Methods in Enzymology

Volume 117
ENZYME STRUCTURE
Part J

METHODS IN ENZYMOLOGY

EDITORS-IN-CHIEF

Sidney P. Colowick   Nathan O. Kaplan

Methods in Enzymology

Volume 117

Enzyme Structure

Part J

EDITED BY

C. H. W. Hirs

DEPARTMENT OF BIOCHEMISTRY, BIOPHYSICS, AND GENETICS
UNIVERSITY OF COLORADO MEDICAL CENTER
DENVER, COLORADO

Serge N. Timasheff

GRADUATE DEPARTMENT OF BIOCHEMISTRY
BRANDEIS UNIVERSITY
WALTHAM, MASSACHUSETTS

1985

ACADEMIC PRESS, INC.

Harcourt Brace Jovanovich, Publishers

Orlando   San Diego   New York   Austin
London   Montreal   Sydney   Tokyo   Toronto

ACADEMIC PRESS, INC.
Orlando, Florida 32887

O 80890

United Kingdom Edition published by
ACADEMIC PRESS INC. (LONDON) LTD.
24–28 Oval Road, London NW1 7DX

LIBRARY OF CONGRESS CATALOG CARD NUMBER: 54-9110

ISBN 0–12–182017–3

PRINTED IN THE UNITED STATES OF AMERICA

85 86 87 88   9 8 7 6 5 4 3 2 1

Table of Contents

Section I. Size, Shape, and Polydispersity of Macromolecules

v

Section II. Interaction of Macromolecules with Ligands and Linkages

Contributors to Volume 117

Article numbers are in parentheses following the names of contributors.
Affiliations listed are current.

José Manuel Andreu (18), *Unidad de Biomembranas, Centro de Investigaciones Biologicas C.S.I.C., Velazquez 144, 28006 Madrid, Spain*

T. Arakawa (5), *Protein Chemistry, Amgen, Thousand Oaks, California 91320*

B. Chu (15), *Department of Chemistry, State University of New York, Stony Brook, New York 11794*

Thomas G. Consler (2), *Department of Biochemistry, Saint Louis University School of Medicine, St. Louis, Missouri 63104*

Lawrence C. Davis (9), *Department of Biochemistry, Kansas State University, Manhattan, Kansas 66506*

H. S. Dhadwal (15), *Department of Electrical Engineering, State University of New York, Stony Brook, New York 11794*

J. R. Ford (15), *Department of Chemistry, State University of New York, Stony Brook, New York 11794*

Susan G. Frasier (16), *Department of Pharmacology, University of Virginia, Charlottesville, Virginia 22908*

Marina J. Gorbunoff (20), *Graduate Department of Biochemistry, Brandeis University, Waltham, Massachusetts 02154*

Paul J. Hagerman (13), *Department of Biochemistry, Biophysics, and Genetics, University of Colorado Health Sciences Center, Denver, Colorado 80262*

J. T. Harmon (6), *American Red Cross Biomedical Research Laboratories, Bethesda, Maryland 20814*

Lyndal K. Hesterberg (8), *Syngene Products and Research, Inc., Fort Collins, Colorado*

B. L. Horecker (17), *Graduate School of Medical Sciences, Cornell University Medical College, New York, New York 10021*

Michael L. Johnson (16), *Department of Pharmacology, University of Virginia, Charlottesville, Virginia 22908*

E. S. Kempner (6), *National Institute of Arthritis, Diabetes, and Digestive and Kidney Diseases, National Institutes of Health, Bethesda, Maryland 20205*

Hiroshi Kido (17), *Department of Enzyme Chemistry, Institute for Enzyme Research, Tokushima University School of Medicine, Tokushima, Japan*

Thomas F. Kumosinski (11, 14), *Eastern Regional Research Center, Agricultural Research Service, U.S. Department of Agriculture, Philadelphia, Pennsylvania 19118*

James C. Lee (2, 8, 21), *Department of Biochemistry, Saint Louis University School of Medicine, St. Louis, Missouri 63104*

Darrell R. McCaslin (3), *Department of Anatomy, Duke University Medical Center, Durham, North Carolina 27710*

George C. Na (24, 25), *U.S. Department of Agriculture, Agricultural Research Service, Eastern Regional Research Center, Philadelphia, Pennsylvania 19118*

Lawrence W. Nichol (12), *Office of the Vice-Chancellor, University of New England, Armidale, New South Wales 2351, Australia*

T. B. Nielsen (6), *Metabolic Research Branch, Naval Medical Research Institute, Bethesda, Maryland 20814*

ROBERT W. OBERFELDER (2, 21), *Department of Biochemistry and Molecular Biology, The University of Texas Health Science Center at Houston, Houston, Texas 77025*

HELMUT PESSEN (11, 14), *Eastern Regional Research Center, Agricultural Research Service, U.S. Department of Agriculture, Philadelphia, Pennsylvania 19118*

ROBERT J. POLLET (1), *Departments of Medicine and Biochemistry, University of South Florida and Veterans Administration Medical Centers, Tampa, Florida 33612*

V. PRAKASH (4), *Protein Technology Discipline, Central Food Technological Research Institute, Mysore 570013, India*

GARY A. RADKE (9), *Department of Biochemistry, Kansas State University, Manhattan, Kansas 66506*

JACQUELINE A. REYNOLDS (3), *Department of Physiology, Duke University Medical Center, Durham, North Carolina 27710*

ROBERT A. SCOTT (23), *School of Chemical Sciences, University of Illinois, Urbana, Illinois 61801*

THEODORE J. SOCOLOFSKY (9), *American Bell Telephone Co., Denver, Colorado 80233*

ALKIS J. SOPHIANOPOULOS (19), *Department of Biochemistry, Emory University, Atlanta, Georgia 30322*

JUDITH A. SOPHIANOPOULOS (19), *Department of Chemistry, Emory University, Atlanta, Georgia 30322*

PHIL G. SQUIRE (10), *Department of Biochemistry, Colorado State University, Fort Collins, Colorado 80523*

SERGE N. TIMASHEFF (4, 5, 24, 25), *Graduate Department of Biochemistry, Brandeis University, Waltham, Massachusetts 02254*

ALBERTO VITA (17), *Laboratoria Biochimica Applicata, Universita Camerino, 62032 Camerino (MC), Italy*

LARRY D. WARD (22), *Graduate Department of Biochemistry, Brandeis University, Waltham, Massachusetts 02254*

DONALD J. WINZOR (12), *Department of Biochemistry, University of Queensland, St. Lucia, Queensland 4067, Australia*

BRUNO H. ZIMM (7), *Department of Chemistry, University of California, San Diego, La Jolla, California 92093*

Preface

"Enzyme Structure," the eleventh volume of *Methods in Enzymology,* was published 18 years ago. Supplements appeared in 1972, 1973, 1978, and 1979. A large part of these volumes was devoted to the presentation of physical techniques used in protein studies. Part I, which appeared recently, is concerned primarily with chemical techniques. This volume and Parts K and L, which are now in preparation, deal in detail with physical methods. It is hoped that they will bring up-to-date coverage of techniques currently available for the study of enzyme conformation, interactions, and dynamics.

As in the past, these volumes present not only techniques that are currently widely available but some which are only beginning to make an impact and some for which no commercial standard equipment is as yet available. In the latter cases, an attempt has been made to guide the reader in assembling equipment from individual components and to help find the necessary information in the research literature.

In the coverage of physical techniques, we have departed somewhat in scope from the traditional format of the series. Since, at the termination of an experiment, physical techniques frequently require much more interpretation than do organic ones, we consider that brief sections on the theoretical principles involved are highly desirable as are sections on theoretical and mathematical approaches to data evaluation and on assumptions and, consequently, limitations involved in the applications of the various methods.

The organization of the material is similar to that of the previous volumes, with Part J being devoted primarily to techniques related to molecular weight measurements and interactions with ligands, the latter a topic that has gained prominence in recent years.

We wish to acknowledge with pleasure and gratitude the generous cooperation of the contributors to this volume. Their suggestions during its planning and preparation have been particularly valuable. The staff of Academic Press has provided inestimable help in the assembly of this volume. We thank them for their many courtesies.

C. H. W. HIRS
SERGE N. TIMASHEFF

METHODS IN ENZYMOLOGY

EDITED BY

Sidney P. Colowick and Nathan O. Kaplan

VANDERBILT UNIVERSITY
SCHOOL OF MEDICINE
NASHVILLE, TENNESSEE

DEPARTMENT OF CHEMISTRY
UNIVERSITY OF CALIFORNIA
AT SAN DIEGO
LA JOLLA, CALIFORNIA

METHODS IN ENZYMOLOGY

EDITORS-IN-CHIEF

Sidney P. Colowick and Nathan O. Kaplan

VOLUME XXXIII. Cumulative Subject Index Volumes I–XXX
Edited by MARTHA G. DENNIS AND EDWARD A. DENNIS

VOLUME XXXIV. Affinity Techniques (Enzyme Purification: Part B)
Edited by WILLIAM B. JAKOBY AND MEIR WILCHEK

VOLUME XXXV. Lipids (Part B)
Edited by JOHN M. LOWENSTEIN

VOLUME XXXVI. Hormone Action (Part A: Steroid Hormones)
Edited by BERT W. O'MALLEY AND JOEL G. HARDMAN

VOLUME XXXVII. Hormone Action (Part B: Peptide Hormones)
Edited by BERT W. O'MALLEY AND JOEL G. HARDMAN

VOLUME XXXVIII. Hormone Action (Part C: Cyclic Nucleotides)
Edited by JOEL G. HARDMAN AND BERT W. O'MALLEY

VOLUME XXXIX. Hormone Action (Part D: Isolated Cells, Tissues, and
Organ Systems)
Edited by JOEL G. HARDMAN AND BERT W. O'MALLEY

VOLUME XL. Hormone Action (Part E: Nuclear Structure and Function)
Edited by BERT W. O'MALLEY AND JOEL G. HARDMAN

VOLUME XLI. Carbohydrate Metabolism (Part B)
Edited by W. A. WOOD

VOLUME XLII. Carbohydrate Metabolism (Part C)
Edited by W. A. WOOD

VOLUME XLIII. Antibiotics
Edited by JOHN H. HASH

VOLUME XLIV. Immobilized Enzymes
Edited by KLAUS MOSBACH

VOLUME XLV. Proteolytic Enzymes (Part B)
Edited by LASZLO LORAND

VOLUME XLVI. Affinity Labeling
Edited by WILLIAM B. JAKOBY AND MEIR WILCHEK

VOLUME 61. Enzyme Structure (Part H)
Edited by C. H. W. HIRS AND SERGE N. TIMASHEFF

VOLUME 62. Vitamins and Coenzymes (Part D)
Edited by DONALD B. MCCORMICK AND LEMUEL D. WRIGHT

VOLUME 63. Enzyme Kinetics and Mechanism (Part A: Initial Rate and Inhibitor Methods)
Edited by DANIEL L. PURICH

VOLUME 64. Enzyme Kinetics and Mechanism (Part B: Isotopic Probes and Complex Enzyme Systems)
Edited by DANIEL L. PURICH

VOLUME 65. Nucleic Acids (Part I)
Edited by LAWRENCE GROSSMAN AND KIVIE MOLDAVE

VOLUME 66. Vitamins and Coenzymes (Part E)
Edited by DONALD B. MCCORMICK AND LEMUEL D. WRIGHT

VOLUME 67. Vitamins and Coenzymes (Part F)
Edited by DONALD B. MCCORMICK AND LEMUEL D. WRIGHT

VOLUME 68. Recombinant DNA
Edited by RAY WU

VOLUME 69. Photosynthesis and Nitrogen Fixation (Part C)
Edited by ANTHONY SAN PIETRO

VOLUME 70. Immunochemical Techniques (Part A)
Edited by HELEN VAN VUNAKIS AND JOHN J. LANGONE

VOLUME 71. Lipids (Part C)
Edited by JOHN M. LOWENSTEIN

VOLUME 72. Lipids (Part D)
Edited by JOHN M. LOWENSTEIN

VOLUME 73. Immunochemical Techniques (Part B)
Edited by JOHN J. LANGONE AND HELEN VAN VUNAKIS

VOLUME 86. Prostaglandins and Arachidonate Metabolites
Edited by WILLIAM E. M. LANDS AND WILLIAM L. SMITH

VOLUME 87. Enzyme Kinetics and Mechanism (Part C: Intermediates, Stereochemistry, and Rate Studies)
Edited by DANIEL L. PURICH

VOLUME 88. Biomembranes (Part I: Visual Pigments and Purple Membranes, II)
Edited by LESTER PACKER

VOLUME 89. Carbohydrate Metabolism (Part D)
Edited by WILLIS A. WOOD

VOLUME 90. Carbohydrate Metabolism (Part E)
Edited by Willis A. Wood

VOLUME 91. Enzyme Structure (Part I)
Edited by C. H. W. HIRS AND SERGE N. TIMASHEFF

VOLUME 92. Immunochemical Techniques (Part E: Monoclonal Antibodies and General Immunoassay Methods)
Edited by JOHN J. LANGONE AND HELEN VAN VUNAKIS

VOLUME 93. Immunochemical Techniques (Part F: Conventional Antibodies, Fc Receptors, and Cytotoxicity)
Edited by JOHN J. LANGONE AND HELEN VAN VUNAKIS

VOLUME 94. Polyamines
Edited by HERBERT TABOR AND CELIA WHITE TABOR

VOLUME 95. Cumulative Subject Index Volumes 61–74 and 76–80
Edited by EDWARD A. DENNIS AND MARTHA G. DENNIS

VOLUME 96. Biomembranes [Part J: Membrane Biogenesis: Assembly and Targeting (General Methods; Eukaryotes)]
Edited by SIDNEY FLEISCHER AND BECCA FLEISCHER

VOLUME 97. Biomembranes [Part K: Membrane Biogenesis: Assembly and Targeting (Prokaryotes, Mitochondria, and Chloroplasts)]
Edited by SIDNEY FLEISCHER AND BECCA FLEISCHER

VOLUME 111. Steroids and Isoprenoids (Part B)
Edited by JOHN H. LAW AND HANS C. RILLING

VOLUME 112. Drug and Enzyme Targeting (Part A)
Edited by KENNETH J. WIDDER AND RALPH GREEN

VOLUME 113. Glutamate, Glutamine, Glutathione, and Related Compounds
Edited by ALTON MEISTER

VOLUME 114. Diffraction Methods for Biological Macromolecules (Part A)
Edited by HAROLD W. WYCKOFF, C. H. W. HIRS, AND SERGE N. TIMASHEFF

VOLUME 115. Diffraction Methods for Biological Macromolecules (Part B)
Edited by HAROLD W. WYCKOFF, C. H. W. HIRS, AND SERGE N. TIMASHEFF

VOLUME 116. Immunochemical Techniques (Part H: Effectors and Mediators of Lymphoid Cell Functions)
Edited by GIOVANNI DI SABATO, JOHN J. LANGONE, AND HELEN VAN VUNAKIS

VOLUME 117. Enzyme Structure (Part J)
Edited by C. H. W. HIRS AND SERGE N. TIMASHEFF

VOLUME 118. Plant Molecular Biology (in preparation)
Edited by ARTHUR WEISSBACH AND HERBERT WEISSBACH

VOLUME 119. Interferons (Part C) (in preparation)
Edited by SIDNEY PESTKA

VOLUME 120. Cumulative Subject Index Volumes 81–94, 96–101

VOLUME 121. Immunochemical Techniques (Part I: Hybridoma Technology and Monoclonal Antibodies) (in preparation)
Edited by JOHN J. LANGONE AND HELEN VAN VUNAKIS

VOLUME 122. Vitamins and Coenzymes (Part G) (in preparation)
Edited by FRANK CHYTIL AND DONALD B. McCORMICK

Section I

Size, Shape, and Polydispersity of Macromolecules

[1] Characterization of Macromolecules by Sedimentation Equilibrium in the Air-Turbine Ultracentrifuge[1]

By ROBERT J. POLLET

The characterization by hydrodynamic methods of minute quantities of partially purified functional proteins from the rate of zonal transport of their specific biochemical activities has been somewhat limited by the empirical basis of the available techniques such as sedimentation through sucrose gradients[2,3] and gel filtration chromatography,[4-6] and by required assumptions regarding molecular asymmetry and hydration. While denaturation of the protein and reduction of disulfide bonds, as in sodium dodecyl sulfate–polyacrylamide gel electrophoresis,[7,8] can minimize the ambiguities of conformation, hydration, and charge, the denaturation process abolishes assayable specific biochemical activity and in some cases these empirical dynamic techniques may still exhibit anomalous behavior.[7,9,10]

The development of sedimentation equilibrium in the air-driven preparative ultracentrifuge (Airfuge, Beckman Instruments, Palo Alto, Calif.) has provided an independent, nonempirical method to characterize partially purified macromolecules within heterogeneous preparations, based on the equilibrium distribution of their biochemical or other specific activities as a function of radial distance in a centrifugal field.[11] Since this method is based on a theoretical thermodynamic description of the system in sedimentation–diffusion equilibrium and is independent of molecular weight standards and zonal transport, the difficulties outlined above are effectively avoided. In addition, the method has opened this area of investigation to the extensive existing literature concerning the use of

[1] This work was supported by National Institutes of Health Grant AM 18608 and by the Veterans Administration.

[2] R. G. Martin and B. N. Ames, *J. Biol. Chem.* **236,** 1372 (1961).
[3] K. S. McCarty, D. Stafford, and O. Brown, *Anal. Biochem.* **24,** 314 (1968).
[4] P. Andrews, *Biochem. J.* **96,** 595 (1965).
[5] H. S. Warshaw and G. K. Ackers, *Anal. Biochem.* **42,** 405 (1971).
[6] L. M. Siegel and K. J. Monty, *Biochim. Biophys. Acta* **112,** 346 (1966).
[7] K. Weber and M. Osborn, *in* "The Proteins" (H. Neurath and R. L. Hill, eds.), Vol. 1, Chapter 3. Academic Press, New York, 1975.
[8] T. B. Nielsen and J. A. Reynolds, this series, Vol. 48, p. 3.
[9] C. Tanford and J. A. Reynolds, *Biochim. Biophys. Acta* **457,** 133 (1976).
[10] Y. Nozaki, N. M. Schechter, J. A. Reynolds, and C. Tanford, *Biochemistry* **15,** 3884 (1976).
[11] R. J. Pollet, B. A. Haase, and M. L. Standaert, *J. Biol. Chem.* **254,** 30 (1979).

sedimentation equilibrium to determine the molecular weights and interactions of proteins in solution, which had been previously restricted to highly purified protein preparations detected optically in the analytical ultracentrifuge.[12-16] The rationale for utilizing the air-driven ultracentrifuge chiefly involves the small size (3.8 cm outside diameter) and fixed angle orientation (tubes 18° from vertical) of its rotor, both of which contribute to very short radial distances from the meniscus to the outermost (centrifugal) portion of the solution (0.3–0.4 cm). Since the time required to approach to within a specific deviation from true sedimentation equilibrium is proportional to the square of this radial distance,[17] short distances are necessary for achievement of experimental sedimentation equilibrium within a reasonable time. Furthermore, the nonsectorial geometry of the tube walls initially creates convective transport of solute[18] which contributes to achievement of experimental sedimentation equilibrium within 24 hr.

Instrumentation and Methodology

Sedimentation equilibrium is performed in the Airfuge, a small ultracentrifuge in which a 3.8-cm-diameter fluted aluminum rotor is lifted and driven by pressure-regulated compressed air at rotational velocities up to 100,000 rpm, generating a maximal centrifugal field of up to 160,000 g. The rotor has positions for 6 cellulose propionate tubes at a fixed angle of 18° to the vertical which may contain up to 175 μl during centrifugation. This instrument and rotor are shown in Fig. 1.

The frequency of rotation is determined stroboscopically by an attachment available from Beckman Instruments, or by a stroboscope (Model 1538A available from the General Radio Company, Concord, Mass.) capable of frequencies up to 150,000/min with an accuracy of ±1%. The rotational velocity is stable after 2 hr to within ±1% from 20,000 to 100,000 rpm.

The protein of interest is dissolved in buffer which is density stabilized[18,19] by the inclusion of 5 mg/ml of the protein itself or a suitable

[12] T. Svedberg and K. O. Pedersen, "The Ultracentrifuge." Oxford Univ. Press (Clarendon), London and New York, 1940.
[13] H. K. Schachman, "Ultracentrifugation in Biochemistry," p. 181. Academic Press, New York, 1959.
[14] H. Fujita, "Mathematical Theory of Sedimentation Analysis," p. 235. Academic Press, New York, 1962.
[15] K. E. Van Holde, in "The Proteins" (H. Neurath and R. L. Hill, eds.), 3rd ed., Vol. 1, p. 225. Academic Press, New York, 1975.
[16] D. C. Teller, this series, Vol. 27, p. 346.
[17] K. E. Van Holde and R. L. Baldwin, J. Phys. Chem. **62**, 734 (1958).
[18] M. A. Bothwell, G. J. Howlett, and H. K. Schachman, J. Biol. Chem. **253**, 2073 (1978).
[19] G. J. Howlett, E. Yeh, and H. K. Schachman, Arch. Biochem. Biophys. **190**, 809 (1978).

Fig. 1. The table top air-turbine ultracentrifuge (Beckman Airfuge). (A) The rotor chamber is shown in the open position so that the small aluminum rotor may be seen. (B) The 6-position rotor with turbine fluting on the bottom.

macromolecule with which the protein does not interact, such as bovine serum albumin[11] or dextran (average M_r = 40,000 or 70,000). The spontaneous concentration gradient of these materials formed in the centrifugal field provides sufficient density stabilization to prevent significant convective disturbance of the solution during centrifugation, deceleration, reorientation, and fractionation.

Each of the untapered cellulose nitrate or cellulose propionate tubes is filled with 100–150 μl of solution and centrifuged at an appropriate rotational velocity for 22–24 hr to achieve sedimentation equilibrium. Centrifugation may be performed at 5–25°,[11,20] with the lower temperatures achieved in a cold room with precooling of the flowing compressed air.[21-23] The solution temperature following centrifugation is usually 3–5° above that of the compressed air entering the centrifuge.

After centrifugation, the driving air pressure is discontinued, followed 2 min later by engagement of the orientation pin for the final 1 min of deceleration. The centrifuge tubes are then removed from the rotor and oriented vertically. The solution is sequentially fractionated from the meniscus with a Hamilton pipet controller having threaded fine control and stabilized by a micromanipulator (Kopf Model 1260[11] or Narishige IM-3[20]), utilizing 10-μl disposable capillary pipets[11] or polyethylene tubing (0.45 mm i.d.).[20] Sequential fractionation may also be performed with a Beckman microtube fractionator with the barrel sealed with vacuum grease. Fractionation takes approximately 10 min per tube during which time the remaining sedimentation equilibrium gradients remain stable. The fractions are diluted appropriately with buffer and the concentration of the protein of interest is determined for each fraction by enzymatic activity, radioactivity, or other specific biochemical activity[11] or by protein absorbance for pure preparations. Minor corrections are made for any small amount of residual nonsedimentable material in a given preparation of protein with $M_r > 30,000$ by subtracting from every data point the protein content remaining at the meniscus of a parallel tube after sedimentation at 100,000 rpm for an additional 6 hr.[11] Alternatively, the fraction of small nonsedimentable material may be estimated independently by ultrafiltration[24] or gel filtration, especially for proteins of $M_r < 30,000$ which would be only partially depleted from the meniscus at the maximum speed of 100,000 rpm.

[20] N. Ueno, H. Miyazaki, S. Hirose, and K. Mirakami, *J. Biol. Chem.* **256,** 12023 (1981).
[21] G. J. Howlett and G. Markov, *Arch. Biochem. Biophys.* **202,** 507 (1980).
[22] R. J. Pollet, B. A. Haase, and M. L. Standaert, *J. Biol. Chem.* **256,** 12118 (1981).
[23] R. F. Vogt, Jr. and D. L. Fiendt, *Clin. Chem.* (*Winston-Salem, N.C.*) **28,** 1490 (1982).
[24] P. E. Bock and H. R. Halvorson, *Anal. Biochem.* **135,** 172 (1983).

Theoretical Considerations

For sedimentation equilibrium of an ideal solution of a macromolecule of molecular weight M at an angular velocity ω and temperature T, the macromolecular concentration $c(r)$ at radial distance r is determined by the differential equation

$$\frac{d \ln c(r)}{dr^2} = \frac{M(1 - \bar{v}\rho)\omega^2}{2RT} = \sigma \tag{1a}$$

or its integrated form

$$c(r) = c(r_m)e^{\sigma(r^2 - r_m^2)} \tag{1b}$$

irrespective of tube geometry, where $\bar{v}$ is the partial specific volume of the macromolecule, ρ is the density of the solution, σ is termed the reduced molecular weight of the macromolecule, and r_m is the radial distance of the meniscus or other reference point in the solution.

Knowledge of the solution geometry is required to transform the experimental concentration profile obtained as a function of solution volume to a concentration gradient as a function of the square of the radial distance during centrifugation as given in the equation above. The centrifuge tube is oriented at a fixed angle θ from the vertical axis and is composed of hemispherical and cylindrical portions with an internal radius a as shown in Fig. 2. During centrifugation the cross-sectional area $A(r)$ of the solution perpendicular to the radial path is the sum of a variably truncated ellipse plus a variably truncated circle whose radius varies, and is given by the following.

For $r \leq r_c$,

$$A(r) = \pi a^2 \csc \theta \tag{2}$$

For $r_c < r < r_c + a(1 + \cos \theta)$,

$$A(r) = a^2 \left[\left(\frac{\pi}{2} + \alpha \sqrt{1 - \alpha^2} + \arctan \alpha \right)(\csc \theta) \right.$$
$$\left. + \left(\frac{\pi}{2} - \beta \sqrt{1 - \beta^2} - \arctan \beta \right)(1 - \alpha^2 \cos^2 \theta) \right]$$

For $r \geq r + a(1 + \cos \theta)$

$$A(r) = \pi a^2(1 - \alpha^2 \cos^2 \theta)$$

where

$$\alpha = 1 - \left(\frac{r - r_c}{a} \right) \sec \theta \quad \text{and} \quad \beta = \frac{\alpha \sin \theta}{\sqrt{1 - \alpha^2 \cos^2 \theta}}$$

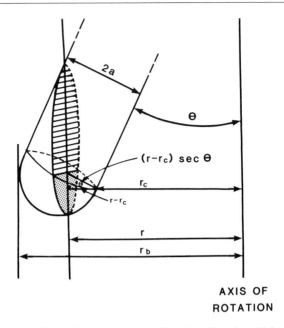

FIG. 2. The geometry of the solution during centrifugation. The tube with internal radius a is oriented at angle θ from the vertical axis. r_m, r_b, and r_c are the respective radial distances from the axis of rotation of the meniscus, the outermost (centrifugal) portion of the solution, and the inner (centripetal) edge of the plane separating the cylindrical and hemispherical portions of the tube.

and r_m, r_b, and r_c are the respective radial distances from the axis of rotation of the meniscus, the outermost portion of the solution, and the inner edge of the plane between the cylindrical and hemispherical portions of the tube (Fig. 2). Beckman specifications yield $\theta = 0.314$ radians, $a = 0.252$ cm, $r_b = 1.47$ cm, and $r_c = 0.98$ cm. The volume $V(r)$ of solution from the meniscus to a plane perpendicular to the radial path at a distance r from the axis of rotation is given by

$$V(r) = \int_{r_m}^{r} A(r) \, dr \tag{3}$$

$V(r)$ may be evaluated numerically,[11] or integrated analytically[24] to yield the graphs shown in Fig. 3. Assuming that the density-stabilized solution reorients without mixing during deceleration of the rotor, Eq. (3) permits transformation of concentration data, obtained as a function of volume from the meniscus of the reoriented solution, to a concentration profile expressed as a function of radial distance from the center of rotation during centrifugation. Direct confirmation of this solution reorientation

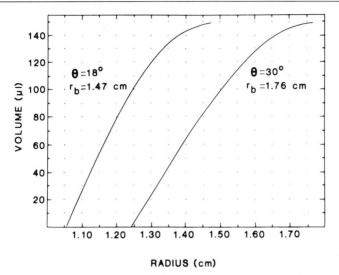

RADIUS (cm)

FIG. 3. Graphs of the volume $V(r)$ from the meniscus as a function of the radial distance from the center of rotation for fixed angle rotors with tubes oriented 18° (A100/18 rotor) and 30° (A100/30 rotor) from the vertical axis. These values were calculated by numerical integration of Eq. (3) for solution volumes of 150 μl, and therefore are applicable to all solutions with total volume $V_T \leq 150$ μl by subtracting $(150 - V_T)$ from all $V(r)$ values. Small corrections have been introduced to accommodate the curvature of the meniscus and volume elements of the solution.[26]

has been obtained by comparison of protein absorption scans of quartz tubes following sedimentation equilibrium in a fixed angle rotor in an air-turbine ultracentrifuge and in a swinging bucket rotor in a preparative ultracentrifuge which does not involve solution reorientation.[25]

The fact that the meniscus and subsequent volume elements are more rigorously cylindrical surfaces with respect to the center of rotation which curve outward (central portion) and inward (edges) from the average planes assumed above leads to small corrections as outlined by Charlwood,[26] which can be well approximated by increasing the radial distance of each volume element by 0.01 cm due to its curvature. The graphs corrected for curvature are given in Fig. 3.

Finally, it should be noted that special relationships which are peculiar to the sector-shaped solution geometry of the analytical ultracentrifuge, e.g., the solution "hinge point," are not applicable to the solution geometry described above.

[25] A. K. Attri and A. P. Minton, *Anal. Biochem.* **133**, 142 (1983).
[26] P. A. Charlwood, *J. Phys. Chem.* **84**, 3122 (1980).

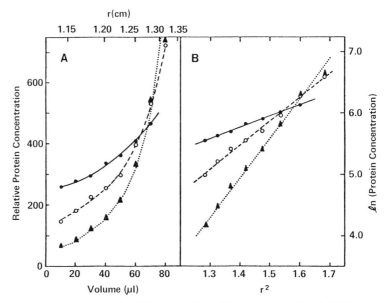

FIG. 4. (A) Sedimentation equilibrium profiles of bovine serum albumin (^{14}C-methylated, 10,000 cpm) (●———●), human γ-globulin (○———○), and beef liver catalase (▲———▲), after centrifugation at 20,000 rpm for 24 hr at 10° in 100 μl of buffer (0.15 M NaCl, 0.02 M sodium phosphate, pH 7.2) at protein concentrations of 5 mg/ml. Bovine serum albumin was measured by ^{14}C radioactivity, human γ-globulin was measured by absorbance at 280 nm, and catalase activity was measured enzymatically. (B) Sedimentation equilibrium data plotted as ln c versus r^2. From the slopes of these lines, the calculated molecular weights for these proteins are 72,000, 153,000, and 244,000, respectively.

Applications

Molecular Weight Determinations

The major application of this sedimentation equilibrium method is the determination of the molecular weights of minute quantities of partially purified proteins. The results for three typical proteins are shown in Fig. 4.[11] At a given angular velocity, the concentration profiles (A) are steeper for the higher molecular weight proteins. As shown in B, log $c(r)$ is a linear function of r^2 and the slope yields the molecular weight of each protein in excellent agreement with its known value. Our results, as well as those of several other research groups, applying this technique to proteins of known molecular weight are presented in Table I. The molecular weights determined by this sedimentation equilibrium method are in good agreement with their known values over a range of 6000–600,000. The concentration profiles which yield ln $c(r)$ as a function of r^2 may be measured by absorbance, enzymatic activity, radioactivity, or any other specific biochemical activity. Thus, this sedimentation equilibrium method yields the molecular weight associated with a specific biochemical activity within a heterogeneous mixture and is based on a thermodynamic description of the system without reference to molecular weight standards.

For sedimentation equilibrium of preparations in which the protein of interest is itself heterogeneous with respect to size, the resulting curvilinearity of ln c as a function of r^2 is often difficult to detect even in the analytical ultracentrifuge,[27,28] and therefore may not be apparent from the more limited number of data points available in the air-turbine ultracentrifuge. As in the analytical ultracentrifuge, the molecular weight derived from $d \ln c/dr^2$ represents the weight-average molecular weight (M_w) of the population of proteins detected by assays which reflect mass concentration (e.g., ultraviolet absorbance). However, for size heterogeneity of functional molecules where the assay of specific biochemical activity in some cases reflects the *molar* concentration of each species, the molecular weight derived from $d \ln c/dr^2$ represents the number-average molecular weight (M_n).

Macromolecules whose characterization has largely depended on sedimentation equilibrium in the air-turbine ultracentrifuge are listed in Table II. It is illustrative to note the use of this method to identify the active molecular species in self-associating enzyme systems. For example, previous studies of purified preparations of the enzyme regulating cholesterol synthesis, 3-hydroxy-3-methylglutaryl coenzyme A reductase, including sedimentation equilibrium in the analytical ultracentrifuge, had yielded a molecular weight of 323,000.[29] However, the hydrodynamic properties of the enzymatic activity appeared inconsistent with this molecular weight[30] and sodium dodecyl sulfate–polyacrylamide electrophoresis suggested subunit molecular weights of 50,000–60,000.[31] This problem was resolved by comparing the radial dependencies of the enzymatic activity and protein absorbance after sedimentation equilibrium of this purified enzyme in the air-driven ultracentrifuge. This method yielded molecular weights of 104,000 for the active enzyme and 320,000 for the protein,[32] indicating the presence of enzyme self-association to form larger inactive species. Subsequently, subunit studies in the presence of

[27] R. K. Dyson and I. Isenberg, *Biochemistry* **10**, 3233 (1971).
[28] P. Munk and D. J. Cox, *Biochemistry* **11**, 687 (1972).
[29] Z. H. Beg and H. B. Brewer, Jr., *Curr. Top. Cell. Regul.* **20**, 139 (1981).
[30] P. A. Edwards, D. Lemongello, J. Kane, J. Schacter, and M. W. Fogelman, *J. Biol. Chem.* **255**, 3715 (1980).
[31] M. Sinensky, R. Torget, and P. A. Edwards, *J. Biol. Chem.* **256**, 11774 (1981).
[32] Z. H. Beg and J. C. Osborne, *Fed. Proc., Fed. Am. Soc. Exp. Biol.* **41**, 1143 (1982); J. C. Osborne, personal communication.

TABLE I

PROTEIN MOLECULAR WEIGHTS BY SEDIMENTATION EQUILIBRIUM IN THE AIR–TURBINE ULTRACENTRIFUGE

Protein	Reference	Rotational velocity (rpm)	Temperature	Partial specific volume	Sed. equil. molecular weight	Molecular weight
Insulin, ^{125}I-labeled (porcine)	11	96,000	10°	0.725	5,600	5,800[a]
Myoglobin (horse)	11	66,500	10°	0.731	17,900	17,500[b]
μ_1-Acid glycoprotein						
^{125}I-labeled (human)	53	35,800	9°	0.704	37,000	36,000[c,d]
^{125}I-labeled (rat)	53	34,900	10°	0.704	40,900	39,600[e]
Ovalbumin, ^{14}C-methylated (hen)	11	35,000	9°	0.733	42,000	44,600[f]
	20	45,000	25°	0.74	40,000	
	24	Not specified	27°	0.742	46,900	
Serum albumin						
[U-^{14}C]Palmitate (human)	21	37,357	8°	0.733	67,800	66,200[g]
^{14}C-methylated (bovine)	11	36,000	10°	0.719	68,000	66,200[h]
	11	20,100			72,000	66,700[i]
	20	38,000	25°		68,000	68,000[h]
^{125}I-labeled (rat)	53	Not specified	9°	0.735	68,500	68,000[j]

Transferrin, ^{125}I-labeled (rat)	53	Not specified	10°	0.725	79,000	79,600[k]
γ-Globulin (human)	11	20,100	10°	0.729	153,000	156,000[l]
Catalase (bovine)	11	20,300	10°	0.720	244,000	248,000[m]
Thyroglobulin (porcine)	11	13,800	10°	0.713	630,000	660,000[n]

[a] H. Brown, F. Sanger, and R. Kitai, *Biochem. J.* **60**, 556 (1955).

[b] M. Dautrevany, V. Boulanger, K. Han, and G. Bisert, *Eur. J. Biochem.* **11**, 267 (1969).

[c] K. Schmidt, H. Kaufmann, S. Isemura, F. Bauer, J. Emura, T. Motoyama, M. Ishiguro, and S. Nanno, *Biochemistry* **612**, 2711 (1973).

[d] B. Fournet, J. Montreuil, G. Streckler, L. Dorland, J. Haverkamp, J. F. G. Vliegenhart, J. P. Binette, and K. Schmidt, *Biochemistry* **17**, 5206 (1978).

[e] G. J. Howlett, P. W. Dickson, H. Birch, and G. Schreiber, *Arch. Biochem. Biophys.* **215**, 309 (1982).

[f] F. J. Castellino and R. Barter, *Biochemistry* **7**, 2207 (1968).

[g] P. Q. Behrens, A. M. Spiekerman, and J. R. Brown, *Fed. Proc., Fed. Am. Soc. Exp. Biol.* **34**, 591 (1975).

[h] J. R. Brown, *Fed. Proc., Fed. Am. Soc. Exp. Biol.* **34**, 591 (1975).

[i] P. G. Squire, P. Moser, and C. T. O'Konski, *Biochemistry* **7**, 4261 (1975).

[j] J. Urban, A. S. Inglis, K. Edwards, and G. Schreiber, *Biochem. Biophys. Res. Commun.* **61**, 494 (1974).

[k] R. T. W. MacGillivray, E. Mendez, J. G. Shewale, S. K. Sinha, J. Lineback-Zins, and K. Brew, *J. Biol. Chem.* **258**, 3543 (1983).

[l] J. L. Oncley, G. Scatchard, and A. Brown, *J. Phys. Chem.* **51**, 184 (1974).

[m] J. B. Sumner and N. Gralen, *J. Biol. Chem.* **125**, 33 (1983).

[n] I. J. O'Donnell, R. L. Baldwin, and J. W. Williams, *Biochim. Biophys. Acta* **28**, 294 (1958).

TABLE II

PROTEINS CHARACTERIZED BY SEDIMENTATION EQUILIBRIUM IN THE AIRFUGE

Protein	Rotational velocity (rpm)	Temperature	M_r
Nucleolar ribonuclease[a] (Ehrlich tumor cells)	32,800	4°	38,500
Lipoprotein lipase (bovine)[38]	Not specified		76,000
Chorismate mutase/prephenate dehydrogenase (*E. coli*)[51]	30,000	8°	78,800[a]
Dihydroorotase[c]	41,800	8°	80,900
HMG CoA reductase[32]	Not specified		
(enzymatic activity)			104,000
(UV absorbance)			323,000
Renin–renin binding protein complex[20]	38,000	25°	113,000
cAMP Phosphodiesterase[d] (rat liver)			
Fraction B	20,900	5°	174,000
Fraction C	20,900	5°	85,000
[125]I-labeled neuraminidase[55] (A/Tokyo3/67 virus)	26,900	10°	196,000
Ribonucleotide reductase[c] (Ehrlich tumor cell)			
ADP reductase activity	18,400	10°	254,000
CDP reductase activity	16,400	10°	304,000
Nitrate reductase[f] (*Chlorella*)	19,100	5°	360,000

[a] D. C. Eichler and T. F. Tator, *Biochemistry* **19**, 3016 (1980).

[b] This M_r was later confirmed as 78,000 in the analytical ultracentrifuge [G. S. Hudson, V. Wong, and B. E. Davidson, *Biochemistry* **23**, 6240 (1984)].

[c] M. W. Washabaugh and K. D. Collins, *J. Biol. Chem.* **259**, 3293 (1984).

[d] T. Yamamoto, F. Lieberman, J. C. Osborne, V. C. Manganiello, M. Vaughn, and H. Hidaka, *Biochemistry* **23**, 670 (1984).

[e] J. G. Cory and A. E. Fleischer, *Arch. Biochem. Biophys.* **217**, 546 (1982).

[f] W. D. Howard and L. P. Solomonson, *J. Biol. Chem.* **257**, 10243 (1982).

protease inhibitors[33,34] and the determination of the nucleotide sequence of the enzyme-specific mRNA have confirmed an actual postprocessing molecular weight of 97,000 for this single polypeptide enzyme,[35,36] consis-

[33] G. C. Ness, S. C. Way, and P. S. Wickham, *Biochem. Biophys. Res. Commun.* **102**, 81 (1981).

[34] D. J. Chin, K. L. Luskey, R. G. W. Anderson, J. R. Faust, J. L. Goldstein, and M. S. Brown, *Proc. Natl. Acad. Sci. U.S.A.* **79**, 1185 (1982).

[35] L. Liscum, R. D. Cummings, R. G. W. Anderson, G. N. DeMartino, J. L. Goldstein, and M. S. Brown, *Proc. Natl. Acad. Sci. U.S.A.* **80**, 7165 (1983).

[36] D. H. Chin, G. Gil, D. W. Russell, L. Liseum, K. L. Luskey, S. K. Basu, H. Okayama, P. Berg, J. L. Goldstein, and M. S. Brown, *Nature (London)* **308**, 613 (1984).

tent with the active form of the enzyme by sedimentation equilibrium in the air-turbine ultracentrifuge.

The molecular weight of the active species in another self-associating enzyme system, bovine lipoprotein lipase with a monomer molecular weight of 41,700,[37] has also been determined by similar methodology. Sedimentation equilibrium in the analytical ultracentrifuge had indicated the presence of reversible oligomer formation,[37] while nonoverlapping molecular weight profiles as a function of initial total enzyme concentration suggested the presence of irreversible aggregation as well. However, while sedimentation equilibrium in the air-turbine ultracentrifuge confirmed the presence of reversible and irreversible aggregation of this enzyme by protein absorbance, the profiles of enzyme activity in these solutions yielded a molecular weight of 76,000 for the active species independent of initial protein concentration.[38] Therefore, in this case the protein dimer represents the functional species in this self-associating enzyme system. These examples demonstrate the unique advantage of the air-turbine ultracentrifuge methodology in the structural–functional characterization of purified complex systems through the comparison of their sedimentation equilibrium behavior with respect to total protein concentration and the corresponding specific biochemical activity.

Protein Self-Association

Sedimentation equilibrium has been the method of choice for the study of self-associating protein systems, and the air-turbine ultracentrifuge provides the opportunity to study these interactions directly without the constraints imposed by optical detection techniques. In the absence of significant nonideality (see below), at sedimentation equilibrium the radial distribution of each molecular species in the self-association scheme is determined by Eq. (1), and at each radial distance the concentration of a given molecular species is determined by the equilibrium between lower and higher order species. Therefore, information concerning the molecular weights, stoichiometry, and equilibrium constants of a self-associating system may be extracted from the experimentally measured total concentration of all species as a function of radial distance at different initial protein concentrations.[15,16,39–41] This may be performed through calculation of the several moment-average molecular weights of the associating

[37] T. Olivercrona, G. Bengtsson, and J. C. Osborne, Jr., *Eur. J. Biochem.* **124**, 629 (1982).
[38] J. C. Osborne, Jr., G. Bengtsson, N. S. Lee, and T. Olivecrona, personal communication.
[39] E. T. Adams, Jr., "Fractions No. 3." Spinco Div., Beckman Instruments, Inc., Palo Alto, California, 1967.
[40] H. Kim, R. C. Deonier, and J. W. Williams, *Chem. Rev.* **77**, 659 (1977).
[41] K. C. Aune, this series, Vol. 48, p. 163.

system, such as M_w and M_n, as functions of the total concentration of all species and relating these functions to the above parameters for various models for self-association, including open models involving an unlimited series of associations having equal molar equilibrium constants (isodesmic association). These determinations are considerably simplified if the molecular weight of the monomer (protomer) is known, in which case the monomer concentration as a function of total protein concentration may be calculated from the sedimentation equilibrium data[42,43] and related to the stoichiometry and equilibrium constants for a selected set of association schemes to be examined.

An example of this type of approach utilizing sedimentation equilibrium in the air-driven ultracentrifuge is the study of insulin at neutral pH, a strongly self-associating system which can be accurately examined in the analytical ultracentrifuge[44] only at concentrations above the dimerization range. However, in the air-turbine ultracentrifuge the sedimentation equilibrium behavior of ^{125}I-labeled insulin may be directly examined at 1000-fold lower concentrations, where insulin is monomeric even at neutral pH. Sedimentation equilibrium of ^{125}I-labeled insulin at 10^{-10} M yields a monomer molecular weight of 5600 and increasing weight-average molecular weights at higher total insulin concentrations.[11] The weight-average molecular weight/monomer molecular weight as a function of increasing total insulin concentration is shown in the Fig. 5, and is consistent with a reversible association scheme[44] involving dimerization of monomers ($K_D = 1.4 \times 10^5$ M^{-1}) and unlimited isodesmic association of the dimers ($K_I = 1.8 \times 10^4$ M^{-1}). As shown in the figure, these data represent a continuous extension of previous analytical ultracentrifuge data[44] downward into the dimerization range, yielding an improved estimation of the dimerization constant.

Alternatively, analysis of self-associating systems may be performed directly from the radial distribution of the total concentrations of all associating species following sedimentation equilibrium. A general approach[45–47] to this problem is the development, from a point by point analysis of the smoothed radial concentration distribution data, of a set simultaneous equations, each being the sum for all molecular species of terms having the form of Eq. (1b) and therefore nonlinear with respect to the unknowns. For the assumed association model, the results of this complex procedure are the molecular weights and meniscus concentra-

[42] R. F. Steiner, *Arch. Biochem. Biophys.* **39**, 333 (1952).
[43] B. K. Milthrope, P. D. Jeffrey, and L. W. Nichol, *Biophys. Chem.* **3**, 169 (1975).
[44] P. D. Jeffrey, B. K. Milthrope, and L. W. Nichol, *Biochemistry* **15**, 4660 (1976).
[45] E. T. Adams, Jr. and J. W. Williams, *J. Am. Chem. Soc.* **86**, 3454 (1964).
[46] P. W. Chun and S. J. Kim, *J. Phys. Chem.* **74**, 899 (1970).
[47] R. H. Haschemeyer and W. F. Bowers, *Biochemistry* **9**, 435 (1970).

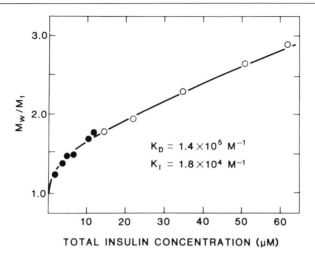

FIG. 5. Weight-average molecular weight/monomer molecular weight, M_w/M_1, for ^{125}I-labeled insulin (●) as a function of total concentration of insulin, as determined by sedimentation equilibrium following centrifugation at 96,000 rpm for 24 hr at 10°. These data are continuous with previous sedimentation equilibrium data[44] (○) at higher insulin concentrations performed in the analytical ultracentrifuge. The results for insulin self-association are well-described by insulin dimerization ($K_D = 1.4 \times 10^5 M^{-1}$) and isodesmic association of dimers ($K_1 = 1.8 \times 10^4 M^{-1}$) to form an open series of higher order species,[44] as shown by the theoretical curve (——).

tions of the associating species, from which their association constants at the meniscus are calculated. The adequacy of the assumed model may be tested for self-consistency of the association constants throughout the radial concentration distribution. While sedimentation equilibrium in the air-turbine ultracentrifuge does not usually yield a sufficient number of data points to permit such a general approach, simple associating systems with known species molecular weights have been examined by a modification[48] of a procedure which resolves the total concentration distribution into a linear combination of theoretical distributions [Eq. (1b)] for each of the known molecular species.[49] An example of this method of analysis applied to air-turbine ultracentrifuge data is provided by Howlett *et al.* who examined the sedimentation equilibrium behavior of ^{125}I-labeled concanavalin A,[48] a dimeric plant lectin which exhibits association into tetramers[50] but is closed to further association. As shown in Fig. 6A, utilizing the known molecular weight of 51,000 for the dimeric concanavalin A,[50] the experimental total concentration distribution (solid circles) is consistent with the theoretical distribution (solid line) for dimer to tetramer

[48] G. J. Howlett, P. J. Roche, and G. Schreiber, *Arch. Biochem. Biophys.* **224,** 178 (1983).
[49] K. C. Aune and M. F. Rohde, *Anal. Biochem.* **79,** 110 (1977).
[50] D. F. Senear and D. C. Teller, *Biochemistry* **20,** 3076 (1981).

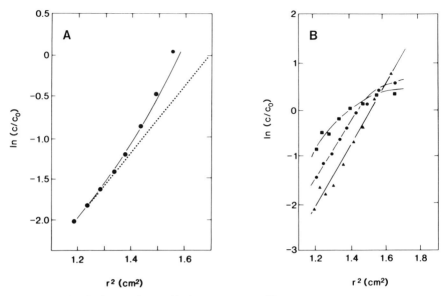

FIG. 6. (A) Sedimentation equilibrium behavior of ^{125}I-labeled concanavalin A (2 mg/ml) in 0.26 M potassium phosphate pH 7.0, 5 mg/ml bovine serum albumin following centrifugation for 22 hr at 36,670 rpm at 10°. The theoretically predicted sedimentation equilibrium curves for pure dimeric concanavalin A (M_r = 51,000) (– – –) and for reversible self-association of dimers to form tetramers (K = 1.6 × 10^4 M^{-1}) (——) are also shown. These data are from Howlett et al.[48] (B) Sedimentation equilibrium behavior of ^{125}I-labeled bovine serum albumin in 0.15 M NaCl, 0.04 M phosphate, pH 7.0, following centrifugation for 22 hr at 39,100 rpm at 10°. Initial protein concentrations were 5 mg/ml (▲), 40 mg/ml (●), and 80 mg/ ml (■), respectively. The downward curvature of the plots at the higher initial protein concentrations indicates the presence of solution non-ideality. The data are from Howlett et al.[53]

association with K = 2 × 10^4 M^{-1}.[48] A similar analysis of the sedimentation equilibrium behavior of chorismate mutase/prephenate dehydrogenase in the air-turbine ultracentrifuge[51] demonstrated dimer–tetramer association (K_{app} = 2.3 × 10^6 M^{-1}) in the presence of NAD[+] and tyrosine, an end-product inhibitor of the enzyme.

Nonideality

Proteins in solution are strictly three-component interacting systems consisting of solvent, protein, and electrolytes which may be well described in a two-component formalism by redefining the component of interest (component 2) as the protein with its associated electrolytes fol-

[51] G. S. Hudson, G. J. Howlett, and B. E. Davidson, J. Biol. Chem. 258, 3114 (1983).

lowing dialysis against the electrolyte solution (component 1), so that the chemical potential of the electrolytes is equal for the new components 1 and 2.[52] For sedimentation equilibrium of very low concentration of proteins in the air-turbine ultracentrifuge, there is also the requirement that the protein of interest must not interact with the macromolecular solute (e.g., bovine serum albumin or dextran) used for density stabilization at a concentration of 5 mg/ml. In addition, the effects of these concentrations of macromolecular solutes with respect to solution nonideality must be considered.

Nonideality of protein solutions is largely secondary to excluded volume effects and repulsive electrostatic effects, both of which are generally negligible at the protein concentrations (<10 mg/ml) and buffer ionic strengths (<0.1) employed in the air-turbine ultracentrifuge. Howlett and colleagues have examined nonideality effects at higher protein concentrations by sedimentation equilibrium in the air-driven ultracentrifuge.[53] Figure 6B shows the concentration profiles and ln c vs r^2 plots of ^{125}I-labeled bovine serum albumin at initial concentrations of 5, 40, and 80 mg/ml. While ln c is a linear function of r^2 at an initial concentration of 5 mg/ml, the graphs become curvilinear (concave downward) at the higher initial concentrations, consistent with nonideality effects. In fact, employing sedimentation equilibrium data at initial bovine serum albumin concentrations of 20, 40, 60, and 80 mg/ml,[53] the colligative second virial coefficient may be calculated as 3×10^{-5} cm^3/g, in reasonable agreement with previous results obtained with other methods.[54] It is likely that significant solution nonideality effects are present at the extreme centrifugal portion of the density-stabilized solution where the bovine serum albumin concentration may be above 50 mg/ml, and these effects may be more pronounced for density stabilization with dextran.

Ligand–Protein Binding

The characterization of reversible ligand binding to proteins largely involves the determination (e.g., by equilibrium dialysis) of the concentration of free and bound ligand following equilibration of the protein with known total concentrations of ligand. This may also be accomplished in the air-turbine ultracentrifuge by selectively depleting the meniscus of protein–ligand complexes by short- or long-term centrifugation of the

[52] G. J. Howlett, P. W. Dickson, H. Birch, and G. Schreiber, *Arch. Biochem. Biophys.* **215,** 309 (1982).

[53] G. J. Howlett, P. W. Dickson, H. Birch, and G. Schreiber, *Arch. Biochem. Biophys.* **215,** 309 (1982).

[54] G. Scatchard, A. C. Batchelder, and A. Brown, *J. Am. Chem. Soc.* **68,** 2320 (1946).

binding systems at maximum speed for large ($M > 100,000$) or smaller (M 18,000–100,000) proteins, respectively.[19] For small nonsedimenting ligands, the meniscus ligand concentration then represents the free ligand concentration of the binding system at equilibrium. This methodology has been used to characterize [^{14}C]CTP binding to aspartate transcarbamoylase,[19] ^{14}C-labeled 5'-AMP binding to pancreatic ribonuclease,[19] and [1-^{14}C]octanoate binding to bovine serum albumin.[21] The reversible binding of a nonsedimenting ligand to a protein may also be indirectly estimated from the sedimentation equilibrium concentration profile of the ligand throughout the protein solution. For example, from the sedimentation equilibrium profile of $1.4 \times 10^{-6} M$ [1-^{14}C]octanoate in the presence of $2.3 \times 10^{-5} M$ bovine serum albumin, Howlett and Markov determined that 18% of the [1-^{14}C]octanoate must be free in order that the concentration profile of the remaining [1-^{14}C]octanoate be consistent with the known molecular weight of 66,000 for the bovine serum albumin to which it is bound.[21] While these concentrations of bound and free ligand represent only one binding data point, they are consistent with the binding affinity constant of $2 \times 10^5 M^{-1}$ estimated by the meniscus depletion method outlined above.

In the case of larger ligands which themselves redistribute in the centrifugal field, if the protein remains significantly larger than the ligand, sedimentation equilibrium can be performed at a rotational velocity which still selectively depletes the meniscus of protein–ligand complexes. In this case, the meniscus concentration of ligand is proportional to the free ligand concentration of the entire binding system before centrifugation and this ratio is determined by centrifugation of the ligand alone under the same conditions. This method has been used to characterize the antibody-antigen binding of ^{125}I-labeled Fab fragments ($M = 48,000$) of monoclonal antibodies directed against influenza virus neuraminidase ($M = 196,000$).[55]

For ligands which redistribute in the centrifugal field, examination of the sedimentation equilibrium concentration distributions throughout the solution has great potential for more fully characterizing ligand–protein binding systems. In the analytical ultracentrifuge, only one of the concentration distributions is generally measurable (usually total protein plus total ligand), so that extraction of the molecular weight, binding stoichiometry, and binding affinity of the acceptor protein must be derived from the methods developed for the analysis of mixed associating systems.[56]

[55] D. C. Jackson, G. J. Howlett, A. Nestorowicz, and R. G. Webster, *J. Immunol.* **130**, 1313 (1983).

[56] E. T. Adams, *Ann. N.Y. Acad. Sci.* **164**, 226 (1969).

The approaches to this problem are similar to those outlined above for analysis of self-associating systems, and involve the calculation of moment-average molecular weights[57] or protomer concentrations[58,59] as a function of total protein concentration or alternatively, the development of a set of simultaneous equations from a point by point analysis of the available radial concentration distribution.[46] Sedimentation equilibrium in the air-turbine ultracentrifuge is uniquely suited to simplifying this difficult problem by permitting the determination of the concentration distributions of both the total ligand and the total protein in a mixed system at binding and sedimentation equilibrium. The concentration profile of total ligand is dependent upon the properties of the binding protein and (1) the molecular weight of the free ligand, (2) the progressive increase in the fraction of binding sites occupied due to the increased free ligand concentration in the centrifugal portions of the solution, and (3) the contribution of the bound ligand to the molecular weight of the protein–ligand complex. A comparison of the sedimentation equilibrium profiles of the protein and ligand, separately and together, would provide binding data at a continuum of free ligand concentrations resulting in a complete characterization of the equilibrium binding system. If the sedimentation equilibrium concentration distributions of the total ligand and total protein can be determined for the mixture at equilibrium, and the ligand is sufficiently small relative to the protein to permit selective meniscus depletion of complexes, then sedimentation equilibrium yields the desired equilibrium binding curve (bound ligand/protein as a function of free ligand concentration) in straightforward fashion.[60]

Characterization of Detergent-Solubilized Proteins

The widespread use of nonionic detergents to solubilize integral membrane proteins, with retention of their structural and functional integrity, has been an extremely effective biochemical technique.[61] However, considering the ambiguities of detergent content, axial ratio, and degree of hydration, the physical characterization of the resulting functional detergent–protein complexes by hydrodynamic methods presents difficulties,[61–63] and is subject to the systematic errors inherent in the empirical

[57] G. J. Howlett and L. W. Nichol, *J. Biol. Chem.* **248**, 619 (1973).
[58] L. W. Nichol, P. D. Jeffrey, and B. K. Milthorpe, *Biophys. Chem.* **4**, 259 (1976).
[59] P. D. Jeffrey, L. W. Nichol, and R. D. Teasdale, *Biophys. Chem.* **10**, 379 (1979).
[60] I. Z. Steinberg and H. K. Schachman, *Biochemistry* **5**, 3728 (1966).
[61] C. Tanford and J. A. Reynolds, *Biochim. Biophys. Acta* **457**, 133 (1976).
[62] C. Tanford, Y. Nozaki, J. A. Reynolds, and S. Makino, *Biochemistry* **13**, 2369 (1974).
[63] S. P. Grefath and J. A. Reynolds, *Proc. Natl. Acad. Sci. U.S.A.* **71**, 3913 (1974).

use as molecular weight standards of water-soluble globular proteins which bind little or no detergent.[64,65]

These difficulties in characterizing detergent–protein complexes may be effectively circumvented by sedimentation equilibrium in the air-driven ultracentrifuge in buffers with varying concentrations of D_2O and $D_2^{18}O$.[22,66] The resulting perturbations of the solution density and corresponding changes in the experimentally determined $(d \ln c/dr^2)$ allow calculation of the molecular weight M_c and partial specific volume $\bar{v}_c$ of the complex through the relation $(2RT/\omega^2)(d \ln c/dr^2) = M_c(1 - \bar{v}_c\rho)$. To accommodate substitution of deuterium for exchangeable hydrogen atoms in the macromolecule,[67] this equation may be modified to yield[22]

$$(2RT/k\omega^2)(d \ln c/dr^2) = M_c[1 - \bar{v}_c(\rho/k)] \qquad (4)$$

where k is the ratio of the molecular weight of the deuterium-substituted complex to that of the unsubstituted complex, and is dependent upon the mole fraction of deuterium in the solvent and the nature of the complex. For proteins, k at each solvent density may be taken as 1.0155 multiplied by the mole fraction of deuterium in the solvent.[68] For the nonionic detergents, there is often only a single exchangeable hydrogen atom, so that deuterium substitution may be neglected for this component. Since these corrections are relatively small, an initial estimate of the weight fraction of protein in the complex (see below) without consideration of deuterium exchange ($k = 1.000$) serves as the basis for setting k at each density equal to 1 + (weight fraction of protein) (0.0155) multiplied by the mole fraction of solvent deuterium at that density.

It is clear from Eq. (4) that $(2RT/k\omega^2)(d \ln c/dr^2)$ is a linear function of (ρ/k) with a slope of $-M_c\bar{v}_c$ and an abscissa intercept of $1/\bar{v}_c$. This graphic approach and the inclusion of $D_2^{18}O$ may offer some advantage relative to the long extrapolation implicit in the algebreic approach of Edelstein and Schachman developed for the analytical ultracentrifuge.[67] Figure 7A shows $(2RT/k\omega^2)(d \ln c/dr^2)$ as a function of (ρ/k) for 7 S ^{14}C-methylated human γ-globulin and micellar *phenyl*-^{3}H-labeled Triton X-100 at 5°. The linear data yield M and $\bar{v}$ values of 158,000 and 0.73 cm³/g for 7 S γ-globulin and 47,000 and 0.91 cm³/g for micellar Triton X-100, respectively, in good agreement with their known values.[62,69] Since this method appears valid for the characterization of these models of the protein and

[64] R. N. Fraule and D. Rodbard, *Arch. Biochem. Biophys.* **171**, 1 (1975).
[65] A. H. Maddy, *J. Theor. Biol.* **62**, 315 (1976).
[66] G. J. Howlett, H. Birch, P. W. Dickson, and G. Schreiber, *Biochem. Biophys. Res. Commun.* **105**, 895 (1982).
[67] S. J. Edelstein and H. K. Schachman, *J. Biol. Chem.* **242**, 306 (1967).
[68] A. Hvidt and S. O. Nielsen, *Adv. Protein Chem.* **21**, 287 (1966).
[69] J. R. Cann, *J. Am. Chem. Soc.* **75**, 4213 (1953).

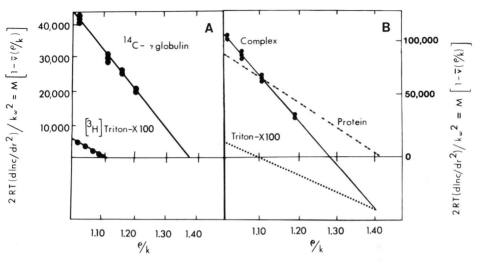

Fig. 7. (A) Density dependence of the sedimentation equilibrium behavior of human γ-globulin and Triton X-100. *phenyl*-³H-labeled Triton X-100 (10,000 cpm) was added to 1% Triton X-100 and 0.5% bovine serum albumin in phosphate-buffered saline containing varying concentrations of D_2O and $D_2{}^{18}O$ and centrifuged at 5° for 24 hr at 50,000 rpm. For each buffer density, $(2RT/k\omega^2)(d \ln c/dr^2) = M_c[1 - \bar{v}_c(\rho/k)]$ is plotted as a function of ρ/k in the above figure. The reciprocal of the abscissa intercept of the linear extrapolation of the data yields a partial specific volume of 0.91 for Triton X-100 and 0.73 for human γ-globulin. (B) Density dependence of the sedimentation equilibrium behavior for the cross-linked ¹²⁵I-labeled insulin–receptor in 0.1% Triton X-100 and 0.5% bovine serum albumin. Covalently cross-linked ¹²⁵I-labeled insulin–receptor complexes (0.2 mg of protein/100 μl) (●———●) were centrifuged for at least 36 hr at 5° in phosphate-buffered saline containing varying concentrations of D_2O and $D_2{}^{18}O$. The sedimentation equilibrium results are analyzed and presented as described above. The reciprocal of the abscissa intercept of linear extrapolation of $(RT/k\omega^2) d(\ln c)/dr^2 = M_c[1 - \bar{v}_c(\rho/k)]$ as a function of ρ/k yields a partial specific volume of 0.78 for the detergent–receptor–hormone complex. The contributions of the receptor protein ($\bar{v}_p = 0.71$) (– – –) and Triton X-100 ($\bar{v}_d = 0.91$) (· · ·) to the sedimentation equilibrium behavior of the complex are shown schematically in the figure as a function of buffer density.

detergent components of detergent-solubilized proteins, the method was applied to characterizing the insulin receptor of the cultured IM-9 lymphoblastoid cell solubilized in Triton X-100. The sedimentation equilibrium behavior of this complex in buffers of increasing density, as shown in Fig. 7B, indicates that $M_c = 475,000$ and that $\bar{v}_c = 0.78$ for this detergent–protein complex.

For sedimentation equilibrium of a multicomponent system consisting of a macromolecular protein and detergent or other components which interact with the protein, the system may be described by[9,62]

$$M_c = M_p(1 + \delta_d + \Sigma_i\delta_i) \tag{5}$$
$$\bar{v}_c = (\bar{v}_p + \delta_d\bar{v}_d + \Sigma_i\delta_i\bar{v}_i)/(1 + \delta_d + \Sigma_i\delta_i) \tag{6}$$

where δ represents the grams of component bound per gram of protein and the subscripts p, d, and i refer to the protein, detergent, and other bound components, respectively. If we assume that (1) detergent extraction of the membrane protein substitutes detergent molecules for the bound lipid and other components ($\delta_i = 0$),[70] (2) $\bar{v}_d$ of the bound detergent is equal to the partial specific volume of micellar detergent[62] ($\bar{v}_d = 0.91$ cm³/g), and (3) $\bar{v}_p$ for the receptor glycoprotein at 5° is approximately 0.71 cm³/g, then Eqs. (5) and (6) yield $M_p = 310,000$ for the receptor glycoprotein alone and $\delta_d = 0.54$ g of Triton X-100 bound per g of receptor protein. The separate contributions of the glycoprotein (dashed line) and detergent (dotted line) to $M_c[1 - \bar{v}_c(\rho/k)]$ are also shown schematically in Fig. 7. At buffer densities near 1.1, the detergent contribution to sedimentation equilibrium is negligible and the behavior of the system is determined by $M_p[1 - \bar{v}_p(\rho/k)]$. This has been the basis for an alternative method of determining M_p in the analytical ultracentrifuge.[71]

Similar methodology utilizing D₂O and an algebreic approach[67] has been applied to characterizing γ-glutamyltranspeptidase solubilized from rat renal membranes with Triton X-100, yielding $M_p = 79,000$ and $\delta_d = 1.2$ g bound detergent per g protein.[66] These methods have also been employed[24] to determine the molecular weight and partial specific volume of glycoproteins, where the carbohydrate component may comprise a large fraction of the molecule. The results of density perturbation of the solvent with D₂O during sedimentation equilibrium in the air-turbine ultracentrifuge are summarized in Table III.

New Instrumentation

The development of the new Beckman TL-100 induction-drive refrigerated tabletop ultracentrifuge and its series of small rotors should extend the applicability of the sedimentation equilibrium methodology described in this chapter. These rotors are considerably smaller than standard preparative ultracentrifuge rotors and have tube capacities of 0.2 to 2.2 ml. Thus, the fixed angle (30°) and swinging bucket rotors for this ultracentrifuge will allow solution columns of <0.5 cm and consequent achievement of sedimentation equilibrium within approximately 24–48 hr.

[70] S. Clarke, *J. Biol. Chem.* **250**, 5459 (1975).
[71] J. A. Reynolds and C. Tanford, *Proc. Natl. Acad. Sci. U.S.A.* **73**, 4467 (1976).

TABLE III

CHARACTERIZATION BY AIRFUGE SEDIMENTATION EQUILIBRIUM OF PROTEINS,
DETERGENTS, AND PROTEIN COMPLEXES IN H_2O, D_2O, AND $D_2^{18}O$

	Sedimentation equilibrium		Reference values cited	
Macromolecule	M_r	$\bar{v}$	M_r	$\bar{v}$
Kininogen light chain-fluorescein (human)[24]	30,500	0.660	—	—
[125]I-labeled human α-acid glycoprotein[66]	40,700	0.718	36,000	0.704
Micellar *phenyl*-^{3}H-labeled Triton X-100[22]	47,000	0.91	50,000	0.91
Ovalbumin-FITC[24]	46,900	0.742	44,600	0.748
[125]I-labeled rat serum albumin[66]	67,600	0.746	65,900	0.735
7 S *methyl*-^{14}C-labeled γ-globulin (human)[22]	158,000	0.73	156,000	0.729
γ-Glutamyltranspeptidase–Triton X-100 complex[66]	175,000	0.819	169,000	0.806
[125]I-labeled insulin–receptor–Triton X-100 complex[22]	475,000	0.78	—	—

Empirical Partitioning Method

Pita and Muller demonstrated the feasibility of partitioning techniques for sedimentation-equilibrium of protein solutions in capillary tubes.[72] Schachman and colleagues[73] developed an empirical practical method of determining molecular weights by partitioning of the solution following long-term centrifugation in the air-turbine ultracentrifuge. After centrifugation of 100 μl of the protein solution for 18 hr, the ratio (F) of the protein concentration remaining after centrifugation to that before centrifugation is determined for the centripetal (nearest the meniscus) 40-μl portion of the solution. The logarithm of F was found to be approximately linearly related to the reduced molecular weight σ of the macromolecule, providing an empirical standard curve for molecular weight determinations.

The theoretical ratio (F) of the average concentration of the macromolecule in the portion of the solution centripetal to a given radial distance of partition r_p to the average concentration in the entire solution is given by

$$F = \frac{\int_{r_m}^{r_p} c(r)A(r)\,dr \Big/ \int_{r_m}^{r_p} A(r)\,dr}{\int_{r_m}^{r_b} c(r)A(r)\,dr \Big/ \int_{r_m}^{r_b} A(r)\,dr} = \frac{V_{m,b}\int_{r_m}^{r_p} A(r)e^{\sigma(r^2 - r_m^2)}\,dr}{V_{m,p}\int_{r_m}^{r_b} A(r)e^{\sigma(r^2 - r_m^2)}\,dr} \tag{7}$$

[72] J. C. Pita and F. J. Muller, *Anal. Biochem.* **47**, 408 (1972).
[73] M. A. Bothwell, G. J. Howlett, and H. K. Schachman, *J. Biol. Chem.* **253**, 2073 (1978).

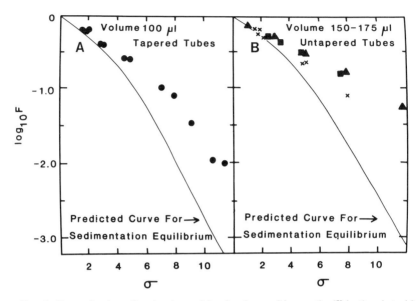

FIG. 8. Determination of molecular weights by the partition method[73] in the air-turbine ultracentrifuge. Following centrifugation of 100 μl (A) or 150 μl (B) of protein solution, the protein concentration remaining in the centripetal 40 or 50 μl of solution, respectively, is measured and divided by the protein concentration before centrifugation to yield the ratio F. The results are expressed as $\log_{10} F$ as a function of the reduced molecular weight $\sigma = M\omega^2(1 - \bar{v}_p)/2 RT$ for a series of proteins in a total volume of 100 μl (●)[73] or 150–175 μl in tubes (▲, ■, ×).[74–76] The theoretically predicted curves of sedimentation equilibrium under these conditions is shown (——) for the 100 μl in tapered tubes[73] and for 150 μl in untapered tubes by numerical integration of Eq. (7).

where $c(r)$ is determined by the integrated form of Eq. (1), $A(r)$ is the cross-sectional area of the solution during centrifugation as a function of r, and $V_{m,b}$ and $V_{m,p}$ are the volumes of the total and centripetal portions of the solution, respectively. Equation (7) may be evaluated by numerical integration for cylindrical tubes where $A(r)$ is given by Eq. (2), and was also evaluated for the tapered tubes used in the original report.[73]

The experimental results and theoretical log F values as a function of σ are presented in Fig. 8[74–76] for the partitioning of the centripetal 40 μl of a 100-μl solution in tapered (left panel) tubes, and for the centripetal 50 μl of a 150-μl solution in cylindrical tubes (right panel). The data from all laboratories show deviations from the results expected for sedimentation equilibrium, such that the F values are 2- to 50-fold larger than those

[74] J. L. Garwin, A. L. Klages, and J. E. Cronan, Jr., *J. Biol. Chem.* **255**, 11949 (1980).
[75] C. O. Rock, J. E. Cronan, Jr., and I. M. Armitage, *J. Biol. Chem.* **256**, 2669 (1981).
[76] R. G. Clarke, G. R. Eagle, and G. J. Howlett, *Aust. J. Biol. Sci.* **32**, 187 (1979).

expected for sedimentation equilibrium in reorienting solutions. These deviations may be caused by various experimental difficulties including solution mixing, incomplete solution reorientation, insufficient centrifugation time, presence of non-sedimentable material, and nonideality effects. Nevertheless, the partitioning data provide a standard curve relating log F and σ which permits molecular weight estimations for unknown samples. Therefore, while this method has become less frequently used, in appropriate circumstances the partitioning method offers a rapid practical technique for molecular weight estimation.

[2] Measurement of Changes of Hydrodynamic Properties by Sedimentation

By ROBERT W. OBERFELDER, THOMAS G. CONSLER, and JAMES C. LEE

Introduction

One often encounters evidence of conformational changes in the study of macromolecules. These structural changes are frequently documented by spectroscopic techniques. Although methods such as fluorescence, UV absorbance, and circular dichroism are highly sensitive spectroscopic techniques, the information yielded may reflect only localized structural perturbations. While these local conformational changes are certainly important, they do not necessarily reflect the *overall* changes that may be occurring in the macromolecule. In order to define accurately the conformational change in question, and to differentiate between local and global structural change, the overall picture of the macromolecule must be examined. One approach which may be used to obtain such information is to study the hydrodynamic properties of the macromolecule by sedimentation, specifically difference sedimentation velocity.[1] This technique monitors only those changes that perturb the hydrodynamic properties of the macromolecule; thus localized perturbations which do not affect the size and/or shape of the macromolecule will not be detected. Positive results using chemical or spectral probes do not necessarily indicate a global structural change, and additional experiments must be performed to ascertain the nature of the perturbation.

Conventional sedimentation studies yield information about the size and shape of macromolecules. Frequently, the sedimentation coefficient

---

[1] In this chapter, the term "difference sedimentation velocity" is used broadly. It refers to sedimentation velocity techniques which measure small changes in the sedimentation coefficient of a macromolecule.

of a macromolecule changes little as a result of the perturbation, so that the effect cannot be described quantitatively using conventional sedimentation. However, using the difference technique one can define accurately small changes in the hydrodynamic properties of the macromolecule in a manner analogous to the technique of difference spectroscopy. By comparing the sedimentation behavior of a perturbed macromolecule to that of a reference, the difference in the sedimentation coefficient may be determined directly as a result of changes in the hydrodynamic properties. Differences as small as 0.01 S have been measured with an accuracy of about ±0.0005 S using this method.

Sedimentation coefficients may be measured with a precision of greater than ±1% and accuracy of about 1%.[2] Hence, in order to measure small changes in the hydrodynamic properties of macromolecules, the difference sedimentation technique is essential. The sources of the 1% limitation in accuracy are numerous. Inaccuracies arise from the measurement of boundary positions, variations in thermal gradients within the rotor, inability to regulate and measure the temperature of the run precisely, and also inability to regulate precisely rotor speed and precession. These variations are minimized by the technique of difference sedimentation velocity. This is accomplished by using simultaneously two cells— one cell contains a solution of the macromolecule serving as the reference while the second cell contains a solution of the macromolecule with the perturbant. Any fluctuation in temperature, rotor speed, or precession will be experienced by both the reference and sample solutions. Thus, inaccuracy due to fluctuations in external variables will be minimized.

Practical Aspects

Difference sedimentation experiments are carried out using a Beckman-Spinco Model E analytical ultracentrifuge. Two-holed rotors (AnD or AnE) are employed for such experiments. Depending upon the exact chosen experimental procedure, either one or two double-sector cells are used. The choice of an optical system depends upon the desired sensitivity of the experiment. Interference optics yield higher sensitivity than schlieren optics, but require more careful optical alignment and experimental preparation.

Sample Preparation

This method requires that two samples be prepared for each experiment. One sample, used as a reference, reflects the unperturbed state of

[2] G. Kegeles and F. J. Gutter, *J. Am. Chem. Soc.* **73**, 3770 (1951).

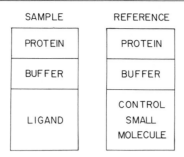

Fig. 1. Diagrammatic scheme for sample preparation.

the macromolecule. The other is the experimental sample with some perturbation of the macromolecule from the reference state. As in most experiments, careful sample preparation is crucial in order to obtain reliable and quantitatively useful data. Since the absolute value of the sedimentation coefficient depends on the concentration of the macromolecule, temperature, ionic strength, solvent density, and viscosity, it is essential that extra attention should be devoted to assure that these variables in the reference and sample solutions are as equivalent as possible. The sample and reference solutions must contain identical concentrations of the macromolecule and they must be equilibrated against the same buffer solution to minimize any fluctuations in solvent density or viscosity.

An aliquot of the ligand will be added to the sample solution, thus, perturbing the solvent density, viscosity, and ionic strength. In order to correct for these nonspecific effects on the sedimentation behavior of the sample macromolecule, noninteracting ligands must be introduced into the reference solution. Tetramethylammonium chloride was employed, for example, in the study of pyruvate kinase to compensate for the nonspecific effects of phenylalanine.[3] The basic procedure for sample preparation is summarized in Fig. 1.

Selection of Optical Systems

The choice of an optical system is dictated principally by the accuracy and precision desired by the investigator. Schlieren optics can measure accurately a change in the sedimentation coefficient (Δs) of ± 0.1 S. However, for a Δs in the range of ± 0.01 S, interference optics must be employed. Each optical system requires a different set of experimental procedures. Those that are adopted for the schlieren optical system are

[3] R. W. Oberfelder, L. L.-Y. Lee, and J. C. Lee, *Biochemistry* **23**, 3813 (1984).

FIG. 2. Sedimentation patterns for rabbit muscle pyruvate kinase in 50 mM Tris, 72 mM KCl, 7.2 mM MgSO$_4$ at pH 7.5 and 4°.

much simpler to execute; hence, it should be employed for initial exploratory tests.

Schlieren Optics. If schlieren optics are chosen, two double-sector cells are used. One of the two cells is assembled using a wedge window (usually 1 or 2°). This allows for the image of one cell to be displaced on the photographic plate, such that each photograph shows two separate schlieren patterns, as in Fig. 2, each representing one of the two cells. This allows for ready comparison of the two samples and permits the determination of the difference sedimentation coefficient. One cell contains the reference macromolecule solution in one sector and the matching reference buffer solution in the other. Both solutions are made up using a substance that balances the solution properties with those of the experimental solutions that contain the ligand. The other cell contains the macromolecule solution in one sector and the buffer solution in the other sector. The basic set up of the experiment is depicted schematically in Fig. 3. Care should be taken to fill each sector to exactly the same level, so that the menisci are superimposed, allowing the easiest analysis

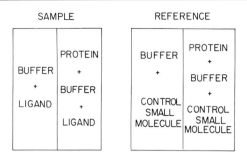

FIG. 3. Diagrammatic scheme for various solutions in each compartment of the two double-sector cells.

of the results. The cells are placed in opposing holes of the two-holed rotor and carefully aligned before starting the run. The reference hole in the rotor is used to mark the reference edge in the photograph, so the plugs must be removed from the rotor in this method. Schlieren patterns are recorded as a function of time on Kodak metallographic plates and analyzed on a microcomparator.

Analysis of these results can be performed in either of two ways. First, calculations of sedimentation coefficients are based on the movement of the maximum refractive index gradient as measured with the microcomparator. All sedimentation coefficients are calculated from the slope of the straight line of the logarithm of the boundary position vs time, as shown in Fig. 4. The observed sedimentation coefficient must be corrected to val-

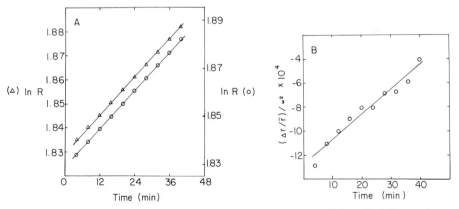

FIG. 4. Effect of 5 mM phenylalanine on the sedimentation coefficient of rabbit muscle pyruvate kinase at pH 7.5 and 4°. (A) Plots of ln (radial distance) versus time. Symbols and experimental conditions are ($\triangle$) 5 mM phenylalanine, and ($\bigcirc$), 5 mM tetramethylammonium chloride. (B) Analysis of the same data points in accordance to Eq. (2).

ues of $s_{20,w}$.[4] The value for Δs is the difference between the observed sedimentation coefficient of the sample and reference solutions. A typical set of data is shown in Fig. 4A. The effect of 5 mM phenylalanine on rabbit muscle pyruvate kinase was tested. The differences in the slopes of these plots indicates a Δs_{obs} of -0.087. Alternatively, Δs can be computed using the following equation[5]

$$\frac{1}{\omega^2} \frac{d(\Delta r/\bar{r})}{dt} = \Delta s \tag{1}$$

where ω is the angular velocity, $\Delta r = r_2 - r_1$, $\bar{r} = (r_1 + r_2)/2$, and t is time. r_2 and r_1 are the radial positions of the schlieren peaks of the sample and reference solution, respectively. The same set of data for pyruvate kinase was analyzed in accordance with Eq. (1) and the results are shown in Fig. 4B. The slope of the plot yields a Δs_{obs} of -0.087. Thus, the two procedures can produce equivalent results. Analysis of data by Eq. (1) expands the data efficiently, thus facilitating evaluation.

Interference Optics. Utilizing this optical system results of high accuracy and precision can be obtained, although much more effort is required for optical alignment and data analysis.[6,7] It is likely that the extra effort and skill required has limited the utilization of this procedure; however, it is one of the few available procedures which allows measurement of precise changes in the hydrodynamic properties of macromolecules.

For determination of a difference sedimentation rate using interference optics, one double-sector cell is used. Kirschner and Schachman[6] reported that unfilled epoxy centerpieces proved advantageous for such experiments because the two sectors have essentially the same slope and almost equal volumes. Centerpieces made from epoxy resin filled with aluminum or charcoal are less satisfactory due to uneven shrinkage during the molding process. These authors further suggested that the lower window holder should have slits of 1 mm width and the upper window holder 4-mm openings in order to facilitate the alignment of the centerpiece walls along radii from the center of the rotor. The cell is assembled as for conventional interference applications. One of the compartments must contain the reference macromolecule solution and the other compartment the sample solution. The two compartments should be filled so that there is a meniscus separation of 0.1 ± 0.05 mm to facilitate subsequent measurements. In this procedure, the cell contains both of the solutions, so the appropriate counterbalance is used in the other hole of the rotor.

[4] T. Svedberg and K. O. Pedersen, "The Ultracentrifuge." Oxford University, London and New York, 1940 (Johnson Reprint Corp., New York).
[5] G. H. Howlett and H. K. Schachman, *Biochemistry* **16**, 5077 (1977).
[6] M. W. Kirschner and H. K. Schachman, *Biochemistry* **10**, 1900 (1971).
[7] M. W. Kirschner and H. K. Schachman, *Biochemistry* **10**, 1919 (1971).

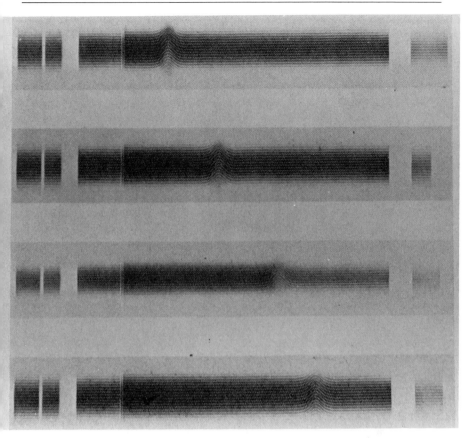

FIG. 5. Difference sedimentation patterns for bushy stunt virus (BSV). Both compartments were filled with BSV at a concentration of 5 mg/ml in 0.2 M potassium phosphate buffer at pH 6.8; the rotor speed was 17,000 rpm. One compartment contained 0.03 ml/ml of D_2O. The photographs were taken at 48, 64, 76, and 88 min after reaching speed. The third photograph was taken with achromatic light using flat glass in place of the Wratten 77A filter. (Reprinted with permission from Kirschner and Schachman.[6] Copyright 1971 American Chemical Society.)

The sedimentation difference between the two samples will be reflected directly on the photographic plate. When a difference in the sedimentation rate exists between the two samples, the interference pattern will show a series of lines, with a peak appearing at the point where there is a difference in the boundary positions of the two solutions. This peak appears similar to those generated by conventional schlieren optics, i.e., it reflects a difference in the refractive indices of the two solutions, as shown in Fig. 5. If there had been no difference between the two samples, the interference pattern would show only a series of straight parallel lines.

Analysis of the interferograms in the difference sedimentation experiment is accomplished by determination of the first moment of the area under the "schlieren-like" profile,[6] since

$$\frac{\int_{\bar{r}_m}^{r_p} \Delta C r dr}{\bar{r}_m^2 C_0} = \left[\left(\frac{r_p}{\bar{r}}\right)^2 \ln \frac{\bar{r}}{\bar{r}_m} \frac{\Delta s}{\bar{s}}\right] - \frac{\Delta \bar{r}_m}{\bar{r}_m} \tag{2}$$

where C_0 and ΔC are the initial macromolecular concentration and difference in concentration between the sample and reference solutions, respectively, $\bar{r}$ is average radial distance, r_p, $\bar{r}_m$ are the radial distance at the plateau region and average radial distance of the meniscii, respectively, $\Delta \bar{r}_m$ is the distance between the two menisci, and $\Delta s/\bar{s}$ is the fractional difference in sedimentation coefficient. Hence, $\Delta s/\bar{s}$ can be obtained directly as the slope of a plot of

$$\frac{\int_{\bar{r}_m}^{r_p} \Delta C r dr}{\bar{r}_m^2 C_0} \quad \text{vs} \quad \left(\frac{r_p}{\bar{r}}\right)^2 \ln \frac{\bar{r}}{\bar{r}_m}$$

C_0 can be obtained in separate low-speed ultracentrifuge experiments with a double-sector synthetic boundary cell. The other parameters are obtainable from the difference patterns where they are resolved into x and y components in the microcomparator. Values of r can be obtained from the x coordinates of these patterns in a manner identical to that in conventional sedimentation experiments. The first moment of the difference concentration distribution, i.e.,

$$\int_{\bar{r}_m}^{r_p} \Delta C r dr$$

is calculated by summing the ordinates, ΔC, measured at equal increments of the square of the radial distance from the axis of rotation.[8] Values of ΔC are proportional to the vertical displacement of the fringes in the difference patterns. A typical set of data for determining the first moment is shown in Table I. Hence, using difference patterns obtained at different times during centrifugation one can tabulate a number of determinations on the first moment and $\Delta s/\bar{s}$ can be determined, as shown in Fig. 6.

Kirschner and Schachman[6] estimated the accuracy of the two sedimentation procedures in measuring hydrodynamic changes and the results are shown in Table II. It is evident that the procedure employing interference optics is more accurate for monitoring small changes in the hydrodynamic properties of macromolecules.

[8] E. G. Richards, Ph.D. Thesis, University of California, Berkeley (1960).

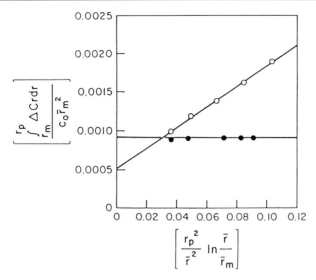

Fig. 6. Difference sedimentation of BSV in H_2O–D_2O solutions. The filled circles represent experiments with identical solutions of BSV at a concentration of 5 mg/ml. The open circles represent data from the experiment shown in Fig. 5 ($\Delta s/\bar{s} = 0.0114$). The data are plotted according to Eq. (2) and the slopes provide a direct measure of $\Delta s/\bar{s}$. (Reprinted with permission from Kirschner and Schachman.[6] Copyright 1971 American Chemical Society.)

Parameters That Cause Changes in Hydrodynamic Properties

There are several parameters which might account for the observed changes in sedimentation coefficients, since

$$s = \frac{M(1 - \bar{v}\rho)}{Nf} \tag{3}$$

where M and $\bar{v}$ are the weight average molecular weight and partial specific volume of the macromolecule, respectively, ρ is the solvent density, N is Avogadro's number, and f is the frictional coefficient. The observed ligand-dependent change in the sedimentation coefficient might arise from a change in M, $\bar{v}$, or f.

Besides promoting conformational changes, ligands may also induce or alter the propensity of the macromolecule to undergo association–dissociation. It is, therefore, essential to obtain independent measurements on the concentration dependence of the sedimentation coefficient. This can be accomplished by conducting a series of sedimentation velocity experiments as a function of macromolecular concentration in the presence and absence of ligand. Sedimentation equilibrium experiments

TABLE I
DETERMINATION OF THE FIRST MOMENT OF A DIFFERENCE
CONCENTRATION DISTRIBUTION[a]

r^b (mm) (1)	y^c (mm) (2)	$y_0{}^d$ (mm) (3)	$(y - y_0)^e$ (mm) (4)	Interval number (5)
13.762	4.791			
14.016	4.793			
14.269	4.793			
14.522	4.792			
14.774	4.792	4.792	0.000	1
15.026	4.773		−0.019	2
15.277	4.726		−0.066	3
15.528	4.626		−0.016	4
15.778	4.468		−0.324	5
16.028	4.308		−0.484	6
16.243[f]	4.325		−0.557	6.9
16.277	4.242		−0.550	7
16.526	4.362		−0.430	8
16.775	4.521		−0.271	9
17.023	4.681		−0.111	10
17.271	4.784		−0.008	11
17.518	4.859		+0.067	12
17.765	4.881		+0.089	13
18.011	4.894	4.894	−0.102	14
18.257	4.894			
18.502	4.896			
18.747	4.894			

Meniscus position $(\bar{r}_m)$, 7.645 mm
Maximum ordinate position $(\bar{r})$, 16.243 mm
Position in plateau region, 18.011 mm

The trapezoidal sum of the ordinates, $\sum\limits_{i=1}^{n} \Delta c_i$, is 2.229 mm

$$\int_{r_m}^{r_p} \Delta c r \, dr \cong \Delta \frac{r^2}{2} \sum_{i=1}^{n} \Delta(c_i - c_i, \text{baseline})$$

$$\cong \Delta \frac{r^2}{2} \sum_{i=1}^{n} \Delta c_i - \frac{n}{2} \Delta c_p$$

$$\cong \frac{15}{2} 2.229 - \frac{14}{2}(-0.102)$$

$$\cong 22.07 \text{ mm}^3$$

[a] Both compartments of a double-sector cell were filled with BSV at a concentration of 5 mg/ml in 0.1 M potassium phosphate buffer, pH 6.80. One compartment contained 3 ml/ml of D_2O.
[b] The radial position in mm is measured from the reference wire. The

TABLE II
COMPARATIVE ACCURACY IN MEASURING
HYDRODYNAMIC CHANGES

	Error in measurement (%)	
Change (%)	Schlieren optics[a]	Interference optics[a]
Volume		
30	4	5
5	25	5
0.3	—	5
Axial ratio		
50	3	5
5	30	5
0.5	—	5

[a] Using procedures outlined in this chapter.

can also yield useful information on this issue. A more efficient procedure is to conduct differential sedimentation experiments.[9] These measurements can yield information on s^0, the sedimentation coefficient at infinite dilution and k, the parameter representing the fractional decrease in the sedimentation coefficient with concentration. If the ligand does not affect the association–dissociation equilibrium of the macromolecule then k is expected to remain the same both in the presence and in the absence of the ligand. The value of k should change if the degree of association is ligand dependent. The values of s^0 should also yield information on the structure of the macromolecule.

The additional contribution of weight by the ligand upon formation of a macromolecule–ligand complex should yield a higher sedimentation

[9] R. Hersh and H. K. Schachman, *J. Am. Chem. Soc.* **77**, 5228 (1955).

interval between each position is at equal increments of the square of the radial distance (15 mm²).

[c] The y ordinate represents the vertical position of the zero-order fringe as read directly on the microcomparator in mm.

[d] y_0 is the average ordinate in millimeters for the supernatant and plateau regions. The readings at low r values correspond to the supernatant and those at high values of r represent the plateau region.

[e] $y - y_0$, where y_0 represents the reading in the supernatant; the absolute value of $y - y_0$ provides a direct measure of Δc_i in millimeters of fringe displacement.

[f] This position corresponds to the maximum ordinate of the difference pattern.

coefficient. One should make an estimate of the maximum contribution of this parameter to the change in $s_{20,w}$ in order to determine whether or not the change in $s_{20,w}$ can be accounted for by this small additive effect.

It is possible that a ligand-dependent change in $\bar{v}$ might arise simply through the additive effect of complex formation, without involving structural changes in the protein. Values for the $\bar{v}$ of the ligand can be calculated according to the method of Traube[10] and McMeekin et al.[11] By assuming that the effect of the ligand on the $\bar{v}$ of the macromolecule is additive, it is possible to estimate the contribution of a $\bar{v}$ change on the observed change in $s_{20,w}$.

Having corrected for the weight and $\bar{v}$ of bound ligands the residual change in $s_{20,w}$ can be attributed to changes in the hydrodynamic properties of the macromolecule. Sedimentation data alone do not provide enough information to resolve the relative contribution of a volume change and an axial ratio change to the observed change in $s_{20,w}$. Additional experiments such as small angle X-ray scattering should be conducted in order to probe the contributions of these factors.

Applications

The major impact of difference sedimentation velocity is in resolving models of allosteric regulation of enzymes.[12,13] In these models it is proposed that an enzyme can exist in various states, each having different affinities for substrates and effectors. The binding of allosteric ligands affects the equilibrium between these states. Kinetic and equilibrium binding data alone do not provide enough information to determine whether the two-state[12] or the sequential model[13] is the best explanation for the data. Differentiation between the two models requires that ligand-dependent changes in the global macromolecular structure be evaluated. According to the two-state model, the completion of the structural change should precede the saturation of the macromolecule with ligands, whereas the sequential model makes no such prediction. In the latter case the structural change will occur in phase with ligand saturation.

One example of this type of application is the interaction of rabbit muscle pyruvate kinase with L-phenylalanine.[3] L-Phenylalanine exerts an allosteric effect upon the activity of this glycolytic enzyme by shifting the steady-state kinetics from hyperbolic to sigmoidal. Using the technique of

[10] J. Traube, Samml. Chem. Chem.-Tech. Vortr. **4**, 255 (1899).
[11] T. L. McMeekin, M. L. Groves, and N. J. Hipp, J. Am. Chem. Soc. **71**, 3298 (1949).
[12] J. Monod, J. Wyman, and J.-P. Changeau, J. Mol. Biol. **12**, 88 (1965).
[13] D. E. Koshland, Jr., G. Némethy, and D. Filmer, Biochemistry **5**, 365 (1966).

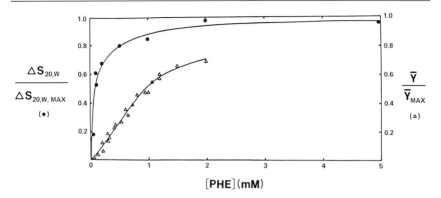

FIG. 7. Comparison of the phenylalanine (PHE) binding isotherm of rabbit muscle pyruvate kinase with the induced structural change, in the presence of KCl and MgSO₄. The lines are the best fit by visual inspection. The symbols are defined as shown in the figure.

difference sedimentation velocity, a global structural change is detected. L-Phenylalanine was shown to produce a titratable structural perturbation, reflecting nearly a 3% (Δs of -0.24 S) change in the sedimentation coefficient. In conjunction with ligand binding data, the sedimentation information can be used to suggest possible mechanistic details, as shown in Fig. 7. It is evident that the completion of the state change precedes the saturation of the enzyme with ligand. At 0.5 mM phenylalanine the state change is greater than 80% complete while the protein is only 25% saturated with the ligand. From this evidence, the two-state model seems to be the most likely. A very similar observation was reported for aspartate transcarbamylase from *Escherichia coli.*[5]

The interaction between NAD$^+$ and rabbit muscle glyceraldehyde-3-phosphate dehydrogenase is a good example of the sequential model. Fuller Noel and Schumaker[14] showed that the fractional change in sedimentation coefficient increases *linearly* as a function of ligand saturation, as shown in Fig. 8. An examination of the concentration dependence of the sedimentation coefficients of the apo- and holoenzymes excludes the possibility of changes in the state of aggregation of the enzyme with ligand binding. The data are most consistent with the sequential mechanism of conformational changes. However, Conway and Koshland[15] reported that all the sulfhydryl groups become exposed upon binding of the first molecule of NAD$^+$. Based on the latter study the data could have been misinterpreted as being evidence for a concerted two-state model. These obser-

[14] J. K. Fuller Noel and V. N. Schumaker, *J. Mol. Biol.* **68,** 523 (1972).
[15] A. Conway and D. E. Koshland, Jr., *Biochemistry* **7,** 4011 (1968).

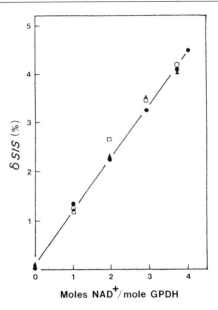

FIG. 8. A plot of the fractional change in the sedimentation coefficient of rabbit muscle glyceraldehyde-3-phosphate dehydrogenase as a function of NAD^+ saturation. Difference sedimentation studies were performed on titrated samples of apoenzyme in D_2O buffer at 5° (○), 10° (□), 15° (▲), and 20° (●), using the apoenzyme as a standard. The solid line is the least-squares fit to the experimental data. (Reprinted with permission from Fuller Noel and Schumacher.[14])

vations clearly demonstrate the pitfalls of using results from techniques that are sensitive to local structural changes to fit models that require global structural information.

In conclusion, difference sedimentation velocity is a powerful technique to obtain information on the global structural change in macromolecules. Using interference optics and associated procedures, a very sensitive and accurate technique is available to monitor changes in the hydrodynamic properties of macromolecules.

Acknowledgments

This work was supported in part by NIH Grants NS-14269 and AM-21489.

[3] Determination of Protein Molecular Weight in Complexes with Detergent without Knowledge of Binding

By JACQUELINE A. REYNOLDS and DARRELL R. McCASLIN

Introduction

The molecular mass of a functioning protein unit or its constituent polypeptides is a physical parameter of fundamental importance in structure–function investigations. Rigorous thermodynamic methods for determining this parameter include light scattering, osmotic pressure, and sedimentation equilibrium. Each of these techniques has unique advantages and drawbacks, and each has limitations on the minimum and maximum molecular mass that can be determined with accuracy. This chapter is concerned with the use of sedimentation equilibrium, a procedure that is generally applicable to proteins in the molecular weight range 10,000 to 500,000 and can be used conveniently for protein–amphiphile complexes without interference from non-protein-containing detergent micelles.

In this chapter we shall present a summary of the theoretical treatment of sedimentation equilibrium in a centrifugal field for systems containing two or more components and discuss the effects of specific interactions between the protein and solute molecules. We shall then proceed to a description of experimental procedures and present examples of the application of this technique to protein–amphiphile complexes. Partial specific volumes are tabulated for amino acid residues, carbohydrates, and a number of detergents and lipids. Atomic and group partial molar volumes that are used to calculate partial specific volumes for chemical compounds are also provided.

Our primary consideration is the determination of protein molecular weight in a *homogeneous* system by a rigorous thermodynamic method, and we will not discuss in detail the mathematical treatment of reversibly (or irreversibly) associating macromolecular particles.

Theory

The molecular weight of a protein in *any* solution can be determined unambiguously by sedimentation equilibrium if the interactions of the protein with solvent components (e.g., H_2O, glycerol, amphiphiles) are explicitly treated.[1,2] In the ultracentrifuge the distribution of protein con-

[1] E. F. Casassa and H. Eisenberg, *Adv. Protein Chem.* **19**, 287 (1964).
[2] J. A. Reynolds and C. Tanford, *Proc. Natl. Acad. Sci. U.S.A.* **73**, 4467 (1976).

METHODS IN ENZYMOLOGY, VOL. 117

centration, c, as a function of the distance from the center of rotation, r (cm), is related to the anhydrous *protein* molecular weight, M_p by the following equation.

$$(2RT/\omega^2)(d \ln c/dr^2) = M_p(1 - \phi'\rho) \tag{1}$$

where ω is the radial velocity in radians/sec, ρ is the *solvent* density in g/cm^3, and $(1 - \phi'\rho)$ is the buoyant density factor which includes contributions from the protein and all bound solvent components. T is the temperature in degrees Kelvin and R is the gas constant, 8.3144×10^7 ergs deg^{-1} mol^{-1}. $(d \ln c/dr^2)$ is the slope of a plot of the log of the protein concentration as a function of the square of the radial distance.

For a two-component system consisting of only protein dissolved in water, the buoyant density factor is

$$(1 - \phi'\rho) = (1 - \bar{v}_p\rho) + \delta_{H_2O}(1 - \bar{v}_{H_2O}\rho) \tag{2}$$

where ρ is the density of water, $\bar{v}_p$ is the partial specific volume of the protein, and δ_{H_2O} is the grams of bound water per gram of protein. The partial specific volume is a thermodynamic parameter defined by the following equation:

$$\bar{v}_p = (\delta V/\delta g_p)_{T,P,g_j} \tag{3}$$

It is simply the change in the volume of a solution on the addition of g_p grams of protein at constant mass of solvent, g_j, and constant T and P.

Since the partial specific volume of water is the inverse of its density, the second term in Eq. (2) is zero and the buoyant density factor is simply $(1 - \bar{v}_p\rho)$ for the system protein–water. (Note that δ_{H_2O} is *not* zero, but bound water does not contribute to the buoyant density term due to the relationship between $\bar{v}_{H_2O}$ and ρ.)

In a multicomponent system where there are interactions between the protein and other solutes the buoyant density factor includes not only the partial specific volume of the protein, but also the mass of ligands bound to the protein and their partial specific volumes. Thus, in analogy to Eq. (3)

$$\phi' = (\delta V/\delta g_p)_{T,P,\mu_i} \tag{4}$$

The difference between this expression and Eq. (3) is that all solutes other than protein are at constant *chemical potential,* μ_i, i.e., they are at equilibrium across a semipermeable membrane.

While in principle ϕ' can be measured directly (see below), a more practical and convenient procedure is to calculate $(1 - \phi'\rho)$ from the following equation.[1]

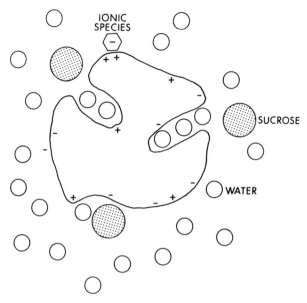

FIG. 1. Schematic diagram of a macromolecular particle with a rough surface surrounded by solvent molecules of varying types and sizes.

$$(1 - \phi'\rho) = (1 - \bar{v}_p\rho) + \delta_{H_2O}(1 - \bar{v}_{H_2O}\rho) + \sum_i \delta_i(1 - \bar{v}_i\rho) \qquad (5)$$

where $\bar{v}_p$ is the protein partial specific volume, $\bar{v}_i$ is the partial specific volume of each bound component, i, and δ_i is the grams of component i bound per gram of protein. As noted in the discussion of Eq. (2), when H_2O is the principal solvent present at 55.5 M, the second term on the right side of Eq. (5) is essentially zero.

The term, δ_i, is the preferential interaction of other solutes with the protein,[3] i.e., it is a measure of the *excess* (or *depletion*) of solutes in the domain of the protein relative to the composition of the bulk solvent. It is apparent that preferential interaction can be either positive or negative since solutes can be preferentially excluded or preferentially "bound" to the macromolecule under investigation. Figure 1 is a schematic diagram of a protein molecule surrounded by water, small ions, and sucrose. Note that water may be occluded ("bound") in clefts that are inaccessible to the larger solute, sucrose. Thus sucrose is preferentially excluded in this case and water molecules preferentially "bound."

[3] J. C. Lee, K. Gekko, and S. N. Timasheff, this series, Vol. 61, p. 26.

The determination of δ_i is often not feasible for a variety of experimental reasons, but it is clear from Eq. (5) that adjustment of the solvent density such that it is the reciprocal of $\bar{v}_i$ will negate the contribution of component i to the buoyant density factor. Since the various isotopes of H_2O are expected to be isomorphous, exchanging freely with bound H_2O, it is a straightforward procedure to increase the solvent density using $H_2{}^{18}O$, ^2H_2O, or $^2H_2{}^{18}O$ and thus maintain the system in a state where the *isotopic* composition of bound *principal solvent* remains the same as that of bulk *principal solvent*. Solutes such as sucrose and CsCl cannot be used to increase the density of the solvent since the high concentrations of these compounds that are required to increase solvent density will alter the activity of water significantly and lead to preferential exclusion (or inclusion) of solvent components in the domain of the protein. In those cases it becomes necessary to know the magnitude of δ, g solute/g protein, for H_2O and other solutes, i, since the terms $\bar{v}_{H_2O}$ and $\bar{v}_{CsCl}$ or sucrose are not equal to the reciprocal of the principal *solvent* density.

It should be noted that there is an alternative approach to the formulation of appropriate mathematical expressions for the sedimentation behavior of protein–amphiphile complexes. It is possible to consider these complexes as a single component such that

$$M_c = M_p \left(1 + \sum_i \delta_i \right) \tag{6}$$

and

$$\bar{v}_c = \left(\bar{v}_p + \sum_i \bar{v}_i \right) \Big/ \left(1 + \sum_i \delta_i \right) \tag{7}$$

From sedimentation equilibrium data at several solvent densities it is possible to determine $\bar{v}_c$ as the reciprocal of the density at which no sedimentation occurs, i.e., $(1 - \bar{v}_c\rho)$ is zero. M_c is then obtained from sedimentation data at a density such that $\bar{v}_c$ is not equal to $1/\rho$ and M_p together with δ_i are calculated from Eqs. (6) and (7).

Experimental Methods

Instrumentation

A Model E analytical ultracentrifuge equipped with a photoelectric scanner is used to obtain the equilibrium distribution of protein–amphiphile complexes as a function of distance from the center of rotation. The concentration of protein is customarily determined by measuring the absorption of light at 280 nm. Unfortunately, some amphiphilic ligands also have significant intrinsic absorption at this wavelength (e.g., the Triton series of detergents) or are contaminated with ultraviolet-absorbing compounds, and these should be avoided in studies of protein molecular weights if at all possible. Some proteins, of course, possess prosthetic groups with absorption bands at wavelengths removed from those of the amphiphilic ligands, and in these cases interference can be avoided by monitoring the protein concentration at the characteristic wavelength of the prosthetic group. The molecular weight of bacteriorhodopsin was determined in Triton X-100 using this procedure.[4] A similar problem occurs if the molecular weight of a protein–amphiphile complex is to be determined in the presence of a substrate which also absorbs at 280 nm (e.g., Ca^{2+}-ATPase detergent in the presence of ATP).

In general, several rotor speeds and initial protein concentrations should be examined and plots of $\ln c$ vs r^2 tested for linearity. When a single sedimenting species is present, $\ln c$ vs r^2 is a straight line at all speeds and initial concentrations. However, if the system is heterogeneous with respect to particle weight or if the particle is self-associating, curvature in these plots will be observed.

Since the binding of a ligand to protein is an equilibrium process, free ligand must always be present to assure the integrity of the complex, and any alteration in the concentration of free ligand will result in changes in binding. It is, therefore, essential that an appropriate centrifuge speed be chosen such that unbound micelles of amphiphiles do not sediment appreciably. If the aggregation number, initial concentration, and $\bar{v}$ of the amphiphile are known, the expected distribution at equilibrium can be calculated directly.

Several scans of the protein distribution at equilibrium are collected and $\ln$ (optical density) calculated as a function of r^2. If the sample is homogeneous, $\ln$ OD vs r^2 is a straight line and the slope can be determined by a least-squares fit to the data. Nonlinear relationships between $\ln$ OD and r^2 can result from self-association, either reversible or irreversible. For a reversible, self-associating system the dependence of $M_p(1 - \phi'\rho)$ on concentration should be independent of the angular velocity and the initial loading concentration. Details of the mathematical treatment of associating systems are beyond the scope of this chapter and may be found in a number of appropriate references.[5,6]

In order to determine how much of the initial protein concentration can be accounted for by the observed sedimentation data, it is essential to

[4] J. A. Reynolds and W. Stoeckenius, *Proc. Natl. Acad. Sci. U.S.A.* **74**, 2803 (1977).
[5] R. H. Haschemeyer and A. E. V. Haschemeyer, *in* "Proteins," p. 181. Wiley, New York, 1973.
[6] R. E. Steiner, *Arch. Biochem. Biophys.* **39**, 333 (1952).

calculate the total recovery of protein in the centrifuge cell at equilibrium. For a nonassociating component

$$c_i = c_0 \frac{B(r_b^2 - r_a^2)}{\exp[B(r_b^2 - r_a^2)] - 1} \exp[B(r_i^2 - r_a^2)] \tag{8}$$

where c_i is the concentration at any r_i, c_0 is the concentration of protein initially in the cell, r_a is the distance of the meniscus from the center of rotation, r_b is the distance of the bottom of the cell from the center of rotation, and $B = M(1 - \phi'\rho)\omega^2/2RT$. Equation (8) applies to any component of molecular weight M with partial specific volume, $\bar{v}$, and can be used to calculate the distribution of non-protein-containing micelles in a specific centrifugal field as discussed previously.

Sample Preparation

Convenient initial optical densities for sedimentation equilibrium using the photoelectric scanner are 0.05 to 0.25, and it is customary to use 0.1-ml samples in order to attain equilibrium rapidly. Samples containing potentially oxidizable substances such as unsaturated fatty acids or non-ionic detergents with polyoxyethylene head groups should be protected with a blanket of argon, and charcoal-filled Epon centerpieces should be used rather than aluminum-filled Epon.

Amphiphilic solutes and lipid-binding proteins partition into fluorocarbon, so a layer of this latter substance cannot be used in the bottom of the centrifuge cell. The absence of this fluorocarbon layer limits the accumulation of data at the bottom of the solution column.

Determination of $\bar{v}$'s and ϕ'

In principle, ϕ' can be measured directly for any solution containing protein(–ligand) complexes by determining the density of these solutions as a function of protein(–ligand) concentration. The solutions must be dialyzed to equilibrium to assure that all dialyzable solvent components are at constant chemical potential. A plot of $\rho_{solution}$ vs c (grams protein/cm³) is used to obtain the limiting slope at infinite dilution which is equal to $(1 - \phi'\rho_0)$, where ρ_o is the density of pure solvent. In practice, however, this procedure requires large amounts of protein and very precise measurements of the concentration of the protein solution. Additionally, many amphiphilic ligands do not dialyze in reasonable periods of time, particularly when the critical micelle concentration is lower than 10^{-4} M.

Similarly, $\bar{v}$'s for single chemical entities can be determined experimentally by measuring the increase in density of the solution as a function of increasing solute concentration. As described for ϕ', the limiting slope

TABLE I
PARTIAL SPECIFIC VOLUMES OF DETERGENTS AND LIPIDS[a]

	$\bar{v}(cm^3/g)$
Normal alkyl detergents	
Octyl glucoside	0.859
Dodecyltrimethylammonium chloride	1.083
Dodecyltrimethylammonium bromide	0.958
Sodium dodecyl sulfate	0.864
Dodecyl octaethyleneglycol monoether	0.973
Tetradecyltrimethylammonium chloride	1.110
Tetradecyltrimethylammonium bromide	0.985
Sodium tetradecyl sulfate	0.890
Brij 56 (16 carbons, 10 oxyethylenes)	0.955
Brij 58 (16 carbons, 20 oxyethylenes)	0.919
Normal alkenyl detergents	
Brij 96 (18 : 1 carbons, 10 oxyethylenes)	0.973
Tween 80 (18 : 1 carbons, 20 oxyethylenes	
attached to sorbitan)	0.896
Substituted phenyl detergents	
Triton X-100 (octylphenyl, 9.7 oxyethylenes)	0.908
Triton N-101 (nonylphenyl, 9.5 oxyethylenes)	0.922
Bile acids	
Deoxycholate	0.778
Glycodeoxycholate	0.77
Taurodeoxycholate	0.76
Chenodeoxycholate	0.78
Cholate	0.771
Glycocholate	0.75
Taurocholate	0.75
Lipids and steroids	
Cholesterol	
Monomer in water	0.988
Micelle	0.949
Monomer in benzene	1.021
Egg yolk lecithin (bilayer in water)	0.984

[a] More extensive tabulations of $\bar{v}$'s can be found in J. C. H. Steele, C. Tanford, and J. A. Reynolds, this series, Vol. 48, p. 11, from which these data were taken. The values for octyl glucoside and cholate were measured by D. R. Mc-Caslin.

at infinite dilution of a plot of $\rho_{solution}$ vs c in grams/cm^3 is $(1 - \bar{v}\rho_0)$. A tabulation of $\bar{v}$'s for a variety of amphiphiles is provided in Table I. (Note that $\bar{v}$ is not necessarily independent of concentration of solute and the tabulated values are obtained from data extrapolated to infinite dilution or to the critical micelle concentration.)

TABLE II
GROUP PARTIAL MOLAR VOLUMES

	Experimental[a] (cm³/mol)	Calculated[b] (cm³/mol)
C	9.1	9.9
H	3.0	3.1
N		1.5
O′		0.4–2.3
O″		5.5
S		15.5
NH_2		7.7
OH	12.0	8.6
CH_2	16.9	16.1
CH_3	17.3	19.2
SH	14.4	18.6
$CH(CH_3)_2$	47.6	51.4
$CH_3CH(CH_3)_2$	64.5	67.5
CH_2SH	30.3	34.7
$CH_2C{\overset{\displaystyle \nearrow O}{\underset{\displaystyle \searrow OH}{}}}$	30.6	35.0
$\underset{H_2N-CNH(CH_2)_3}{\overset{HN \diagdown}{}}$	84.2	88.4
(OCH_2CH_2)	37.1	37.7
H^+	−5.8	
Na^+	−6.6	
K^+	3.6	
Rb^+	8.7	
NH_4^+	13.0	
NH_3^+	14.0	
Ca^{2+}	−28.6	
Cl^-	23.6	
Br^-	30.5	
OH^-	1.8	
SO_4^{2-}	24.8	
HPO_4^{2-}	18.5	

[a] F. J. Millero, A. L. Surdo, and C. Shin, *J. Phys. Chem.* **82,** 784 (1978); H. L. Friedman and C. V. Krishnan, *in* "Water" (F. Franks, ed.), p. 1. Plenum, New York, 1973.

[b] Calculated from E. J. Cohn and J. T. Edsall, "Proteins, Amino Acids, and Peptides." Van Nostrand-Rheinhold, Princeton, New Jersey, 1943.

TABLE III
CALCULATED PARTIAL SPECIFIC VOLUMES OF
AMINO ACIDS AND SUGARS[a]

Residue	$\bar{v}(cm^3/g)$
Amino acid	
Asp	0.60
Asn	0.62
Thr	0.70
Ser	0.63
Glu	0.66
Gln	0.67
Pro	0.76
HyPro	0.68
Gly	0.64
Ala	0.74
Cys	0.61
Val	0.86
Met	0.75
Ile	0.90
Leu	0.90
Tyr	0.71
Phe	0.77
Lys	0.82
His	0.67
Arg	0.70
Trp	0.74
Hexose	0.613
n-Acetylhexosamine	0.666
Methyl pentose	0.678
n-Acetylneuraminic acid	0.548
n-Glycolloylneuraminic acid	0.557

[a] The *residue* partial specific volumes and weights do not contain the molecule of water than is eliminated on formation of a peptide bond between two amino acids or an oligosaccharide linkage between two sugar molecules.

In general, $\bar{v}$ can be calculated with reasonable accuracy from knowledge of the atomic constituents of the molecule.[7]

[7] E. J. Cohn and J. T. Edsall, "Proteins, Amino Acids, and Peptides." Van Nostrand-Reinhold, Princeton, New Jersey, 1943. This equation does not include the "co-volume" term which is 14.1 cm^3/g mol and must be added to the total volume obtained by summation of atom and group volumes for each molecule. For a protein or other macromolecule of large molecular weight, this term is negligible. For charged molecules in aqueous solution a term arising from electrostriction of the solvent must also be included. For α amino acids this number is -13.3 cm^3/mol.

TABLE IV
DENSITY OF WATER AND WATER ISOTOPES[a,b]

	F	d_{max}	t_{max}
H_2O	1	1	3.986
^2H_2O	1.0555	1.10602	11.230
$H_2^{18}O$	1	1.11255	4.298
$^2H_2^{18}O$	1.0555	1.21691	11.458

[a] F. Steckel and S. Szapiro, *Trans. Faraday Soc.*
59, 331 (1963). $\tau = t - t_{max}$. t_{max} and t in °C. d_{max}
and d in g/cm³.
[b] $1 - (d/d_{max}) = F\tau^2(1.74224 + 482.502/(\tau + 77.861))10^{-6}$.

$$\bar{v} = \left(\sum_i \bar{v}_i\omega_i\right)\bigg/\sum_i \omega_i \qquad (9)$$

Molar volumes of atoms and some chemical groups are given in Table II, and the $\bar{v}$'s of amino acid and selected carbohydrate residues are provided in Table III. All of these data refer to aqueous solutions since $\bar{v}$ contains terms representative of solvent–solute interactions and hence differs from one solvent to another. $\bar{v}$'s for amphiphiles presented in Table I are those for the micellar state since it is not unreasonable to assume that this state is more akin to that of a protein–amphiphile complex than is the monomeric state in which the amphiphile is completely surrounded by water.

Procedures in the Absence of Knowledge of Ligand Binding

When the determination of δ_i, the amount of bound ligand/gram protein, is not feasible the contribution of any *one* bound ligand to the buoyant density term can be negated as described in the theoretical section of this chapter. If heavy water is used to adjust the solvent density such that $\rho = 1/\bar{v}_i$, both the molecular weight of the protein and $\bar{v}_p$ are altered in a reciprocal fashion by isotopic substitution of exchangeable protons on the protein(–ligand) moiety. The magnitude of this alteration can be calculated if the amino acid composition of the protein is known. The use of ^{18}O water would from this standpoint be more desirable, but the cost is somewhat greater than for heavy water.

Figure 2 shows an example of data obtained for the major polypeptide from human high-density serum lipoprotein complexed with two different detergents and centrifuged in solvents of differing density. In this case, the "blank out" density of the detergent is in an accessible range using ^2H_2O to increase the solvent density. If the detergents possessed $\bar{v}$'s

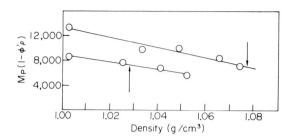

FIG. 2. Sedimentation equilibrium data for AI from human high-density lipoprotein complexed with Lubrol WX (upper line) and dodecyl octaethyleneglycol monoether (lower line). Total detergent concentrations were 5 mM and protein concentration was 5×10^{-6} M. Arrows indicate the value of $\rho = 1/\bar{v}$ for each detergent. The AI polypeptide contains 440 exchangeable protons per molecule which would increase the molecular weight by 1.5% for *complete* isotopic substitution in pure ²H₂O and decrease $\bar{v}_p$ to 0.724 cm³/g from 0.736 cm³/g. M_p in dodecyl octaethyleneglycol monoether is 29,000 and in Lubrol WX is 32,000. The sequence molecular weight is 28,400.

outside of the accessible density range, the data could have been extrapolated outside the region of experimental points with a consequent loss in accuracy. This polypeptide has been sequenced and it should be noted that the molecular weight obtained in the homogeneous detergent, dodecyl octaethyleneglycol monoether, is within 2% of the correct value. However, the heterogeneous detergent, Lubrol WX, gives a molecular weight that is too high by 11%, a result that has been attributed to binding of specific components in the heterogeneous ligand and a consequent incorrect value of $\bar{v}_i$, i.e., the average $\bar{v}$ was assumed, which will be

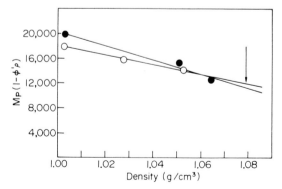

FIG. 3. The AI polypeptide from human serum high-density lipoprotein in didecanoyl phosphatidylcholine at two different lipid binding levels. The point of intersection of the two lines showing $M_p(1 - \phi'\rho)$ as a function of ρ is the experimentally determined value of $1/\bar{v}_{\text{lipid}}$. The arrows show the calculated value of $1/\bar{v}_{\text{lipid}}$.

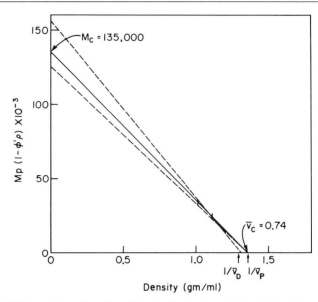

FIG. 4. $M_c(1 - \bar{v}_c\rho)$ as a function of density for bovine rhodopsin in sodium cholate. The experimentally determined ranges of $M_c(1 - \bar{v}_c\rho)$ for samples prepared in H_2O and ^2H_2O are indicated by short vertical bars at $\rho = 1.01$ and 1.11 g/cm^3, respectively. The dashed lines are the limiting lines when constrained to pass through the experimental data at the two densities and to have a ρ axis intercept between $1/\bar{v}_{\text{cholate}}$ and $1/\bar{v}_{\text{rhodopsin}}$. The solid line is the best fit line when the protein–detergent complex is also required to contain an integral number of rhodopsin polypeptide chains ($M = 41,745$) and to have a positive value for δ_{cholate}.

incorrect if there is preferential binding of specific subpopulations of the heterogeneous detergent. This hypothesis could have been tested if data had been obtained for two levels of Lubrol WX binding to the AI polypeptide in solvents of increasing density. In this case, two separate straight lines of differing slope would have been observed for plots of $M_p(1 - \phi'\rho)$ vs ρ for the two different samples, and these two lines would have intersected at a density equal to the reciprocal of the partial specific volume of the *bound* detergent. An example of this type of experiment is shown in Fig. 3 in which the ligand is didecanoyl phosphatidylcholine and the protein is again the AI polypeptide. The arrow marks the reciprocal of the calculated ligand partial specific volume. Note that the two lines intersect at a slightly different value, namely, 0.941 rather than 0.927 cm^3/g. Using the former measured value for $\bar{v}$ of dodecanoyl phosphatidylcholine, the protein molecular weight is 58,000 corresponding to 2.04 polypeptide chains per particle. If the calculated $\bar{v}$ is used, the molecular weights are 55,500 and 52,500 for the two samples, again corresponding to 1.85 and

1.95 copies per particle. It is clear that these results are all within experimental error equal to 2 polypeptides/complex.

One of the more serious problems that presents itself in the determination of protein molecular weight in a protein–amphiphile complex occurs when the $\bar{v}$ of the amphiphile is close to the $\bar{v}$ of the protein and the binding is sufficiently weak that δ_i cannot be determined with a sufficient degree of accuracy. This situation occurs for rhodopsin in sodium cholate (v_p is 0.744 g/cm^3 and v_{cholate} is 0.771 g/cm^3) and Fig. 4 provides an example of one means to deal with it. $M_c(1 - \bar{v}_c\rho)$ was determined as a function of solvent density using $^2\text{H}_2\text{O}:\text{H}_2\text{O}$ mixtures [see Eqs. (6–7)]. For a detergent–protein complex $M_c(1 - \bar{v}_c\rho)$ is a linear function of density if the effects of deuterium exchange are negligible. Since $1/\bar{v}_c$ as defined by Eq. (7) must lie between $1/\bar{v}_p$ and $1/\bar{v}_{\text{cholate}}$, the point of intersection at $M_p(1 - \phi'\rho) = 0$ of the line fitted to the experimental data is bounded as shown in Fig. 4, placing an additional constraint on the system. The two broken lines represent the limiting fits to the data with the requirement that the x intercept must lie between $1/\bar{v}_p$ and $1/\bar{v}_{\text{cholate}}$. The solid line is the best fit with two additional constraints, namely, the known subunit molecular weight of rhodopsin (41,745) and the requirement that δ_{cholate} be positive.

[4] Calculation of Partial Specific Volumes of Proteins in 8 M Urea Solution

By V. Prakash and Serge N. Timasheff

The partial specific volume, $\bar{v}$, of a protein is a quantity which plays a major role in the accurate determination of the molecular weight of proteins by sedimentation equilibrium and in the correct interpretation of small-angle X-ray scattering results. A commonly used procedure to determine the number of subunits or to characterize the association–dissociation process in a protein is to measure the molecular weight of the native associated protein in the presence of low ionic strength buffers, followed by a measurement of the molecular weight of the dissociated subunits in the presence of high concentrations of a denaturant, such as guanidine hydrochloride *or* urea. Lack of knowledge of the correct value of $\bar{v}$ in the presence of the denaturant can lead to erroneous conclusions on the stoichiometry of subunit composition in the assembly of macromole-

METHODS IN ENZYMOLOGY, VOL. 117

cules.[1,2] Even though a method is available for the calculation of the partial specific volume of a protein in the presence of guanidine hydrochloride,[3] no method is available for the estimation of $\bar{v}$ in the presence of urea. Due to the lack of such methods, it is a common practice to estimate this quantity in the presence of urea, especially when the protein is available in only a small quantity.

The partial specific volumes of native proteins in dilute buffer are usually calculated according to the procedure of Cohn and Edsall,[4] which yields reasonably correct values of $\bar{v}$. However, in the presence of high concentrations of denaturants, such as 8 M urea, this method of calculation neglects the contribution of the protein–solvent interactions which can be considerable, since in a three-component system, namely, macromolecule, denaturant, and water, the macromolecule generally interacts with one or both of the solvent components in the bulk solvent. Such preferential interactions introduce a correction into the calculated value of $\bar{v}$, and unless this correction is applied, the calculated $\bar{v}$ may be in error by as much as 0.02 ml/g. Our recent study on the interactions of proteins with urea by density measurements, however, shows that the magnitude of the change in the presence of urea may be much higher. An error of such magnitude in the partial specific volume may lead to very serious errors in the calculated molecular weight, when this value of $\bar{v}$ is used in conjunction with sedimentation equilibrium in 8 M urea. Even though there is information in the literature on the variation of $\bar{v}$ of some proteins and the binding of urea,[5] no procedure for its calculation was available until recently.

Based on the results of our study,[6,7] a simple method has been proposed[7] to calculate the apparent partial specific volumes of proteins in 8 M urea, which greatly reduces the uncertainties inherent in the previously used procedures. The ony information required for the computation of $\bar{v}$ in urea is the amino acid composition of the protein. The methodology stems from the study of the interactions of proteins with urea and the calculation is based on the fact that the total binding of urea to proteins can be calculated from the experimentally measured preferential interac-

[1] E. Reisler and H. Eisenberg, *Biochemistry* **8**, 4572 (1969).

[2] D. L. Barker and W. P. Jencks, *Biochemistry* **8**, 3879 (1969).

[3] J. C. Lee and S. N. Timasheff, *Biochemistry* **13**, 257 (1974).

[4] E. J. Cohn and J. T. Edsall, "Proteins, Amino Acids and Peptides," p. 370. Van Nostrand-Reinhold, Princeton, New Jersey, 1943.

[5] J. Span and S. Lapanje, *Biochim. Biophys. Acta* **295**, 371 (1973).

[6] V. Prakash, C. Loucheux, S. Scheufele, M. J. Gorbunoff, and S. N. Timasheff, *Arch. Biochem. Biophys.* **210**, 455 (1981).

[7] V. Prakash and S. N. Timasheff, *Anal. Biochem.* **117**, 330 (1981).

tion parameter, along with the calculated hydration of proteins at a known pH according to the procedure of Kuntz.[8]

Theory

The molecular weight, $M_{2,app}$, determined by sedimentation equilibrium in the three component system consisting of water, protein, and urea is related to the molecular weight, M_2, after extrapolation to zero protein concentration, by

$$M_{2,app} = \frac{2RT}{(1 - \bar{v}_2\rho)_{m_3}\omega^2} \frac{d \ln C_2}{d(r^2)}$$

$$= M_2 \left[1 + \frac{(1 - \bar{v}_3\rho)_{m_2}}{(1 - \bar{v}_2\rho)_{m_3}} \left(\frac{\partial g_3}{\partial g_2}\right)_{T,P,\mu_3} \right] \tag{1}$$

where water, protein, and urea are components 1, 2, and 3, respectively, following the standard notation of Scatchard[9] and Stockmayer,[10] ρ is the density of the solvent, C_2 is the protein concentration in any units, R is the universal gas constant, T is the temperature in degrees Kelvin, P is pressure, ω is the angular acceleration, r is the distance from the center of rotation, g_i is the concentration of component i in grams per gram of water in the system, $\bar{v}$ is its partial specific volume, and μ_3 is the chemical potential of component 3. The subscripts m_j indicate that the partial specific volume of component i is measured at conditions at which the molality, m_j, of component j ($j \neq i$) is identical in the bulk solvent and in the solution of component i. Hence, the value of the apparent molecular weight obtained experimentally contains a contribution from the preferential interactions between the protein and solvent components denoted by the term $(dg_3/dg_2)_{T,P,\mu_3}$.[11]

What is the effect of urea denaturation on the buoyancy of a sedimenting protein molecule? The buoyancy is affected in two ways. First, the volume of the protein may change upon denaturation; second, the interactions of the protein with solvent components may affect the buoyancy of the kinetic particle. Experimental evidence indicates that the major change in buoyancy results from the interactions of the denatured protein with the solvent components and that the change in volume of a protein upon denaturation in 6 M GuHCl or 8 M urea is generally small and its

[8] I. D. Kuntz, *J. Am. Chem. Soc.* **93**, 514 (1971).
[9] G. Scatchard, *J. Am. Chem. Soc.* **68**, 2315 (1946).
[10] W. H. Stockmayer, *J. Chem. Phys.* **18**, 58 (1950).
[11] S. N. Timasheff, *Adv. Chem. Ser.* **125**, 327 (1973).

effect on the partial specific volume of the protein is negligible.[3,6] Thus, in practice one may assume that ϕ_2, the partial specific volume at infinite dilution of the protein, measured at conditions at which the molalities of the solvent components in the protein solution and in the reference solvent are identical, is equal to $\bar{v}_2$, the partial specific volume of the native protein in dilute buffer.

With the above concept in mind, the true molecular weight, M_2, may be obtained from standard sedimentation equilibrium data, if the partial specific volume of the protein is measured under conditions at which the chemical potential of component 3 in the protein solution is identical with that in the bulk solvent. Thus, at protein concentration extrapolated to zero,

$$M_2 = \frac{2RT}{(1 - \phi_2'\rho)} \frac{d \ln C_2}{d (r^2)} \tag{2}$$

where ϕ_2' is the apparent partial specific volume of the protein in chemical equilibrium with the solvent.

Hence the purpose now is to evaluate theoretically ϕ_2' in 8 M urea from a knowledge of the amino acid composition of the protein. At infinite dilution of the macromolecular species, the quantity ϕ_2' is related to the preferential interaction parameter,

$$(\partial g_3/\partial g_2)_{T,P,\mu_3} \simeq (\partial g_3/\partial g_2)_{T,\mu_1,\mu_3} \equiv \xi_3$$

by the relation

$$\phi_2' = \phi_2 - \xi_3(1/\rho - \bar{v}_3) \tag{3}$$

since,[12,13]

$$\xi_3 = [(\partial\rho/\partial g_2)_{\mu_3} - (\partial\rho/\partial g_2)_{m_3}]/(\partial\rho/\partial g_3)_{m_2} \tag{4}$$

The preferential interaction parameter is related, in turn, to the total amounts of solvent components bound to the protein by,[14]

$$\xi_3 = A_3 - g_3 A_1 \tag{5}$$

where A_3 is the total solvation expressed as grams of urea per gram of protein, A_1 is the total hydration in grams of water per gram protein, and g_3 is the solvent composition, expressed as grams of urea per gram of water. Combining Eqs. (3) and (5) results in

$$\phi_2' = \phi_2 - (1/\rho - \bar{v}_3) (A_3 - g_3 A_1) \tag{6}$$

[12] E. F. Casassa and H. Eisenberg, Adv. Protein. Chem. **19**, 287 (1964).
[13] G. Cohen and H. Eisenberg, Biopolymers **6**, 1077 (1968).
[14] H. Inoue and S. N. Timasheff, Biopolymers **11**, 737 (1972).

Examination of Eq. (6) shows that ϕ_2', the apparent partial specific volume of a protein in chemical equilibrium with solvent, can be calculated with a knowledge of ϕ_2, A_1, A_3, g_3, ρ, and $\bar{v}_3$. Values of ρ and $\bar{v}_3$ are available in the literature for a number of solvent systems and, even if not available, they can be determined easily. For 8 M urea, $\rho = 1.1152$ g/ml at 20° and $\bar{v}_3 = 0.763$ ml/g.[6] The parameter g_3 is determined by solvent composition; for 8 M urea the value of g_3 is 0.752 g urea/g water.[6] The only unknown parameters at this stage are A_1, A_3, and ϕ_2. A_1 can be calculated[3,7,15] for each protein from the hydration values of its constituent amino acids. We have selected the method of Kuntz.[8] A_3 is calculated by using the model of one urea molecule being bound to each pair of peptide bonds in the protein and to each aromatic amino acid side chain including histidine.[6] ϕ_2 is set equal to $\bar{v}_2^0$ which, in turn, can be calculated from the amino acid composition of the protein by the method of Cohn and Edsall.[4] Substitution of all the above values into Eq. (6) yields the value of ϕ_2'.

Method of Calculation

The method of calculation will be presented, using chymotrypsinogen as the example.[16] The amino acid composition of the protein has been arbitrarily expressed as residues per 10^5 of protein.

Calculation of $\bar{v}_2^0$

The partial specific volume of the protein is calculated from its amino acid composition,[4] using values of the partial specific volumes, $\bar{v}_i$, of the constituent amino acids, by[4]

$$\bar{v}_2 = \sum N_i(W_i\bar{v}_i) \Big/ \sum N_iW_i \qquad (7)$$

where N_i is the number of residues of amino acid of type i and W_i is its residue weight, i.e., the molecular weight reduced by 18.0. The individual values of W_i and $\bar{v}_i$ are listed in the table for the amino acids. The amino acid composition of chymotrypsinogen is also listed in the table. Using these values and Eq. (7), a value of $\bar{v}_2^0 = 0.731$ ml/g is calculated for chymotrypsinogen. The measured value in dilute buffer is 0.733 ml/g.[17,18]

[15] J. C. Lee and S. N. Timasheff, *Arch. Biochem. Biophys.* **165,** 268 (1974).
[16] This calculation has been reproduced from Prakash and Timasheff.[7]
[17] J. C. Lee, K. Gekko, and S. N. Timasheff, this series, Vol. 61, p. 49.
[18] J. Skerjanc, V. Dolecek, and S. LaPanje, *Eur. J. Biochem.* **17,** 160 (1970).

CALCULATION OF THE PARTIAL SPECIFIC VOLUME FROM HYDRATION AND AMINO ACID DATA FOR CHYMOTRYPSINOGEN A[a]

Amino acid	W_i[b] (g/mol)	$\bar{v}_i$[c] (ml/g)	$W_i\bar{v}_i$ (ml/mol)	N_i (residues/ 10^5g protein)	N_iW_i	$N_iW_i\bar{v}_i$	H_i[d] (hydration mol H_2O/ mol residue)	H_iN_i
Lys	128.2	0.82	105.1	55	7051	5781.8	4.5	247.5
His	137.2	0.67	91.9	8	1097.6	735.4	4	32
Arg	156.2	0.70	109.3	16	2499.2	1749.4	3	48
Asp	115.1	0.60	69.1	35	10359	6219	2	70
Asn	—	—	—	55			2	110
Thr	101.1	0.70	70.8	90	9099	6369.3	2	180
Ser	87.1	0.63	54.9	109	9493.9	5981.2	2	218
Glu	129.1	0.66	85.2	20	7617.9	5026.8	2	40
Gln	—	—	—	39			2	78
Pro	97.1	0.76	73.8	35	3398.5	2582.9	3	105
Gly	57.1	0.64	36.5	90	5139	3289.0	1	90
Ala	71.1	0.74	52.6	86	6114.6	4524.8	1.5	129
Cys	103.2	0.61	63.0	39	4024.8	2455.1	1	39
Val	99.1	0.86	85.2	90	8919	7670.3	1	90
Met	131.2	0.75	98.4	8	1049.6	787.2	1	8
Ile	113.2	0.90	101.9	39	4414.8	3973.3	1	39
Leu	113.2	0.90	101.9	74	8376.8	7539.1	1	74
Tyr	163.2	0.71	115.9	16	2611.2	1854.0	3	48
Phe	147.2	0.77	113.3	23	3385.6	2606.9	0	0
Trp	186.2	0.74	137.8	31	5772.2	4271.4	2	62
Total				958	100422.7	73416.9		1707.5

[a] $A_3 = 0.334$ g/g; $\rho = 1.1152$ g/ml; $\bar{v}_3 = 0.763$ ml/g; $g_3 = 0.751$ g/g; $A_1 = 0.307$ g/g; $\bar{v}_2^0 = 0.73$ ml/g. ϕ_2' (calc) $= 0.717$ ml/g; ϕ_2' (exp) $= 0.720$ ml/g.

[b] $W_i = M_i - 18.0$, where M_i is molecular weight of neutral amino acid.

[c] $\bar{v}_i$ = specific volume of amino acid residue.

[d] The hydration values are at pH 4.0, obtained from Kuntz.[8]

Calculation of A_1

The total hydration of a protein, A_1, is given by the relation,

$$A_1 = \sum H_i N_i \tag{8}$$

where H_i is the hydration of amino acid species i. Kuntz[8] has determined the values of H_i for all the amino acids at several pH values and the values at pH 4.0 are listed in the table. According to the data of Kuntz,[8] aspartate, glutamate, and tyrosyl residues show a strong pH dependence of their hydration. Since in the determination of the partial specific volumes of various proteins in urea solution[6] and the subsequent development of a model for urea binding to proteins, the pH of the system was 4.0, we have used values of 2.0 and 2.0 for the hydration of aspartic and glutamic acids. The total hydration of the protein was calculated to be 1707.5 mol of H_2O/ 10^5 g protein, i.e., $A_1 = 0.307$ g of H_2O/g protein.

Calculation of A_3

The extent of urea binding to proteins in 8 M urea solution is given by[6]

$$\frac{\text{moles of urea}}{\text{moles of protein}} = \frac{\sum N_i - 1}{2} + \sum N_{\text{aromatic}} \tag{9}$$

Using the above equation, A_3 can be calculated by arbitrarily setting the molecular weight of the protein equal to 10^5.

$$A_3 = \frac{\text{moles of urea}}{\text{moles of protein}} \times \frac{M_r \text{ urea}}{M_r \text{ protein}}$$

$$= \frac{60}{100,000} \left(\frac{\sum N_i - 1}{2} + N_{\text{aromatic}} \right) \tag{10}$$

The arbitrary molecular weight, 10^5, gives for chymotrypsinogen 958 residues per 10^5 g, and a total of 957 peptide bonds. Since one molecule of urea binds to each pair of peptide bonds,[6] 479 molecules of urea are bound to the backbone of chymotrypsinogen. In addition, one molecule of urea is bound to each aromatic side chain, giving another 78 molecules of urea bound to the protein. This results in a total of 554 mol urea/10^5 g of chymotrypsinogen, or $A_3 = 0.334$ g urea/g protein.

Calculation of ϕ_2'

Substitution of the above values of $\bar{v}_2^0$, A_1, and A_3 into Eq. (6) yields directly ϕ_2'. From such a calculation, a value of $\phi_2' = 0.717$ ml/g was obtained for chymotrypsinogen. This should be compared to the experimentally determined value of 0.720 ml/g.[6]

General Comments

While the presently described method of calculating ϕ_2' is generally applicable in the determination of molecular weights in 8 M urea, it must be applied with caution, since other parameters, such as volume changes upon denaturation and the presence of carbohydrate, lipids, and nucleic acids may interfere with the calculations. As has been discussed earlier, any volume change which occurs when a protein system is transferred from water to 8 M urea solution should also manifest itself in the partial specific volume change. In such a case, ϕ_2 is not equal to $\bar{v}_2^0$, but the deviation should not be great since usually these volume changes are small.

In some cases, however, the volume decrease is significant. For example, transfer from water to 8 M urea of β-lactoglobulin, papain, and α-chymotrypsin leads to volume changes of -240, -440, -300 ml/mol, respectively.[6] On the other hand, for the proteins β-lactalbumin, lysozyme, and chymotrypsinogen A, the volume decrease is 30, 70, and 80 ml/mol, respectively.[6] At present there is no definite set of rules for predicting the volume change which accompanies the unfolding of any given protein. The magnitude of this volume change can be expected, however, to be similar in various denaturants, so that knowledge of the volume change upon unfolding in any strong denaturant can be taken as an indication of the extent to which it will affect the overall calculation of ϕ_2' in 8 M urea. Finally, this method of calculation of ϕ_2' in 8 M urea should not be used for glycoproteins, lipoproteins, and proteins associated with nucleic acids.

Acknowledgments

This work was supported in part by NIH Grants GM-14603 and CA-16707.

[5] Calculation of the Partial Specific Volume of Proteins in Concentrated Salt and Amino Acid Solutions[1]

By T. ARAKAWA and S. N. TIMASHEFF

The physicochemical characterization of proteins, at times, requires working in solvents which contain high concentrations of low-molecular-weight compounds. For example, organisms adapted to low water poten-

[1] This work was supported in part by NIH Grants GM-14603 and CA-16707.

tial accumulate low-molecular-weight compounds at high concentrations to raise the osmotic pressure of the cytoplasm.[2] These include salts, such as NaCl and KCl, amino acids, and related compounds. Since the proteins of these organisms frequently are unstable in dilute salt,[3] the characterization of their physicochemical properties, such as molecular weight, must be made in concentrated solution of these compounds.[4] Parameters which play an important role in sedimentation equilibrium and sedimentation velocity experiments, density gradient sedimentation, as well as in X-ray scattering, X-ray diffraction, and some column techniques are the preferential hydration of the proteins and the apparent partial specific volume, $\phi_2^{\prime \circ}$, of the protein measured at dialysis equilibrium, i.e., under conditions where the chemical potentials of the additive are identical in the protein solution and the bulk solvent. This last quantity can be greatly different from the true partial specific volume, ϕ_2°, of the protein, reflecting the preferential interactions of the protein with the solvent components.[5-8] For example, $\phi_2^{\prime \circ}$ has been found to be higher than ϕ_2° by as much as 0.05 ml/g in 1 M Na_2SO_4 for BSA[7,9] and this difference may become even greater for smaller proteins. Whereas ϕ_2° can be calculated from the amino acid composition of a protein with reasonable accuracy, the isopotential apparent partial specific volume, $\phi_2^{\prime \circ}$, can be obtained only experimentally, e.g., by densimetry, after equilibrating the protein solution with the reference solvent by dialysis. This procedure, however, requires relatively large amounts of protein and extensive dialysis, which may not be practical for many proteins because of stability problems.[10] In view of these difficulties it would seem useful to have a simple method of estimating the isopotential apparent partial specific volume in such concentrated solvents. We have developed such a procedure for concentrated salt and amino acid solutions.

Lee and Timasheff[11] and Prakash et al.[12] have developed a method for calculating $\phi_2^{\prime \circ}$ in 6 M guanidine–HCl and 8 M urea solutions. This

[2] P. H. Yancey, M. E. Clark, S. C. Hand, R. D. Bowlus, and G. N. Somero, *Science* **217**, 1214 (1982).

[3] J. K. Lanyi, *Bacteriol. Rev.* **38**, 272 (1974).

[4] R. Jaenicke, *Annu. Rev. Biophys. Bioeng.* **10**, 1 (1981).

[5] K. C. Aune and S. N. Timasheff, *Biochemistry* **9**, 1481 (1970).

[6] T. Arakawa and S. N. Timasheff, *Biochemistry* **22**, 6536 (1982).

[7] T. Arakawa and S. N. Timasheff, *Biochemistry* **22**, 6545 (1982).

[8] T. Arakawa and S. N. Timasheff, *Arch. Biochem. Biophys.* **224**, 169 (1983).

[9] Abbreviation used: BSA, bovine serum albumin.

[10] J. C. Lee, K. Gekko, and S. N. Timasheff, this series, Vol. 61, p. 26.

[11] J. C. Lee and S. N. Timasheff, *Anal. Biochem.* **117**, 330 (1981).

[12] V. Prakash, C. Loucheux, S. Scheufele, M. J. Gorbunoff, and S. N. Timasheff, *Arch. Biochem. Biophys.* **210**, 455 (1981).

method is based on the fact that the extents of denaturant binding and hydration are functions of the amino acid composition.[12,13] Since the patterns of preferential interaction for salts and amino acids with proteins are completely different from those for the denaturants, that method cannot be applied. We have found, however, that in these systems there exists a correlation between the preferential interaction and the protein surface area[14] which renders possible the calculation of the preferential interactions for various proteins from the data obtained with other proteins. This method is described in the present chapter.

Calculation

Defining component 1 = water, component 2 = protein, and component 3 = additive, the isopotential apparent partial specific volume, $\phi_2^{\prime\circ}$, can be calculated from the isomolal partial specific volume, ϕ_2°, and the preferential hydration of the protein in the particular solvent, $\xi_1 = (\partial g_1/\partial g_2)_{T,\mu_1,\mu_3}$, by

$$\phi_2^{\prime\circ} = \phi_2^\circ + g_3\xi_1[(1/\rho_0) - \bar{v}_3]$$ (1)

where g_i is the concentration of component i in grams per gram of water, T is the thermodynamic (Kelvin) temperature, μ_i is the chemical potential of component i, ρ_0 is the density of the solvent, and $\bar{v}_3$ is the partial specific volume of the additive. The value of $\bar{v}_3$ can be either measured densimetrically[16] or calculated from the plot of the specific volume vs weight fraction of components using published data.[7,8,16-18] The value of ϕ_2° can be either determined by densimetry[10] or calculated from the amino acid composition by the method of Cohn and Edsall.[19] The additive concentration, g_3, is calculated from C_3/C_1, where C_1 and C_3 are the concentrations of water and additive in g/ml, or approximated by

$$g_3 \simeq \frac{C_3\bar{v}_1}{1 - C_3\bar{v}_3}$$ (2)

with $\bar{v}_1$ set equal to 1.0.

[13] J. C. Lee and S. N. Timasheff, *Biochemistry* **13**, 257 (1974).
[14] This characteristic property of ξ_1 arises from the mechanism of the preferential interactions in these solvent systems, namely, from the preferential exclusion of these substances from the protein surface due to the surface tension effect.[6,15]
[15] J. C. Lee and S. N. Timasheff, *J. Biol. Chem.* **256**, 7193 (1981).
[16] T. Arakawa and S. N. Timasheff, *Biochemistry* **23**, 5912 (1984).
[17] "International Critical Table," Vol. 3. McGraw-Hill, New York, 1928.
[18] L. A. Dunn, *Trans. Faraday Soc.* **62**, 2349 (1966).
[19] E. J. Cohn and J. T. Edsall, "Proteins, Amino Acids and Peptides," p. 372. Van Nostrand-Reinhold, Princeton, New Jersey, 1943.

The preferential hydration parameter, ξ_1, is related to the preferential interaction of component 3 with the protein, $\xi_3 = (\partial g_3/\partial g_2)_{T,\mu_1,\mu_3}$, by

$$\xi_1 = -\frac{1}{g_3}\left(\frac{\partial g_3}{\partial g_2}\right)_{T,\mu_1,\mu_3} \tag{3}$$

The calculation of ξ_1 is based on the observation[7,8] that, for a number of additives, this parameter is (1) proportional to the specific surface area of proteins, s_2 (except at extremes of pH) and (2) close to independent of the concentration of component 3. (The specific surface area is defined as s_2/M_2 where s_2 and M_2 are the surface area and molecular weight of the proteins.) This is true for aqueous solvent systems where component 3 is one of a number of salts (Na_2SO_4, NaCl, $MgSO_4$, and CH_3COONa), sugars (glucose, sucrose, lactose), or amino acids (glycine, alanine, and betaine). As a result, for these systems it becomes possible to calculate the value of ξ_1 for any protein B from that determined for a specific protein A by

$$\xi_1^B = \xi_1^A(s_2^B/s_2^A) \tag{4}$$

where s_2^B and s_2^A are the specific surface areas of proteins B and A, respectively. Since for globular proteins, the surface area of a protein molecule is proportional, within a close approximation, to $M_2^{2/3}$, we obtain $(s_2^B/s_2^A) = (M_2^A/M_2^B)^{1/3}$ and, hence,

$$\xi_1^B = \xi_1^A(M_2^A/M_2^B)^{1/3} \tag{5}$$

As is evident, ξ_1 is not very sensitive to molecular weight, permitting the use of approximate values of molecular weights in this calculation. Such approximate values may be estimated from, e.g., a sedimentation equilibrium experiment with the use of ϕ_2^o, instead of $\phi_2'^o$, or a gel permeation chromatography experiment.

The preferential interactions have been measured for a number of amino acids[8] and salts.[7] The following experimentally determined values of ξ_1 can be used to calculate the preferential hydrations and isopotential partial specific volumes of proteins. Values of ξ_1 obtained for lysozyme ($M_2 = 14,300$) were 0.55 g/g for glycine and β-alanine, 0.50 g/g for α-alanine, 0.45 g/g for betaine, and 0.55 g/g for CH_3COONa. Values of ξ_1 measured for BSA ($M_2 = 68000$) were 0.45 g/g for Na_2SO_4, 0.25 g/g for NaCl, and 0.38 g/g for $MgSO_4$. These values may be used as ξ_1^A in Eq. (5). For KCl, one may use the ξ_1 value determined for NaCl.[20]

---

[20] Values of ξ_1 have also been tabulated by Kuntz and Kauzmann,[21] who have expressed them as the hydration parameter in three component solutions.

[21] I. D. Kuntz and W. Kauzmann, *Adv. Protein Chem.* **28**, 239 (1974).

Sample Calculations

For lysozyme in 1 M NaCl, $\phi_2'^\circ$ was calculated to be 0.722 ml/g from the ξ_1 value measured with BSA ($\xi_1^{BSA} = 0.25$), with $\rho_0 = 1.04$ g/ml, $g_3 = 0.0597$ g/g, $\bar{v}_3 = 0.331$ ml/g, and $\phi_2^\circ = 0.707$ ml/g: the experimental value is 0.723 ml/g. Other values calculated in this manner were β-lactoglobulin in 1 M MgSO$_4$, $\phi_2'^\circ$, calculated $= 0.788$ ml/g ($\phi_2'^\circ$ experimental $= 0.785$ ml/g); lysozyme in 1 M MgSO$_4$, 0.748 ml/g (0.747 ml/g); BSA in 1 M CH$_3$COONa, 0.748 ml/g (0.747 ml/g); BSA in 1 M β-alanine, 0.745 ml/g (0.744 ml/g); BSA in 1 M α-alanine, 0.744 ml/g (0.744 ml/g).

The Case of Ammonium Sulfate

The use of $(NH_4)_2SO_4$ as a protein precipitant and medium for X-ray crystallographic studies suggests that it would be useful to know $\phi_2'^\circ$ values of proteins in this solvent, although no ξ_1 data are available. This can be calculated from the data of Tuengler et al.,[22] who have determined $\phi_2'^\circ$ as a function of $(NH_4)_2SO_4$ concentration for two proteins. Rearranging Eq. (1) and introducing the values of ϕ_2° [set equal to $\phi_2'^\circ$ in the absence of $(NH_4)_2SO_4$] and $\phi_2'^\circ$, obtained from lobster tail lactate dehydrogenase ($M_2 = 136,000$) in 1.2 M $(NH_4)_2SO_4$, and g_3, calculated approximately from their data, we obtain

$$\xi_1[(1/\rho_0) - \bar{v}_3] = (\phi_2^\circ - \phi_2'^\circ)/g_3 = 0.215 \text{ ml/g} \qquad (6)$$

Assuming that $[(1/\rho_0) - \bar{v}_3)]$ is constant, this quantity can be used in place of ξ_1. This approach was tested by calculating $\phi_2'^\circ$ at various $(NH_4)_2SO_4$ concentrations for pig heart lactate dehydrogenase ($M_2 = 145,300$) and comparing with the experimental values.[22] The results were calculated value $= 0.742$ ml/g (experimental value $= 0.740$ ml/g), 0.751 ml/g (0.749 ml/g), 0.760 ml/g (0.756 ml/g), and 0.769 ml/g (0.764 ml/g) at 0.3, 0.6, 0.9, and 1.2 M salt, respectively. In all cases, the calculated values agreed with the measured ones within the experimental error.

A calculation of the parameter of Eq. (6) for BSA, after correction for the molecular weight difference, gave a value of 0.27 ml/g, which is smaller than that for Na$_2$SO$_4$ (~ 0.35 ml/g) but greater than that for NaCl (~ 0.17 ml/g), indicating that the preferential interaction of $(NH_4)_2SO_4$ with proteins is intermediate, between those of Na$_2$SO$_4$ and NaCl, provided that $\bar{v}_3$ for $(NH_4)_2SO_4$ is not very different from the value for these two salts.

[22] P. Tuengler, G. L. Long, and H. Durchschlag, *Anal. Biochem.* **98,** 481 (1979).

Comments

Although, in general, the agreement is good between values of $\phi_2'^\circ$ calculated according to the procedure of this chapter and experimental values, deviations can be expected when proteins have highly asymmetric shapes or in the case of lipoproteins and glycoproteins. Nevertheless, the calculations described in this chapter give estimates of $\phi_2'^\circ$ which can be used in analyzing sedimentation equilibrium results. On the other hand, the use of ϕ_2° leads to serious errors, in particular when the salts are Na_2SO_4 and $(NH_4)_2SO_4$ and for smaller proteins. It should be mentioned also that the effect of the concentrated salts on $\phi_2'^\circ$ is not simply due to high ionic strength, since salts such as KSCN and $MgCl_2$ have either little effect on $\phi_2'^\circ$ or even lower it at high salt concentrations.[7,16] Therefore, it is totally incorrect to attempt such calculations for one salt using interaction data obtained for another salt.

[6] Molecular Weight Determinations from Radiation Inactivation

By J. T. HARMON, T. B. NIELSEN, and E. S. KEMPNER

Introduction

Biological molecules damaged by ionizing radiation (X rays, γ rays, or high energy electrons) lose activity. The activity remaining after irradiation of lyophilized or frozen samples can be analyzed by target theory, yielding the molecular weight of the functional unit independent of the molecular shape or volume. The technique of radiation inactivation depends on the deposition of large amounts of energy directly in the biologically active structure under conditions where radiation products from other molecules are without effect on the measured property. Such direct radiation damage is independent of the surrounding environment. It follows that purification is not required; if the starting material has sufficient activity, molecular weights can be determined after irradiation of intact cells, seeds or frozen tissues.

Each interaction of high energy radiation with a molecule results in the deposition of massive amounts of energy, of the order of 1500 kcal/mol. This severely damages the molecular structure and results in a complete loss of biological activity. The energy is dissipated by complex mechanisms which result in the large scale breakage of covalent bonds. After low doses of radiation higher mobility polypeptides are observed by SDS gel electrophoresis, indicating complete fragmentation of the molecules. These and previous results[1] suggest that damage may occur in the backbone far from the initial site of interaction. Molecules which have not suffered these interactions are undamaged and fully functional. This can be seen most clearly in cases of ligand binding to irradiated receptors: analysis by the method of Scatchard[2] of binding data obtained from control and irradiated samples reveals a decrease in number of binding sites with no change in affinity.[3-5] In the case of enzymes, a double reciprocal plot (v^{-1} vs s^{-1}) shows a change in V_{max} but no change in K_m.[6,7] It is an important feature of radiation inactivation that there is no evidence for partially active molecules such as a structurally intact polymer with reduced catalytic function. Similarly, radiation-damaged molecules are not active; that is, there are no fragments of the original peptide which retain enzymatic potential.

Microwave, visible, and ultraviolet quanta have relatively low energies and cannot cause a single large energy deposition in biologically active molecules. Many of these quanta do not cause even a single ionization per event and are inappropriate for target analysis studies. Not only must the radiation be ionizing, the ionizations must also occur randomly throughout the mass of the sample; this feature is characteristic of hard X rays, γ rays, and high energy electrons, but not of protons, α particles, and other "high LET"[8] radiations.

The major limitation of this technique is the requirement that loss of biological activity be independent of radiation damage to other molecules which are not involved in the measured function. Exposure conditions must ensure that the only damage to biochemically active structures will be due to the action of radiation directly on these molecules ("direct action of radiation"). Irradiation of liquid solutions results principally in degradation products from the solvent molecules. Many of these are chemically reactive, and after diffusion to solute molecules they can inactivate by simple chemical interaction ("indirect action of radiation"). Thus target analysis cannot be applied to radiation inactivation of mate-

[1] D. L. Aronson and J. W. Preiss, *Radiat. Res.* **16**, 138 (1962).
[2] G. Scatchard, *Ann. N.Y. Acad. Sci.* **51**, 660 (1949).
[3] C. J. Steer, E. S. Kempner, and G. Ashwell, *J. Biol. Chem.* **256**, 5851 (1981).
[4] T. L. Innerarity, E. S. Kempner, D. Y. Hui, and R. W. Mahley, *Proc. Natl. Acad. Sci. U.S.A.* **78**, 4378 (1981).
[5] A. Doble and L. I. Iverson, *Nature (London)* **295**, 522 (1982).
[6] S. J. Adelstein and L. K. Mee, *Biochem. J.* **80**, 406 (1961).
[7] D. J. Fluke, *Radiat. Res.* **51**, 56 (1972).
[8] LET = linear energy transfer, the amount of energy deposited in matter per unit length of particle trajectory.

rials in liquid solutions or cells in the wet state. These complications can be avoided by use of lyophilized samples or frozen materials. At low temperatures the principal molecular effect is restriction of diffusion. When irradiations are carried out at very low temperatures, the solvent radiation products react locally and do not reach the solute molecules. The requirement that radiation effects be directly on the biochemically active molecules is then met, and target analysis permits calculation of the size of the functionally active unit.

The concept of a functional unit implies a collection of one or more molecules, all of which are required in order for the measured reaction to proceed. From target analysis alone there is no implication of spatial relationship between the molecules, only that they all must have escaped radiation destruction. The target size is the mass equal to the sum of the masses of all the necessary structures. For simple enzymes the target size has been found to accurately correspond to the molecular weight of a single polypeptide chain or to the mass of an entire molecule composed of several subunits. The physical basis for these observations is hypothesized to be general destruction of any polymer in which radiation causes a primary ionization; the loss of one chain may result in a conformational instability of other subunits with subsequent loss of activity. Alternatively, if a catalytic process involves the presence of several subunits, the loss of any one can terminate the reaction. This distinction is illustrated by glutamate dehydrogenase, a hexamer of identical subunits. The target size for activity corresponds to the size of the hexamer,[9] suggesting that destruction of a single subunit destabilizes the whole complex. In other enzymes, a subunit has been found to be the functional unit, apparently not requiring the remainder of the molecule for activity.[10]

Two caveats should be mentioned pertaining to the definition of the structure involved in the functional unit. The first relates to the transmission of destructive energy across a disulfide bond. The molecule ricin is composed of two polypeptide chains linked by a single disulfide bond. When the activity associated with the A chain (inactivation of the protein-synthesizing capacity of eukaryotic ribosomes) is examined in the intact ricin molecule, the target size reflects both chains.[10a] When the A chain is irradiated separately, the target size is that of the A chain only. Thus although the A chain functions separately and independently, if it is linked by a disulfide bond to the B chain the functional size is increased by the size of the B chain; a primary ionization occurring in one chain causes

[9] E. Blum and T. Alper, *Biochem. J.* **122,** 677 (1971).
[10] E. S. Kempner and W. Schlegel, *Anal. Biochem.* **92,** 2 (1979).
[10a] H. T. Haigler, D. Woodbury, and E. S. Kempner, *Proc. Natl. Acad. Sci. U.S.A.* **82,** 5357 (1985).

destruction via a disulfide bond to the second chain. The second caveat relates to the functional size of glycoproteins. It appears that the carbohydrate portion of a glycoprotein does not transfer destructive energy to the polypeptide chain. This was shown with the enzyme invertase.[10b] This enzyme contains 50% carbohydrate which can be totally removed without loss of enzymatic activity. The functional size of this enzyme is unaltered on removal of the carbohydrate portion of the molecule. Similar observations have also been made with other glycoproteins. However, it is not yet known whether the carbohydrate would add to the functional size if it were required for activity. The mechanism of radiation destruction of glycoproteins has not yet been fully examined.

The target size which is observed represents the minimal assembly of polymer chains which produce activity under the assay conditions. If the conditions of sample preparation (lyophilization, salts, detergents, etc.) affect the physical state of the molecules in the assay mix, then the target size may be altered accordingly. By analogy, gel electrophoresis of a disulfide-linked dimer will yield different results depending on the presence of reducing agents during sample preparation.[11] For radiation inactivation these preparative procedures would have to alter the molecules— such as by destruction of interchain disulfide bridges or the functional activity expressed in the assay.

It is important to remember that hydrodynamic, electrophoretic, and chromatographic techniques measure parameters different from those measured by target analysis. For example, sedimentation equilibrium and gel permeation chromatography are used to measure the molecular weight and excluded volume, respectively, both of which are dependent on the "particle" size, i.e., the entire assembly of polypeptide chains. Similarly, gel electrophoresis in SDS usually measures the molecular weight of isolated polypeptide chains.[11] Target size analysis relies on the relationship between function and the native structure of the enzyme.

There is as yet no theoretical limitation on the types of activities which can be studied by radiation inactivation. Any function which survives lyophilization or freezing can be analyzed for a radiation-sensitive target. The procedures we describe in this report have been applied successfully to a wide variety of organisms, cell and tissue preparations, and biological activities. Other techniques are also reported by others to be useful in some irradiation situations and reference will be made to them in appropriate sections of this chapter.

[10b] M. E. Lowe and E. S. Kempner, *J. Biol. Chem.* **257,** 12478 (1982).
[11] T. B. Nielsen and J. A. Reynolds, this series, Vol. 48, p. 3.

Basis of Target Analysis

The analysis of target size is well founded empirically[10,12,13] and is based on the probability of interaction between the incident radiation and the orbital electrons of the irradiated matter. The probability of n ionizations occurring in an irradiated sample is given by the Poisson formula[10]:

$$P(n) = x^n e^{-x}/n! \qquad (1)$$

where x is the average number of ionizations in that mass.

From gel electrophoresis of irradiated proteins (Fig. 1) it is clear that a single primary ionization or "hit" results in large-scale destruction of structure. The activity remaining after exposure to ionizing radiation results from units which have not been "hit."

The probability of no ionizations in Eq. (1) is $P(0) = e^{-x}$. The average number of ionizations in a given mass due to a radiation dose D is $x = KD$ where K is a parameter dependent on the biochemical system studied. Thus, the surviving biochemical activity (A) will decrease as a simple exponential function of the radiation dose:

$$A = A^0 e^{-KD} \qquad (2)$$

where A^0 is the initial (nonirradiated) activity. The radiation dose D is in units of rads, defined as the absorption of 100 ergs/g material. The value of K may thus be obtained as the slope of the line $\ln(A/A^0) = -KD$. The molecular weight of the biological functional unit (M_r) is proportional to K and is given by the empirical equation:

$$M_r = -cKS_t \qquad (3)$$

where c is a constant (6.4×10^{11} rad daltons) and S_t reflects the effect of temperature on the apparent target size[14] (Table I). For example, at $-135°$, $S_t = 2.8$ and Eq. (3) becomes[15]

$$M_r = 1.79 \times 10^{12} K \qquad (4)$$

If there is more than one structure (1, 2 . . . of activity A^0_1, A^0_2 . . .) and each of these structures is independently responsible for a fraction of the measured biological function, the loss of function upon irradiation may be described as an additive series of exponentials[10]:

[12] D. E. Lea, "Actions of Radiations on Living Cells," 2nd ed. Cambridge Univ. Press, London and New York, 1955.
[13] G. R. Kepner and R. I. Macey, *Biochim. Biophys. Acta* **163**, 188 (1968).
[14] E. S. Kempner and H. T. Haigler, *J. Biol. Chem.* **257**, 13297 (1982).
[15] T. B. Nielsen, P. M. Lad, M. S. Preston, E. Kempner, W. Schlegel, and M. Rodbell, *Proc. Natl. Acad. Sci. U.S.A.* **78**, 722 (1981).

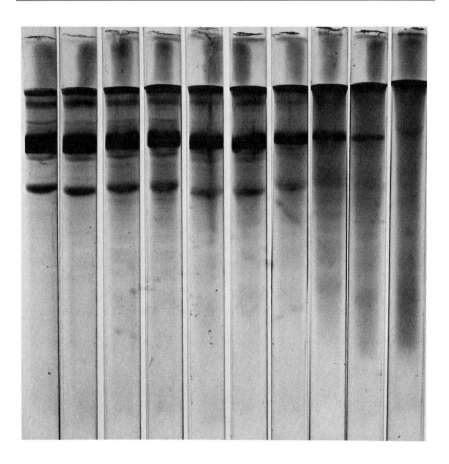

FIG. 1. SDS–PAGE of irradiated proteins. Fluorescamine-coupled glutamate dehydrogenase[19] was irradiated frozen at $-135°$ with 10 MeV electrons. After exposure to radiation, equal quantities were applied to each tube. The doses were (left to right): 0–1–3–6–9–12–24–48–72–96 Mrads. Coomassie Blue staining reveals a progressive decrease in stain intensity of the original band and the appearance of lower molecular weight species in radiation-damaged samples (E. S. Kempner and J. H. Miller, unpublished observations, 1981).

$$A = A^0_1 e^{-K_1 D} + A^0_2 e^{-K_2 D} + \cdots \qquad (5)$$

where $K_1, K_2 \ldots$ are related to the mass of components $1, 2 \ldots$, each component with activity $A^0_1, A^0_2 \ldots$. For two targets of sufficiently different sizes, the survival curve [Eq. (5)] can be resolved into two

TABLE I
CHANGE OF RADIATION SENSITIVITY OF
ENZYMES WITH TEMPERATURE

°C	Factor[14]
−200	4.41
−180	3.87
−160	3.40
−140	2.98
−120	2.62
−100	2.30
−80	2.02
−60	1.77
−40	1.56
−20	1.37
0	1.20
+30	0.99
+70	0.76
+100	0.63
+150	0.45

simple exponential functions from which both molecular weights can be calculated. By this means the fraction of the total activity resulting from each component may be obtained from the intercepts of each of the resolved curves on the ordinate.

It is useful to empirically evaluate the validity of the hypothesis that a loss in function results from a change in the number of units, rather than a change in affinity for substrate or cofactor. A method suitable for hormone receptors is to evaluate the binding affinity and number of binding sites by determination of the binding isotherm after various doses of radiation. Enzymes may be evaluated by a double reciprocal analysis, as for example Fig. 2 for hexokinase. Notice that the apparent affinity (K_m), indicated by the x intercept ($-1/K_m$), of the surviving enzymes for the substrate was unchanged, suggesting substantial molecular integrity for these units. Similar results are obtained for other enzymes and receptors studied in these ways.[4,7]

Irradiation

Permanent destructive effects on macromolecules can be observed after exposure to a variety of radiations. For purposes of target analysis, only X rays, γ rays, and high energy electrons (>1 MeV) are useful in determination of molecular weight, while protons, deuterons, and α parti-

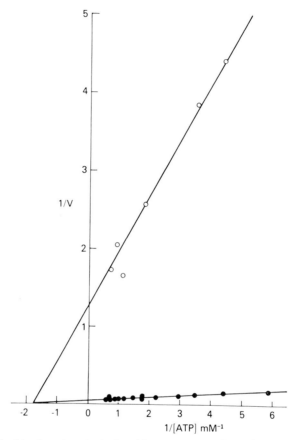

FIG. 2. The kinetics of rat brain hexokinase represented as Lineweaver–Burk plots, measured after a low dose (12 Mrads, ●) and a high dose (72 Mrads, ○) of irradiation with 10 MeV electrons. Reduced V_{max} is seen with increasing radiation dose but the same K_m (0.60 ± 0.12 mM) is observed in control and all irradiated samples (M. Suarez and S. Ferguson-Miller, unpublished observations, 1983).

cles find utility in describing the cross-sectional area of the target.[16] Other radiations (e.g., ultraviolet and neutrons) have effects which are not useful for these kinds of studies, but can reveal other properties of matter. Low energy electrons have limited ability to penetrate condensed matter; this property has been used advantageously to localize the depth of biologically active units in cells.[17] The lower energy electromagnetic radia-

[16] E. C. Pollard, W. R. Guild, F. Hutchinson, and R. B. Setlow, *Prog. Biophys.* **5,** 72 (1955).
[17] J. W. Preiss, *Arch. Biochem. Biophys.* **75,** 186 (1958).

tions are absorbed by specific atoms (viz. soft X rays) or chemical groups (e.g., ultraviolet, and also visible and infrared light, although these latter two are not ionizing and do not normally cause permanent and irreversible effects).

To be useful for target analysis the ionizing radiations must not only cause permanent damage in organic molecules, these effects must also occur randomly throughout the mass of the sample. The mean distance between ionizations must be large compared to the size of the molecules so that two independent events do not occur in the same polymer. It is precisely for this reason that α particles and other heavily ionizing radiations are not appropriate agents; these ionizations occur so close together that there is essentially continuous damage along the trajectory of the particle. Ionizations caused by high energy electrons occur at an average spacing of 2500 Å along the electron trajectory, a distance quite large with respect to proteins and other biomolecules. Similar distances are found with X rays and γ rays.

The absorption of a beam of particles occurs within a certain distance which is a function of the nature of the absorbing material and the type and energy of the particle. This distance is the maximum range of the particle in a given material. The absorption of electromagnetic radiation occurs randomly throughout the thickness of the target material; there is an exponential decrease in the number of photons with distance which varies with the nature of the absorber. The characteristic "half-value" thickness is used to describe the absorption of X rays and γ rays by a particular material.

Each ionization caused directly by γ rays or electrons is called a "primary ionization." It is characterized by the average transfer of ~65 eV of energy from the radiation to the target molecule.[13,18] The energy is dissipated throughout the polymer in which the primary ionization occurs.[19] The energy is absorbed by processes such as bond breakage and excitations[20] and lower energy transfers. Because of the large amount of energy, many different effects occur in each molecule and some dissipative processes cause massive damage.

Irradiation of biological material not only destroys biological structures, but there can also be reactions in the buffers, containers, and gas phases of the sample. In general these do not perturb target analysis, although there are two special conditions which must be considered. One is the induction of artificial radioactivity, especially by high energy elec-

[18] A. M. Rauth and J. A. Simpson, *Radiat. Res.* **22,** 643 (1964).
[19] E. S. Kempner and J. H. Miller, *Science* **222,** 586 (1983).
[20] R. E. Lapp and H. L. Andrews, "Nuclear Radiation Physics," 2nd ed., p. 128. Prentice-Hall, Englewood Cliffs, New Jersey, 1954.

trons of 15 MeV and greater and also by more massive atomic projectiles. Although damage to the sample from these processes may be small compared to that directly from the electron or γ-ray beam, radiations from these new isotopes may make the sample unsafe to handle. Usually these isotopes have short half-lives so that a day is sufficient to allow their decay, but it is more facile to avoid the problem where possible by using radiations which do not cause these effects. The other special condition concerns absorption of a significant fraction of the radiation beam by the container so that appreciably reduced doses are delivered to the sample.

Sources of Ionizing Radiation

In view of the restrictions on the radiation for molecular weight determinations by target analysis, only three types of radiation sources will be considered: X rays, γ rays, and high energy electrons. The amount of radiation necessary in most studies is quite large. The dose of radiation is limited principally by the radiation flux from the source; exposure times can vary from seconds to days or even weeks. With isotope-produced γ rays from even the largest ^{60}Co source, long irradiations are required. Nevertheless, for stable lyophilized samples it is possible to perform radiation studies successfully at low dose rates.[21] Although ^{60}Co decays with an uncomfortably short (5.3 years) half-life, this certainly represents the cheapest, most compact, most reliable radiation source with the smallest requirements on operating budgets. No other isotope source yields sufficient dose rates to be practical. Large beam currents of electrons can be produced by a variety of accelerators providing much higher radiation fluxes which can reduce exposure times to minutes and hours. Such brevity is particularly advantageous for studies with frozen samples. X-Ray sources are readily available but produce dose rates two orders of magnitude lower than linear electron accelerators[22] and are, therefore, limited in usefulness.

Electron accelerators are of three major types: Van der Graaff, Febetron, and linear accelerators (LINACS). Of these, the highest dose rate is available from the Febetrons, but only as a short (50 nsec) pulse and the total accumulated dose is low. The Van der Graaff accelerators are widely available (although most are proton accelerators, not electron) and can produce high dose rates from electrons which have been accelerated to energies of 1 to 4 MeV. The linear accelerators can produce electrons with 10 to 100 times more energy. However, they are large machines

[21] G. Beauregard and M. Potier, *Anal. Biochem.* **122**, 379 (1982).
[22] T. S. Nikitina, E. V. Zhuravskaya, and A. D. Kuzminsky, "Effect of Ionizing Radiation on High Polymers," p. 4. Gordon & Breach, New York, 1963.

TABLE II
RANGE IN WATER OF IONIZING RADIATIONS

Energy (MeV)	Range[a] (cm)	Recommended maximum sample thickness (cm)
Electrons		
10	5.2	2
5	2.5	1
3	1.7	0.7
1	0.4	0.15
X rays and γ rays		
10	32	5
3	18	2.8
1	9.7	1.5
0.3	5.8	1.0
0.1	4.2	0.7

[a] From "Radiological Health Handbook," rev. ed., pp. 122, 123, 133. DHEW-PHS Bur. Radiol. Health, Washington, D.C., 1970. Range for X rays and γ rays is the half-value layer. Recommended thickness is for 90% transmission. Absorption due to sample container wall has not been considered.

requiring extensive space and personnel. Moreover, only a few of these are "high beam current" machines capable of producing the large radiation doses required.

An important factor in consideration of radiation sources is the energy of the photon or electron. For ^{60}Co γ rays, of course, the energy is fixed at 1.3 MeV. Many electron accelerators are adjustable to different energy beams. Because of the absorption characteristics of these particles, it is prudent to select an electron energy for which the range is appreciably longer than the thickest sample to be irradiated. For samples as thick as 1 cm, electrons of 10 MeV (range in water approximately 5 cm[23]) are useful. Table II lists the maximum thickness recommended for several different energy radiations. These should be interpreted cautiously; it is the energy necessary for entrance to the biochemical substance. It is the same as the energy of the original radiation source only when there is no significant absorption of the beam due to air, box or the individual sample containers.

[23] A. B. Brodsky, "CRC Handbook of Radiation Measurement and Protection," Sect. A, Vol. I. CRC Press, West Palm Beach, Florida, 1978.

The most desirable radiation source is an electron accelerator. The ability to be turned on and off offers inherent safety features. A high beam current machine allows the production of high dose rates. A minimal energy of 5 MeV permits a reasonable thickness for samples, while a maximum energy of the order of 15–20 MeV avoids the induced isotope radioactivity which can create logistical problems.

We have been most successful with a LINAC producing a large current of 13 MeV electrons. Because the beam emerges from the machine as a narrow beam of a few centimeters diameter, only a few samples could be exposed to significant doses of radiation at the same time. A useful technique is to spread the beam over a larger area. Although this decreases the dose delivered per square centimeter, it allows more preparations to be exposed simultaneously. We use ~2 cm water to scatter the beam over a 225 cm^2 area at 1 m from the beam port. There is a concurrent loss of beam energy, so that the electrons actually enter the sample with an average energy of 10 MeV. Dose rates of 30 Mrads/hr are obtained from electron pulses 4 μsec long at a repetition rate of 30/sec. It is important that the dose rate be uniform over the entire area of the exposure plane. One way to accomplish this is to make the experimental array of samples relatively small compared to the radiation field. If the array were placed close to the accelerator beam port, there would be restrictions in the number of samples that could be simultaneously exposed; far from the beam port there would be a considerable reduction in dose rate and beam energy. Our experimental conditions allow the variation in dose rate to be up to 10% across the array. An alternative approach to obtain different doses is to use the variation in electron beam intensity over the exposure field.[24] A very similar procedure has been used with samples placed at different distances from a ^{60}Co γ-ray source.[25] All samples are exposed for the same time period, but at different positions in the field.

Measurement of Radiation Dose

Residual biochemical activity decreases exponentially with radiation exposure. Accordingly, inaccuracies in dose measurement are reflected by large changes in measurable function. Determination of radiation dose is therefore of paramount significance in these experiments. With care, routine measurements can be made accurate to almost 5%. However, several important aspects of dosimetry must be considered. No matter which of several possible methods is utilized, reproducibility can often be

[24] F. Lübbecke, D. R. Ferry, H. Glossman, E.-L. Sattler, and G. Doell, *Naunyn-Schmiederberg's Arch. Pharmacol.* **323,** 96 (1983).

[25] S. Uchida, K. Matsumoto, K. Takeyasu, H. Higuchi, and H. Yoshida, *Life Sci.* **31,** 201 (1982).

an obstacle. In choosing among the different methods, the first criterion is the kind of ionizing radiation used—electrons, γ rays, etc. Some methods show an energy dependence in which differing dose measurements can be obtained, for example, with electrons varying only in their energy. The response of some dosimeters varies with the rate of radiation delivery (e.g., ionization chambers saturate at high dose rates). Others show significant dependence on temperature (e.g., radiochromic dye films). There may be a time factor between exposure of the detector and the time the reading is made (e.g., X-ray films). The dose delivered may depend on the thickness of the sample; although this is discussed in the section on sample preparation, it is clear that the question of dosimetry can be deeply involved in this problem. It would be desirable to have both a measure of the cumulative dose delivered as well as an instantaneous determination of the dose rate which could be performed simultaneously with sample exposure.

We have found that thermoluminescent dosimeters (TLDs) are the most useful. TLDs are crystalline materials such as LiF. When irradiated, electrons are dislodged from the crystal structure, creating a "hole." Both holes and electrons can be trapped in lattice imperfections. Upon subsequent heating, the holes or electrons escape from these traps and emit light. TLDs show good reproducibility and accuracy, and have little time, temperature, or energy dependence. Dose rate is not a problem and they can be used with electrons or γ rays, so they may often be the method of choice. Unfortunately their very nature precludes their use as a continuous monitor, and they are sensitive in the range of a kilorad. With a pulsed radiation source such as a linear accelerator, beam intensity can be determined from a few pulses before and after sample irradiation. With a stable machine, the dose rate varies little (less than 5% change in rads per pulse over a 6-hr period). In our experiments, several TLDs are placed in glass sample vials which are irradiated at the four corners and center of the sample array; in the course of a single radiation experiment, 60 independent dose determinations are made.

Other methods of dosimetry exist. Several different classes of radiation measurements are based on a dose-dependent change in optical density. These include the standard ferrous–ferric sulfate conversion (the Fricke dosimeter) as well as techniques utilizing photography, polymethylmethacrylate[26] ("Perspex"), colored cellophane,[9,27] and radiochromic dye films.[14,28] There are also ionization chambers, calorimetric

[26] R. J. Berry and C. H. Marshall, *Phys. Med. Biol.* **14,** 585 (1969).
[27] J. Cuppoletti, C. Y. Jung, and F. A. Green, *J. Biol. Chem.* **256,** 1305 (1981).
[28] W. L. McLaughlin, J. C. Humphreys, B. B. Radak, A. Miller, and T. A. Olejnik, *Radiat. Phys. Chem.* **14,** 535 (1979).

methods, and even biologic materials with known radiation sensitivities,[29-31] all of which have been used to calibrate radiation exposures. The biological technique is by far the most cumbersome method of dose determination, and yet is among the best. It offers the measurement of dose actually delivered to the sample since the virus or enzyme used as a dosimeter can be intimately mixed with the experimental material. In many cases endogenous enzymes, which are found naturally in membrane or microsomal preparations being studied, can be used as true "microdosimeters." The difficulties involve absolute knowledge of the molecular weight, the radiation sensitivity of these markers and also the need to assay their surviving activity in the same samples which are the object of the investigation. The temperature dependence is the same in both unknown and marker, and therefore this procedure self-corrects for the effect of temperature of irradiation.

Other dosimetric techniques have inherent shortcomings. Calorimetry and photographic techniques are not practical for routine use. The Fricke dosimeter is normally reserved as a calibration standard for other dosimeters. Ionization chambers offer promise of being able to continuously monitor radiation, but cannot be put directly into the sample environment. The radiochromic dye films can most easily and directly reveal the entire radiation field and detect inhomogeneities; unfortunately they are temperature sensitive and UV sensitive (even from fluorescent lights!).

Caution must be exercised to establish that the dosimeter measures the radiation dose actually absorbed by the sample. Common errors are to forget the effects due to the sample container, the temperature of the sample, or variation of dose with depth in the sample. For biochemists, radiation dosimetry is the most unfamiliar aspect of target analysis studies.

Irradiation Conditions and Apparatus

Shielding and Ozone. All variations on the radiation inactivation method involve a common factor of very large radiation exposures (1–100 Mrads). This is 10^4–10^6 times greater than the lethal exposures for humans (LD_{50} ~300 rads). Personnel must be protected from radiation by shielding, distance, or time. A further hazardous situation occurs because of the effects of ionizing radiation on the environment of the sample and the exposure room. Induced radioactivity in samples and containers can occur with some types of radiation and might be sufficiently great to be of

[29] D. J. Fluke, *Radiat. Res.* **28,** 336 (1966).
[30] J. W. Preiss, M. Belkin, and W. G. Hardy, *J. Natl. Cancer Inst. (U.S.)* **27,** 1115 (1961).
[31] M. M. S. Lo, E. A. Barnard, and J. O. Dolly, *Biochemistry* **21,** 2210 (1982).

concern for health and safety (see Sources of Ionizing Radiation). Indirect effects of radiation are also of major significance. Just as radiolytic products occur in an aqueous environment, radiation beams also interact with the atmosphere and these ionizations lead to new chemical species. Due to the high sensitivity of olfactory detection, this is most obvious in the production of ozone. Especially high concentrations of ozone are found whenever a high dose rate source is in contact with the atmosphere. The extreme chemical reactivity of ozone is a problem for three different reasons: the potential for explosion, the health hazard, and the possibility of ozonolysis of the experimental samples. Ozone boils at $-112°$ (and oxygen at $-183°$). Experiments at lower temperatures offer the possibility of ozone condensation which may be suspected from both smell and the distinctive blue color of the liquid. Evaporating solutions of ozone are extremely unstable and likely to explode. Even low dose rate machines, such as isotope sources, cause ozone formation which can accumulate to moderate levels in an enclosed volume. Appropriate ventilation by exhaust systems to safely dispose of the gas should always be used, especially before personnel enter the sample area.

Experimental samples may be adversely affected by the highly reactive ozone. Any materials in contact with an oxygen-containing atmosphere during radiation exposures are susceptible. Thus irradiation of samples *in vacuo* or sealed under a nitrogen atmosphere offers insurance against ozonolysis. The methods we have used have retained a small amount of air with the samples in a sealed vial. Ozone smell can be detected upon opening those vials which had been exposed to electron radiation, but this ozone is blown off before thawing of frozen samples. In other experimental designs, samples which are open to the atmosphere of the room during irradiation are exposed to much higher amounts of ozone. This problem may be exacerbated even further where the sample has a large surface area as in the case of a thin film. Addition of an antioxidant such as butylated hydroxytoluene after irradiation may be of some merit in these conditions.[32]

Temperature Measurement and Control. Thermistor measurement of temperature works well at temperatures as low as that of Dry Ice ($-78°$). Lower temperatures are best determined with platinum resistance probes, such as Model 4150, Type 539 "PT," a four-wire lead probe from the Yellow Springs Instrument Company. Resistance measurements were made with a Model 172A digital multimeter from Keithley Instruments. Good results were also obtained with a platinum resistance probe made by

[32] D. A. Thompson, M. Suarez-Villafane, and S. Ferguson-Miller, *Biophys. J.* **37**, 285 (1982).

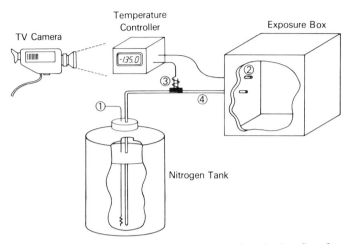

Fig. 3. Temperature measurement and control system. Samples in a Styrofoam box are exposed to a radiation beam. Temperature is maintained by a variable flow of cold nitrogen gas from a large nitrogen tank; an electrical heater (1) boils the liquid nitrogen and maintains a positive pressure on the liquid in the closed system. Temperature is detected by a platinum resistance probe (2) in the exposure box. A digital readout on the temperature controller is picked up by a TV camera for remote monitoring. When the temperature rises above a preset point, a solenoid valve (3) is opened, permitting the flow of nitrogen gas through an insulated delivery tube (4) to the exposure box.

Omega Engineering Model #K2028, but avoiding the Teflon-insulated wire. This polymer is readily degraded to a powder by high radiation fields with resultant electrical shorts. Both of these probes function reliably and accurately at temperatures as low as that of liquid nitrogen ($-196°$). These probes contain metallic wires and some are encased in metallic sleeves. Since these materials will cause appreciable scatter of electrons, inhomogeneities of the radiation field and radiation "shadows" can occur behind these probes. We therefore avoid placing the probes directly in front of the samples, but exercise caution that the probe placement permits faithful temperature determination of the samples.[14]

Sample temperature is maintained by a stream of cold nitrogen gas (Fig. 3). A large container (31 liters) of liquid nitrogen is fitted with a heater-withdrawal device (Model CE-8, Frigitronics of Connecticut). An electrical heater immersed in the liquid boils the nitrogen in a tightly closed system. The heater is turned off when a pressure of 0.7 kg/cm^2 (10 lb/in.^{2}) has built up in the gas phase above the liquid nitrogen. During subsequent removal of liquid nitrogen the gas phase pressure remains constant at 0.5 kg/cm^2 without additional heating. Alternatively, a tank of nitrogen gas can be used to provide pressure for liquid nitrogen flow.

Access to the closed nitrogen system is achieved by a solenoid valve (Model #91C89C7A, Valcor Engineering). This solenoid valve also contains Teflon-coated wire which should be replaced or shielded from the radiation field. When the solenoid valve is opened the pressure in the nitrogen container forces liquid nitrogen up the withdrawal tube where it rapidly boils at −196°. This provides a strong stream of cold nitrogen gas which is directed through an insulated flexible tube to the irradiation box. The high flow rate of nitrogen gas insures rapid mixing throughout the volume of the box and uniformity of temperature. The gas escapes through a loosely fitting lid as well as several small holes in the sides of the box. It is important that one hole be placed at the floor of the Styrofoam box to allow drainage of any gases which might condense.

The Omega probe can be used in conjunction with a temperature controller, Model 4202 (Omega Engineering). This device can be set to any desired temperature between −198 and +196°. When the detected temperature rises above that set value, the controller opens the solenoid valve. This requires wiring the controller such that the controlling relay is normally open. When the temperature is above the set point, the relay closes, thereby energizing the solenoid to allow nitrogen flow. In operation the system maintains the set temperature of −135° within 2°. Heating of the samples due to the radiation field is not more than 0.3°, and the same temperature is determined in the frozen samples as is routinely measured in the bulk gas phase.[14]

Other authors have attempted to control temperature by different means. Air streams at various temperatures are blown over the samples during radiation exposure in the most popular method.[33] For many years it was common to place samples in good thermal contact with a large temperature-controlled mass such as a metal slab or a block of Dry Ice; sand baths have also been used to this end. Reliance on the ambient air temperature is commonly reported, and may be adequate except at high dose rates where there is a possibility of heating effects in the samples.

Irradiation Box and Sample Holder. During radiation exposure, samples should be held in a fixed position in the field under conditions where accurate dose determinations[33a] and temperature control are possible. One reported variation is to cyclically move the samples in the beam during exposure[34,34a] which could minimize effects of inhomogeneities in the radiation field. For irradiations at low temperatures, our sample vials

[33] S. R. Levinson and J. C. Ellory, *Biochem. J.* **137**, 123 (1974).
[33a] G. Beauregard, S. Giroux, and M. Potier, *Anal. Biochem.* **132**, 362 (1983).
[34] C. Y. Jung, T. S. Hsu, J. S. Hah, C. Cha, and M. N. Haas, *J. Biol. Chem.* **255**, 361 (1980).
[34a] C. Y. Jung, *in* "Molecular and Chemical Characterization of Membrane Receptors" (C. Venter, ed.), pp. 193–208. Liss, New York, 1984.

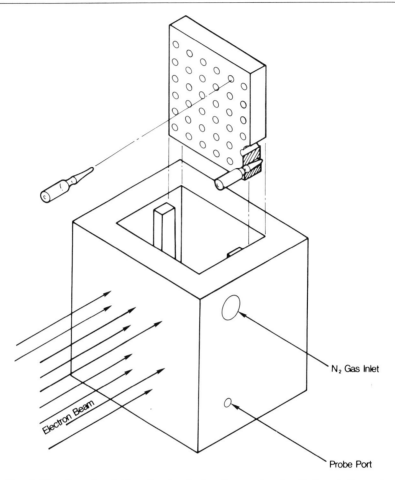

FIG. 4. Styrofoam irradiation chamber for low temperature irradiation studies and sample holder. Chamber cover not shown. Sample array is lowered into tracks in the chamber during radiation exposure.

are held in a Styrofoam block, and the entire assembly of block and vials are placed in a Styrofoam box (Fig. 4). This material is an excellent insulator, permitting appropriate temperature control, and is almost transparent to high energy electron radiation. Dose measurements made in the presence and absence of the Styrofoam box are virtually identical. The limited absorption of radiation by Styrofoam precludes significant damage, and the box can be reused for many experiments until the material begins to powder and crumble (total dose of ~1000 Mrads).

A Styrofoam block 17 × 18 × 3.8 cm is drilled with 121 holes, each 7 mm in diameter. The tapered neck of the sample vials are inserted into the 11 × 11 array so that the frozen sample in the bottom of the vial is facing the beam (see Fig. 4). The Styrofoam box is 25.5 × 26 × 21 cm, with 3.3-cm-thick walls and floor. A loosely fitting Styrofoam cover is placed over the top of the box. Several ports in the side walls of the box are used for temperature measurement and control.

This exposure box is placed on a wooden table and held in place with polyethylene "bookends." The box and table can be maneuvered into position and aligned with a laser coaxial with the electron beam such that the beam center falls on the midpoint of the sample array.

When the entire assembly has reached thermal equilibrium at −135°, the electron beam is turned on, entering the front face of the Styrofoam box at a distance of approximately 80 cm from the beam port of the linear accelerator. After a predetermined radiation exposure, the beam is turned off and the accumulated ozone in the exposure room is withdrawn by exhaust fans. The sample array is removed and warmed to −80° on a bed of Dry Ice. Some irradiated samples are removed and replaced with new sample vials. The radiation exposure process is repeated until all samples have received the desired cumulative radiation doses.

Radiation exposure at room temperature is performed in an identical manner, except that the Styrofoam box is replaced with a similar box which has had the front and side faces removed. A squirrel-cage fan (Dayton Electric; Model 1C939) blows room temperature air over the vials during the irradiation and the temperature, detected by a small probe placed among the sample vials, is monitored remotely.

Sample holders and supports of other materials have been used, but each possess certain technical disadvantages. We avoid aluminum, steel, and other metals near the sample vials because of the significant radiation backscatter that metals cause in the radiation beam. This can lead to inhomogeneities in the radiation field and make dosimetry extremely complex. The use of certain plastics (acrylics, fluorocarbons) is inadvisable because of their mechanical failure at relatively low radiation exposure as well as difficulties with temperature control.

Dose Range to Examine. In the simplest cases, when only one component is responsible for the total activity, remaining biochemical activity is found to be a single exponential function of radiation dose. In more complex systems the survival curve appears concave or convex (see the section on Analysis and Interpretation of Target Data). Since the interpretation by target analysis depends directly on the nature of this curve, it is necessary to establish definitely the activity–dose relationship. It would be desirable to determine biological activity over several log decrease of the activity, to 1 or 0.1% that of the control. The dose required to inactivate the function to this degree is related to the radiation sensitivity of the process being measured. This is commonly indicated by the D_{37}, the radiation dose required to reduce the activity to e^{-1} that of the control. Three times larger doses ($3 \cdot D_{37}$) leaves 5% of the activity remaining. Table III gives the percent surviving activity as multiples of the D_{37}. These values of course are for a single exponential decay. For complex inactivation curves, detailed analysis is possible only when the loss of activity approaches a simple exponential function at sufficiently high radiation doses. To firmly establish the exponential loss of activity in simple or complex curves requires a large range of doses. For example, an exponential loss which reached 20% after 10 arbitrary radiation dose units can be compared with a linear loss which reached the same 20% at 10 units (Fig. 5); at no point do the two curves differ by more than 15% of the initial activity. On a semilogarithmic plot, the two models clearly differ only after 12 arbitrary units (when the exponential survival has reached 14.5%). The major difficulty then becomes the reliability of the biochemical assay when only a small fraction of the initial function remains. When nonspecific binding or background enzyme reaction is high and the survival curve cannot be determined below 20%, the application of target analysis becomes uncertain.

Sample Preparation

Requirements of Sample Size and Concentration

Since the functional size is determined by the loss of activity with increasing radiation dose, it is imperative that the investigator be cognizant of the assay being used to measure the activity. If the functional size of an enzyme, for example, is being evaluated, it is necessary to know the reaction steps involved (i.e., a single-step reaction or a coupled reaction) and to verify that the activity measured is dependent on the concentration of enzyme under investigation. Many artifacts can enter the data if the investigator is not careful. For example, if an enzyme requires a cofactor for activity, then the cofactor must be present in sufficient quantity so as not to be rate limiting at any time during the course of the reaction.

An advantage of radiation analysis is that the functional size obtained is a reflection of the structural state of the macromolecules at the time of the irradiation. Lyophilization[35] or freezing[35a] in phosphate buffer may

---

[35] D. Parkinson and B. A. Callingham, *Radiat. Res.* **90,** 252 (1982).
[35a] J. Harmon and E. Kempner, unpublished observations.

TABLE III
EXPONENTIAL LOSS OF OBSERVABLE ACTIVITY
WITH RADIATION EXPOSURE

Radiation dose in multiples of D_{37}	Activity remaining as percentage of control
0	100.
1	36.79
2	13.53
3	4.98
4	1.83
5	0.67
6	0.25
7	0.09

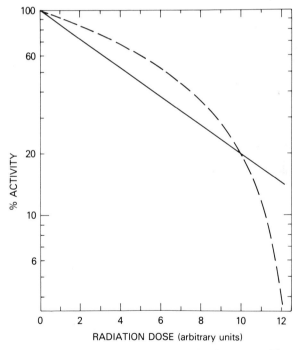

FIG. 5. Comparison of two theoretical radiation inactivation curves. The solid line represents a simple exponential decay which is set to 20% survival at 10 arbitrary radiation units. The dashed line is the expected survival of a material which is destroyed in direct proportion to radiation exposure (linear with dose), also set at 20% at 10 units.

alter the inactivation curves of some enzymes. The presence of a detergent in the sample may also alter results; this effect is very complicated and cannot be predicted in advance.[3,36] For whole-cell irradiation, several investigators have included a cryoprotective reagent in the sample.[27,37] Since most of these reagents act by preventing ice-crystal formation, one must be aware that the effects of these reagents on the results have not been examined.

Another parameter which must be considered prior to preparation of samples for irradiation is the stability of biological activity to freeze-thaw or lyophilization. Radiation inactivation experiments can still be performed on systems which undergo a reproducible loss in activity. However, the results obtained in these systems will reflect only the remaining activity which has survived these processes.

A wide range of protein concentrations may be used, but very low concentrations (less than 1 μg/ml) should be avoided due to the possibility of adsorption onto the glass vial. Added carrier proteins will generally prevent this adsorption. The maximum amount of protein should be limited in those cases where aggregation and/or precipitation of denatured protein occurs with increasing radiation dose.

The volume of the sample will be determined by the absorption characteristics of the radiation being used. Specifically, the thickness of the sample and its container should be predetermined as indicated in the section on Sources of Ionizing Radiation (Table II). If a 10 MeV electron source is used, 2 cm is the recommended maximum thickness of samples in thin-walled glass containers (Kimble #12012).

The protein concentration and volume of the sample must also conform with the requirements of the assay system under investigation. It is essential that sufficient activity be present in the sample vial to allow the accurate determination of 1–10% of the initial activity (unirradiated control). The initial activity to be considered is defined as the activity remaining after freeze-thaw or lyophilization and reconstitution treatments which may be required for the radiation source being used. This requirement for sufficient activity in the sample assists the investigator in determining the number of functional components with considerably more accuracy. It has been our experience that if the sufficient activity requirement is not met there is considerable error associated with the evaluation of the inactivation data. For example, if the sensitivity of an enzyme assay is accurate to 0.1 nmol product/min and the activity before

[36] H. Noël, G. Beauregard, M. Potier, G. Bleau, A. Chapdelaine, and K. D. Roberts, *Biochim. Biophys. Acta* **758**, 88 (1983).

[37] F. A. Green, C. Y. Jung, J. Cuppoletti, and N. Owens, *Biochim. Biophys. Acta* **648**, 225 (1981).

irradiation is 0.2 nmol/min then one can accurately determine the radiation inactivation to only 50% of the initial or nonirradiated control activity. Any collection of data at less than 50% initial activity will obviously be inaccurate. Data of this type may suggest a two-component system (due to the inability to detect lower enzyme activities) when, in fact, there is only one component associated with this activity.

An internal standard (e.g., an enzyme of known target size) should be included in the same irradiation vial as the sample under investigation. This serves as an indicator of malfunction in the experiment. The assay of an internal standard may elucidate several kinds of problems: first, it can offer direct evidence that a particular irradiation vial was not handled properly, e.g., the sample may have undergone multiple freeze-thaw cycles due to a freezer malfunction or due to problems in transit. Second, the internal standard can be used to verify the temperature at which the irradiation was performed. This is especially useful if accurate temperature measurements are unavailable. Finally, the internal standard can be used to verify the radiation dose to which a particular vial has been exposed. In fact, this procedure has been used by several investigators[30,31] as an internal dosimeter. An internal standard may already exist in the sample if an impure system (e.g., cell or membrane) is being irradiated. Thus, determination of 5'-nucleotidase on membranes can be used as an endogenous internal standard. If the sample to be irradiated is a pure enzyme, then an exogenous standard should be added. Care must be taken to ensure that the exogenous standard does not interfere with the activity measurements of the unknown sample and vice versa.

Preparation of Irradiation Vials

There are a number of procedures in the literature for the preparation of samples and containers for different radiation sources.[34,38] Rather than reviewing these procedures, we will present the protocol for preparation of irradiation vials which we have used for irradiations with high energy electrons.

The vials used are 2-ml untreated ampoules (Kimble #12012). If desired, these vials can be treated with commercial products, for example, Prosil-28 (PCR Research Chemicals), to decrease the adsorption of proteinaceous material. After treatment, the vial is labeled with an appropriate identification code for that particular set of sample vials. The labeling is done with a permanent marking pen such as Sanford's Sharpie or Alpco #300, marker. In order to retain the identification code while handling the vials, it is necessary to cover this code with clear tape (Tucktape #205,

[38] G. Beauregard and M. Potier, *Anal. Biochem.* **132**, 362 (1983).

Tuck Industries). Many commercially available tapes will not withstand the low temperatures and/or the high doses of radiation. The identification code is covered with at least 1.5 turns of tape around the vial. Care is taken to prevent overhandling the tape, which will decrease its effectiveness.

After the vials are labeled and taped, the samples can be added with a 2-in. needle attached to an appropriately sized syringe or with Lang-Levy glass pipettes. The purpose is to deliver the entire liquid volume below the constricted neck of the ampoule. Now the samples are ready for freezing. The particular method of freezing the samples [fast freeze in liquid N_2 ($-195°$) or Dry Ice ($-80°$), or slow freeze in $-20°$ freezer] is determined by whichever procedure will be less detrimental to the activity of interest. After freezing the samples can be lyophilized if necessary.

Once the samples are frozen, lyophilized (if necessary) and equilibrated at $-80°$, they should be sealed quickly with an oxygen-gas torch without thawing the samples. To accomplish this, use a reasonably hot flame, hold the open end of the vial with forceps and the frozen end with protective gloves; rotate the vial in the flame to soften the glass at a point approximately 1.5 cm from the open end of the vial. When the glass begins to soften, pull gently on both ends of the vial until the length of the vial has increased by approximately 1 cm. Leave the vial in the flame just until the two pieces of the vial melt apart. Place the sealed vial on Dry Ice being careful not to touch the hot end.

There is a considerable amount of earlier literature on the effect of the contents of the gas phase in the top of the vial.[39,40] In our studies we have not observed any differences in results obtained with samples sealed either under vacuum, nitrogen,[40a] oxygen,[40b] or air as long as the ozone produced in air-containing vials was removed prior to thawing the samples as described in the section on Handling Samples after Irradiation.

Storage of Samples

Once the samples are prepared they should be stored at $-60°$ or below to preserve activity. If the samples must be stored at $-196°$ to preserve activity, then care must be taken in warming the glass vials. Improperly

[39] J. A. V. Butler and A. B. Robins, *Radiat. Res.* **17**, 63 (1962).
[40] W. Günther and H. Jung, *Z. Naturforsch., B: Anorg. Chem., Org. Chem., Biochem., Biophys., Biol.* **22B**, 313 (1967).
[40a] M. D. Suarez, A. Revzin, R. Narlock, E. S. Kempner, D. A. Thompson, and S. Ferguson-Miller, *J. Biol. Chem.* **259**, 13791 (1984).
[40b] T. Goldkorn, G. Rimon, E. S. Kempner, and H. R. Kaback, *Proc. Natl. Acad. Sci. U.S.A.* **81**, 1021 (1984).

sealed vials may explode due to liquid nitrogen which may enter the vial through a pinhole. Storage conditions after the samples have been irradiated are the same as before irradiation.

After irradiation the samples are stored at temperatures below $-60°$ to maintain activity. The time between irradiation of the sample and assaying residual activity can often be quite substantial. However, it is possible that the irradiated samples may have a shorter storage time than the control. If a particular function or activity has a very short freezer life, the assay must be performed as quickly as possible.

Measurement of Activity

Handling Samples after Irradiation

On the day when the assay will be performed, the irradiated samples plus at least one unirradiated control are removed from frozen storage. Immediately, the tops of the vials are removed and any ozone in the top of the sample chamber is removed by purging with air or nitrogen gas. This is done prior to sample thawing to prevent any ozone from possibly oxidizing the sample. Once aqueous samples have thawed, a visual inspection should be made to ascertain that all the samples have remained homogeneous and no precipitate is in evidence. If a precipitate is present, the samples must be handled accordingly. If the precipitate does not interfere with the assay, it may be best not to remove it. However, if the assay involves a light absorption technique then the precipitate must be removed to prevent light scattering. Usually, when a precipitate is formed it occurs in the unirradiated control sample as well as in the irradiated sample. Thus, it would behoove the investigator to prevent any precipitate from forming in the initial sample preparation step. When bovine serum albumin is used as a carrier protein (to decrease adsorption of an enzyme to the glass vial) it may begin to precipitate after exposure to radiation doses greater than 60 Mrad.

When lyophilized samples are resuspended after irradiation, a homogeneous resuspension is required. This can usually be obtained by aspirating the sample through a small bore needle (22-gauge) into a syringe. Some soluble proteins may be very difficult to reconstitute. This particular problem will become more serious at higher doses of radiation and in the presence of carrier protein. Although sonication or homogenization may increase the solubility of the sample, the effect of these procedures on sample activity must be carefully examined.

It is necessary to measure the protein concentration in the irradiated samples by either the Lowry procedure[41] or the Biuret procedure.[42] The fluorescamine procedure[43] may be used only if no quenching agents are produced in the samples by irradiation. A protein assay which uses Coomassie Blue to quantitate the protein (Bradford procedure[44]) cannot be used for irradiated samples because color formation is not proportional to protein content after irradiation.

Assay of Irradiated Samples

Assuming the initial sample preparation was performed correctly, the irradiated samples can be assayed directly. When an assay is so cumbersome that only several irradiated samples can be assayed at any one time, it is necessary to have at least one unirradiated control for each group of irradiated samples. This should adjust for between-assay variability.

When there is significant assay-to-assay variation in the data, it may be necessary to irradiate multiple samples at each dose of radiation in order to obtain an accurate functional size. Thus, after repetitions of the radiation experiment the variation in the functional size will contain the variation in the sample preparations, the variation from assay-to-assay and the variation from one irradiation run to the next.

The effect of radiation should be on the maximum velocity (V_{max}) of the functional unit to perform its function rather than on the ability of the functional unit to associate with its substrate (K_m). However, it is necessary to determine whether the K_m or K_D changes with radiation dose. Target analysis predicts that only maximum velocity (V_{max}) changes will occur. Several complex systems have been presented in the literature[45,46] to indicate that apparent affinity constants can change after irradiation. Thus, the investigator must examine these parameters on the irradiated samples and compare them to the control unirradiated samples. This determination need not be performed at all the radiation doses used to obtain an estimate of the functional size but rather at radiation doses selected by an educated examination of the inactivation curve.

[41] O. H. Lowry, N. J. Rosebrough, A. L. Farr, and R. J. Randall, *J. Biol. Chem.* **193**, 265 (1951).

[42] A. G. Gornall, C. J. Bardawill, and M. M. David, *J. Biol. Chem.* **177**, 751 (1949).

[43] P. Bohlen, S. Stein, W. Dairman, and S. Udenfriend, *Arch. Biochem. Biophys.* **155**, 213 (1973).

[44] M. M. Bradford, *Anal. Biochem.* **72**, 248 (1972).

[45] J. T. Harmon, C. R. Kahn, E. S. Kempner, and W. Schlegel, *J. Biol. Chem.* **255**, 3412 (1980).

[46] R. J. Turner and E. S. Kempner, *J. Biol. Chem.* **257**, 10794 (1982).

Analysis and Interpretation of Target Data

Two goals in the analysis of target data are to distinguish single component [Eq. (2)] from multiple component [Eq. (5)] curves, and to obtain the best fit of the data for the calculation of molecular weight. The first step in the analysis is, for ligand binding data, the calculation of the ratio of bound/free ligand (B/F). This calculation is required of any equilibrium binding study in which the amount of ligand bound is a significant fraction of the total amount added. Under these circumstances, the total ligand concentration is not equivalent to the free ligand concentration and therefore "pseudo-first-order" kinetics cannot be assumed.[47] A corresponding correction is not necessary for measurements of enzyme activity when the concentration of substrate is large (compared to the K_m or apparent affinity of the enzyme for the substrate) and does not change appreciably during the course of the reaction. Next, the data should be normalized to the initial activity or binding ratio and the results graphed according to Eq. (2). That is, dose versus the natural logarithm of normalized activity. It will often be clear from the graph whether or not the data fit a simple exponential [i.e., Eq. (2), see Fig. 6a]. In the simplest case the data may be analyzed by a regression analysis,[48] by a regression analysis weighted according to the uncertainty in the data, or by a regression analysis constrained to intercept the initial (nonirradiated) value of the biological function.[49] If there is a question as to whether data fit a simple exponential, a statistical test should be applied to check whether a two-line model fits the data better than one-line.[50] The error in the sum of squares for the two-line model [SSE(F)] is equal to the sum of the SSEs of each component line. Similarly, a regression analysis of the one-line model will yield an error in the sum of squares, SSE(R). The two models may be compared by use of the F statistic:

$$F = \frac{SSE(R) - SSE(F)}{2} \div \frac{SSE(F)}{D_f} \tag{7}$$

where the degree of freedom associated with SSE(F) is the sum of the degrees of freedom associated with the SSE for each line, or $D_f =$

[47] D. Rodbard, in "Receptors for Reproductive Hormones" (B. W. O'Malley and A. R. Means, eds.), p. 289. Plenum, New York, 1973.
[48] C. Fewtrell, E. Kempner, G. Poy, and H. Metzger, Biochemistry 20, 6589 (1981).
[49] E. S. Kempner, J. H. Miller, W. Schlegel, and J. Z. Hearon, J. Biol. Chem. 255, 6826 (1980).
[50] J. Neter and W. Wasserman, "Applied Linear Statistical Models," p. 160. Richard D. Irwin Inc., Homewood, Illinois, 1974.

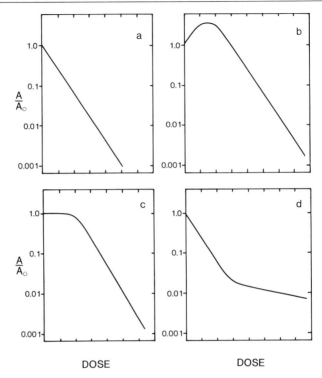

FIG. 6. Idealized types of complex curves observed after radiation inactivation.

$(n_1 - 2) + (n_2 - 2)$, and n_1, n_2 are the number of observations in each line of the two-line model. Small values of F mean that the data are best fit by the one-line model. Conversely, large values of F mean that the data are better fit by a two-line model and either the slopes or the intercepts of these two lines are different. These possibilities can be resolved with further analysis[50] if necessary. Proper analysis will require assays on replicate samples over a wide dose range, so that less than 5% of the activity remains at the highest dose. Even so, the uncertainty in molecular weights calculated from complex curves is often considerable because the available dose range may be restricted.

Several types of complex curves (i.e., not simple exponentials) may be considered. In each of the following cases, the interpretation of the data was consistent with the biochemical nature of the enzyme or receptor, but other interpretations are possible. Curves like Fig. 6b which show an *increase* in the measured function after irradiation at low doses and then decrease at higher doses have been interpreted as indicating the presence of a radiation-sensitive inhibitor[45] or a radiation-induced transition from a low activity to a high activity form.[51] Curves like Fig. 6c which remain at the initial activity before decreasing exponentially are consistent with either redundant sites (as in immunoglobulin M when a hemolytic assay is used[52]) or as a special case of the curve in Fig. 6b.[45,53] Curves like Fig. 6d which show different limiting slopes at high and low doses of radiation have been interpreted as consistent with the presence of two or more independent structures responsible for the same biological function[54] or a possible result in a system regulated by association-dissociation.[55,56,57]

Several experimental factors may also affect target analysis data. Chief among these are the effects of detergent[3] or buffer[34] on target size (see the section on Sample Preparation) and the possible effect of the presence of immunological domains (within a protein structure) on the measured target size.[48,52] In addition, particular experimental systems may be subject to specific problems. In the study of carrier-mediated transport in membrane vesicles it is clear that irradiation increases passive diffusion rates.[46] Although some increased membrane permeability may be tolerated by measurement of changes in passive diffusion, it may be necessary to reconstitute the carrier into fresh, unirradiated vesicles.[40b] The possibility of rearrangement of the components of a regulated system should be considered.[54] In the case of tryptophan synthetase, conditions were established for the recombination of a regulatory component with the catalytic component by irradiation of the purified and mixed components.[49] A similar experiment may not be feasible for an impure enzyme. The effect of a radiation-induced change in the kinetic constants should be considered (see the section on Measurement of Activity). Lastly, as mentioned under "Measurement of Activity," the solubility of the enzyme may decrease upon irradiation. If as a result the amount of enzyme in the assay is diminished, the consequent calculation of mass would be in error.

When the analysis of the data is completed, a value or set of values for the target size should have resulted. How should these be interpreted? In the simple case the target size may be correlated with subunit size or with

[51] R. L. Kincaid, E. Kempner, V. C. Manganiello, J. C. Osborne, Jr., and M. Vaughan, *J. Biol. Chem.* **256**, 11351 (1981).

[52] W. Rosse, H. J. Rapp, and T. Borsos, *J. Immunol.* **98**, 1190 (1967).

[53] J. T. Harmon, C. R. Kahn, E. S. Kempner, and M. L. Johnson, in "Current Views on Insulin Receptors" (D. Andreani, R. De Pirro, R. Lauro, J. Olefsky, and J. Roth, eds.), p. 37. Academic Press, London, 1981.

[54] W. Schlegel, E. S. Kempner, and M. Rodbell, *J. Biol. Chem.* **254**, 5168 (1979).

[55] P. Simon, S. Swillens, and J. E. Dumont, *Biochem. J.* **205**, 477 (1982).

[56] A. S. Verkman, K. Skorecki, and D. A. Ausiello, *Proc. Natl. Acad. Sci. U.S.A.* **81**, 150 (1984).

[57] M. Potier and S. Giroux, *Biochem. J.* **226**, 797 (1985).

the hydrodynamic molecular weights. However, the target size is that of the functional unit and may correspond to the subunit size, a combination of subunits, or the particle or hydrodynamic mass. Thus, a function may require the presence of multiple subunits as in glutamate dehydrogenase.[9] On the other hand, for enzymes composed of unlike subunits with multiple measurable functions (e.g., adenylate cyclase[15] or of tryptophan synthetase[49]), each functional unit may correspond to a different combination of subunits. Any assignment of structure or composition to the functional units derived from target analysis thus must also be based on biochemical information about the receptor or enzyme.

Conclusions

We have described in detail the procedures used in radiation inactivation experiments and the reasons for technical steps. With each refinement we achieve increases in accuracy, reproducibility and (we hope) validity. Comparison with known molecular sizes indicates that with our present techniques we often attain agreement within 10%; a survey of the older literature[10] indicated agreement better than 15%.

Certain caveats should always be borne in mind. This method reveals the mass of a functional unit as it exists in the physical state determined by sample preparation and irradiation conditions. The meaning of a functional unit depends on the particular assay which is used. And finally, the target size is a sum of all the necessary parts.

It is possible to determine target sizes for any measurable function which can survive freezing and thawing. If the function can also survive lyophilization, methodological procedures may be somewhat simpler. There is no requirement for purification, and the active unit may be studied *in situ* in membranes and even in intact cells. But the best feature of all is that it works.

[7] Measurement of Molecular Weights of Elongated Molecules by Viscoelastometry

By BRUNO H. ZIMM

The viscoelastic method can be thought of as an extension of the measurement of intrinsic viscosity. With the viscosity of a macromolecular solution is necessarily associated a relaxation time, a time during which the conformations of the molecules come to steady state after the sudden application of stress. During this time the system is able to store some of the energy absorbed from the applied stress. In other words, it shows elasticity as well as viscosity; hence the phenomenon is called viscoelasticity. In dilute solutions the relaxation time depends strongly on molecular shape and size and offers a way of measuring these quantities in a nondestructive manner. The method has been applied principally to very large DNA molecules, where the relaxation times are conveniently long, of the order of seconds or greater, and where other size and shape methods become difficult. When the relaxation times are shorter than about 1 sec, which is the case with most proteins, viscoelasticity of dilute solutions can be measured accurately only by very special equipment, such as that developed by Ferry and his collaborators,[1] and in this case the phenomenon is mainly of interest for studying molecular flexibility.[2]

With large linear-chain DNA molecules the relaxation time varies about as $M^{1.67}$, where M is the molecular weight, and is therefore a sensitive measure of the latter. Actually, the relaxation time measures the cube of the hydrodynamic radius of the molecule; so it is sensitive to extension changes, such as those induced by changing salt concentration, as well as to molecular weight. However, under constant solution conditions it is a good measure of molecular weight, in the same way that intrinsic viscosity or sedimentation constant is. It is attractive as a measure of molecular weight for several reasons: it is more sensitive to molecular weight than most other hydrodynamic properties; its measurement produces a curve of relaxation against time which contains information about the distribution of molecular weights, and the nature of this curve is such that the largest species present are emphasized, which is most useful when one desires to look at an undegraded long-chain material in the presence of degradation products or smaller impurities. The last is often the case with chromosomal DNA.

The application of this method to DNA was first described by Chapman et al.[3] in 1969. It was soon applied successfully to very large DNA by Klotz[4] working with bacteria and by Kavenoff working with Drosophila.[5,6] The Drosophila work showed for the first time that DNA molecules were large enough to run unbroken from one end of the chromosome to the other.

[1] J. L. Schrag and R. M. Johnson, Rev. Sci. Instrum. 42, 224 (1971).
[2] S. Hvidt, J. D. Ferry, D. L. Roelke, and M. L. Graeser, Macromolecules 16, 740 (1983).
[3] R. E. Chapman, L. C. Klotz, D. S. Thompson, and B. H. Zimm, Macromolecules 2, 637 (1969).
[4] L. C. Klotz and B. H. Zimm, J. Mol. Biol. 72, 779 (1972).
[5] R. Kavenoff and B. H. Zimm, Chromosoma 41, 1 (1973).
[6] R. Kavenoff, L. C. Klotz, and B. H. Zimm, Cold Spring Harbor Symp. Quant. Biol. 38, 1 (1973).

The apparatus used for such measurements is essentially a refined concentric rotating cylinder, or Couette, viscometer. (For a full description see Klotz and Zimm[7] and Bowen and Zimm.[8]) The rotor of this viscometer is a Cartesian diver, suspended by bouyant forces in the middle of the solution without physical contact with the walls of the chamber; its height is usually maintained by a servo mechanism actuated by a beam of light that senses the position of the rotor. The rotor is turned by a rotating magnetic field acting on a metal ring in the rotor. To measure the elastic relaxation of a solution, the magnetic drive is started and the rotor allowed to spin until it reaches a steady rate; the drive is then abruptly turned off. If the solution is viscoelastic, the stored elastic energy causes the rotor to reverse its direction of rotation, and finally to come gradually to rest. The final part of the decay to rest is exponential, and the time constant of this part, which is usually obtained from the slope of a plot of the logarithm of the rotor's angular position against time, is the longest relaxation time of the solution. The angular position of the rotor is measured by an optical system that observes a piece of sheet polarizer in the rotor. Relaxation times vary from seconds for DNA of molecular weight about 10^8 (large viruses) to hours for DNA above 10^{10} (chromosomal DNA of higher organisms).

As with sedimentation or other hydrodynamic measurements, the relaxation time depends on the DNA concentration, so it is necessary to measure at low concentrations. For accurate work it is desirable to plot the time against the concentration and extrapolate to infinite dilution.

The semilog plot of position against time is most useful when the sample is monodisperse with respect to molecular weight or when the distribution of molecular weights cuts off sharply at the high end, since in these cases the relaxation curve has a prominent longest relaxation time corresponding to the largest species. (Even when there is only one species, shorter relaxation times are also present, corresponding to complicated internal motions of the molecule, but these contribute only a few percent to the relaxation and are not of much consequence.)

The longest relaxation times of DNA molecules can also be measured by other methods. Electrooptical effects (electrobirefringence and linear electrodichroism) are particularly useful for small DNA, where the relaxation times are of the order of micro- or milliseconds, times which are easily measurable with electronic techniques. Here the main applications have been in studying molecular structure and flexibility.[9,10] For large

[7] L. C. Klotz and B. H. Zimm, *Macromolecules* **5,** 471 (1972).

[8] B. C. Bowen and B. H. Zimm, *Biophys. Chem.* **9,** 133 (1979).

[9] P. J. Hagerman, *Biopolymers* **20,** 1503 (1981).

[10] M. Hogan, N. Dattagupta, and D. M. Crothers, *Proc. Natl. Acad. Sci. U.S.A.* **75,** 195 (1978).

DNA these effects become complicated and are not yet well understood, but the related stress-optical effect (transient flow birefringence) has been used successfully[11]; its main drawback in comparison to viscoelasticity is the need for larger amounts of solution.

[11] D. S. Thompson and S. J. Gill, *J. Chem. Phys.* **47,** 5008 (1967).

[8] Measurement of Hydrodynamic Properties of Active Enzyme by Sedimentation

By LYNDAL K. HESTERBERG and JAMES C. LEE

Introduction

Why Active Enzyme Centrifugation?

An entirely novel approach to enzyme structure–function analysis was opened about 1963 with the introduction of active enzyme centrifugation (AEC) by Cohen and co-workers.[1-3] The methods combined sedimentation and the enzymatic reaction in such a way that the macromolecule is *not* observed, but rather the reaction catalyzed by the enzyme is. For the first time, it was possible to simultaneously measure the hydrodynamic properties of the fully active complex and the enzymatic activity. With previous techniques it has usually been impossible to determine the hydrodynamic properties of the fully active enzyme molecule, especially if the molecule existed only in a partially purified state. Such knowledge, however, is of the utmost importance when trying to understand the enzymatic reaction mechanism at a molecular level.

This method allows one to obtain the hydrodynamic parameters, sedimentation coefficient, s, and diffusion coefficient, D, while the enzyme–substrate complex is fully active. Therefore, one can compare the conformational or oligomeric state of the active enzyme with that of the inactive enzyme. In addition the structural information provided by AEC will allow a correlation of the physical state of the enzyme with the enzymatic activity in that state. This is extremely valuable in associating systems where activity may change depending on the polymeric state of the enzyme.

[1] R. Cohen, *C.R. Hebd. Seances Acad. Sci.* **256,** 3513 (1963).
[2] R. Cohen and M. Mire, *Eur. J. Biochem.* **23,** 267 (1971).
[3] R. Cohen, B. Giraud, and A. Messiah, *Biopolymers* **5,** 203 (1967).

Advantages/Disadvantages

One of the advantages offered by AEC is that the method is rapid and direct, and the hydrodynamic parameters obtained are actually those of the active unit within the same ranges of protein concentration generally used for steady-state kinetic studies, i.e., μg/ml or lower. In addition, since the method monitors only the enzyme's activity profiles, no enzyme concentration effects need to be considered in the analysis. AEC also allows for a high degree of specificity in the experimental design. Since only the particular substrate or product of interest is measured, one can monitor the enzyme of choice without interference from other proteins.

It has also become apparent that this technique will work well in crude preparations of the enzyme. Enzymes may behave very differently when surrounded by the cellular milieu as compared to the purified state. For example, certain enzymes which participate in the same metabolic pathway may aggregate or in some manner influence the conformational states of the others.[4,5] AEC will allow a detailed inspection of the structural properties of specific enzymes in a complex mixture of proteins and thereby will allow the study under conditions as close to the cellular environment as possible.

One of the experimental disadvantages of AEC is the high pressures generated during the length of the AEC run. Since pressures on the order of 100–500 atmospheres are generated at the base of a centrifuge cell at high rotor speeds, even small changes in the partial specific volume of the protein can lead to marked effects in the sedimentation behavior. In general, the polymerization of a protein will lead to changes in the volume. It is well known that pressures in the range of 70–400 atmospheres can produce transformations in the aggregation state of various polymeric systems.[6,7] Specifically, the apparent association constant for polymerization can be dramatically shifted by the pressure gradient developed in an ultracentrifuge cell. Detection of pressure-induced effects and the required correction factors have been thoroughly outlined by Harrington and Kegeles.[8]

Other Techniques

Other methods have been developed with the advantage of running at atmospheric pressure to avoid perturbation of pressure-sensitive associa-

[4] M. R. Kuter, C. J. Masters, and D. J. Winzor, *Arch. Biochem. Biophys.* **225,** 384 (1983).
[5] R. S. Liu and S. Anderson, *Biochemistry* **19,** 2684 (1980).
[6] R. P. Frigon and S. N. Timasheff, *Biochemistry* **14,** 4567 (1975).
[7] R. Josephs and W. F. Harrington, *Proc. Natl. Acad. Sci. U.S.A.* **58,** 1587 (1967).
[8] W. F. Harrington and G. Kegeles, this series, Vol. 27, p. 306.

tion equilibria. One of these is the column scanning method of Ackers.[9,10] An active enzyme HPLC system has also been described by Furman and Neet[11] which measures product formation directly. This technique avoids both the problems of enzyme reassociation during assay inherent in conventional studies and the pressure perturbations of AEC. However, this latter system has the disadvantage that the enzyme-containing fractions are diluted with substrate solution after separation on a column. For concentration-dependent protein systems, this could alter the results significantly. Nevertheless, the speed with which the experiment can be performed is a distinct advantage.

Scope

The scope of this chapter will be limited to AEC and will include (1) some of the practical aspects of designing and performing the experiments, (2) data analysis, (3) interpretation of the data to establish a model, and (4) theoretical simulation of data leading to a confirmation of the model obtained from experimentally observed data.

The object of this chapter is to provide an outline of the steps necessary to obtain reliable, accurate data without the complications provided by various artifacts. The analysis of the data is discussed, although this is, in most cases, a very simple matter. The interpretation of the data is crucial and potentially the most complex step. However, an investigator who is aware of the potential problems and artifacts in data interpretation and who makes use of the resources available for computer simulations will be able to provide the theoretical confirmation of the proposed model, leading to a better understanding of the mechanism of action of an enzyme.

Basic Principles of AEC

The basic technique of AEC as pointed out by Cohen and Mire[2] is band sedimentation of an enzyme solution. This involves the layering of a small volume of enzyme sample over a large volume of buffer in a rotating centrifuge cell. This buffer contains all of the substrates and cofactors necessary for activity. As the thin band of enzyme sediments through the substrate solution, it catalyzes the reaction and yields products. The protein band cannot be observed directly because of the small concentration

[9] G. K. Ackers, *in* "Methods of Protein Separation" (N. Catsimpoolas, ed.), Vol. 2, p. 1. Plenum, New York, 1976.
[10] L. C. Davis and G. A. Radke, *Biophys. Chem.* **18,** 241 (1983).
[11] T. C. Furman and K. E. Neet, *J. Biol. Chem.* **258,** 4930 (1983).

of enzyme used. Therefore, the sedimentation and diffusion coefficients of the enzyme are calculated from the reaction catalyzed by the enzyme band. There is the requirement, therefore, that the reaction can in some way be monitored spectrophotometrically. Ideally, the disappearance of substrate or the appearance of product could be directly observed with the scanner. However, in many enzyme reactions this is not possible. Therefore, other systems must be used. If the reaction releases hydrogen ion, a pH-dye coupled assay system is possible. Also, the reaction of many enzymes can be coupled to other enzymes which in turn catalyze a reaction which can be monitored spectroscopically. The use of either a pH-dye system or the coupled enzyme system requires a series of control experiments and may introduce a variety of artifacts into the experiment. These are discussed in detail in the following section. Nevertheless, the AEC technique is valid for any assay that can be monitored spectroscopically.

One of the other requirements for AEC experiments is an adequate positive density gradient in the assay mixture to counteract the negative gradient associated with the leading edge of the sedimenting band. During acceleration of the ultracentrifuge rotor, a small amount of enzyme is layered as a thin film onto the contents of the centrifuge cell. It is crucial that the formed boundary be sharp and that no enzyme be present anywhere else in the cell except at the meniscus before the desired ultracentrifugal speed is reached. This requires that the density of the substrate solution in the cell be somewhat higher than that of the enzyme solution.

In general, this may be achieved by dissolving the enzyme in 0.01 M buffer, whereas the substrates are contained in 0.1 M buffer. In some cases, however, it may be necessary to increase the density of the substrate solution by the addition of NaCl, glycerol, or any other suitable compound. The suitability of such an additive is dependent mainly on three criteria: (1) it should in no way affect the enzymatic system to be tested, (2) it should not react with substrates or any other products that might shift the equilibria from normal assay conditions, and (3) the molecular weight of the additive should be small enough to prevent sedimentation or gradient formation in the centrifugal field where it will be used.

The technique of AEC, in addition, has inherent assumptions, one of which is that every enzyme molecule in the sedimenting band acts on its substrate with the same velocity. This occurs when the substrate concentration remains at the saturating level throughout the entire distribution of the enzyme. One must also assume that the reverse reaction does not take place to any significant extent. The experimental steps which one must undertake to ensure the validity of these assumptions are discussed below.

Practical Aspects

Fundamentals

Active enzyme sedimentation studies require a Beckman-Spinco Model E analytical ultracentrifuge equipped with a UV-scanner and a RTIC unit to measure and regulate the temperature. For AEC studies, a charcoal-filled Epon, Type I (Vinograd type) double-sector band-forming centerpiece with sapphire windows is recommended. Both sectors are filled with 0.32 ml of the desired assay mixture containing substrates, cofactors, and the necessary material to form a stabilizing gradient. The capillary chamber of the sample sector is filled with 5–15 μl of assay buffer containing the desired concentration of enzyme minus the substrates. The chamber of the reference sector is filled with the same volume of assay buffer. The calculations for the required rpm of the rotor have been outlined by Cohen and Claverie.[12] Basically, the sedimentation coefficient of the enzyme and the rotor speed required are ~7 S, >50,000 rpm, ~10 S, >42,000 rpm, and ~15 S, >34,000 rpm. The basic methods for data collection and calculations have been described by Kemper and Everse.[13]

Density Gradient

The success or failure of an individual AEC run depends heavily on the formation of an adequate positive density gradient to support the layered band of enzyme. If an adequate positive density gradient is not established, the enzyme band reaches a region of negative density gradient and then sinks until it reaches a region of sufficiently positive density gradient. Generally, solvents or salts are used to provide a sufficient positive density gradient. The selection of the particular one depends on the enzyme system used. However, D_2O, glycerol, sucrose, and a variety of neutral salts are widely used.

In addition to these, much can be done by manipulation of experimental procedure to enhance the formation of a gradient.

Rotor Acceleration/Sinking. First, rotor acceleration at the start of an AEC run should be very gradual up to 10,000 rpm. Other considerations in the start-up were pointed out by Wei and Deal.[14] Worth special consideration at higher enzyme concentration is the incorporation of a 5- to 20-min "holding" period at 5000 rpm. This allows for the complete layering of the enzyme onto the surface, as well as for the formation of the positive

[12] R. Cohen and J. Claverie, *Biopolymers* **14**, 1701 (1975).
[13] D. L. Kemper and J. Everse, this series, Vol. 27, p. 67.
[14] G. J. Wei and W. C. Deal, Jr., *Biochemistry* **18**, 1129 (1979).

density gradient. The exact speed and length of the holding period and start-up procedure are best determined empirically. If the gradient has not formed sufficiently, the enzyme will sink rapidly.

Extreme sinking can be visualized in two ways. Using schlieren optics, one can observe the layering and stability of the banded enzyme solution. During layering over a sufficient positive density gradient, increasing amounts of the enzyme can be seen at the meniscus giving rise to a pattern similar to that seen in the early phase of a schlieren sedimentation velocity experiment. As the enzyme band sediments into the substrate mixture, the band will diffuse and slowly diminish. However, in the event of enzyme sinking, the band at the meniscus will be visible only shortly, if at all. Instead, one sees a curvature of the schlieren pattern at the bottom of the cell.

For ultracentrifuges with only uv scanner optics, it is more difficult to detect sinking immediately, but it becomes flagrantly apparent 10 to 20 min into the run. The first indication, based on our experience, is an uneven or skewed appearance of product near the meniscus as well as increasing amounts of product at the bottom of the cell. In cases where the enzyme does not sink the entire distance to the bottom, the total amount of substrate will disappear in a very uneven and irregular manner.

If the operator fails to observe enzyme sinking visually, it can be detected in the data analysis. In cases of severe sinking, large increases in the sedimentation rate are indicative of the problem. In the event of sinking which is not so severe, nonlinear log R vs time plots will also indicate the lack of a sufficient positive density gradient. To correct the problem of sinking, the positive density gradient in the substrate solution must be increased and be allowed time to form. Slow rotor acceleration combined with a holding period with higher concentrations of the solvent of choice should provide the necessary conditions to prevent sinking.

Loading Volume. The volume of the layered enzyme is also an important consideration in the choice of buffers and the formation of the gradient. As pointed out by others it is advisable to use the minimum volume possible.[15] A volume of 5 to 15 μl would be the least which can be practically used as anything less will likely give inconsistent layering results. Significant errors, due to volume or failure of the density gradient, are likely to occur at loading volumes larger than 30 μl.

Solvent System. An additional factor must be considered when solvent systems are employed for the formation of the positive density gradient. The introduction of salts, D_2O, glycerol, etc., to protein solutions may lead to unexpected results if the interaction of protein and solvent is not

[15] D. J. Llewellyn and G. D. Smith, *Arch. Biochem. Biophys.* **190**, 483 (1978).

considered. The basic equation for obtaining the sedimentation coefficient under standard conditions is

$$s_{20,w} = s_{obs} \left(\frac{\eta_{t,w}}{\eta_{20,w}}\right)\left(\frac{\eta_{solvent}}{\eta_{t,w}}\right)\left[\frac{(1 - \bar{v}_2^0\rho)_{w,20}}{(1 - \phi_2^0\rho)_{solvent,20}}\right] \quad (1)$$

where η is viscosity, ρ_w and $\rho_{solvent}$ are the densities of water and the solvent, respectively, and $\bar{v}_2^0$ and ϕ_2^0 are partial specific volumes of the protein in water and in the presence of organic solvents, respectively. Observed sedimentation coefficients in the presence of D_2O must be corrected for the deuteration of the protein also. Equation (1) becomes

$$s_{20,w} = s_{obs} \left(\frac{\eta_{t,w}}{\eta_{20,w}}\right)\left(\frac{\eta_{solvent}}{\eta_{t,w}}\right)\left[\frac{(1 - \bar{v}_2^0\rho)_{20,w}}{(1 - \phi_2^0(\rho/k)_{t,solvent,20}}\frac{1}{k}\right] \quad (2)$$

where k, the correction factor for the deuteration of the protein, is equal to 1.008 for a solvent with 50% D_2O.

In the presence of organic solvents, the correction factors due to protein–solvent interactions are much more significant, since

$$(\delta g_3/\delta g_2)_{T,\mu_1,\mu_3} = \frac{(\delta\rho/\delta c_2)_{\mu_3} - (\delta\rho/\delta c_2)_{m_3}}{(\delta\rho/\delta c_3)_{m_2}} \quad (3)$$

where $\delta g_3/\delta g_2$ is the preferential interaction parameter between protein (component 2) and solvent (component 3). $\delta\rho/\delta c_i = 1 - \phi_i\rho$ for component i, and the subscripts μ_3 and m_3 indicate that the measurements are obtained at constant chemical potential and constant chemical composition, respectively.[16] Thus, the correct equation to calculate for $s_{20,w}$ is

$$\frac{1}{s_{20,w}} = \frac{1}{s_{obs}}\frac{\eta_{20,w}}{\eta_{t,w}}\frac{\eta_{t,w}}{\eta_{solvent}}\left[1 + \frac{1 - \bar{v}_3\rho}{1 - \bar{v}_2\rho}(\delta g_3/\delta g_2)\right] \quad (4)$$

It is evident that, depending on the magnitude for $\delta g_3/\delta g_2$, failure to consider such preferential interaction parameters may lead to significant differences in the value of sedimentation coefficients. Since $\delta g_3/\delta g_2$ is a measure of the amount of solvent molecules (component 3) present in the immediate domain of the protein (component 2) over its concentration in the bulk solvent, it may assume either a positive or negative value. A positive value of this parameter indicates an excess of solvent molecules in the domain of the protein; a negative value means a deficiency of solvent molecules or an excess of water (component 1) in the immediate domain of the protein. A positive value of $\delta g_3/\delta g_2$ would lead to a lower corrected value for $s_{20,w}$ than if the preferential interaction was not taken into consideration. Conversely, a negative value of $\delta g_3/\delta g_2$ would yield a

[16] E. F. Casassa and H. Eisenberg, Adv. Protein Chem. **19**, 287 (1964).

higher corrected value for $s_{20,w}$. It is evident that determination of the preferential interaction parameter is essential. Details of the principles of multicomponent theory have been outlined by Lee *et al.*[17]

The presence of organic solvents may not only induce preferential interaction of protein with solvent but it may also affect the association constants of protein–protein interaction.

D_2O is known to enhance the self-association of some protein systems[18-21] and the possible mechanism is that deuterium bonding is of greater strength than hydrogen bonding.[22,23] The effects of glycerol and sucrose on protein self-association have been demonstrated in phosphofructokinase and microtubule assembly.[24-26] The proposed driving force is derived from a thermodynamically unfavorable interaction between protein and the solvent system. Thus, to help circumvent the effects of solvents, the use of multiple solvent systems is recommended, along with the correction for preferential solvent interactions.

Assumptions

The most crucial assumption is that every enzyme molecule is reacting with the same velocity. The validity of the data rests on this assumption, and, therefore, on the ability of the investigator to test for the experimental conditions required to insure that the assumption is true. When the conditions are not fulfilled, an enzyme molecule in the leading edge of the enzyme band produces more products in a given time than an enzyme molecule in the trailing part of the band. Since the band widens because of diffusion during centrifugation, a much higher and erroneous value is obtained for $s_{20,w}$.

This can occur under any condition which leads to substrate depletion. This can most often be seen simply by the loading of an excess amount of the enzyme. To determine if this artifact occurs, one performs a series of centrifugations with varying amounts of enzyme. A plot of the sedimentation value versus enzyme concentration should be linear with zero slope

[17] J. C. Lee, K. Gekko, and S. N. Timasheff, this series, Vol. 61, p. 26.
[18] S. N. Timasheff, *Protides Biol. Fluids* **20**, 511 (1973).
[19] S. Paglini and M. A. Lauffer, *Biochemistry* **7**, 1827 (1968).
[20] J. J. Lee and D. S. Berns, *Biochem. J.* **110**, 465 (1968).
[21] P. A. Baghurst, L. W. Nichol, and W. H. Sawyer, *J. Biol. Chem.* **247**, 3198 (1972).
[22] S. N. Timasheff and R. Townend, *Protides Biol. Fluids* **16**, 33 (1969).
[23] G. Némethy and H. A. Scheraga, *J. Chem. Phys.* **41**, 680 (1964).
[24] J. C. Lee and S. N. Timasheff, *Biochemistry* **16**, 1754 (1977).
[25] S. N. Timasheff, R. P. Frigon, and J. C. Lee, *Fed. Proc., Fed. Am. Soc. Exp. Biol.* **35**, 1886 (1976).
[26] L. K. Hesterberg and J. C. Lee, *Biochemistry* **19**, 2030 (1980).

when no artifact is present. Whereas when the conditions are inadequate, the sedimentation values will increase with increasing enzyme concentration. Care must be taken, as a concentration-dependent associating system can also give a similar result. Steady-state kinetics are also very valuable in assessing under which conditions too much enzyme activity may be present, and are crucial for differentiating between overloading and an associating system.

The degree to which the substrate present at any point in the cell is converted to product depends also on the length of time the enzyme band takes to pass that point. At low rpm, the enzyme will take longer to pass than at higher rpm. So conditions which do not show overloading at 60,000 rpm may give rise to this artifact at 30,000 rpm. It is important to bear this in mind when choosing the experimental conditions.

Depletion of substrate can lead to dramatic deviation from the actual sedimentation coefficient of the active component. Cohen and Claverie[12] demonstrated that with an initial concentration of substrate equivalent to the Michaelis constant, K_m, and with a high consumption of substrate by the passage of a band of enzyme, the observed $s_{20,w}$ would be 10% higher than the actual value. However, with an initial substrate concentration of 10-fold of K_m and with over 50% consumption of substrate, the observed value for $s_{20,w}$ may only be ~1–2% higher.

One of the major concerns in active enzyme sedimentation experiments is denaturation of the enzyme in the course of these experiments. Since the area enclosed by the sedimentation boundary represents the amount of product in the cell, it is possible to monitor the activity of the enzyme in the course of a sedimentation experiment. If no denaturation of the enzyme or no dissociation into inactive aggregates during the sedimentation procedure has occurred, then the total amount of product in the cell should increase linearly with time.

Accordingly, if the total amount of product, expressed in arbitrary units of area, is plotted as a function of the time of sedimentation a linear plot should be produced, as shown in Fig. 1. Any deviation should be investigated further.

Another assumption made is that the substrate or product being measured does not diffuse significantly between two consecutive scans. This has been shown to hold true, so long as a minimum angular velocity is maintained. This minimum value depends on the size of the enzyme being assayed and has been outlined thoroughly by Cohen and Claverie.[12] The rule of thumb outlined is for an enzyme sedimenting at ~7 S, >50,000 rpm is necessary, for ~10 S, >42,000 rpm, and for ~15 S, >34,000 rpm is required. The minimum S for which AEC can be used reliably is ~3 S.

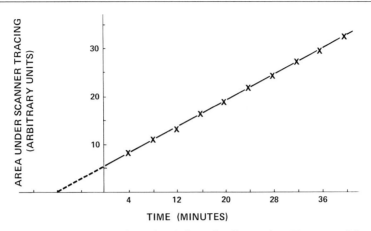

TIME (MINUTES)

Fig. 1. Amount of product formed and time of sedimentation. Enzyme activity is assumed to be proportional to the area enclosed by the scanner tracing. The time shown is from the start of scans, while the actual layering time, as observed visually, occurred 10.75 min prior to the initial scan at 0 min. (Reprinted with permission from Hesterberg and Lee.[26] Copyright 1980 American Chemical Society.)

Methods in Enzyme Assays

The most desirable assay system would be to spectrophotometrically measure the disappearance of substrate or the appearance of product for the specific enzyme's reaction. If suitable absorption wavelengths are available such as the case for dehydrogenase enzymes and NADH, then the simplest system is the better. In practice, it is always more desirable to be able to follow the appearance of product, rather than substrate disappearance. In this way, meniscus effects can be minimized. However, many enzyme's substrates or products cannot be monitored spectroscopically. In these cases, more complex assay systems are required, and in turn more artifacts are possible for the unwary investigator. One widely used assay system is commonly referred to as the coupled enzyme assay. The concept is that the reaction products of the enzyme of interest, enzyme A, are linked to another enzyme, B, whose product is detectable spectroscopically. In this assay system, enzyme B is included in the assay mixture. To use this system for AEC, two new conditions must be met. First, the sedimentation coefficient of the coupling enzyme B must be lower than the smallest active unit of enzyme A. Second, the concentration of enzyme B must always be sufficient so that the intermediate product is transformed immediately into the final product. When low concentrations of enzyme B are used, artificially low values of sedimentation will be observed for enzyme A. The correct value is determined from a series of centrifugations with varying concentrations of enzyme B, then extrapolation of the value to an infinite concentration of enzyme B. In most cases, the line should be flat once sufficient amounts of enzyme B are present.

The second condition requires that the coupling enzyme, enzyme B, does not denature or lose activity during the duration of the AEC run. This can be tested in a manner analogous to the denaturation test for enzyme A discussed in the preceding section.

Another assay system has also been used in cases where a H^+ is released during the reaction. The utility of such a pH-dependent, dye-linked assay system has been demonstrated and is valid for AEC studies even on systems in rapid association–dissociation.[27] The underlying concept for this assay is that a spectrophotometric change in an acid–base indicator dye can be observed as it is titrated by the H^+ produced during the reaction. However, this assay has some negative aspects. First, very low buffer and salt conditions are required, so as to maximize the sensitivity of the dye assay. If too high a concentration of the buffer is present, the dye will not be sensitive to the enzyme's reaction. Also, the dye and the buffer should be carefully selected so they have the same pK_a. If these vary too widely, a nonlinear response will result. In any event, the assay mix should be titrated to confirm a linear response in the absorbance of the dye upon increasing amounts of H^+. In general, a coupled enzyme system will afford higher sensitivity than a pH-dependent dye-linked assay.

Data Analysis

Sedimentation Values

The rate of sedimentation of the active enzyme can be obtained by analyzing the scanner-derived tracings from the individual run. The mechanics of measurement have been outlined by Kemper and Everse.[13] Cohen and co-workers[3] have described four methods of analysis based on the differences between successive scans. The most accurate method is a rigorous (and tedious) method of moments. This requires the use of computer facilities and is rather time consuming. Interested readers are referred to the papers of Cohen et al.[3] However, two less rigorous methods are most often used. These are, first, the method of the midpoint of the scanner traced boundary and, second, the difference curve method. The method of the midpoints is to simply measure the midpoints of the leading edges of the zones directly. These midpoints at various time intervals provide the necessary information for the log R vs time plots from which

[27] J. P. Shill, B. A. Peters, and K. E. Neet, *Biochemistry* **13**, 3864 (1974).

one obtains values for $s_{20,w}$. The latter method requires measuring the rate of sedimentation of the migration of the maxima in the difference curve. The difference curve method is derived directly from the rigorous methods, but with several assumptions being made. The most crucial assumption is that the diffusion of the product is negligible during the time of the scan.

The difference curves are very simply constructed by superimposing two consecutive scans and measuring the difference in absorbance between the first and second scan at a given radius, as shown in Fig. 2. When the differences at a number of radii are measured, a difference curve can be constructed. The difference in absorbance is then directly related to the change in concentration of either product or substrate between those two time points.

Direct comparisons of these two methods have shown that the method of the midpoints is equal to, if not better than, the difference curve method,[15,26,28] as shown in Fig. 2. However, for associating protein systems much more information can be gleaned from the difference curve method than from the method of midpoints and will be discussed below. The lower limit of these two methods requires that the enzyme sediments at ≥3 S.[15] For enzymes sedimenting more slowly than this, the more rigorous methods must be used.[1]

The error in the data measurements increases dramatically as either the enzymes s value drops below 3 S or the angular velocity of the rotor decreases below the optimum.[12] Llewellyn and Smith[15] have simulated product distribution curves for enzymes sedimenting both above and below the 3 S limit. They have graphically demonstrated the effects of both sedimentation and diffusion coefficients on the time necessary for the enzyme to move from the meniscus region and away from the complicating edge effects.[1,3] Any steps which can increase the rate of sedimentation will minimize the errors inherent in the less rigorous methods of analysis.

Diffusion Coefficient

The diffusion coefficient of the active enzyme species can also be calculated from the data obtained from the difference curve analysis. This has been outlined previously.[12,29] The total area of the difference curve, A, divided by the peak of the curve, H, when squared is equal to the function of $[\exp(2s\omega^2 t) - 1]$ where s is the sedimentation coefficient, ω is the angular velocity of the rotor, and t is the time. The slope of a plot of $(A/H)^2$ versus $[\exp(2s\omega^2 t) - 1]$ gives the value of the diffusion coefficient.

[28] B. L. Taylor, R. E. Barden, and M. F. Utter, *J. Biol. Chem.* **247**, 7383 (1972).
[29] J. Claverie, *Arch. Biochem. Biophys.* **202**, 160 (1980).

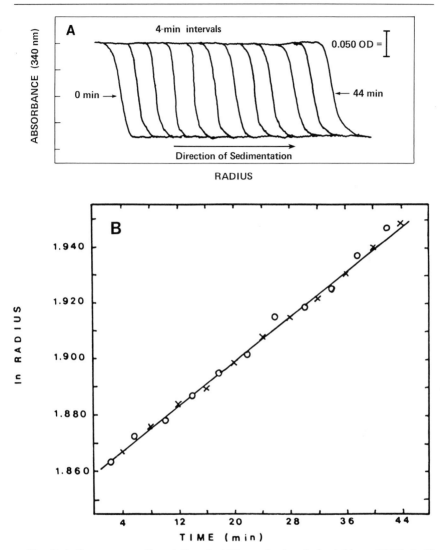

Fig. 2. Active enzyme sedimentation of rabbit muscle phosphofructokinase (PFK) at pH 8.55 and 23°. The coupled enzyme assay system was utilized in 50% D_2O. The conditions for centrifugation were as follows: rotor speed, 60,000 rpm; scan speed, fastest; chart speed, 25 mm/sec; noise suppression, off; wavelength, 340 nm; optical density range, 0–1.0; 7.5 µl of PFK at 3.15 µg/ml was layered onto the assay mixture. (A) Successive scanner tracings superimposed on a single frame of reference. (B) Relation between ln R and time for data obtained from the inflection point of the boundary (×) and the difference curve (○). The time of the difference curve was taken to be the mean time of the scans. (C) Difference curves constructed from the scanner traces. (Reprinted with permission from Hesterberg and Lee.[26] Copyright 1980 American Chemical Society.)

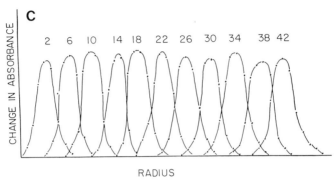

FIG. 2. (continued)

However, even under conditions of simulated data, the inherent error in the diffusion coefficient has been reported to be about 10 times that of the sedimentation coefficient. As pointed out by Cohen and Claverie,[12] if only one active species is considered, if there are no problems in experimental design (centrifuge speed, substrate depletion, etc.), and if a scanner in combination with a medium-sized computer is used, then one can apply the rigorous methods of data analysis and get s and D. However, a simple visual comparison of simulated curves with experimental difference curves should provide a reasonable D value. The inherent error arises from the fact that the observed activity curves are not in all cases identical to the protein profile.[29]

As pointed out previously, the collection and analysis of AEC data are usually straightforward and simple. The relationship between the changing product or substrate distribution and the sedimentation of the enzyme can however be complex. It is the interpretation of the data which is more complicated, especially when associating protein systems are involved.

Data Interpretation

For an associating system the value of $s_{20,w}$ may be strongly concentration dependent and cannot be quantitatively analyzed because of the varying and unknown concentration of the enzyme in the sedimenting zone. However, the presence of multiple active forms can be detected.

Correlation between AEC and Steady-State Kinetics

Before the interpretation of data can proceed, one of the initial necessary experiments is a correlation between the enzymatic activity ob-

COMPARISON OF PHOSPHOFRUCTOKINASE ACTIVITY: STEADY-STATE KINETICS
VS ACTIVE ENZYME SEDIMENTATION[a]

| Assay system | Stabilizing gradient solvent | Observed average $\Delta OD/min$[b] | |
		Steady state	AEC
Coupled enzyme	50% D_2O	5.1×10^{-2}	5.3×10^{-2}
Coupled enzyme	10% sucrose	2.3×10^{-1}	2.1×10^{-1}
Coupled enzyme	10% glycerol	2.4×10^{-1}	2.5×10^{-1}
pH dye	10% glycerol	9.2×10^{-3}	8.2×10^{-3}

[a] Reprinted with permission from Hesterberg and Lee.[26] Copyright 1980 American Chemical Society.
[b] Observed ΔOD is normalized to 1 μg of PFK, 1 ml volume, and 1.0 cm path length.

served in AEC and that from steady-state kinetics under identical conditions. Since the area under each difference curve represents the total activity of the enzyme for the time between two consecutive scans, it is possible to measure the activity of the enzyme during the centrifugation experiment. Since the exact amount of the enzyme present at any point in the centrifuge cell is not known, the activity must be calculated by assuming that all of the protein added to the capillary chamber was layered and sedimented in the cell. Also, the observed base width and height of the difference curve must be corrected for the intrinsic magnification factor, calibration of the absorptivity vs pen deflection, and the light path of the centrifuge cell. The enzyme activity expressed as the change in absorption per minute is calculated by dividing the total area calculated for a given difference curve by the number of minutes between two successive scans. For manual data analysis, the experimental uncertainty is ±10%.[26] With computer-assisted data analysis, the experimental uncertainties would be somewhat better. It is obvious that if the corrected activity observed per minute in AEC is in good agreement with steady-state kinetic data, then one knows that the species observed in the AEC are the full complement of active forms, as shown in the table for rabbit muscle phosphofructokinase. However, if this correlation is very poor, or if activity is lost with time, a variety of problems may be occurring. First, the experimental design may be such that enzyme is being inactivated or lost. Another possibility is a pressure-induced loss of activity, due to the high fields present in the ultracentrifuge. In the case of concentration-dependent associating systems, the dilution effects during AEC may favor the formation of inactive, smaller forms of the enzyme. In such a case, not only would the activity not correlate well, but it would also appear to diminish with time. Studies, therefore, should be initiated to identify the causal factor(s) of the discrepancy between the AEC and steady-state kinetic activities.

Detection of Multiple Active Components

Multiple active components in an enzyme system may consist of either interacting (associating–dissociating) or noninteracting components. The simpler case is when multiple noninteracting forms are present. What cannot be overemphasized is the tremendous utility of the difference curves in detecting, analyzing, and quantitating the various active species. The presence of *separate* peaks of activity sedimenting through the centrifuge cell will be clearly present in a difference curve, while possibly hidden in the reaction boundary. Cohen and Claverie[12] and Llewellyn and Smith[15] have simulated sedimentation profiles for different combinations of multiple active forms, both self-associating or complexing between active and inactive forms. For a simpler system of an interacting active monomer–active dimer, depending on the equilibrium relaxation time a bimodel or unimodel sedimentation pattern can be detected as shown in Fig. 3. With infinite relaxation time (i.e., two independent species) or magnitude of the length of the experiment, then biomodel patterns will be observed. Otherwise, a unimodel pattern is expected, as dictated by the Gilbert theory.[30] Cohen and Claverie[12] have further simulated sedimentation patterns of an unusual case involving $E + R \rightleftharpoons C$, where E is the free active enzyme, R is a more massive and inactive molecule, and C is the complex of these two components. Depending on whether C is active or inactive, a set of patterns was generated, as shown in Fig. 4. In the case of active C, two peaks are clearly visible. The elevation of baseline between the two peaks is crucially important, since it indicates interaction between the two species. In the case of inactive C, an elevation of baseline is still clearly visible although only one peak can be detected. Such an observation signals the existence of an invisible and fast sedimenting particle interacting with the free enzyme E. The elevation of baseline is due to the dissociation of the complex while it sediments and thus it yields active E molecules, which have sedimented further than the free E molecules not complexed with R.

In practice, a variety of multiple active forms have been detected for pyruvate carboxylase[31] and transcarboxylase.[32]

---

[30] G. A. Gilbert, *Discuss. Faraday Soc.* **20**, 68 (1955).
[31] B. L. Taylor, W. H. Frey, II, R. E. Barden, M. C. Scrutton, and M. F. Utter, *J. Biol. Chem.* **253**, 3062 (1978).
[32] E. M. Potto and H. G. Wood, *Biochemistry* **16**, 1949 (1977).

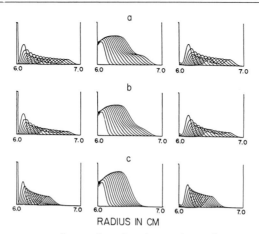

RADIUS IN CM

Fig. 3. Active enzyme sedimentation of an interacting active monomer–active dimer system. Here $s_M = 6$ S, $D_M = 7.07$ (in 10^{-7} cm²/sec); $s_D = 10$ S, $D_D = 5.9$. (a) Equilibrium relaxation time is infinite, i.e., two independent species. (b) Equilibrium is slow, its relaxation time, 2000 sec, being of the same order of magnitude as the length of the experiment. (c) Equilibrium is fast compared to the sedimentation (relaxation time 200 sec). The left-hand curves show the invisible total distributions of the monomer plus dimer species themselves (including the just layered band at $t = 0$), the center ones, the observed product distributions, and the right-hand ones the corresponding differences curves. Successive distributions were calculated at $t = 500$ sec and then every 200 sec, up to 3100 sec (and for the corresponding differences curves at 600 sec, and then every 200 sec, up to 3000 sec). The centrifugation speed was 60,000 rpm. The equilibrium constant was taken as $K = 1$ in the three simulations; at the time of the band layering ($t = 0$), the equilibrium between the two species was reached (i.e., in mass units $C_M = C_D$). (Reprinted with permission from Cohen and Claverie.[12] Copyright 1975 John Wiley and Sons, Inc.)

An additional consideration is whether or not the enzyme equilibrium is sensitive to ligand-induced association or dissociation. As pointed out by Neet,[33] the oligomeric states of a wide variety of enzymes are very sensitive to ligand-induced changes. These can also dramatically alter the association parameters, thereby changing the state of the enzyme as it sediments through the cell.

Heterogeneity

Molecular heterogeneity or microheterogeneity refers to differences in the molecular composition of the active unit of an enzyme. In practice, testing for this has been avoided, neglected, or forgotten. The aversion has been due to the large amount of meticulous calculation which is required to effectively show heterogeneity of an enzyme sample on a

[33] K. E. Neet, *Bull. Mol. Biol. Med.* **4,** 101 (1979).

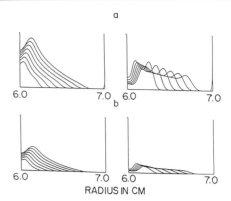

FIG. 4. An unusual case: the equilibrium $E + R = C$. At $t = 0$, the three species E, R, C, were at their equilibrium concentrations in the layered band (in mass units $C_E = 1$, $C_R = C_C = 10$, i.e., the number of molecules of each species was thus about the same at $t = 0$). The equilibrium relaxation time was taken as 200 sec. Centrifuge speed 45,000 rpm; the product distributions (left curves) were calculated at 500 sec, and then every 200 sec up to 2900 sec; the right curves are the corresponding differences ones. For the other parameters, see Fig. 3. (a) Both E and C are active. (b) Only E is active. R is always inactive. (Reprinted with permission from Cohen and Claverie.[12] Copyright 1975 John Wiley and Sons, Inc.)

molecular level. However, two groups[34,35] have shown the format for testing such a system using sedimentation studies.

In both cases, very small changes in the s or D values were observed during the sedimentation of the sample for which there was no other obvious explanation. The use of simulated data obtained by computer modeling was entirely consistent with the conclusion that molecular heterogeneity was present.

As pointed out by Gilbert and Gilbert,[35] the confirmation of heterogeneity is possible through a simulation test even when there is no independently evaluated diffusion or sedimentation coefficient. This is a consequence of the fact that spreading of the enzyme boundary or band by diffusion follows a different mathematical relationship than spreading due to heterogeneity. Hence, experimental and simulated data may be significantly different if the simulated data assume only spreading by diffusion. Any differences, therefore, will be an indicator of heterogeneity.

Simulation of Experimental Data

The AEC methods allow a simple and precise determination of both the sedimentation and diffusion coefficients. As mentioned earlier, for the

[34] B. Schmitt and R. Cohen, *Biochem. Biophys. Res. Commun.* **93**, 709 (1980).
[35] G. A. Gilbert and L. M. Gilbert, *J. Mol. Biol.* **144**, 405 (1980).

simplest case of a homogeneous protein with a single active species, it does not require an extreme degree of sophistication to arrive at a good value for s and D. Unfortunately, the real world often complicates the situation. The existence of enzyme systems which are capable of associating or dissociating to a series of polymeric and possibly active forms poses some questions: What is the stoichiometry of components in the complex and what are the equilibrium constants? As the complexity of the situation becomes apparent, it becomes essential to resort to the comparison of experimental and simulated curves. While deriving an empirical method from computer simulations may appear at first glance as artificial, it has been repeatedy shown to be of great value in analyzing polymerizing systems.[12,34,35] Simulations have been used by others to demonstrate experimental possibilities for AEC.[12,15] However, it must be remembered that the simulation procedure itself must be carefully checked for its realism and reliability.

Nevertheless, the use of simulated data provides the final test necessary to confirm a model. The raw data first provide a series of possible models and these are then used to find the most likely interpretations of the data. The best model is the one which consists of the minimum number of variables, yet can generate information that fit the raw data collected initially. In complex systems, such an analysis provides the only method to confirm that the interpretation of the data is consistent with the observed physical parameters of the system.

Concluding Remarks

AEC is a simple and precise method to obtain both sedimentation and diffusion coefficients for the active species. As with any technique, there are a series of assumptions, limitations, and possible artifacts which must be taken into consideration in the experimental design. However, to the informed and alert investigator, AEC provides an excellent means to directly observe the hydrodynamic properties of the fully active enzyme–substrate complex in either a purified or crude state. The technique also allows for the investigation of enzyme interactions with other cellular components such as filaments, or nucleic acids.

The ability to simulate AEC data is important not only because it allows the detection of possible artifacts, but it allows one, having completed an experiment, to subsequently test the interpretation of the data to ensure that it is consistent with the experimentally observed results. The potential of AEC for analyzing crude enzyme mixtures in a cellular environment with sensitivity and specificity is unmatched, and in combination with theoretical simulation will provide a valuable tool for future investigations.

[9] Computer-Controlled Scanning Gel Chromatography

By LAWRENCE C. DAVIS, THEODORE J. SOCOLOFSKY, and
GARY A. RADKE

Historical Background

Direct scanning gel chromatography (DSGC) was developed in the laboratory of Ackers about 15 years ago.[1] Several reviews of the technique are available.[2-4] At that time, minicomputers were scarcely in existence and microcomputers had not even been considered. With the ever-increasing power and decreasing prices of small computers the potential for computer control of DSGC has greatly increased. Scanners are presently limited by optical considerations discussed below, but in this area also there are likely to be rapid advances within the next few years. One difficulty this presents an author is that the system to be described is already obsolete, or at least not state-of-the-art. Comments describing likely places for improvement will be included in the description of the system.

The basic principle of DSGC is the relative movement of a light source and receiver past a gel column packed with some light transmitting and liquid-permeable matrix. Thus far, Sephadex has proved most useful primarily because it scatters light mostly in the forward direction where it can be captured efficiently even in the ultraviolet. Choice of matrix for the scanner depends on being able to get enough light through in a short time so that one need observe one position in the column for only a short time. With a more highly scattering matrix, gel equilibration and measurement of partition coefficients can be done using photon counting but successful use of scanning requires that the time for each single observation (point) be relatively short, if column flow rates are to be reasonable.

The first scanning chromatograph[1] was designed around a stationary optical system with a column that moved past this source and receiver. Although this made the optics simple, there were some difficulties with the column settling over time from vibration during its movement and the Teflon column guides sometimes produced streaks on the column.[5] We

[1] E. E. Brumbaugh and G. K. Ackers, *J. Biol. Chem.* **243**, 6315 (1968).
[2] G. K. Ackers, *Adv. Protein Chem.* **24**, 343 (1970).
[3] G. K. Ackers, *in* "The Proteins" (H. Neurath and R. L. Hill, eds.), 3rd ed., Vol. 1, p. 1. Academic Press, New York, 1975.
[4] G. K. Ackers, *in* "Methods of Protein Separation" (N. Catsimpoolas, ed.), Vol. 2, p. 1. Plenum, New York, 1976.
[5] G. K. Ackers, personal communication.

METHODS IN ENZYMOLOGY, VOL. 117

therefore modified the scanner design to keep the column stationary and move the light source and receiver past it (Fig. 1). This has resulted in column lifetimes of several months with no detectable change in the packing attributable to the use of the scanner. This arrangement, however, introduces noise into the photomultiplier signal as a result of vibration during its motion. We have recently installed a mid-ultraviolet fiber optic bundle purchased from Maxlight, Inc., so that the photomultiplier tube can remain stationary during the scans while light is carried from the column to the PM tube via the fiber optic bundle. E. E. Brumbaugh and G. K. Ackers have built a system in which fiber optic bundles transmit light to and from the column while column and optical system remain stationary.[5] All of our examples from our published work cited below have been obtained with a moving photomultiplier.

Brumbaugh et al.[6] described use of a system basically the same as that originally developed but with the addition of a NOVA Z/2 minicomputer for data processing. Their data processing program was a relatively simple one that allowed addition, subtraction, multiplication by a constant, inversion, and calculation of natural logs. Hard copy of data was obtainable by paper tape or X-Y plotter on-line. Data manipulations were visualized on a storage oscilloscope. Powell et al.[7] used a modified version of the original scanner with a Hewlett-Packard minicomputer to digitize the data and an IBM 370 to analyze the data.

Scanning gel chromatography has been applied to measurement of the partition coefficient of proteins such as hemoglobin that undergo association and dissociation reactions[2,4,7] and micelle formation by fatty acyl CoA.[7] It has been used also for active enzyme chromatography both in continuous flow of small zones[8] or with flow stoppage[9,10] or for broad zones.[11] Because of the interests of this laboratory, the emphasis throughout this chapter will be on active enzyme chromatography applications of the scanning system.

Design of the System

Figure 1 provides a block diagram of our scanner as it is presently constructed.[9] The heart of the scanner is a precision syringe drive ram

[6] E. E. Brumbaugh, E. E. Saffen, and P. W. Chun, *Biophys. Chem.* **9,** 299 (1979).
[7] G. L. Powell, J. R. Grothusen, J. K. Zimmerman, C. A. Evans, and W. W. Fish, *J. Biol. Chem.* **256,** 12740 (1981).
[8] M. M. Jones, J. W. Ogilvie, and G. K. Ackers, *Biophys. Chem.* **5,** 339 (1976).
[9] T. J. Socolofsky, G. A. Radke, and L. C. Davis, *Anal. Biochem.* **125,** 307 (1982).
[10] L. C. Davis and G. A. Radke, *Anal. Biochem.* **125,** 315 (1982).
[11] L. C. Davis and G. A. Radke, *Biophys. Chem.* **18,** 241 (1983).

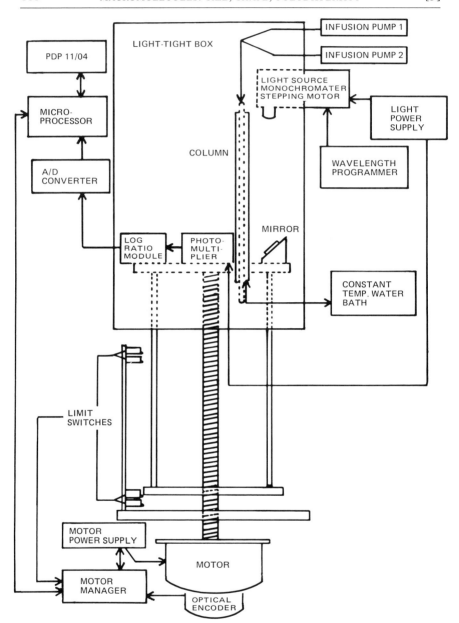

FIG. 1. Direct scanning gel chromatography system. Details are provided in the text. (From ref. 9, by permission.)

from Update Instruments, Madison, WI. This was chosen because it had a programmable velocity and distance profile, was capable of moving a relatively heavy load (in excess of 25 kg), obtained a constant velocity rapidly, stopped quickly, and was highly reproducible. The key feature of the syringe ram is a high-power low-mass DC motor with a photodiode optical encoder. The photodiode output from this is fed to a motor controller which regulates velocity and distance. The same signal can be directed to a computer or microprocessor to provide information for triggering A/D conversions by the optical scanning system. If the load to be carried is relatively small, as, for instance, fiber optic bundles, one could use a much simpler system based on a stepping motor with an encoder to record position accurately,[5] or one could use a simple motor with a linear "timing fence" such as is used in line printers, mounted parallel to the column being scanned. Stepping motors will drift, and it is useful to have a discrete end point such as the switches shown in Fig. 1. These ensure that every scan is initiated at exactly the same point.

The design of the syringe ram is such that scanning in the upward direction only is the most convenient scanning arrangement. Each scan is initiated when the motor moves above the microswitch at the bottom. We have chosen to have the point spacing a constant number (variable, but generally 100) of encoder pulses between samplings of the output of the photomultiplier by the A/D converter. The pitch of the drive screw is not metric so the distance usually used comes out to 1.016 mm/point of the scan with 194 points per scan of 19.5 cm. Each point is a simple average of 32 A/D conversions taken at equal intervals over 1/60 sec. The number 32 is chosen for simplicity of binary arithmetic because the observations can simply be cumulated and then shifted to give an average. This constrains us to a maximum velocity of 60 points (about 6 cm) per second which gives a scan time of slightly over 3 sec for the usual 20 cm column. The interval of 1/60 sec was chosen to permit active filtering of the signals to remove stray ground loop signals or other extraneous sources of low frequency noise. With careful design it may be possible to eliminate more of these sources without the filter, but we have no need for greater scanning speeds and this provides the simplest means of filtering out 60-cycle noise. This same technique was applied to rapid sampling of a Gilford spectrophotometer output to give very clean signals. The primary source of noise in either case seems to be the A.C. component of the tungsten lamp power supply which results in light source intensity fluctuations.

As mentioned above, A/D conversions are triggered in response to the optical encoder pulses of the motor. A SYM-1 microprocessor (Motorola 6502 chip) is used as the sensor and triggering device. It also does the necessary arithmetic to make an average point and store it away in a file.

A counter is included to keep track of the points per file and of the file number. It then "reloads" to await initiation of another scan when the microswitch at the bottom of the scan is passed. Economical use of the scanner requires that scans be taken only as frequently as necessary; otherwise one will drown in data. The SYM-1 is programmable either directly from its keypad or via a PDP-1104 (to be discussed below). Using the clock on the SYM-1, times for the next scan, the interval between scans and the delay after this triggering time until actual initiation of a scan are all programmed in. The SYM-1 was provided with enough RAM memory to store 100 scans.

At the completion of each scan the SYM-1 inquires of the PDP-11 whether it is in a mode to receive the file. If it is, the file is shipped to the PDP-11 which then stores it on a floppy disk for later data manipulation. If the PDP-11 is otherwise engaged, the SYM-1 transfers the file to its memory and transfers it to the PDP-11 at the next opportunity. We could have carried out the entire operation using only the PDP-11, but it is a somewhat slower machine than the SYM-1 and is far too powerful to be waiting around to do scans. This decision to install a microprocessor was an important one, because data manipulation and presentation take nearly as long as the scanning process, and if the computer is tied up doing a series of scans it cannot be used to put data into presentable form. The relatively low price of microprocessors makes this "duplication" quite reasonable. A typical run takes several hours to set up, in order to get the column equilibrated with appropriate buffers, and then a few more hours to do the actual chromatography of interest. To get more than one complete run per day requires full-time use of the scanning system to make baselines, follow the solute of interest, and take final baselines. It is then necessary to subtract baselines from all the scans, find rates of migration or partition coefficients, make plots or graphs, etc. This also takes several hours, even with the assistance of programs to be described below.

Thus far we have chosen to restrict our studies to the visible region of the spectrum, in part because of difficulty in getting a sufficiently intense, inexpensive light source to use in the UV, but primarily because nitrogenase, the enzyme of initial interest to us, requires the presence of dithionite for stability and this absorbs intensely in the UV. For most of our work a 6-V, 25-W tungsten bulb, such as is found in a DU or Gilford, has proved to be adequate. When working at 340 or 350 nm, a 120-V tungsten projection lamp (GE, #CDJ used in Klett colorimeters) can be substituted in the mounting position of a hydrogen lamp to provide greater light intensity. With this lamp, water cooling of the lamp housing is necessary to prevent burnout of the lamp. The Klett colorimeter lamp is designed to operate at line voltage in the AC mode. Use of the active filter on the scanner is essential in this case because it eliminates the cyclical change in voltage (and hence of light intensity) that occurs in AC operation. The actual voltage to the bulb is regulated by a Variac and is usually kept at 80–85% of maximum. At this voltage the lifetime of the bulb is greatly extended over what it is at 100% of rated voltage. Obviously, halogen lamps could be used to provide higher light intensities.

A J-Y Optics Double H-10 monochromator was used to provide a highly monochromatic light source with good throughput of light. It has a stepping motor-controlled wavelength drive which is easily interfaced to the PDP-11 so that changing wavelengths between scans is relatively simple. This allows one the capability of doing scans at multiple wavelengths in a single experiment. Thus one can follow the migration of two different constituents through a column simultaneously. When studying interacting systems this allows one to calculate profiles for both constituents and thereby to derive association constants. We have used the dual wavelength method successfully with model systems.

Light from the monochromator passes through a quartz collimating lens upon entry into the chamber housing the scanner. Because this is a simple, single lens, it is impossible to perfectly collimate light of all wavelengths. We have used 525 nm as a convenient collimating test wavelength and at that wavelength we are able to obtain a good image of the bulb filament over most of the length of the active scan, showing effective collimation.

The collimated image is reflected from a front-silvered mirror mounted at a 45° angle so that the beam is directed at a 1-mm slit in front of the column to be scanned. This angle is not terribly critical if the distance from mirror to slit is short. We have adjusted the mirror by insuring that with the column removed light passes directly through the inlet and a 1-mm outlet slit. Various heights of outlet slit have been tested and a 2-mm slit has been found to improve performance because it allows collection of more scattered light. Ray tracing shows that this introduces a negligible error into the position observed and pathlength through the column because the outlet slit is 2 cm away from the column. Increasing the slit height beyond 2 mm was not found to gain much further increase in light intensity. The outlet slit serves mainly as a stray-light shield for the photomultiplier which is painted black on sides and back. There is ~1-cm space for an interference filter immediately between slit and photomultiplier tube, but we have not had occasion to use one. At the levels of photomultiplier signal and light absorption that we have been using, stray light (from the room outside) has not been a problem.

Pathlength of the column is obviously important. Small-diameter columns will transmit more light but require higher concentrations of absorbing solute to give the same absorption. There are also light scattering and refraction effects of the column itself. For instance, a column of less than 0.5 cm diameter, unless very cleverly masked, would tend to have excess stray light, not passing through the column solution but around the walls. Columns of several centimeters diameter would consume larger volumes of the absorbing solution and have high blank absorbancies (actually light scattering) so that finding a high enough intensity light source could become a problem. We have chosen to use quartz or glass columns of 1 cm effective pathlength although for particular applications other sizes could be used. Short pathlength flat columns might be used for highly scattering matrices[12] but the flow characteristics of these have not been investigated.

The angle of collection of light from the column is important for maximizing the light throughput. For Sephadex G-200 the intensity of scattered visible light (436 nm) is about 1/10 as great at 30° as it is straight through a 1-cm column.[13] The relative intensity of scattered light at larger angles increases as the wavelength decreases. Brumbaugh et al.[5] found that at 220 nm the intensity at 30° was about half that straight through. Thus a very wide angle of collection is needed to capture all of the light. This can be managed in the horizontal dimension but light scattered along the length of the column cannot be collected without significantly diminishing resolution. With our present column configuration and a fiber optic bundle 2×10 mm placed 1 cm from the column, the light loss at 405 nm is about 60% (0.75 A) comparing transmission through an empty vs a filled column. The light loss by scattering (apparent absorbance) increases as $1/\text{wavelength}^2$; so at 210 nm it would be expected to be about 3 A.[13] In the fiber optic system used by Brumbaugh et al.[5] the reported absorbance of G-200 at 220 nm is 2 A perhaps because their horizontal collection angle is ±40°. Brumbaugh and Ackers (cited in 2) found an apparent absorbance of <2 for Sephadex G-200, using an end-on photomultiplier close to a scanned column. The acceptance angle for that design was not specified but must have been at least ±30°.

The entire column is fitted closely into a solid brass (2.5×5 cm) jacket that was machined to provide entrance and exit masking of the column as well as temperature control. The jacket was machined in two pieces and then covered with black plastic electrician's tape to prevent reflections from the sides of the column. Over the region of the column to be scanned the brass is beveled away from the column at 45° so that the passage of light is not hindered. On the inlet side the vertical slit formed by the two brass pieces is 4 mm wide, while at the outlet side it is 9 mm wide. The

[12] L. C. Davis, Anal. Biochem. 120, 95 (1982).
[13] L. C. Davis, Biophys. Chem. 10, 55 (1979).

central cylindrical hole was made 14 mm in diameter to take a column of 12.5 mm outside diameter allowing for the black masking. One-fourth-inch refrigeration tubing was brazed to channels along the edges of the brass jacket to provide water circulation for temperature control. We have operated only at ambient temperature or above, so that condensation on the column has not been a problem. Because there is relatively little electronic gear inside the scanner box, there has been little heat to remove. For operation at low temperatures it would be necessary to monitor the actual column temperature and to provide dry nitrogen in the scanner box to prevent condensation.[5]

The inside of the scanner box is painted black, and black felt is fitted around the entries of wires, etc., to prevent light from getting in. We are unable to detect any stray light signal at the maximum voltage used (650), with or without room lights on.

The photomultiplier tube chosen for the scanner is the 1P28 which is standard in many spectrophotometers, including the Beckman DU. We have used a number of individual tubes and found that the highest sensitivities are obtained with those scavenged from abandoned DU electrometers. Regular production run tubes from commercial suppliers had considerably less sensitivity and a poorer signal-to-noise ratio (typically 100–400 A/Lm and 0.5–5 nA dark current). For work in the ultraviolet one should consider a "solar-blind" plug-in replacement for the 1P28, such as Hamamatsu R166 or R166 UH. We have not tested this type extensively. Sensitivity of the R166 is typically about 20-fold less than one of the better 1P28 tubes at 280 nm, but it is *very* much less responsive to visible light. When mounted in the scanner unmasked, removing one side of the box elicited little response with natural daylight in the room, relative to a D_2 lamp, when the monochromator was set for 220 nm. The maximum permissible anode current is only 10 μA, 10-fold lower than the 1P28, but monochromatic D_2 sources probably cannot exceed this light intensity anyway with a column in place. Our best R166 has a rated dark current of 6 pA at 1000 V and a measured gain at 220 and 280 nm equal to our best 1P28 tube. Three others have ~20-fold greater noise (based on manufacturers' ratings of individual tubes) and 2- to 20-fold lower (measured) gain at 650 V. The sensitivity of any PM tube could be greatly increased by increasing voltage to the rated maximum of 1250 V.

Once a beam of light has passed through the column and reached the photomultiplier tube, the process of data collection and manipulation begins. The 1P28 produces a current proportional to input light intensity over a wide range of light intensities. At high light intensity it is subject to saturation and fatigue. Continuous operation at an output of 1 μA is acceptable with little decrease in responsiveness observed over many hours. Levels of 3 μA for short periods of time cause no difficulties and brief excursions to 10 μA are acceptable. The rated maximum is an average of 100 μA over 30 sec, a limit we have never tested. We have arranged the scanner so that the PM tube is not exposed to light while the scanner is "idling" between scans. Light intensities are adjusted so that baseline scans in the absence of absorbing solute do not exceed about 3 μA. Most scans are obtained with output signals in the range of 100 nA to 1 μA. If one were scanning at two different wavelengths, matching light intensities to keep signals in this range could prove troublesome, but we have done so successfully.

An alternative to matching light intensities is to vary the PM voltage. We built a digital voltage controller based on the output of an 8 bit D/A converter. This did not give a sufficiently steady voltage over the time of the scans, presumably because of the approximation error of the 8 bit D/A converters. An electrometer designed around a Gilford system that controls PM voltage could be used and would provide an adequately stable signal if the active filter were used. Otherwise, even with a D_2 lamp, there is a distinct 60-cycle ripple in the output of the older model Gilford system, which was intended for use with a regular recorder, not a fast acquisition system.

The PM signal must somehow be converted to a digital form and amplified for data manipulation. Relatively recently A/D converters have become commercially available with precision up to 16 bits which allows resolution of 1/64,000. In 1980 the largest available A/D converter was 14 bits. If one uses absolute signal intensities and a linear amplifier and then takes the log digitally, after A/D conversion, one is restricted to a resolution equivalent to 1 nA in 64 μA with a 16-bit converter. On the other hand, if one takes the log of an analog signal and then converts it to a digital form, one can in principle work (with, e.g., the 757 P) over a range from 1 nA to 1 mA, though most PM tubes are not linear over that whole range. With a typical output voltage of 10 V from a linear amplifier and using a 16-bit A/D with zero absorbance set at 9 V, an absorbance of 3 A gives a signal of 9 mV. It should be possible to go from 0 to 3 absorbance units with a precision of 1/64 at 3 A using a 16 bit A/D converter. However, to get 1% precision at 4 A would require a 20 bit A/D converter, which is not yet commercially available.

A nontrivial consideration in using an A/D converter is the amount of digital data generated. A 20-bit A/D converter requires twice the space of a 10-bit one to store the same amount of information. With a 12-bit converter one can pack each data point into 1.5 words of an 8-bit microprocessor. With a 16-bit converter one needs to use two words for each point, reducing the available memory by 1/3. Only if it significantly increased precision would the extra effort be worthwhile. With a logarithmic amplifier the resolution by the A/D converter is in principle independent of absorbance, so that a 12-bit A/D converter gives resolution to ~1/4000 even at 3 A. In practice, we have found the log ratio amplifier (757 P of Analog Devices) to be severely nonlinear with signals below 100 nA in our rapid sampling mode (Fig. 2). It has been suggested to us[14] that this is a function of the slow response time of log amplifiers with very small signals. The difficulties come principally from the photomultiplier noise which in a ratio mode gives rise to apparent negative signals (in the difference between reference and measuring signal). Our sampling frequency of ~2000 kHz is equal to the band width of the 757 P at 10 nA, so that the amplifier cannot accurately track the changing signal.

A log ratio amplifier was chosen originally so that we could use a reference PM tube to compensate for fluctuations in light intensity over the course of a scan. We found in practice, however, that the light source is sufficiently stable (<0.01 A/hr drift) for the types of experiments that we are doing, and introducing a second photomultiplier introduces a second source of noise. Now we supply a constant 1 μA current to the reference side so that we are able to monitor the absolute signal intensity of the photomultiplier as a function of light source voltage or wavelength. This is important in protecting the photomultiplier from fatigue. Using a Gilford system one could avoid this problem since in this case the PM voltage is automatically varied and is the measured quantity. Alternatively, one could have automatic slit correction as on a more modern wavelength scanning spectrophotometer than the DU. Probably the ideal arrangement would be to install a fiber optical scanner on the optical bench of a Gilford, or comparable Cary, wavelength scanning spectrophotometer, with a fast signal response time. This approach would not work so well with some of the newer high resolution single-beam spectrophotometers such as the Beckman DU-8 because they have a 1/4 sec signal averaging routine which would limit the scanning rate to 4 points per second or about one scan per minute.

The level of precision and resolution required for the scanner is a function of the type of experiment that one plans to do. We have tested several PM tubes over a range of voltages with or without movement of the scanner. With the scanner stationary, typically observed noise levels (standard deviation of any point in four 194-point scans with each "point" consisting of 32 samplings) are in the range of 3.8 to 8.5 mV, depending on PM tube, voltage, light source, etc. The 12-bit A/D converter in a ±5 V bipolar mode represents each 2.5-mV change by one bit so that observed

[14] M. van Swaay, personal communication.

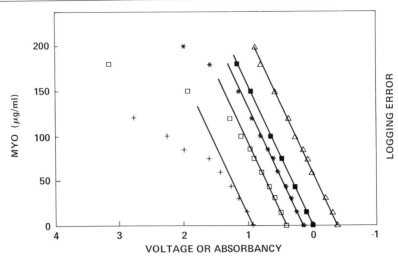

FIG. 2. Variation in apparent absorbance as a function of photomultiplier output current. A series of dilutions of equine metmyoglobin were tested at 415 nm using two different photomultiplier tubes, each at two different voltage settings. Each data point represents the average of a 194-point "scan" of a stationary round cuvette of 1 cm pathlength. The symbols (△) and (□) are for PM tube A at two voltages; the symbols (*) and (+) are for PM tube B at those same voltages. The (■) is for absorbancies determined in a Gilford spectrophotometer. All lines are drawn parallel using the best fit to the Gilford data, offset to the voltage value of the blank at each voltage setting for each PM tube. Voltages are shown relative to a reference of 1 μA but with the sign inverted, equivalent to absorbance with an arbitrary offset. The fall-off in signal response begins in each case when the input current becomes less than 100 nA and shows the same shape of response curve.

error is 1.5 to 3.5 bits. This noise level would be less than one bit with a 10-bit A/D converter, which is why we replaced the 10-bit A/D converter of the PDP-1104 by the 12-bit Analog Devices one. If the voltage change for one absorbance unit is 1 V, the error level is from 0.38 to 0.85%. The rated log conformity of the 757 P is ±0.5% relative to input or ±2.17 mV relative to output over the range of signals under consideration. We are thus operating near or at the limit of precision of the amplifier when the scanner is stationary. A series of measurements (nonmoving "scans") on myoglobin solutions of increasing concentration with two voltage settings and two PM tubes is shown in Fig. 2, which was prepared using the PLOT1 program (see below). Error levels for 2/3 of the "scans" in this series were below 20 mV over a range of input signals from 3 μA to 10 nA. One PM tube had more than half its error levels below 10 mV while the other had >1/2 above 10 mV. Least noise was found between 100 nA and 1 μA input current.

With the scanner running and a column in place the observed noise level is usually ±25 mV or less, depending to some extent on signal intensity, wavelength and uncontrollable external conditions. For instance, a typical set of baseline scans selected at random and taken during a study of catalase migration[14] showed an average standard deviation for four scans of 0.0108 V on a signal equivalent to 630 nA input. The noise error in measuring the catalase concentration was thus less than 2% (simple sum of errors) when a 1-V absorbance scale was used. Recent experiments using fiber optics on the output side of the scanner have given comparable results with average standard deviations of <0.01 V on a 0.5-V signal.

For work with nonenzymatic proteins such as hemoglobin one needs to be able to measure very reliably but fairly slowly over a wide range of protein concentrations so as to derive partition coefficients and then to calculate association constants. Ackers' group has extensively exploited their scanner and photon counting systems for this sort of work[15,16] using gel saturation experiments to measure concentrations accurately at both high and low concentrations. The practical lower limit found by Jones et al.[16] for scanning at 220 nm was in the range of 5 μg/ml (~0.05 A in gel) while for photon counting it was <1 μg/ml.[15] Our system limit approaches that of Jones et al. under optimum operating conditions.

When we first constructed a scanning system we included a "flow cell" at the bottom of the column so that we could monitor the concentration of a solute in the absence of the gel and make calculations of the partition coefficient. This worked reasonably well for standard substances such as hemoglobin or catalase, though a quite broad zone (at least 1 ml) must be passed through the cell to reach the equilibrium concentration. More recently we have focused our interests on enzymes for which active enzyme chromatography is a simpler way to monitor their apparent partitioning in the column.[8] Because most enzymes are active at much lower concentrations than could be practically monitored directly, we have abandoned the use of the flow cell and depend on rate of migration through the column to measure size and state of aggregation.

For determining rates of migration on simple systems it is not necessary to have particularly clean spectra or very stable baselines. So long as one is able to obtain an estimate of the centroid of migration from a number of times during the run one can derive a reliable estimate of the rate of migration. Calibration of the column with standards allows good

[15] G. K. Ackers, E. E. Brumbaugh, S. H. C. Ip, and H. R. Halvorson, *Biophys. Chem.* **4,** 171 (1976).
[16] M. M. Jones, G. A. Harvey, and G. K. Ackers, *Biophys. Chem.* **5,** 327 (1976).

estimation of the rate of migration and hence the size of the unknown relative to those standards.

We have usually packed our columns with more gel than is needed to completely fill the observable region and then placed a porous polyethylene frit on the top of this. Buffer flow rates (typically 70 μl/min or less) are controlled by a Harvard continuous infusion pump with 50-ml disposable syringes. This produces a constant flow rate with no fluctuations detectable and gives no problems with compression of the gel column during several weeks of operation. By having the column slightly longer than the region scanned we can tolerate potential adsorption of solutes to the gel (junk) without distortion of the baseline in the region of the scan. For broad zones, sample is applied directly via the same line and infusion pump system as for buffer flow. For introduction of small pulses of enzyme into a substrate-saturated column we use a second pump and a smaller syringe to give the equivalent flow rate. Simple liquid chromatography valves allow one to switch readily from one line to another. Because the column is slightly longer than the region scanned, we never see a very sharp boundary between solute-free and solute-containing regions. Generally, our columns have been packed about 1 cm longer than the scanned region so there is always dispersion equivalent to this length plus dispersion introduced during the loading process (lines + frit). Initial dispersion is always the most rapid spreading that occurs during chromatography and is the least reproducible. Dispersion coefficients are most reliably measured by simulation beginning with a real profile (see below).

Temperature control of infusion solutions is sometimes desirable, as for instance to slow down an active enzyme reaction prior to its entry into the column.[10] We have fitted jackets of copper pipe wrapped by copper tubing plus a layer of insulation to cool the 50-ml disposable syringes on the infusion pump. The inlet lines from syringe to column are surrounded by Tygon and glass tubing to contain dithionite solutions for work with anaerobic systems. These could be used instead to circulate a temperature-controlled fluid if that were critical to the experiment. Dithionite in Tris base remains stable in the jacketing system for at least a week and will effectively protect anaerobic solutions from oxygen diffusion, except from that which might be adsorbed to the polyethylene inlet lines themselves. Being able to jacket the lines in this simple way is a significant advantage of a stationary column.

Data Processing[17]

The computer-controlled scanning system generates data at a rapid rate. These must be processed and plotted out for visualization indepen-

---

[17] Programs are available from the senior author.

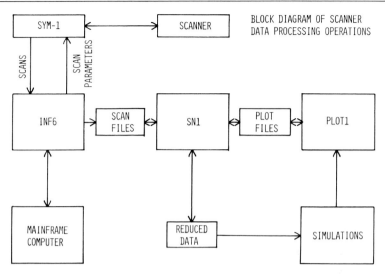

DIAGRAM 1.

dently of the scanning operations (Diagram 1). It was found convenient to divide the computer operations into three parts[18]: one is the interface itself which runs the scanner, the second is the data manipulation section, and the third is the final plotting for permanent visual records. A fourth related function is the simulation of results for matching particular models to the observed spectra.[19] The interfacing routine was built into a more general program that allows communication with the main computer system of the University via a modem, permits printout to a teletype and can communicate with an APPLE II computer.

The PDP-11 is used under 1NF6, the interface program, to initialize the SYM-1 by loading the operating parameters from a stored copy of zero page and then serves as a passive receiver of files from the SYM-1, echoing them to the Tektronix monitor. It also can display the contents of various registers of the SYM-1 that keep track of the number of scans made, the time they were made, when they were shipped to the PDP-11, etc. The PDP-11 is used under SN1, the scanner number cruncher (Table I), to manipulate the data into printable form. It can also send the results to the video monitor for inspection during the process. Both 1NF6 and SN1 have interrupt file storage features so that one can rapidly switch from one to the other without losing data that were partially processed, etc.

[18] T. J. Socolofsky, unpublished programs for M.S. degree in Computer Science, Kansas State University, 1980.
[19] L. C. Davis and M. S. Chen, *Arch. Biochem. Biophys.* **194,** 37 (1979).

TABLE I

MAIN FEATURES OF SN1-DATA PROCESSING PROGRAM

RD = Read a scan file	Identifies options for reading a scan file from disk.
WR = Write a scan file	Identifies options for writing a modified scan file to disk.
PK = Pack a plot file	Allows/instructs user to move reduced data from a scan file to a plot file.
UP = Unpack a plot file	Allows/instructs user to move data from a plot file to a scan file for further manipulation.
RP = Read a plot file	Prompts user in reading a plot file from disk.
WP = Write a plot file	Prompts user in writing a plot file to disk. Plot files are variable size in a fixed size common block in memory.
ED = Edit data sets	Allows/instructs user to edit scan files, once in memory arrays.
DD = Data description	Displays description of last scan file read from disk. Description includes file name and the time it was sent to the PDP-11.
PD = Plot data	Plots data stored in a *scan* file; useful for quickly visualizing a scan or reduced data, but no coordinates can be plotted.
CC,N = Calculator section (divided into six sections because of memory and video screen limitations.)	Reduces data to conventional presentation forms such as derivatives, centroids, etc.

Typing a two-letter response displayed in SN1 main results in a submenu being displayed. The general function of these are noted on the right side above. Submenus contain user prompts to get files transferred and manipulated.

The calculator sections (CC,N) work on the data sets retrieved via the RD option in the main menu. Six data sets may be maintained in fixed size arrays in memory at once. In all examples below 'N' signifies set number.

CC,1:1,   Delete a specified set (clears the array space).

2,   Move—move an entire set of data from one location to another or just move the X or Y values between sets.

3,   Change or look at constants. (Up to six user defined constants may be stored.)

4,   Operation with constant. $XN=XN(op)CN$ where XN denotes the X values of set N, $YN=YN(op)CN$ where YN denotes the Y values of set N and CN is constant value N. (op) = $+, -, *$, and $/$.

5,   Look at set sizes. Useful for identifying truncated sets; found in CC,1–CC,6

CC,2:1,   Add and subtract with Y. $YN=YN-YN$ or $YN=YN+YN$

2,   Switch values of X and Y.

3,   Automatic scanner calculations. Change bits to volts of set N or change bits to volts and subtract a baseline. (The baseline is stored in set 5 and is initially created by averaging two or more user-selected scans.)

CC,3:1,   Average specified sets located in positions 1 to 4. The average is stored in position 5 and the standard deviation is stored in position 6.

2,   Form a three point moving average of set N. (Smooths noisy data.)

3,   Calculate the first derivative of set N.

TABLE I (*continued*)

CC,4:1,   Identify the set you want section four to operate on.

    2,   Delete data over a specified range in X. For example, delete all points less than X(i) or all points greater than X(i) or points between X(i) and X(j) where i and j are user-defined values of X that occur within the set being operated on.

    3,   Find the average of Y over a range in X and calculate the centroid. The user specifies the set of points over a range in X to be averaged for the leading edge, the set to be averaged for the trailing edge, and the set to be used in finding the centroid.

    4,   Put results in set N for drawing square front (equivalent sharp boundary).

CC,5:   Computes the sum of Y values over a user-defined range of X and finds the midpoint of Y over that range (to find midpoint of peak area in time difference scanning chromatography or small zones).

CC,6:   The user identifies the set to be operated on and the set to receive the results, N' and the number of points a group (G) is to contain. This section then sums the Y values of each successive G of X for set N and stores the results as new points in set N' (allows unbiased amplification of small gradients, or compression of a 194 point scan to <100 pts so that it can be used to initiate a simulation).

The PDP-11 is used with PLOT1 (Table II) to take the finally processed data and turn it into labeled graphs with the XY recorder. PLOT1 is also able to use externally generated files to do the same thing (Fig. 2). This feature allows us to use results from simulations and put them on the same coordinates as the raw data for direct comparison.

The basic unit of information obtained by the direct scanning system is a file which consists of a collection of absorbance values for a constant number of points which represent the observable length of the column. The following steps are required for the determining of final absorbance profiles. First, a number of files are accumulated to represent the baseline of the column immediately prior to introduction of the sample. The column scanner, particularly the light source, should be turned on and warmed up for several hours if constant absolute magnitude baselines are important. If only a relative profile of the column packing is needed, this is less important. Typically we take four scans at 5-min intervals immediately prior to the time that the sample profile first appears in the column and use these as the baseline. These four scans are averaged together and the average is stored in a temporary file in the PDP-1104. Files are obtained from the A/D converter and stored in the form of bits and need to be converted to volts before presentation of the data on a monitor or X-Y recorder. All operations beyond the initial acquisition of the data file are

TABLE II

MAIN FEATURES OF PLOT1—DATA PLOTTING ROUTINE

The program presents the following display:

DT = Data type in — Allows data to be entered from the terminal keyboard.

DE = Data edit — Allows the user to insert, delete, or change data points.

DD = Data description — May be used to change or add to the data description provided by the scanner.

DW = Data write to disk — Prompts user to write a file to disk. (The file written will contain all data sets currently in memory.)

DR = Data read from disk — Prompts user to read a specified plotting file from disk.

DH = Data hard copy — Provides hard copy of all data sets currently in memory and the data description as provided by the user or scanner.

DA = Data manipulation options — Performs various operations on data sets: touch return to advance program to the next main menu.

Graph Description

WD = Write description to disk — Write a completed graph description to disk.

RD = Read description from disk — Reads a previously stored graph description from disk.

DA_ = Define X or Y axis — Defines beginning and ending points on plotter for X and Y axis.

DL_ = Define X or Y axis label — Allows user to type in the label and assign a character size. The routine tells the user if the label will fit on the graph and automatically centers it.

DS_ = Define X or Y scale	May be calculated automatically or set by user. Axis offset, scale in units/cm, frequency of tick marks, and size of tick labels.
SS = Specify symbol size	Allows the user to define the size of the symbol to be used for plotting points.
GH = Hard copy of graph description	Prints a hard copy of the above information; touch return to advance program to next main menu description.

Plotting Options

DA_ = Draw axis and tick marks for X and Y	Draws the axis and tick marks only for the axis specified.
DT_ = Draw tick labels for X or Y	Draws the tick labels only for the axis specified.
DL_ = Draw label for X or Y	Draws the label only for the axis specified.
PS_ = Plot points for data set 1–12	Plots points from specified data set.
LS_ = Draw straight lines for data set 1–12	Draws straight lines connecting points for specified data set.
DD = Draw description	The description, its size, and location on the graph are defined by the user in this subroutine only. This is not the same description previously defined, but the figure legend.
AL = Draw everything except straight lines	Automatically does all of the above except draw straight lines plotting the data. Plotting data lines must be performed as a separate step.

Response to each choice is the proper two-letter code and an extension if requested by a_. PLOT1 sends both video monitor and X-Y plotter under a user option, allowing preliminary testing of graphs before hard copy is prepared.

under control of a large, "menu-driven" processing program (Table I). Once the baseline has been established, automatic baseline subtraction and conversion of bits to volts can be done for each subsequent file.

The files are then displayed on the Tektronix video monitor and, if desired, may be plotted on a Houston Instruments X-Y plotter (Table II). While the scan profile is displayed on the monitor, one chooses a region to represent the baseline and one to represent the plateau concentration and between these two regions calculates the centroid of the profile. Calculations of centroid positions can be done with a precision of better than ±0.4 points under favorable operating conditions. For a series of broad zones of potassium chromate the baseline noise level was ±0.011 V and the linear regression estimate of the positions of ~12 centroids for each of three boundaries had a confidence of greater than 0.9995. The SEM of the difference between succeeding centroid positions was less than ±0.2 points (mm). Similar confidence levels were obtained with a series of solutes (Fig. 3). The inset gives an example of regions chosen as baseline and plateau absorbancy and the centroid calculated between these. These data have been smoothed for presentation by use of a three-point moving average, twice. The actual calculations are usually done on the raw data prior to smoothing. If the baseline of the column is relatively flat there is no need to subtract it prior to calculation of the centroid.

Rates of centroid migration are a useful measure of the average size of a protein or other macromolecule. However, for associating systems the whole profile is more informative than the simple centroid. Some years ago we developed a set of simulation programs to run on the PDP-1104, based on those of Cann and Goad,[19,20] to calculate migration profiles for some simpler associating systems. Calculation of dispersion profiles is a trivial subroutine of this for nonassociating solutes. Usually the simulations are begun by imagining a sharp boundary between solvent and solute and then calculating the profile after intervals of dispersion (and transport). Because the loading of a column may not lead to a perfectly symmetrical boundary and cannot lead to a perfectly sharp one, it may be preferable to use real data as the starting point of a simulation. Consequently, we have modified the simulation program to allow us to put in a profile obtained from the scanner as the initial state of the simulation.[21] In this case it is necessary to use a smoothed profile to ensure that there are no artifactual boundary shape effects from random noise in the scans. (The simulation does not deal well with negative profile gradients.)

[20] J. R. Cann, "Interacting Macromolecules." Academic Press, New York, 1970.
[21] G. A. Harvey, Ph.D. Thesis, University of Virginia, Charlottesville (1977).

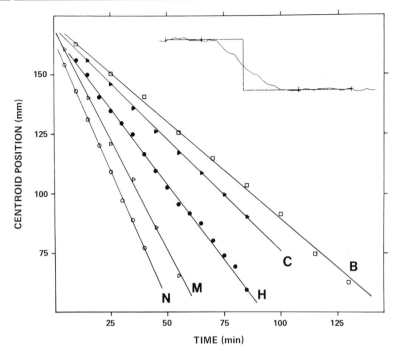

FIG. 3. Rates of centroid migration through a column of Sephadex G-200. Centroid positions were determined after subtracting a baseline scan obtained prior to applying the solute of interest. In the inset a typical profile for catalase is shown along with the calculated centroid. The bracketed straight lines are the baseline and plateau absorbancies used to calculate the central position. In the main figure, Blue Dextran (B) was monitored at 405 nm and a solution of 5 mg/ml was applied. Catalase (C) was monitored at 405 nm at 0.4 mg/ml; human hemoglobin (H) from Sigma was monitored at 400 nm, at 0.1 mg/ml; equine myoglobin (M) was monitored at 410 nm at 0.1 mg/ml. NADH (N) was monitored at 375 nm at a concentration of 0.3 mM. All solutes were dissolved in 10 mM phosphate buffer, pH 6.9. (From Ref. 9, by permission.)

The profiles in Fig. 4 show typical results for several different column flow rates with the same solute, myoglobin. In each case reasonable fits are obtained that fix the dispersion coefficient to within about a factor of 2. As expected, dispersion increases with flow rate.[3,9] In a number of experiments we got fitted estimates of dispersion coefficients accurate to ±30% or so, with a range of solutes and flow rates.[9] In general the error was of constant *relative* magnitude for each flow rate. Improving the precision of the estimation would be possible but serves little purpose since the only use to us of such estimates is to determine whether an

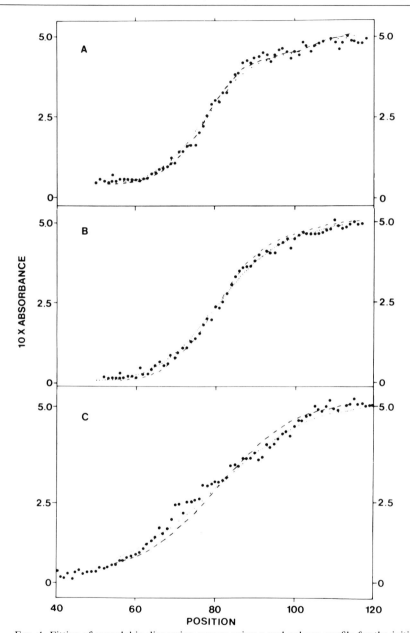

FIG. 4. Fitting of myoglobin dispersion curves using a real column profile for the initial boundary conditions. In each part an initial scan with a centroid near 40 mm into the column was used. The diffusion simulation previously described[19] was applied with appropriate choice of dispersion coefficient for a suitable time. The simulated curves were then com-

unknown protein is showing anomalously large dispersion indicative of heterogeneity. Such specific instances can then be further investigated by more detailed study.

Some examples of such studies have been described for hexokinase, aldolase, and glucose-6-phosphate dehydrogenase.[10] In these three instances we used another option on the scanner, stopping flow after the enzyme had passed part way down the column. Scans were then taken at various times and the rate of conversion of substrates to products was monitored. Rates of product formation were estimated by manual plotting of the absorbance vs time for every third point throughout the length of the column, after subtracting a baseline derived from the time of flow stoppage.

Such calculation by plotting is a very tedious way to extract information from the scans. A much more efficient way is simply dividing the absorbance of an entire scan file by the time from flow stoppage; the rate is produced directly. (This assumes that scanning is fast relative to the enzyme reaction rate.) Four rate files can be held in temporary arrays simultaneously and then averaged together to derive a mean rate with the standard deviation of each point. This is just one example of why a good data processing program is an essential part of any scanning system. The program we have developed is "menu-driven" and "user-friendly" so that a relative novice can work through from beginning to end. It is written in FORTRAN IV and runs on a PDP-1104 with 32K of core memory, but it could be resident in a central computer with a terminal, provided that the rate of data transfer is fast. The major function of the program is file transfer with only trivial mathematical manipulation of the files. Thus, transfer time is the limiting factor in most operations.

Application of the System

Active enzyme chromatography was developed by Ackers' group[8] using small zones of enzyme applied manually to columns equilibrated

---

pared to the real profile after the equivalent development time when the centroid was near 120 mm into the column. (A) Initial centroid position point 160, final point 80 (reference point 0 is the bottom of the column in all cases). The flow rate was ~1 ml/hr and the elapsed time between the two scans shown was 320 min. Simulated dispersion curves with coefficients of 0.9×10^{-5} (———) and 1.6×10^{-5} (· · ·) cm^2/sec are shown. (B) Initial centroid position point 163, final point 82. Flow rate was ~2 ml/hr and elapsed time 160 min. Simulated dispersion curves with coefficients of 2.5×10^{-5} (———) and 5×10^{-5} (· · ·) cm^2/sec are shown. (C) Initial centroid position point 156, final point 79. Flow rate was ~4 ml/hr and elapsed time 80 min. Simulated dispersion curves with coefficients of 1.5×10^{-4} (———) and 2.5×10^{-4} (· · ·) cm^2/sec are shown. (From Ref. 9, modified, by permission.)

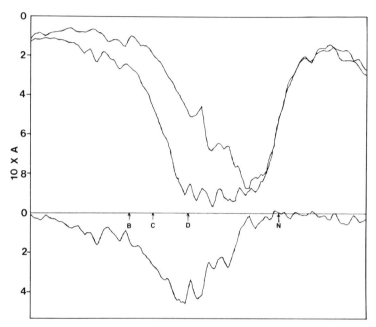

FIG. 5. Time difference spectrum after stopping column flow with glucose-6-phosphate dehydrogenase. Flow was stopped after 70 min following sample application. Scans taken at 5 and 35 min after stopping flow are shown, along with the difference between them. Increasing absorbance is toward the bottom in the original spectra, and in the difference spectrum. The expected positions of Blue Dextran (B), catalase (C), diaphorase (D), and NADH (N) are shown for reference. The bottom of the column is toward the left. (From Ref. 10, by permission.)

with appropriate substrates. Rates of enzyme migration were derived from positions of absorbance profiles as a function of time. We introduced the notion of stopping flow during the passage of the enzyme through the column to provide a profile of the enzyme distribution.[10] In this way the actual dispersion of the enzyme could be calculated rather than having to assume that it followed ideal behavior. In fact, we found that aldolase and glucose-6-phosphate dehydrogenase both showed some association behavior that would not be detectable by conventional active enzyme chromatography, in our scanning system (Fig. 5). In principle the associations should be detectable by elution of the enzyme from a column and careful assay of the profile of enzyme activity. The scanning system merely simplifies this process by doing ~200 assays simultaneously at much lower concentrations of enzyme than are usually used.

One system in which the association behavior was obvious even without the flow stopping option was the association of diaphorase with keto

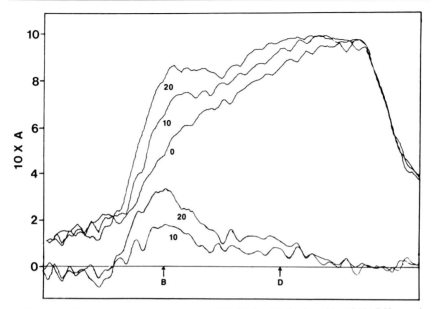

FIG. 6. Active enzyme chromatography of KGDC with lipoic acid and NADH as substrates. After the column was saturated with assay mixture, 5 μl of a 31 mg/ml stock of KGDC was diluted into 500 μl of the assay mixture and 0.150 μl was applied to the column. Chromatography was stopped after 52 min and the development of enzyme activity was monitored. The original spectra, smoothed with two three-point moving averages, are shown above, with the difference below. The expected migration positions of Blue Dextran (B) and diaphorase (D) are indicated below the difference spectra. Absorbance increases toward the bottom of the figure and the top of the column is at the right. (From Ref. 11, by permission.)

acid dehydrogenases.[10] A nearly ideal boundary shape for a "single" component, glucose-6-P dehydrogenase, migrating through the column producing product at a constant rate, is shown in Fig. 5. Notice that the leading boundary, indicative of the enzyme activity, is much broader than the trailing boundary, which is product only. This is a function of both the relative dispersion coefficients for the enzyme and the small molecule and time spent moving through the column. Comparing the profile for diaphorase activity of α-ketoglutarate dehydrogenase (KGDC) with that of glucose-6-P dehydrogenase it is apparent that the diaphorase is much more dispersed than it should be for a molecule of its size (Fig. 6). When flow is stopped and the development of the activity profile is permitted to continue for a period of time the cause of the dispersion becomes apparent. At the concentration of protein applied to the column, the diaphorase is dissociating from the complex during the course of chromatography on the column.

With small zone chromatography it is not possible to make any reasonable estimate of the association constant for the diaphorase and KGDC complex.[22] In order to obtain an estimate of the association constant for diaphorase with KGDC and pyruvate dehydrogenase (PDC) we developed a technique of broad zone active enzyme chromatography.[11] In this case the principle of operation is the same as for the small zone except that enzyme is mixed with substrate and infused continuously at the same rate as column flow in its absence, until a broad zone (3–10 ml) has been applied to the column. Then the column is developed further with assay mixture and the profiles of enzyme activity are monitored. Again the column flow may be stopped and the development of the broad zone enzyme profile monitored.

The particular case that we studied is one described by Gilbert and Jenkins[23] in their theoretical work on mass transport, in which the larger constituent has a migration rate independent of the presence of the smaller constituent. Both KGDC and PDC have molecular weights of several million so that they are totally excluded from Sephadex G-200 whether or not there are a few molecules of the diaphorase component associated. Its molecular weight is about 100,000 and ~6 molecules are thought to be associated with either the KGDC or PDC complexes.

When the rate of migration of the larger constituent is independent of the presence of the smaller one, the relative association can be derived from the rate of migration of the smaller constituent at different starting concentrations. For instance, if diaphorase is fully associated with the complex it will migrate at the complex rate, while if it is fully dissociated it will migrate much more slowly as an M_r 100,000 protein. Partial association will result in a rate of migration intermediate between these two limits, and simply proportional to the percentage associated (Fig. 7).

Using a slow substrate for the diaphorase it was possible to increase the enzyme concentration 50-fold above that used with the regular substrate. Estimates of association constant at these two concentrations were in good agreement: ~12 and ~15 nM, respectively. For PDC we derived an association constant of 26 nM from an experiment at the lower protein concentration. These experiments could be done by conventional elution chromatography in the presence of the substrates but would require many hours of assays at the end. The same type of experiment might in principle be done with high-performance liquid chromatography but thus far the keto acid dehydrogenases have only been chromatographable under conditions of high salt, which severely inhibits enzyme function and leads to dissociation of other constituents of the complex.[24]

[22] J. K. Zimmerman and G. K. Ackers, *J. Biol. Chem.* **346**, 1078 (1971).
[23] G. A. Gilbert and R. C. L. Jenkins, *Proc. R. Soc. London, Ser. A* **235**, 240 (1960).
[24] T. E. Roche, personal communication.

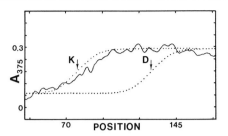

FIG. 7. Broad zone chromatography of KGDC using NADPH as substrate. Assay mix was applied for 4 h, followed by enzyme in assay mix for 1 h. 100 μl of enzyme (31 mg/ml) was diluted into 6 ml for this experiment. The rate shown (——) is the average of four scans from 65 to 80 min taken 5 min apart. The simulated profile ($\cdots$) is that of diaphorase fitted to the appropriate absorbance difference. The arrows indicate the expected migration positions of KGDC (K) and diaphorase (D). The top of the column is to the right; position is shown in mm. (Modified from Ref. 11 by permission.)

Nitrogenase is an associating system of two proteins that are required for enzymatic activity. When we first began development of the scanning system it was common belief that they formed a relatively stable complex. More recent kinetic experiments have indicated that the association–dissociation must be fast and the overall association constant is relatively weak (10^7 liters/mol or less) so that under conditions of enzyme assay (<1 μM) the two proteins would only be partially associated except in the presence of a large excess of one component. We have done active enzyme chromatography of the enzyme using the oxidation of dithionite as a measure of enzyme activity. Indeed we find that the overall enzyme activity behaves as if it migrates at the rate of the smaller component if an equimolar mixture is applied to the column as a broad zone. This result has been confirmed using elution chromatography at various ratios of the protein with specific postcolumn assay for the larger component and the combined activity separately.[25] Because scanning gel chromatography is limited to only one observable substrate with this enzyme, it has not proved to be particularly useful in studies of the association under conditions of enzyme turnover. There were previously reports of association between the proteins under non-turning-over conditions but these proved to be an artifact in the ultracentrifugation experiments.[26]

One useful offshoot of scanning chromatography is a simplified calibration procedure for gel filtration and HPLC columns.[27] The scanner

[25] L. C. Davis, unpublished studies.

[26] J. R. Postgate, R. R. Eady, D. J. Lowe, B. E. Smith, R. N. F. Thorneley, and M. G. Yates, in "Mechanisms of Oxidizing Enzymes" (T. V. Singer and Ondarza, eds.), p. 173. Am. Elsevier, New York, 1978.

[27] L. C. Davis, J. Chromatogr. Sci. **21**, 214 (1983).

produces results in terms of rates of migration of standard proteins or active enzymes. This is equivalent to the reciprocal of the elution volume, or at constant flow rate, reciprocal elution time. A plot of this measure vs Stokes radius gives a straight line over the useful range of a column partitioning ability without having to derive either the void or the included volumes. This calibration is somewhat simpler than the presumably more precise method of Ackers[2] and is more reasonable than using the log of molecular weight which requires all proteins to be alike. As an example, we can consider the case of hexokinase, which is well established to have a monomer molecular weight of 50,000, while its Stokes radius ($\sim$33 Å) in either narrow or broad zone active enzyme chromatography, or elution gel chromatography, is that of a much larger globular protein, nearer 75,000.[10,11,25] Furman and Neet[28] used plots of log molecular weight vs elution time and with the two standards that they chose (ovalbumin and bovine serum albumin) obtained an estimated molecular weight of 50,000 for hexokinase monomers. We used the above-described calibration procedure, as well as that of Ackers, and found the somewhat larger Stokes radius whereas by replotting our data as log M_r vs elution time, we obtained the same result as Furman and Neet. Published crystallographic data indicate a radius of 25 Å which is much smaller than our observed values in chromatography, suggesting that in solution the protein has a rather different conformation from that observed in the crystal.[11]

Acknowledgments

Research from this laboratory has been supported by NIH Grant GM 23039 and by the Kansas Agricultural Experiment Station. This is contribution number 84-133B of the Kansas Agricultural Experiment Station.

[28] T. C. Furman and K. E. Neet, *J. Biol. Chem.* **258**, 4930 (1983).

[10] Hydrodynamic Characterization of Random Coil Polymers by Size Exclusion Chromatography

By PHIL G. SQUIRE

Since its discovery by Porath and Flodin,[1] Sephadex gel filtration has proven to be one of the most powerful methods for the purification of proteins. Studies of the gel filtration of fractionated dextran samples, in

[1] J. Poráth and P. Flodin, *Nature (London)* **183**, 1657 (1959).

comparison with proteins, by Granath and Flodin[2] clearly demonstrated that the elution volume of a macromolecular solute on permeation through a given column was determined by the molecular size of the solute. Since globular proteins are tightly coiled, and are often roughly spherical, their molecular sizes are simply related to their molecular weights, and as a consequence, Sephadex gel filtration has become a widely used method for determining the molecular weights of proteins. For polysaccharides and other random coil polymers, a simple, general relationship between molecular size and molecular weight does not exist, and, as a consequence, the goal of relating the molecular weight of these molecules to their elution volumes has not been easy to achieve. In this chapter, we shall explore some of the problems that have been encountered, and describe current approaches to this goal.

During the past decade, technological advances have resulted in new column packings, some of which can be used at high pressures. As a consequence, a more general term for the general separation process is appropriate. Since the fundamental process that determines the elution volume of a macromolecule is size exclusion, either occurring in the interior channels of the beads, or at their external surfaces,[3] I prefer the term size exclusion chromatography (SEC), but gel permeation chromatography (GPC) and gel filtration chromatography (GFC) are also used to describe the same separation process.

A recent monograph[4] reviews the theory and practice of SEC and describes column packings, as well as methods for determining the molecular weight distribution of polymers, and these important subjects will not be treated here. I also refer the reader to a recent review by Hagnauer[5] for references to size exclusion chromatography of specific random coil polymers and other macromolecules as well as other recent books and symposium editions on the general subject of SEC. Here, I shall first review a method of constructing a universal calibration curve that has proven to be quite successful by high polymer chemists, and probably deserves wider use among biochemists working with flexible biopolymers. This will be followed by a review of my own efforts to relate elution volumes obtained by SEC to the hydrodynamic and statistical parameters of random coil polymers.

[2] K. A. Granath and P. Flodin, *Makromol. Chem.* **48,** 160 (1961).
[3] P. G. Squire, A. Magnus, and M. E. Himmel, *J. Chromatogr.* **242,** 255 (1982).
[4] W. W. Yau, J. J. Kirkland, and D. D. Bly, "Modern Size Exclusion Chromatography." Wiley, New York, 1979.
[5] G. L. Hagnauer, *Anal. Chem.* **54,** 265R (1982).

The Universal Calibration Curve

The concept of a universal calibration curve, as introduced by Benoit et al.[6] is based on the Einstein viscosity law

$$[\eta] = \nu N V_h / M \tag{1}$$

This equation relates the hydrodynamic volume, V_h, of a polymer of molecular weight, M, to its intrinsic viscosity $[\eta]$ in cm^3/g. N is Avagadro's number, and ν is a shape factor which has the value 2.5 for spherical macromolecules. Thus, molecules having the same value of $[\eta]$ M should elute with the same elution volume, V_e. For convenience, the authors suggest that calibration curves be constructed by plotting log($[\eta]$ M) vs elution volume, and that data for all types of spherical macromolecules should fall on the same universal calibration curve. I would point out, however, that these arguments do not predict that the relationship between these parameters should necessarily be linear. In fact, most, but not all of the calibration curves that I have seen in the literature, e.g., Fig. 1,[6a] show definite upward curvature.

Universal calibration curves have been widely used by polymer chemists and this method has become the basis of the ASTM standard test method[7] for measuring molecular weight distributions of polymer samples. While this specific test method is limited to polymers soluble in tetrahydrofuran, the general principles of the method are of broader applicability.

Having selected an SEC column with an appropriate size exclusion range and solvent compatability the approach based on Eq. (1) is rather simple. One must first calibrate the columns with standard preparations with narrow molecular weight distributions and known intrinsic viscosities. (Calibration with standards having a broad distribution of molecular weights is also possible,[4,8] but the procedure is more complicated and the results probably less accurate.) Typically the relationship between molecular weight and intrinsic viscosity of these fractions is given in terms of the Mark–Houwink equation,

$$[\eta] = K M^a \tag{2}$$

Molecular weight standards with narrow distributions are available from commercial sources, both for dextran and polyethylene glycol.

[6] H. Benoit, Z. Grubisic, P. Rempp, D. Decker, and J. G. Zilliox, J. Chim. Phys. 63, 1507 (1966).

[6a] Z. Grubisic, P. Rempp, and H. Benoit, J. Polym. Sci., Part B 5, 753 (1967).

[7] "1980 Annual Book of ASTM Standards," Pt. 35, p. 875, ANSI/ASTM D3593-77. Am. Soc. Test. Mater., Philadelphia, Pennsylvania, 1980.

[8] J. Janca, Adv. Chromatogr. 19, 37 (1981).

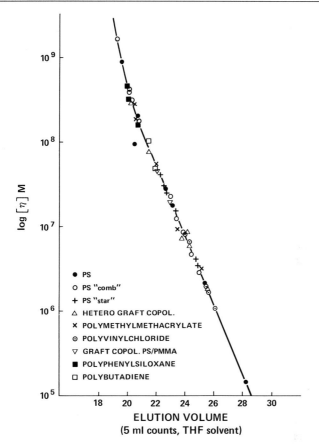

FIG. 1. Universal calibration plot (reprinted with permission from Grubisic *et al.*[6a]).

These standards are appropriate for use in aqueous solvents. Polystyrene in tetrahydrofuran is the standard specified in the ASTM method.[7] The Mark–Houwink constants for a wide range of polymers in tetrahydrofuran are tabulated in the ASTM report.[7] Other values are compiled by Yau et al.[4] The Mark–Houwink constants for dextran have been evaluated at 20° by Granath[9]

$$[\eta] = 0.443 \ M^{0.42} \tag{3}$$

Corresponding data for polyethylene glycol at 25° and for the molecular weight range 200–8000, taken from a review by Bailey and Koleske[10] are

[9] K. A. Granath, *J. Colloid Sci.* **13**, 308 (1958).
[10] F. E. Bailey, Jr. and J. V. Koleske, *in* "Nonionic Surfactants" (M. J. Schick, ed.), p. 794. Dekker, New York, 1967.

$$[\eta] = 0.156\, M^{0.5} \tag{4}$$

The data are then plotted as $\log[\eta]\, M$ vs V_e, as in Fig. 1. When the test sample is examined in the same column, the product $[\eta]\, M$ can be read directly from the calibration curve and the hydrodynamic volume calculated by Eq. (1).

In order to determine the molecular weight of a test sample, either of two approaches may be followed. If the project is of somewhat limited scope, one can merely determine the intrinsic viscosity of the sample, and calculate M from the product $[\eta]\, M$ read from the universal curve. For projects of more extensive scope, it would probably be desirable to fractionate a preparation with a broad distribution of molecular weights into fractions with narrow distributions, possibly by preparative SEC, determine the molecular weights and intrinsic viscosities of these samples and calculate the Mark–Houwink constants. When this is done, then the molecular weight of any sample can be calculated from the $[\eta]\, M$ value read from the graph from the equation $[\eta]\, M = KM^{a+1}$.

Sources of error in using this procedure are discussed and illustrated by Yau et al.[4] and Janca.[8] Briefly, they arise from three sources. Theories of SEC are all based on the assumption that the size exclusion process uniquely determines the elution volume, and yet the possibility of reversible adsorption is difficult to dismiss and, where it occurs, errors in the interpretation may easily result. Experimental errors, including the measurements of viscosity and elution volume as well as inadequate temperature control can also contribute. A third source of error is less obvious, but not less worrisome. Equation (1), which is the basis of the method, relates the hydrodynamic volume of a macromolecule which may be a highly deformable random coil to its intrinsic viscosity. This experimental parameter, when properly measured, corresponds to infinite dilution, and for macromolecules of the type we are discussing here, the results should be extrapolated to zero shear. When we analyze SEC data by the procedure described here, we are applying Eq. (1) under conditions where the concentrations and shear rates may both be substantial.

Since SEC separates molecules according to size, one might inquire as to which parameter of size should be chosen. Hydrodynamic volume is certainly a reasonable choice, but, as pointed out by Janca,[8] other parameters of size have been considered as well, and universal calibration curves can also be constructed based on these parameters. The product $[\eta]\, M$ is related to the radius of an equivalent sphere, R_e by the equation

$$[\eta]\, M = (10N\pi/3)R_e^3 \tag{5}$$

and to the radius of gyration, R_g, by

$$[\eta]\, M = 6^{3/2}\phi R_g^3 \tag{6}$$

Here, ϕ is the universal viscosity constant having the value 2.1×10^{23} when $[\eta]$ is in $cm^3 g^{-1}$.

Finally, $[\eta]$ is related to the square root of the mean square end to end distance, $\langle r^2 \rangle^{1/2}$ by the equation

$$[\eta] M = \phi \langle r^2 \rangle^{3/2} \tag{7}$$

Columns Calibrated with Globular Proteins

About 20 years ago I[11] derived an equation which related elution volumes obtained by Sephadex gel filtration to the molecular weights of proteins and the molecular size of dextran, a random coil polymer. This equation was based on a mathematical model which represented the channels within the beads as an assembly of cones, cylinders and crevices. While this was not a very realistic model for the Sephadex gel, it may be a reasonable approximation to the internal channels of the beads currently used for high-pressure size exclusion chromatography (HPSEC). (Unpublished scanning electron micrographs of Toyo Soda TSK-G3000 SW column packings by M. E. Himmel support this opinion.) More recent versions[12] of the original equation appear in two forms [Eqs. (3) and (4) of Ref. 12] which are equivalent, the choice being dictated by whether the interstitial volume outside the gel beads, or the total volume accessible to solvent could be more precisely measured. We have recently reported[3] an uncertainty in the measurement of V_0 resulting from external size exclusion and conclude that Eq. (4) (of Ref. 12) is less subject to error. This equation has the form

$$\frac{V_e^{1/3} - V_t^{1/3}}{V_o^{1/3} - V_t^{1/3}} = F_{(v)} = \frac{M^{1/3} - A^{1/3}}{C^{1/3} - A^{1/3}} \tag{8}$$

Here, V_e is the elution volume of a protein of molecular weight M, V_0 is the interstitial volume outside the spherical gel beads, estimated as the elution volume of a protein of molecular size too large to enter the gel pores, and V_t is the total volume available to solvent, estimated as the elution volume of a probe of low molecular weight. The function $F_{(v)}$ is calculated from elution volumes according to the first quality. The second equality predicts a linear relationship between $F_{(v)}$ and $M^{1/3}$. This equation also permits one to estimate the upper and lower limits of molecular weights that are separated by the column. At the intercept at $F_{(v)} = 1$, $M^{1/3} = C^{1/3}$, and at the intercept at $F_{(v)} = 0$, $M^{1/3} = A^{1/3}$. Data obtained with three TSK columns having different pore sizes are shown in Fig. 2, and the values of the constants A and C are entered into Table I.[13]

[11] P. G. Squire, *Arch. Biochem. Biophys.* **107**, 471 (1964).
[12] M. E. Himmel and P. G. Squire, *Int. J. Pept. Protein Res.* **17**, 365 (1981).
[13] P. G. Squire, *J. Chromatogr.* **210**, 433 (1981).

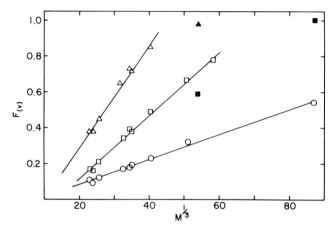

FIG. 2. Data from native proteins[13] plotted according to Eq. (8) for G4000 SW (○), G3000 SW (□), and G2000 SW (△). Data presented as solid symbols refer to γ-globulin, which is retarded due to adsorption, and to a protein used for estimation of V_o. These points were not included in the least-squares analysis. (Reprinted with permission from Ref. 13.)

For use with macromolecules other than globular proteins, we convert the calibration curve to $F_{(v)}$ vs r_p. To do this, we merely multiply the constants $A^{1/3}$, $M^{1/3}$, and $C^{1/3}$ by the factor 0.794×10^{-8}. This follows from a well known equation from hydrodynamics for native proteins:

$$r = \left(\frac{3\,M\bar{v}}{4\,\pi N}\right)^{1/3} \cdot \left(1 + \frac{w}{\bar{v}\rho}\right)^{1/3} = 0.794 \times 10^{-8}\,M^{1/3} \qquad (9)$$

In this calculation the value 0.73 was used for the partial specific volume, $\bar{v}$, and 0.53 g of water per gram of protein as the hydration, w. These are

TABLE I
CALIBRATION CONSTANTS OBTAINED FROM
FIG. 2

Column	A	C	$r_a(\text{Å})$	$r_c(\text{Å})$
G2000 SW	940	91,000	7.8	36
G3000 SW	2460	340,000	10.7	56
G3000 SW[a]	3900	330,000	12.5	55
G4000 SW	551	$3.4 \cdot 10^6$	6.5	120

[a] From Himmel and Squire.[12] All other data from Squire.[13] (Reprinted with permission from Ref. 13.)

mean values calculated from an earlier study[14] of the hydrodynamic properties of 21 globular proteins of known structure. Calibration constants in terms of molecular radii are also given in Table I. These constants also provide an estimate of the limiting values of these parameters for separation by the primary SEC process on the three types of columns.

We have now converted Eq. (8) to the form

$$F_{(v)} = ar_p + b \tag{10}$$

where

$$a = \frac{1}{r_c - r_a} \quad \text{and} \quad b = \frac{-r_a}{r_c - r_a}$$

If we accept the premise that size exclusion chromatography separates macromolecules according to size, we are left with a question. What parameter of size of a random coil polymer is it that is equal to the radius of a hydrated globular protein that has the same elution volume? Various authors have suggested that it may be the radius of gyration, R_g, or the radius of the equivalent hydrodynamic sphere, R_e, but it seems likely that no a priori answer exists. We shall seek an experimental answer. We define this unknown parameter as the "SEC radius" and assume that it is related to the molecular weight by an equation of the form

$$r_p = \text{SEC radius} = gM^z \tag{11}$$

Entering Eq. (11) into Eq. (10), and taking the logarithm, we have

$$\log(F_{(v)} - b) = \log ag + z \log M \tag{12}$$

This equation predicts a linear relationship between $\log(F_{(v)} - b)$ and $\log M$, with slope z. Since a is known [Eq. (10)], the constant g can be calculated. In this way, we generate a regression equation of the form in Eq. (11) which can be compared with equations of the same form which relate any of the statistical or hydrodynamic parameters of size to molecular weight.

This approach was tested with data from Kato et al.[15] who reported the elution volumes for polyethylene glycols and dextran fractions of narrow molecular weight distributions on the same three columns. These data were plotted according to Eq. (12) as seen in Figs. 3 and 4, and the constants, g and Z, obtained by least squares analysis of the data from the linear portion of the curves (open symbols of Figs. 3 and 4), are recorded as "SEC radius" in columns 2, 3, and 4 of Table II. The remaining columns of Table II contain equations for the radius of the equivalent

[14] P. G. Squire and M. E. Himmel, *Arch. Biochem. Biophys.* **196**, 165 (1979).
[15] Y. Kato, K. Komiya, H. Sasaki, and T. Hashimoto, *J. Chromatogr.* **190**, 297 (1980).

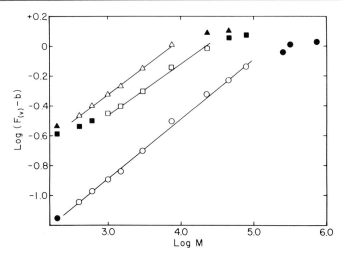

Fig. 3. Data from polyethylene glycol fractions plotted according to Eq. (12) for G4000 SW (○), G3000 SW (□), and G2000 SW (△). Data presented as solid symbols lie outside the linear region. (Reprinted with permission from Ref. 13.)

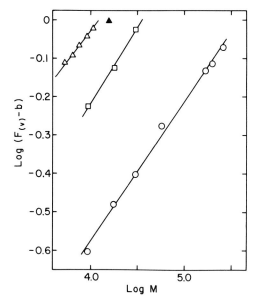

Fig. 4. SEC data for dextran plotted according to Eq. (11). Data reported for G4000 SW (○) and G3000 SW (□) were calculated from the elution volumes of fractions with rather narrow molecular-weight distributions, while the G2000 SW (△) data were obtained from the integral elution curve of a sample with a broad molecular-weight distribution. (Reprinted with permission from Ref. 13.)

TABLE II

COMPARISON FOR "SEC RADII" WITH PARAMETERS MEASURED BY CONVENTIONAL HYDRODYNAMIC METHODS[a]

Polymer	"SEC radius"			Equivalent sphere, R_e		Radius of gyration, R_g		
	G4000 SW	G3000 SW	G2000 SW	From sed. vel.	From $[\eta]$	From sed. vel.	From $[\eta]$	From light scattering
PEG	$0.87\,M^{0.40}$ 21 Å	$1.38\,M^{0.35}$ 23 Å	$1.02\,M^{0.37}$ 20 Å	$0.145\,M^{0.59}$ 16 Å	$0.29\,M^{0.50}$ 16 Å	$0.22\,M^{0.59}$ 25 Å	$0.37\,M^{0.50}$ 20 Å	
Dextran	$0.937\,M^{0.36}$ 38 Å	$0.69\,M^{0.40}$ 43 Å	$1.39\,M^{0.36}$ —	$0.160\,M^{0.54}$ 42 Å	$0.34\,M^{0.47}$ 43 Å	$0.24\,M^{0.54}$ 63 Å	$0.43\,M^{0.47}$ 54 Å	$0.66\,M^{0.43}$ 56 Å

[a] Radii in Å evaluated at $M = 3000$ for polyethylene glycol (PEG) and at $M = 30,000$ for dextran are given below the regression equation.

sphere, R_e, and the radius of gyration R_g. Below the regression equations in each column the values of these parameters corresponding to a molecular weight of 3000 for polyethylene glycol and 30,000 for dextran are given for comparison. Except for dextran on G-2000 SW, these values fall within the linear portions of the curves in Figs. 3 and 4.

The primary data used in the calculations of the regression equations for R_g and R_e were taken from the literature (see Ref. 13 for original sources) and are as follows: For dextran, $s_{20,w} = 2.24 \times 10^{-15} M^{0.46}$, $\bar{v} = 0.593$, $[\eta] = 0.243 M^{0.42}$, and $R_g = 0.66 M^{0.43}$ from light scattering measurements. For PEG, $s_{30,w} = 1.26 \times 10^{-15} M^{0.41}$, $\bar{v} = 0.8392$, and $[\eta] = 1.56 M^{0.5}$. The sedimentation coefficient of PEG, corrected to 25° for the different viscosity and density of water is $s_{25,w} = 1.11 \times 10^{-15} M^{0.41}$, and this relationship was used here.

Light scattering data give the radius of gyration directly and this equation is recorded in the final column. The radius of the equivalent hydrodynamic sphere, R_e, was calculated from the intrinsic viscosity by Eq. (5), and was calculated from the sedimentation coefficient by the equation

$$R_e = \frac{M(1 - \bar{v}\rho)}{6\pi\eta_0 N s_{20,w}} \tag{13}$$

where η_0 is the viscosity of water at 20°. I used the equation $R_e = 0.788 R_g$, from Eqs. (5) and (6), for interconverting R_e and R_g values from viscosity, and $R_e = 0.663 R_g$ for sedimentation data. This follows from equations[16] which express the translational frictional coefficient, f, in terms of R_e and R_g, $f = 6\pi\eta_0 R_e = 5.1 \times 6^{1/2}\eta_0 R_g$.

There is fairly good agreement among the equations relating "SEC radius" to molecular weight, for the three columns with quite different pore characteristics. This is true for both polymers. The values of the "SEC radius" for polyethylene glycol evaluated at $M = 3000$ fall between the two values for the radius of gyration and they are in poor agreement with R_e values calculated either from sedimentation velocity or intrinsic viscosity measurements. On the other hand, the values for dextran, $M = 30,000$, agree better with the values for R_e than with those for R_g calculated from data obtained by either of the three experimental methods.

It is somewhat disturbing that the exponential terms for the "SEC radius" tend to be somewhat lower than the corresponding terms for the Stokes radius and the radius of gyration. This point was emphasized, perhaps overemphasized, in Ref. 13, because of the fact that if the expo-

---

[16] C. R. Cantor and P. R. Schimmel, "Biophysical Chemistry." Freeman, San Francisco, California, 1980.

nential terms in two equations are markedly different, the calculated radii cannot be in good agreement over a wide range of molecular weights. We note, however, that the exponential term for R_e, as well as the radius of gyration varies over a rather broad range depending on whether these parameters were obtained by sedimentation, viscosity, or light scattering measurements. These parameters are probably more accurately determined by light scattering, and the exponent for "SEC radius" agrees with that for light scattering better than does the exponent from sedimentation data and about as well as the one from viscosity. For polyethylene glycol on the other hand, the exponential term for the SEC radius is considerably smaller than the corresponding terms calculated from sedimentation or viscosity measurements. This is likely a consequence of the highly extended nature of this polymer in aqueous solution.

Considering its structure, a hydrocarbon chain interrupted at each third position by an oxygen atom, PEG has unusual properties.[10,17] It is highly soluble in water even at degrees of polymerization of 10^5 or higher. This implies strong interactions with the solvent. In water, the polymer assumes a highly extended structure. This implies a stiffness which has been attributed[17] to a conformational transition from a zig-zag structure to a more compact highly ordered meander conformation which predominates at degrees of polymerization above 9. The viscosity is highly dependent upon shear rate. A 1% solution of PEG with a molecular weight of about 2×10^6 displays[17] a 10-fold change in viscosity within the narrow shear rate range of 0 to 0.2 sec^{-1}. Compared with the data from the SEC of dextran fractions the regression equation for SEC radius of PEG has two unusual features, as pointed out above. The exponential term is significantly lower than the values of the corresponding terms based on sedimentation or viscosity measurements, and the SEC radius corresponds to a larger parameter of size R_g for PEG in contrast to R_e for dextran. It seems likely that both of these unusual features are related to the high shear dependence and the stiffness of the polymer chain.

Acknowledgment

This work was supported by the Colorado State University Experiment Station and is published as Scientific Series #2874.

[17] M. Rösch, in "Nonionic Surfactants" (M. J. Schick, ed.), p. 753. Dekker, New York, 1967.

[11] Structural Interpretation of Hydrodynamic Measurements of Proteins in Solution through Correlations with X-Ray Data

By THOMAS F. KUMOSINSKI and HELMUT PESSEN

Introduction

Analysis of protein structure by X-ray crystallography has revealed irregular surfaces consisting of many clefts, grooves, and protuberances. To understand the overall contribution of these features to protein properties, one must bear in mind that globular proteins generally carry out their biological function in aqueous solution, where they are both fully solvated and in a dynamic state, whereas crystallography observes a static structure, and one in which bound water is nearly undetectable. Significant questions from a biochemical perspective are, therefore, whether the solution structure differs from the crystallographic structure, what the possible differences are, and how they may affect biological activity. Hydrodynamic parameters, which are sensitive to surface characteristics, have been calculated from X-ray crystallographic coordinates and compared with solution values in attempts to answer such questions, but with only marginal success. There appears to be a need for structural information from other nonhydrodynamic sources for use in conjunction with hydrodynamic parameters, so that the combined results may be compared with those from X-ray crystallography. Small-angle X-ray scattering (SAXS) is a method particularly suited to meet that need.

The calculation of hydrodynamic parameters from SAXS data can afford various insights. Comparison with observed values can remove ambiguities regarding the relative contributions of shape and hydration to the frictional ratio. X-Ray data from either diffraction or SAXS can provide a measure of molecular surface roughness. Comparison of diffraction results with those from SAXS can indicate differences between a technique that gives detailed structural information, albeit without regard to hydration, and one that gives less geometric detail, but furnishes information on hydration as well as molecular shape and surface area. This chapter will deal with each of these aspects in turn to show how they can be used to advantage.

At other times it may be of value to have a means for deriving structural parameters from available hydrodynamic data which would not permit this unambiguously by traditional methods. It will be shown how from sedimentation coefficients (the best suited among hydrodynamic parame-

METHODS IN ENZYMOLOGY, VOL. 117

ters for this purpose), in conjunction with general expressions and a knowledge only of molecular weights and partial specific volumes, one can obtain reasonable estimates of a number of useful characteristics. Among these are molecular surface areas, hydrated volumes, axial ratios, and radii of gyration of globular proteins, both in the native state and on undergoing structural changes.

Approaches to the Problem Area

The three hydrodynamic parameters under consideration here are obtained experimentally by the observation of flow under the influence of an applied force. The proportionality factors relating the flow rates to the respective forces are the sedimentation coefficient, s, for a gravitational field; the diffusion coefficient, D, for a concentration gradient; and the viscosity, η, for a shearing force. In solutions, these coefficients will vary with concentration. In the cases of sedimentation (the major focus of this chapter) and diffusion, quantities more characteristic of the solute are the respective parameters s^0 and D^0, obtained by extrapolation to infinite dilution; in the case of viscosity, similar extrapolation of a derived quantity, the reduced viscosity, gives the intrinsic viscosity, $[\eta]$. Each of the first two coefficients can be used individually to characterize a macromolecule with respect to either its molecular weight or its frictional properties, provided the other of these is already known from independent measurement or is held constant. Used jointly, sedimentation and diffusion give molecular weight and frictional information simultaneously. The intrinsic viscosity is closely related to the frictional coefficient (i.e., the ratio of frictional force to relative particle velocity) but, in contrast to flexible and extended molecules, for the globular proteins which are our concern here it is not significantly related to molecular weight.

It thus becomes evident that an understanding of frictional coefficients is central to an interpretation of the three hydrodynamic parameters. Frictional coefficients, however, are not directly accessible by experiment. The experimentally accessible frictional ratio (the ratio of the frictional coefficient of the actual—hydrated and nonspherical—molecule to that of a corresponding theoretical—nonhydrated and spherical—molecule of equal dry volume) by definition combines information on two kinds of properties, not readily separated: hydration and shape (the latter in terms of anisotropy, expressed for convenience as the axial ratio of a hypothetical ellipsoid of revolution). Although neither may be obtained explicitly, a range of reasonable assumptions for one will give a range of possible values for the other. This procedure is feasible in the case of globular proteins, for which both ranges are relatively limited, and has led to the practice of balancing the relative contributions of hydration and shape according to some particular criterion, or else of assuming an average hydration value, such as 0.2 or 0.25 g of water per gram of dry protein, in order to arrive at approximate axial ratios.[1-3] Results have often appeared dubious.[4,5] Such procedures have been criticized on the basis that some of the hydration model assumptions are not appropriate and, instead, simultaneous use of parameters from sedimentation, diffusion, and viscosity measurements on the same solutions has been advocated.[4,6,7] The improvements are still not altogether convincing. The problem may be that frictional coefficients derived from the various hydrodynamic processes are not, in principle, identical, because the different types of forces involved require different hydrodynamic models.[5]

A reason for less than satisfactory results from both of the above approaches, which is worth examining in some detail, is suggested by the accumulated X-ray crystallographic evidence relating to the surface structure of the protein molecule. The assumption of smoothness inherent in the models, like the use of ellipsoidal modeling in the first place, has been no more than a convenience, adopted because these were the only geometrical bodies for which a complete theory predicting frictional ratios was available.[8-11] However, the severe surface irregularities revealed by X-ray diffraction data for globular proteins are certain to affect frictional properties. Earlier calculations of structural frictional coefficients from sedimentation coefficients, based simply on unit-cell parameters from X-ray diffraction[12] and leading to excessively high hydration values, have been improved by rigorous calculations for a number of proteins based on a shell model[13] to take into account the dependence of frictional coefficients on surface roughness, or "rugosity," i.e., wrinkledness. (This concept was introduced without explicit definition; it is quantitated elsewhere in terms of a surface area in excess of the smooth surface of the model.[14])

[1] J. L. Oncley, *Ann. N.Y. Acad. Sci.* **41**, 121 (1941).

[2] E. J. Cohn and J. T. Edsall, "Proteins, Amino Acids and Peptides," pp. 428ff. Van Nostrand-Reinhold, Princeton, New Jersey, 1943.

[3] C. Tanford, "Physical Chemistry of Macromolecules," pp. 358ff. Wiley, New York, 1961.

[4] H. Scheraga and L. Mandelkern, *J. Am. Chem. Soc.* **75**, 179 (1953).

[5] I. D. Kuntz and W. Kauzmann, *Adv. Protein Chem.* **28**, 239 (1974).

[6] M. Stern, *Biochemistry* **5**, 2558 (1966).

[7] P. G. Squire, P. Moser, and C. T. O'Konski, *Biochemistry* **7**, 4261 (1968).

[8] A. Overbeck, *J. Reine Angew. Math.* **81**, 62 (1876).

[9] D. Edwardes, *Q. J. Pure Appl. Math.* **26**, 70 (1893).

[10] R. O. Herzog, R. Illig, and H. Kudar, *Z. Phys. Chem., Abt. A* **167**, 329 (1934).

[11] F. Perrin, *J. Phys. Radium* [7] **7**, (1936).

[12] P. G. Squire and M. E. Himmel, *Arch. Biochem. Biophys.* **192**, 165 (1979).

[13] D. C. Teller, E. Swanson, and C. De Haën, this series, Vol. 61, p. 103.

[14] T. F. Kumosinski, H. Pessen, and H. M. Farrell, Jr., *Arch. Biochem. Biophys.* **214**, 714 (1982).

Using first an approximation due to Kirkwood[15-17] and later a more rigorous theory, Teller *et al.*[13] calculated frictional coefficients from X-ray coordinates. They found that agreement with experimental values was reached only when a single-layer hydration shell was added to the crystallographic model in the first case, and only to charged groups on the protein surface in the second.

This illustrates a further reason for disagreements: X-ray diffraction does not observe quite the same entities as do solution methods. Proteins, as mentioned, function in an environment where they are expected to be hydrated, but X-ray structural analysis can show only a fraction of even the most tightly bound water. Model calculations to derive frictional ratios directly from data-bank X-ray coordinates, therefore, can take no account of bound water, although its presence must moderate the effects of the surface irregularities. In addition, X-ray diffraction observes a static structure, disregarding protein breathing[18] in solution. Yet other differences may be due to electrostriction in the crystal resulting from charged groups in the protein.

To resolve some of these ambiguities, an ideal complement to the hydrodynamic methods would be a method which (1) is not hydrodynamic, and therefore not dependent on frictional ratios; (2) independently gives hydration information, as well as structural information from which frictional ratios can be obtained; and (3) unlike X-ray diffraction, does examine proteins in solution. Small-angle X-ray scattering meets all these requirements.[19] It has the additional qualification that it allows the determination of the surface-to-volume ratio, an exceedingly useful parameter in this context. A comparison of hydrodynamic coefficients obtained by SAXS with empirical values thus will provide an approach to interpreting the latter in a meaningful way.

There are two different expressions that can be used as a starting point for the calculation of axial ratios (and thus frictional ratios[11]) of scattering-equivalent ellipsoids from SAXS data[20]; they give different answers. Method 1 gives an estimate of the overall molecular shape, without regard to rugosity. Method 2 makes use of the surface-to-volume ratio obtained from SAXS and translates it into a hypothetical axial ratio descriptive of the surface area instead of the overall geometry of the molecule. In effect, it provides the molecular model of Method 1 with the required additional surface by stretching or flattening it (depending on whether one deals with

15 J. G. Kirkwood, *Recl. Trav. Chim. Pays-Bas* **68,** 649 (1949).
16 J. G. Kirkwood, *J. Polym. Sci.* **12,** 1 (1954).
17 J. García de la Torre and V. A. Bloomfield, *Q. Rev. Biophys.* **14,** 81 (1981).
18 T. E. Creighton, *Prog. Biophys. Mol. Biol.* **33,** 231 (1978).
19 H. Pessen, T. F. Kumosinski, and S. N. Timasheff, this series, Vol. 27, p. 151.
20 V. Luzzati, J. Witz, and A. Nicolaieff, *J. Mol. Biol.* **3,** 367 (1961).

a prolate or an oblate ellipsoid) and arrives at a frictional ratio reflecting the extra surface presented by the rugosities. One can use either of these axial ratios (denoted 1 and 2), by way of frictional ratios derived from them, to estimate hydrodynamic coefficients, in particular sedimentation coefficients.[20] It turns out that those predicted from the surface-sensitive axial ratios (Method 2) are in excellent agreement with experimental sedimentation coefficients, whereas the more conventional (Method 1) axial ratios are poor predictors.[21]

Establishing Correlations between Hydrodynamic and X-Ray Data

The approach described in the following is largely adapted from the work of the present authors.[21] Its main thrust will be directed toward the utilization of sedimentation coefficients because these are found to furnish the best correlations; diffusion and viscosity correlations will be discussed to the extent the requisite data were available.

The criterion for selection of proteins for the present purpose was the availability of two kinds of data in the literature: (1) the sedimentation coefficient $s_{20,w}^0$, or data which allow it to be calculated, and (2) requisite SAXS data. The latter refers to reported values of (a) the radius of gyration, R_G, and (b) at least two others of the following three parameters: the hydrated volume, V, the surface-to-volume ratio, S/V (required for all proteins in the lower molecular weight range), and the axial ratio or some other shape ratio, depending on the model. These SAXS data are referred to hereafter as the primary parameters for the protein reported, as contrasted to secondary parameters in a particular case, namely, those that could be derived from the primary ones if not reported independently. In addition, the proteins considered here are roughly globular, with no flexibility, as seen by SAXS.

An extensive search of the literature to date has produced a total of 19 globular proteins and 2 spherical viruses that meet the stated criteria (for references, see Table I; this differs from Table I of Ref. 21 by an additional protein, 10a, and a number of corrections and recalculated values, but the original sequence of numbers has been retained). In view of the nearly three decades that the SAXS technique has been available, this is a comparatively small number. It would appear that most SAXS investigators do not determine S/V for solutions of bipolymers on account of severe experimental difficulties caused by the extremely small scattering signal from protein solutions as well as certain instrumental limitations. In fact, only 11 proteins in the data set actually have experimentally determined surface areas. Fortunately for our purpose, 10 of these proteins

[21] T. F. Kumosinski and H. Pessen, *Arch. Biochem. Biophys.* **219**, 89 (1982).

STRUCTURAL AND HYDRODYNAMIC PARAMETERS FROM SAXS

Macromolecule[a]	M[c]	$\bar{v}$,[c] ml/g[d]	Model[b]	R_G, Å	V, Å³	S/V, Å^{-1}	$(a/b)_1$[e]	$(a/b)_2$[f]	$(f/f_0)_1$[g]	$(f/f_0)_2$[h]	s_1, S[i]	s_2, S[j]	From sedimentation $s_{20,w}^0$, S
1. Ribonuclease (bovine pancreas)[k,1]	13,690[l]	0.696[3]	PE	14.8	22,000	0.29	1.87	3.69	1.036	1.161	2.03	1.81	1.78[4]
2. Lysozyme (chicken egg white)[1]	14,310[m]	0.702[3]	PE	14.3	24,200	0.25	1.42	2.92	1.011	1.107	2.07	1.89	1.91[5]
3. α-Lactalbumin (bovine milk)[1]	14,180[n]	0.704[3]	PE	14.5	25,100	0.24	1.43	2.81	1.012	1.099	2.02	1.86	1.92[6]
4. α-Chymotrypsin (bovine pancreas)[7]	22,000[o]	0.736	PE	18.0	37,170[p]	0.157	2.0	2.02	1.044	1.045	2.36	2.36	2.40[8,9]
5. Chymotrypsinogen A (bovine pancreas)[7]	25,000[o]	0.736	PE	18.1	37,790[p]	0.160	2.0	2.12	1.044	1.051	2.67	2.65	2.58[10]
6. Pepsin[11,q]	34,160[12]	0.725[13]	PE	20.5	54,870[p]	0.26	2.0	4.76	1.044	1.234	3.36	2.84	2.88[14]
7. Riboflavin-binding protein, apo (pH 3.0) (chicken egg white)[15]	32,500[o]	0.720[r]	PE	20.6	66,500	0.203	1.63	3.58	1.021	1.153	3.12	2.76	2.76
8. Riboflavin-binding protein, holo (pH 7.0) (chicken egg white)[15]	32,500[o]	0.720[r]	PE	19.8	55,600	0.213	1.76	3.62	1.029	1.156	3.28	2.92	2.92
9. β-Lactoglobulin A dimer (bovine milk)[16]	36,730[s]	0.751[3]	PE	21.6	60,250[t]	0.166[t]	2.13	2.93	1.052	1.108	3.12	2.99	2.87[17]
10. Bovine serum albumin[18]	66,300[u]	0.735[3]	PE	30.6	142,000[w]	0.146	2.90	3.88	1.105	1.174	4.62	4.34	4.30[w]
10a. Lactate dehydrogenase, M₄ (dogfish)[x]	138,320	0.741	OE	34.7[x]	253,300[t]	0.0893[p]	0.409	—	1.069	—	7.50	—	7.54[20]
11. β-Lactoglobulin A octamer (bovine milk)[16]	146,940	0.751[v]	OE	34.4	215,000[t]	0.125[t]	0.347	0.255	1.097	1.162	7.89	7.45	7.38[17]

(continued)

TABLE I (continued)

Macromolecule[a]	Model[b]	Auxiliary parameters M^c; $\bar{v}$, ml/g[d]	SAXS parameters R_G, Å	SAXS parameters V, Å³ / S/V, Å⁻¹	SAXS parameters $(a/b)_1^e$ / $(a/b)_2^f$	Calculated from SAXS $(f/f_0)_1^g$ / $(f/f_0)_2^h$	Calculated from SAXS s_1, S[i] / s_2, S[j]	From sedimentation $s_{20,w}^0$, S
12. Glyceraldehyde-3-phosphate dehydrogenase, apo (bakers' yeast)[21]	HOC	142,870[z] 0.737[z]	32.1	264,200 0.0995[p]	0.636 0.389	1.018 1.078	8.15 7.70	7.6[23]
13. Glyceraldehyde-3-phosphate dehydrogenase, holo (bakers' yeast)[21]	HOC	145,520[aa] 0.737[r]	31.7	250,000 0.1016[p]	0.614 0.384	1.024 1.080	8.46 7.97	8.0[23]
14. Malate synthase (bakers' yeast)[24]	OE	170,000[v,25] 0.735	39.6	338,000 0.0843[p]	0.363 —	1.089 —	8.40 —	8.25[25]
15. Pyruvate kinase, apo[bb] (brewers' yeast)[26]	OEC	190,800[cc] 0.734[27]	43.5	406,000 0.0879[p]	0.321 0.298	1.112 1.127	8.70 8.62	8.70[27]
16. Pyruvate kinase, holo[bb] (brewers' yeast)[26]	OEC	192,160[aa] 0.734[r]	42.5	406,000 0.0855[p]	0.349 0.320	1.096 1.113	8.92 8.80	8.81[27]
17. Catalase (bovine liver)[28]	PC	248,000[29] 0.730[3]	39.8	420,000 0.0752[p]	1.91 2.24	1.038 1.060	12.20 11.96	11.3[29]
18. Glutamate dehydrogenase (bovine liver)[30]	PC	312,000[31] 0.749[31]	47.0	668,000 0.0648[p]	1.98 2.30	1.043 1.064	12.18 11.93	11.4[31]
19. Turnip yellow mosaic virus[32]	S	4.97×10^6[33] 0.666[33]	108[p]	11.49×10^6[p] 0.0214[p]	1.0 —	1.0 —	104 —	106[33]
20. Southern bean mosaic virus[34]	S	6.63×10^6[35] 0.696[35]	111[p]	12.25×10^6[p] 0.0210[p]	1.0 —	1.0 —	124 —	115[35]

a Superscript numerals following entries indicate references as listed below. Tabulated data were taken from the references thus designated in the first column, unless noted otherwise for a particualr parameter.

b Geometric model used to describe scattering particle: PE, prolate ellipsoid; OE, oblate ellipsoid; PC, prolate (elongated) cylinder; OEC, oblate (flattened) elliptical cylinder; HOC, hollow oblate cylinder; S, sphere.

c Molecular weights, by preference, were based on amino acid compositions and sequences wherever available, except in some cases where ...

Partial specific volumes were the cited authors' values or, in some cases, more accurate values found in the literature. Corrections for temperature differences between 25 and 20° were not in general made for $\bar{v}$ because resulting differences in $s_{20,w}^0$ are minimal and do not affect comparisons between the different s values.

[e] From Eq. (3a). Prolate or oblate cylinders were modeled as equivalent prolate or oblate ellipsoids, respectively.
[f] From Eq. (3b) or (3c). Cylinder modeled as in Note [e].
[g] From Eq. (2a) or (2b), based on Eq. (3a).
[h] From Eq. (2a), based on Eq. (3b); or Eq. (2b), based on Eq. (3c).
[i] From Eqs. (1a) and (1b), based on Eq. (3a).
[j] From Eqs. (1a) and (1b), based on Eq. (3b) or (3c).
[k] Key to references:

[1] H. Pessen, T. F. Kumosinski, and S. N. Timasheff, *J. Agric. Food Chem.* **19**, 698 (1971).
[2] M. O. Dayhoff, ed., "Atlas of Protein Sequence and Structure," Vol. 5. Natl. Biomed. Res. Found., Washington, D.C., 1972.
[3] J. C. Lee and S. N. Timasheff, *Biochemistry* **13**, 257 (1974).
[4] D. A. Yphantis, *J. Phys. Chem.* **63**, 1742 (1959).
[5] A. J. Sophianopoulos, C. K. Rhodes, D. N. Holcomb, and K. E. Van Holde, *J. Biol. Chem.* **237**, 1107 (1962).
[6] M. J. Kronman and R. E. Andreotti, *Biochemistry* **3**, 1145 (1964).
[7] W. R. Krigbaum and R. W. Godwin, *Biochemistry* **7**, 3126 (1968).
[8] G. W. Schwert, *J. Biol. Chem.* **179**, 655 (1949).
[9] G. W. Schwert and S. Kaufman, *J. Biol. Chem.* **190**, 807 (1951).
[10] P. E. Wilcox, J. Kraut, R. D. Wade, and H. Neurath, *Biochim. Biophys. Acta* **24**, 72 (1957).
[11] A. A. Vazina, V. V. Lednev, and B. K. Lemazhikin, *Biokhimiya (Moscow)* **31**, 629 (1966).
[12] T. G. Rajagopalan, S. Moore, and W. H. Stein, *J. Biol. Chem.* **241**, 4940 (1966).
[13] T. L. McMeekin, M. Wilensky, and M. L. Groves, *Biochem. Biophys. Res. Commun.* **7**, 151 (1962).
[14] R. C. Williams, Jr. and T. G. Rajagopalan, *J. Biol. Chem.* **241**, 4951 (1966).
[15] T. F. Kumosinski, H. Pessen, and H. M. Farrell, Jr., *Arch. Biochem. Biophys.* **214**, 714 (1982).
[16] J. Witz, S. N. Timasheff, and V. Luzzati, *J. Am. Chem. Soc.* **86**, 168 (1963).
[17] T. F. Kumosinski and S. N. Timasheff, *J. Am. Chem. Soc.* **88**, 5635 (1966).
[18] V. Luzzati, J. Witz, and A. Nicolaieff, *J. Mol. Biol.* **3**, 379 (1961).
[19] G. L. Miller and R. H. Golder, *Arch. Biochem. Biophys.* **36**, 249 (1952).
[20] A. Pesce, T. P. Fondy, F. Stolzenbach, F. Castillo, and N. O. Kaplan, *J. Biol. Chem.* **242**, 2151 (1967).
[21] H. Durchschlag, G. Puchwein, O. Kratky, I. Schuster, and K. Kirschner, *Eur. J. Biochem.* **19**, 9 (1971).
[22] R. Jaenicke, D. Schmid, and S. Knof, *Biochemistry* **7**, 919 (1968).
[23] R. Jaenicke and W. B. Gratzer, *Eur. J. Biochem.* **10**, 158 (1969).
[24] D. Zipper and H. Durchschlag, *Eur. J. Biochem.* **87**, 85 (1978).
[25] G. Schmid, H. Durchschlag, G. Biedermann, H. Eggerer, and R. Jaenicke, *Biochem. Biophys. Res. Commun.* **58**, 419 (1974).

(continued)

References to TABLE 1 (continued)

[26] K. Müller, O. Kratky, P. Röschlau, and B. Hess, Hoppe-Seyler's Z. Physiol. Chem. **353**, 803 (1972).

[27] H. Bischofberger, B. Hess, and P. Röschlau, Hoppe-Seyler's Z. Physiol. Chem. **352**, 1139 (1971).

[28] A. G. Malmon, Biochim. Biophys. Acta **26**, 233 (1957).

[29] J. B. Sumner and N. Gralén, J. Biol. Chem. **125**, 33 (1938).

[30] I. Pilz and H. Sund, Eur. J. Biochem. **20**, 561 (1971).

[31] E. Reisler, J. Pouyet, and H. Eisenberg, Biochemistry **9**, 3095 (1970).

[32] P. W. Schmidt, P. Kaesberg, and W. W. Beeman, Biochim. Biophys. Acta **14**, 1 (1954).

[33] R. Markham, Discuss. Faraday Soc. **11**, 221 (1951).

[34] B. R. Leonard, Jr., J. W. Anderegg, S. Shulman, P. Kaesberg, and W. W. Beeman, Biochim. Biophys. Acta **12**, 499 (1953).

[35] G. L. Miller and W. C. Price, Arch. Biochem. Biophys. **10**, 467 (1946).

[l] From Dayhoff (Ref. 2, this table), p. D-130.

[m] From Dayhoff (Ref. 2, this table), p. D-138.

[n] From Dayhoff (Ref. 2, this table), p. D-136.

[o] Value reported by cited authors (see Note c).

[p] Secondary parameter, calculated with use of indicated model from values of primary parameters of cited authors (see under Selection of Proteins).

[q] Origin of preparation not stated.

[r] Value for apoenzyme used, since $\bar{v}$ for holoenzyme not available.

[s] From M. O. Dayhoff, ed., "Atlas of Protein Sequence and Structure," Suppl. 1, p. S-83. Natl. Biomed. Res. Found., Washington, D.C., 1973.

[t] Unpublished data of authors of Ref. (16), this table (S. N. Timasheff, personal communication).

[u] From M. O. Dayhoff, ed., "Atlas of Protein Sequence and Structure," Suppl. 2, p. 267. Natl. Biomed. Res. Found., Washington, D.C., 1976.

[v] This value of molecular volume appears to be high, as was the molecular weight of 81,200 reported by the listed authors, pointing to the possible presence of aggregation products. For s_1 and s_2, listed in the table, the amino acid sequence molecular weight was used, together with a proportionally adjusted volume of 115,940. The inconsistent use of $V = 142,000$ with $M = 66,300$ would result in $s_1 = 4.43$ and $s_2 = 4.06$.

[w] From Miller and Golder (Ref. 19, this table). The value reported in Ref. 18, this table, is 4.1 at 0.75%. Allowing for the concentration dependence according to Ref. 19, the two values are equivalent.

[x] Unpublished data of H. Pessen, T. F. Kumosinski, G. S. Fosmire, and S. N. Timasheff.

[y] Value for 9 (dimer) used, since $\bar{v}$ for octamer not available.

[z] From Dayhoff (Ref. 2, this table), pp. D-147, D-148.

[aa] Calculated from value for apoenzyme.

[bb] The designations "apo" and "holo," although not strictly correct in this case, are used for brevity. They refer to "native" and "fructose diphosphate liganded," respectively.

have molecular weights less than 100,000 and thus could be expected to show the effects of rugosity. The remaining ones have molecular weights greater than 100,000, where it could be expected that the rugosity would make relatively little contribution to the structural portion of the frictional ratio. Shape information, however, is available for these proteins from SAXS results in the high-angle region. In fact, all the high-molecular-weight proteins have been found to be approximated best by cylinders (either prolate or oblate), with the exception of lactate dehydrogenase, β-lactoglobulin octamer, and malate synthase; these resemble oblate ellipsoids of revolution.

Theory

For the frictional ratio we make use of a decomposition, due to Oncley,[1] of the total ratio, f/f_0, into shape- and hydration-dependent factors, f_e/f_0 and f/f_e, respectively. The form of Svedberg's equation to be used to calculate theoretical sedimentation coefficients $s^0_{20,w}$ from SAXS structural parameter is[22]

$$s^0_{20,w} = \frac{M(1 - \bar{v}\rho)}{(f/f_0)\, 6\pi\eta N r_0} \tag{1a}$$

where the subscript "20,w" denotes reference to water at 20°, M is the anhydrous molecular weight obtained from the amino acid sequence or composition whenever possible, $\bar{v}$ is the partial specific volume of the protein (calculated or, preferably, experimental), ρ is the density and η is the viscosity of water at 20°, and N is Avogadro's number. It should be noted that for our calculations r_0, the Stokes radius (in cm), will be related to the scattering volume V of the hydrated macromolecule (in cm³) instead of the more customary $\bar{v}$, by the relationship

$$r_0 = (3V/4\pi)^{1/3} \tag{1b}$$

Since the scattering volume, in contrast to $\bar{v}$, already reflects the hydrated molecule, the corresponding frictional ratio is really f_e/f_0, although it was written above (and for simplicity will continue to be written in the following) as f/f_0.

The frictional ratio f/f_0, then, is here the structural factor of the total frictional ratio for the hydrated particle. We model all molecules as prolate or oblate ellipsoids of revolution[11]:

$$\frac{f}{f_0} = \frac{(p^2 - 1)^{1/2}}{p^{1/3}\ln[p + (p^2 - 1)^{1/2}]}, \qquad (p > 1, \text{ prolate}) \tag{2a}$$

[22] T. Svedberg and K. O. Pedersen, "The Ultracentrifuge," p. 22. Oxford Univ. Press (Clarendon), London and New York, 1940.

$$\frac{f}{f_0} = \frac{(1 - p^2)^{1/2}}{p^{1/3} \tan^{-1}[(1 - p^2)^{1/2}/p]}, \qquad (p < 1, \text{ oblate}) \qquad (2b)$$

where p equals a/b, b is the equatorial radius, and a is the semi-axis of revolution of the ellipsoid. (The usage of $p = a/b$ is in agreement with that of Luzzati and co-workers[20]; note that this p is the reciprocal of the p defined by Teller et al.[13]) The axial ratios p were determined from SAXS parameters by the method of Luzzati,[20] with the use of either of two dimensionless ratios, r_1 and r_2, defined as follows (where V is the volume of the macromolecule, R_G is the radius of gyration, and S is the external surface area):

$$r_1 \equiv \frac{3V}{4\pi R_G^3} = p \Big/ \left(\frac{p^2 + 2}{5}\right)^{3/2}, \qquad (p \gtrless 1) \qquad (3a)$$

and

$$r_2 \equiv R_G \frac{S}{V} = \frac{3}{2p} \left[1 + \frac{p^2}{(p^2 - 1)^{1/2}} \sin^{-1} \frac{(p^2 - 1)^{1/2}}{p} \right] \left(\frac{p^2 + 2}{5}\right)^{1/2}, \qquad (p > 1) \qquad (3b)$$

or

$$r_2 \equiv R_G \frac{S}{V} = \frac{3}{2p} \left[1 + \frac{p^2}{(1 - p^2)^{1/2}} \tanh^{-1} (1 - p^2)^{1/2} \right] \left(\frac{p^2 + 2}{5}\right)^{1/2}, \qquad (p < 1) \qquad (3c)$$

These equations incorporate the geometric relationships $V = (4/3)\pi ab^2$, $R_G = [(a^2 + 2b^2)/5]^{1/2}$, and the expressions for S in terms of a and b for prolate and oblate ellipsoids of revolution, respectively. The ratios r_1 and r_2 may be seen from the limiting case of a sphere to be subject to the constraints $r_1 \leq 2.152$ and $r_2 \geq 2.324$.

Regarding the evaluation of p from Eq. (2), when f/f_0 is known, or from Eq. (3), when r_1 or r_2 is known, it may be noted that no closed expression for p is available. For this reason the earlier literature (p. 326 of Ref. 3; 12) made use of plots of these functions (readily calculated for p as the independent variable), which then permitted graphical evaluation to a limited precision. Greater precision could be obtained from compilations of these values, involving somewhat voluminous tables or, if more compact (cf. p. 41 of Ref. 22), requiring interpolation. With the availability of computers, or even programmable desk calculators, it is a relatively simple matter to program an iterative algorithm which can rapidly evaluate p to the desired precision for any value of f/f_0, r_1, or r_2 within the domains of these variables.

It is to be emphasized that f/f_0 and r_0 are derived from solution structural parameters without any assumption regarding the contribution of hydration to the frictional ratio; also, no assumption is necessary concerning the symmetric or asymmetric placement of the water molecules, or concerning electrostriction effects, in contrast to the use of three-dimensional X-ray crystallographic structures for correlation with sedimentation data of globular proteins, where such assumptions cannot be avoided.[23]

Correlations between Sedimentation and X-Ray Diffraction Data

To ensure that our special selection criteria have resulted in a set of data not very different from those for globular proteins in general, we first test our set of 21 globular macromolecules against those of Squire and Himmel[12] and Teller *et al.*[13] (selected for a different purpose and according to different criteria), as suggested by the Svedberg relationship[22] for spherical molecules in the form

$$s_{20,w}^0 = [M^{2/3}(1 - \bar{v}\rho)/\bar{v}^{1/3}](3\pi^2 N^2/4)^{-1/3}(6\eta)^{-1} \tag{4}$$

where all parameters have been previously defined. A plot of $s_{20,w}^0$ vs $M^{2/3}(1 - \bar{v}\rho)/\bar{v}^{1/3}$ is shown in Fig. 1 for all 21 macromolecules. Fitting a least-squares straight line with zero intercept to all points gives a slope of 0.00950 ± 0.00002 S cm g^{-1} mol$^{2/3}$. (The 19 proteins alone give a line of slope 0.00931 ± 0.00009.) Also shown in Fig. 1 is the theoretical line for molecules considered as smooth spheres, which constitutes an upper limit of slope 0.0120 in the same units, obtained from Eq. (4) by evaluation of the collection of constants.[13] Squire and Himmel[12] and Teller *et al.*[13] obtained the equivalents of slopes of 0.0108 and 0.010 for their respective sets of proteins. These values are not greatly different from those above. One may assume, therefore, that our set has approximately the same average rugosity as other globular proteins. This statistical correlation is purely empirical and has no structural foundation; frictional coefficients are not explicitly considered.

Frictional ratios may be introduced into this approach by means of the relationships developed by Teller[24] between accessible surface area A_s, packing volume V_p, radius R_p from the packing volume, and molecular weight M, which were derived by calculations based on the X-ray crystallographic structures of a set of proteins first used by Chothia.[25] The relationships are

[23] F. M. Richards, *Annu. Rev. Biophys. Bioeng.* **6,** 151 (1977).
[24] D. C. Teller, *Nature (London)* **260,** 729 (1976).
[25] C. Chothia, *Nature (London)* **254,** 304 (1975).

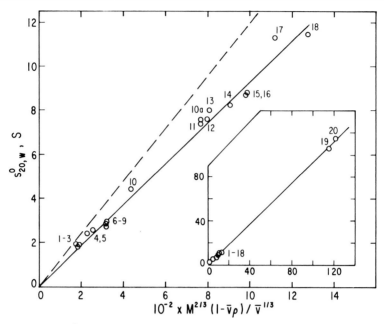

FIG. 1. Plot of $s_{20,w}^0$ vs the function $M^{2/3}(1 - \bar{v}\rho)/\bar{v}^{1/3}$ for the 19 proteins, as numbered in Table I. Solid line: linear least-squares fit, with slope 0.00931 ± 0.00009. Dashed line: theoretical upper limit line expected for proteins considered spherical, with slope 0.0120. Inset: corresponding points and lines for 21 biopolymers, including two viruses in addition to the 19 proteins of the main figure; scales in same units; slope 0.00950 ± 0.00002. (Adapted from Fig. 1 of Ref. 21.)

$$A_s = 11.12 \pm 0.16\ M^{2/3} \qquad (\text{in } \mathring{A}^2) \qquad (5a)$$

and

$$V_p = 1.273 \pm 0.006\ M \qquad (\text{in } \mathring{A}^3) \qquad (5b)$$

(5b) is equivalent to

$$R_p = 0.672 \pm 0.001\ M^{1/3} \qquad (\text{in } \mathring{A}) \qquad (5c)$$

by reason of $V_p = (4/3)\pi R_p^3$. R_p is related to the radius of gyration by $R_G = (3/5)^{1/2}R_p$, as may be verified from Eq. (3a), with $p = 1$.

From these expressions, axial ratios for prolate or oblate ellipsoids of revolution can be calculated by means of Eqs. (3a–c). (S and V here are represented by A_s and V_p, respectively, although it should be realized that these are rough approximations only, and that V_p, in particular, is not a hydrated volume.) The molecular weight cancels out for both $3V/(4\pi R_G^3)$ (the smooth-surface model) and $R_G S/V$ (the rugose-surface model), as it

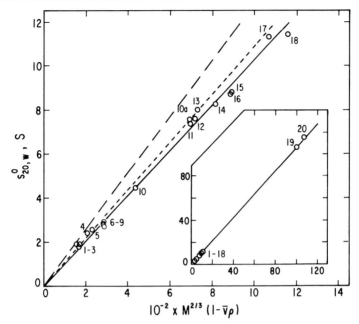

FIG. 2. Plot of $s_{20,w}^0$ vs the function $M^{2/3}$ $(1 - \bar{v}\rho)$ for the 19 proteins, as numbered in Table I. Solid line: linear least-squares fit, with slope 0.01030 ± 0.00010; dashed line: theoretical for smooth-surface models; dotted line: theoretical for rugose-surface models. Inset: corresponding plot and line including two viruses; scales in same units as main figure; slope 0.01079 ± 0.00003. (Adapted from Fig. 2 of Ref. 21.)

must, these expressions being dimensionless. Since Eq. (5c) is based on a spherical model,[24] use of $3V/(4\pi R_G^3)$ here will necessarily result in axial ratios of 1 for the smooth model. The information contained in S, however, is independent of the assumption of such a model and will, therefore, permit calculation of equivalent axial ratios from $R_G S/V$ (prolate: 3.96; oblate: 0.238), and thus frictional ratios from Eqs. (2a) and (2b) (prolate: 1.180; oblate: 1.178). Equations (1a,b), with V again used in place of $\bar{v}$ as a measure of the Stokes radius r_0, then yield

$$s_{20,w}^0 = M^{2/3}(1 - \bar{v}\rho)k \times 10^{-13} \qquad (6)$$

where k, the collection of constants in Eq. (4), becomes 0.01284 for all smooth-surface models and about 0.0109 (prolate: 0.01088; oblate: 0.01090) for rugose-surface models.

Figure 2 is a plot of $s_{20,w}^0$ vs $M^{2/3}(1 - \bar{v}\rho)$ for our 21 macromolecules. A straight line with zero intercept fitted to the experimental data yields a value of 0.01078 ± 0.00003 for k in Eq. (6). (For the 19 proteins alone, k

equals 0.01030 ± 0.00010.) Comparison between these experimental values for k and those above, derived from the X-ray crystallographic structures, shows that a prolate or oblate ellipsoid of revolution with an equivalent S/V ratio (rugose-surface model) describes the hydrodynamic behavior of globular proteins to within about 1%, whereas the smooth model is off by nearly 20%. This is in agreement with the conclusions concerning the rugosity of the surface reached by Teller *et al.*,[13] who, as mentioned in the Introduction, used a more exact calculation of the frictional coefficient, with data from a different set of proteins.

Estimation of Sedimentation Coefficients from SAXS

With these considerations in mind, we turn to using SAXS results in an attempt to predict sedimentation coefficients. In Table I, the radius of gyration, R_G, volume, V, and surface-to-volume ratio, S/V, are listed for our set of 21 macromolecules. Also tabulated are the partial specific volumes, $\bar{v}$, and the anhydrous molecular weights, M (obtained in most cases from the amino acid sequence or composition), as well as indications of the geometric model which best describes the scattering particle as determined by SAXS. It should be noted, however, that experimental values of S/V are available only for proteins 1 through 11. Values for the other macromolecules had to be calculated from their smooth-surface model. Axial ratios calculated for each protein from the SAXS results of Table I and Eq. (3a) for the $3V/(4\pi R_R^3)$ relationship, and Eqs. (3b) or (3c) (as the case may be) for the $R_G S/V$ relationship, are given as $(a/b)_1$ and $(a/b)_2$, respectively.

For each of the proteins for which both values are available, $(a/b)_2$ is larger than $(a/b)_1$ when the model is prolate, the reverse if oblate. The differences become somewhat less as the molecular weight of the protein increases; this would be consistent with the notion that flow lines are influenced by the rugae (which presumably remain of about constant average dimensions) to a lesser extent as the volume of the particle increases. Frictional ratios for $(a/b)_1$ and $(a/b)_2$, calculated from Eqs. (2a) and (2b), are listed as $(f/f_0)_1$ and $(f/f_0)_2$, respectively. It should be recalled that the assumption was made that all proteins can be approximated by spherical, prolate, or oblate ellipsoidal models. This assumption is least exact for proteins 12, 13, and 15–18, which are more nearly cylinders; however, it is still a useful approximation and generally considered reasonable.[18] From the molecular weights, partial specific volumes, and frictional ratios, one can obtain sedimentation coefficients for the smooth-surface (s_1) and rugose-surface (s_2) models by means of Svedberg's equation [Eq. (1a)], with the Stokes radius in this equation calculated from the scattering volume listed in the table. These values as well as the experimentally determined $s_{20,w}^0$ are given in Table I.

It is seen that, whereas s_1 values are consistently larger than $s_{20,w}^0$, s_2 generally is very close to $s_{20,w}^0$, in agreement with Teller's conclusion that the hydrodynamic behavior of proteins is influenced by the rugose accessible surface area.[13] The agreement between s_2 and $s_{20,w}^0$ is particularly remarkable for the holo- and apo-forms of several proteins in this data set, viz. riboflavin-binding protein and glyceraldehyde-3-phosphate dehydrogenase. In these cases, $s_{20,w}^0$ values change owing to some configurational change in the protein, and the calculated s_2 values evidently follow these changes quite faithfully.

It must be noted that in cases 12–20 the differences, if any, between $(a/b)_1$ and $(a/b)_2$ [and consequently between $(f/f_0)_1$ and $(f/f_0)_2$, and between s_1 and s_2] are not due to rugosity since, in the absence of experimental S/V values, the rugosity could not be taken into account. Instead, S/V in cases 10a, 14, 19, and 20 was calculated from models of smooth ellipsoids or spheres, so that the information content of $R_G S/V$ must be identical to that of $3V/(4\pi R_G^3)$, and only one axial ratio and one s is calculated and listed (designated here as s_1, as the designation s_2 would incorrectly imply that an independent S/V was involved). In cases of other smooth bodies, such as cylinders (Nos. 12, 13, 15–18), there will be a difference between $(a/b)_1$ and $(a/b)_2$, and thus between s_1 and s_2, because these bodies have been represented by ellipsoids of equal volume, for the sole reason that frictional ratios for ellipsoids can be readily calculated by means of Perrin's equations. These differences will not, therefore, reflect rugosity but the excess surface due to the difference in model (elsewhere[14] termed S_B, the excess surface due to body shape other than ellipsoidal, as distinguished from S_X, the additional contribution to surface area due to rugose surface texture). To the extent that this additional surface affects hydrodynamic properties, s_2 in these cases also should afford the better estimate of $s_{20,w}^0$.

In a few instances the agreement between s_2 and $s_{20,w}^0$, while still very satisfactory, is less striking than in the majority of the cases. In 4 and 5, the molecular weights reported by the authors were somewhat lower than values from known amino acid composition, so that the possibility of partial autoproteolysis cannot be excluded, with unknown consequences for the SAXS values. In 9, we are dealing with a known dimer, which might be more accurately represented by an elongated, rounded cylinder than by a prolate ellipsoid. Altogether, however, when it is considered that these SAXS data were compiled from scattered and sometimes fragmentary sources ranging over a period of nearly three decades—obtained by a variety of observers, of varying familiarity with the technique, and

TABLE II

VISCOSITY AND DIFFUSION

Protein	$[\eta]$, ml/g	$D^0_{20,w} \times 10^7$	$(a/b)_1$	$(a/b)_{\text{visc}}$	$(a/b)_{\text{diff}}$	$(a/b)_2$
1. Ribonuclease	3.30^a	10.68^i	1.87	2.56	3.46	3.69
2. Lysozyme	2.5^b	10.4^j	1.42	1.55	3.41	2.92
3. α-Lactalbumin	3.01^c	10.57^e	1.43	1.82	3.21	2.81
5. Chymotrypsinogen A	2.5^d	9.5^k	2.0	1.7	2.47	2.12
6. Pepsin	3.93^e	9.0^e	2.0	3.4	1.4	4.76
9. β-Lactoglobulin dimer	3.4^f	7.82^l	2.13	2.7	3.1	2.92
10. Bovine serum albumin	3.69^g	$6.16^{e,m}$	2.55	3.8	3.97	4.18
17. Catalase	3.9^h	4.1^n	1.91	3.1	3.05	2.24

^{a} J. G. Buzzell and C. Tanford, *J. Phys. Chem.* **60**, 1204 (1956).
^{b} J. Léonis, *Arch. Biochem. Biophys.* **65**, 182 (1956).
^{c} D. B. Wetlaufer, *C. R. Trav. Lab. Carlsberg* **32**, 125 (1961).
^{d} C. Tanford, K. Kawahara, S. Lapanje, T. M. Hooker, Jr., M. H. Zarlengo, A. Salahuddin, K. C. Aune, and T. Tagahaki, *J. Am. Chem. Soc.* **89**, 5023 (1967).
^{e} A. Polson, *Kolloid-Z.* **88**, 51 (1939).
^{f} L. G. Bunville, Ph.D. Thesis, State University of Iowa, Ames (1959).
^{g} C. Tanford and J. G. Buzzell, *J. Phys. Chem.* **60**, 225 (1956).
^{h} R. E. Lovrien, Ph.D. Thesis, State University of Iowa, Ames (1958).
^{i} J. M. Creeth, *J. Phys. Chem.* **62**, 66 (1958).
^{j} J. R. Colvin, *Can. J. Chem.* **30**, 831 (1952).
^{k} G. W. Schwert, *J. Biol. Chem.* **190**, 799 (1951).
^{l} R. Cecil and A. G. Ogston, *Biochem. J.* **44**, 33 (1949).
^{m} R. L. Baldwin, L. J. Gosting, J. W. Williams, and R. A. Alberty, *Discuss. Faraday Soc.* **20**, 13 (1955).
^{n} J. B. Sumner and N. Gralén, *J. Biol. Chem.* **125**, 33 (1938).

using different instruments of several different types and different methods of data evaluation—the agreement shown in Table I is all the more remarkable.

Correlation of Diffusion Coefficients and Intrinsic Viscosities with SAXS Data

The above method for calculation of sedimentation coefficients from SAXS results may be useful in the calculation of other hydrodynamic quantities. Table II presents values of axial ratios $(a/b)_{\text{diff}}$ derived from experimental diffusion coefficients and the use of SAXS volumes from Table I, along with the smooth-surface axial ratio, $(a/b)_1$, and rugose-surface axial ratio, $(a/b)_2$, of Table I. Even though large errors frequently exist in experimental diffusion coefficients, the axial ratios derived from them are seen to be mostly closer in magnitude to the rugose-surface than to the smooth-surface axial ratio. [In fact, some are larger than the rugose-surface axial ratio, most likely because of experimental error in the diffusion coefficient. The value for pepsin is very low, probably for the same reason: judging from its molecular weight relative to those of chymotrypsinogen A and β-lactoglobulin dimer (see Table I), the single value of the diffusion coefficient recorded for pepsin in the literature, 9.0, appears much too high; a value nearer 8.2 would be more in accord with those for the other two proteins. This would lead to 2.96 for $(a/b)_{diff}$.] The indication is that linear diffusion depends on the surface characteristics of a particle, as is the case with sedimentation. Determination of axial ratios from scattering volume and the intrinsic viscosity are less straightforward, as seen also in Table II. Here the experimentally derived axial ratio $(a/b)_{visc}$ is closer to the $(a/b)_2$ of Table I for only β-lactoglobulin A (dimer), bovine serum albumin, and catalase. The $(a/b)_{visc}$ for lysozyme, α-lactalbumin, and chymotrypsinogen A is closer to $(a/b)_1$, while ribonuclease and pepsin have $(a/b)_{visc}$ values approximately equidistant between the smooth-surface and rugose-surface axial ratios.

No clear reason for these discrepancies is apparent. However, Kuntz and Kauzmann[5] (pp. 289–306) have also observed that hydration values derived from intrinsic viscosity and sedimentation coefficients show inconsistencies beyond those expected from experimental error. They suggest that discrepancies arise because the hydrodynamic volumes for diffusion and viscous flow are inherently different. While the results presented here do suggest such a difference, resolution of these matters will require the investigation of a larger number of proteins for which SAXS, diffusion, sedimentation, and viscosity data may become available. Meanwhile, when attempting to use sedimentation coefficients in conjunction with intrinsic viscosities to arrive at estimates of axial ratios, surface areas, or hydrated volumes for a particular protein, one should be aware of the possibility that the two methods do not actually measure identical geometric parameters. The applications section of this chapter will present an alternative procedure for estimating the contribution of hydration to the total frictional coefficient in order to obtain the contribution of the axial ratio by itself, and thus estimates of geometric parameters from sedimentation coefficients.

Structural Comparisons of SAXS and X-Ray Diffraction

A comparison of volumes from SAXS with theoretical volumes derived from the X-ray diffraction structure according to Teller[24] is shown in Fig. 3. The SAXS solution volume is seen to be consistently higher than the volume from the crystallographic structure. Fitting a least-squares

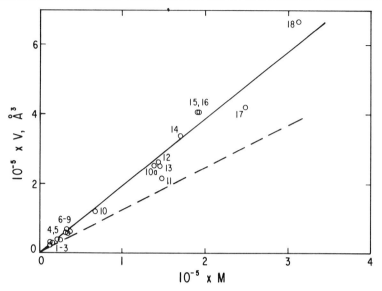

FIG. 3. Plot of scattering volume vs molecular weight for the 19 proteins in Table I. Solid line (SAXS): linear least-squares fit, with slope 1.964 ± 0.045. Dashed line: from X-ray crystallographic structure data [cf. Eq. (5b)], with slope 1.27. (Adapted from Fig. 3 of Ref. 21.)

straight line with zero intercept to the SAXS volume vs molecular weight plot gives a slope of 1.964 ± 0.045, while the corresponding slope for the diffraction data is 1.27.[24]

Further, the SAXS surface area (Fig. 4) can be compared with the accessible surface area according to Teller.[24] Here, the SAXS surface area is slightly lower, and fitting a straight line with zero intercept to the data as a function of $M^{2/3}$ gives a slope of 9.49 ± 0.25, while Teller's value is 11.12. (It may be added that each of the above calculations can also be attempted with a polynominal of degree 2, i.e., with extra terms in M^2 for the volume, and in $M^{4/3}$ for the surface area, but the extra terms are found to result in no statistically significant differences.) The volume of a protein in solution from SAXS, therefore, is found to be larger than the volume from the X-ray crystallographic results, whereas the surface area in solution is slightly lower than the crystallographic accessible surface area. The increase in volume can be expected owing to solvation effects (see this volume [14]); other factors being equal, such an increase would be expected also to yield a correspondingly increased surface area. The contrary decrease in surface area actually observed appears to indicate that the binding of solvent to the macromolecule results in less anisotropy, less rugosity, or a combination of both these effects. In fact, the

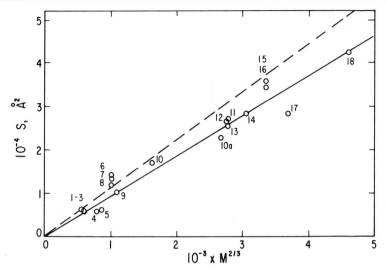

FIG. 4. Plot of surface area from small-angle X-ray scattering vs $\frac{2}{3}$ power of molecular weight for the 19 proteins in Table I. Solid line (SAXS): linear least-squares fit, with slope 9.49 ± 0.25. Dashed line (X-ray diffraction): accessible surface area computed from three-dimensional X-ray structure [cf. Eq. (5a)], with slope 11.12. (Adapted from Fig. 4 of Ref. 21.)

binding sites should lie within some of the rugae or deeper clefts or grooves of the macromolecule. Calculation of $(a/b)_2$ from the fitted SAXS results (i.e., $A_s = 9.49\,M^{2/3}$ and $V = 1.964\,M$), along with the spherical assumption used for X-ray crystallographic data [i.e., $R_G = (3/5)^{1/2}(3V/4\pi)^{1/3}$], yields an average axial ratio for a prolate ellipsoid of revolution of 2, as compared with 3.96 from the X-ray diffraction results. Although this calculation cannot be entirely correct since it can be seen from Table I that the average smooth-surface axial ratio is 1.8 rather than 1, this is the only type of comparison available in view of the lack of literature values for R_G calculated from the X-ray crystallographic structures.

However, the above results with respect to the increase in SAXS volume over the X-ray crystallographic volume could be due also to elec-trostriction of the protein upon crystallization. In fact, the concept of a dynamic alteration of protein conformation in solution ("breathing") has been previously introduced.[18] Whether the observed increase in volume is due to binding of solvent components or to the breathing of the macromol-ecule cannot be resolved without extensive additional studies. These would include sedimentation in $H_2^{17}O$ and $H_2^{18}O$ for increased solvent density, and small-angle neutron scattering using $H_2^{17}O$ to avoid the in-creased hydrophobic interactions shown to occur in D_2O.

Further Applications: Calculation of Solution Geometric Parameters

Calculations from Sedimentation Coefficients without SAXS Data (Table III)

In the absence of access to an SAXS instrument, structural parameters of monomeric proteins in solution can be computed with reasonable accuracy from their sedimentation coefficients, anhydrous molecular weights, and partial specific volumes with the aid of the data shown in Figs. 3 and 4. In these figures surface areas and volumes are shown as functions of molecular weight to the $\frac{2}{3}$ power and to the first power, respectively. Although the statistics for the linear least-square fits to the data appear to be good, the direct use of these functions to evaluate geometric parameters from individual experiments is inadvisable because it can lead to large errors.

It has been pointed out[21] that the increase in hydrodynamic volume (which is equivalent to the SAXS volume) over the partial specific volume, $\bar{v}_2$, at moderate salt concentrations is attributable to the hydration of the protein, and that monomeric proteins of molecular weight less than 100,000 have hydration values close to an average of 0.280 g H_2O/g protein, while oligomeric proteins of molecular weight greater than 100,000 have values close to an average of 0.444 g H_2O/g protein. (Recalculated from emended data of Table I of Ref. 21 for Nos. 1–6, 8–10 and for Nos. 12, 14, 15, 17, and 18, respectively.) These average values, in conjunction with the anhydrous molecular weight, yield a reasonable estimate of the particle volume from the relationship

$$V = (M/N)(\bar{v}_2 + A_1\bar{v}_1) \tag{7}$$

where M is the anhydrous molecular weight, N is Avogadro's number, $\bar{v}_2$ is the partial specific volume of the particle, A_1 is the hydration value, and $\bar{v}_1$, the partial specific volume of bulk water, here may be taken as 1 g/ml.

For the surface area calculation for proteins of molecular weight less than 100,000 one can use the solution value from SAXS, $S = 9.49 \pm 0.25$ $M^{2/3}$. However, on examination of Fig. 4 it is evident that for some of these proteins the surface area might come closer to the accessible surface area value, $11.12\ M^{2/3}$. This discrepancy must be taken into account when shape factors are calculated from sedimentation values. For this reason this analysis is divided into separate procedures: one for globular monomeric proteins of molecular weight less than 100,000 (Table IIIA, where both the accessible surface area and the solution surface area must be included in the considerations), and one for oligomeric proteins of molecular weight greater than 100,000 (where only the solution surface area can be used).

For Monomers. Since monomeric proteins of molecular weight less than 100,000 generally can be represented by prolate ellipsoids of revolution with an average smooth-surface axial ratio of 1.8 (see Table I), the criterion can be adopted that any calculated smooth-surface axial ratio greatly (e.g., more than 25%) in excess of this value is prima facie not reasonable and therefore not to be accepted. Assuming the appropriate average of 0.280 g/g for the extent of hydration, one can estimate the hydrated volume V from $\bar{v}_2$ and M by Eq. (7). From the $s_{20,w}^0$, a frictional coefficient $(f/f_0)_2$ can be estimated by use of Eqs. (1a) and (1b), and from this a rugose axial ratio $(a/b)_{2,\text{K-P}}$ by Eq. (2a). Substitution of this ratio for p in Eq. (3b) gives $r_2 \equiv R_G S/V$. S can be estimated either by Eq. (5a) from the accessible surface area ($S = 11.12\, M^{2/3}$) or from the solution area ($S = 9.49\, M^{2/3}$), and corresponding values of R_G can be calculated from Eq. (3b). Table IIIA shows the results of such calculations performed on proteins 1–10 of Table I (with the exception of apo-RBP at pH 3.0), $R_{\text{K-P}}$ and R_{as} being the radii of gyration thus obtained from solution surface areas (this chapter) and from accessible surface areas, respectively. To determine which radius of gyration is more nearly correct, smooth-surface axial ratios [$(a/b)_{1,\text{K-P}}$, derived from solution surface area, and $(a/b)_{1,as}$, derived from accessible surface area] are evaluated from V and each R_G by Eq. (3a). Since values of $3V/(4\pi R_G^3)$ greater than 2.15 are geometrically meaningless for ellipsoids of revolution (noted under the respective axial ratios in Table III as NO, for "not obtainable"), one chooses the R_G value for which $(a/b)_1$ is not impossible or, if both are possible, the lower value. If both $(a/b)_{1,\text{K-P}}$ and $(a/b)_{1,as}$ have been designated as NO, it is to be assumed that $(a/b)_{2,\text{K-P}}$ does not differ from $(a/b)_{1,\text{K-P}}$ because the molecule has a relatively smooth surface and is best represented by a radius of gyration R_{sm}, calculated from the value of $3V/(4\pi R_G^3)$, obtained in this instance from the rugose axial ratio $p = (a/b)_{2,\text{K-P}}$ by Eq. (3a). This is the case with chymotrypsinogen and α-chymotrypsin, which both have essentially the same $(a/b)_1$ and $(a/b)_2$ values (Table I). The results of this procedure can be verified in Table III, where the calculated radii of gyration in italics are in good agreement with the experimental values from SAXS. (An exception is pepsin, whose surface area from SAXS is also much larger than that from accessible surface.) It is seen also that calculated rugose-surface [$(a/b)_{2,\text{K-P}}$] as well as smooth-surface [$(a/b)_{1,\text{K-P}}$ or $(a/b)_{1,as}$] values are in reasonable agreement with axial ratios determined experimentally by SAXS.

For Oligomers. For oligomeric proteins of molecular weight greater than 100,000, radii of gyration and smooth-surface axial ratios can be calculated as for monomers. Here, however, only the solution surface area relationship, $S = 9.49\, M^{2/3}$, can be used. Whether the particle is a

TABLE III
STRUCTURAL CALCULATIONS FROM $s_{20,w}^0$

Protein	R_{K-P}, Å [a] / R_{as}, Å [b]	R_{sm}, Å [c]	R_{exp}, Å [d]	$(a/b)_{2,K-P}$ [a]	$(a/b)_{2,exp}$ [d]	$(a/b)_{1,K-P}$ [a] / $(a/b)_{1,as}$ [b]	$(a/b)_{1,exp}$ [d]
A. Globular proteins, M < 100,000							
1. Ribonuclease	18.6		14.8	3.93	3.69	3.24	1.87
	15.8					2.24	
2. Lysozyme	15.0		14.3	2.90	2.90	1.83	
	12.7					NO[e]	
3. α-Lactalbumin	13.8		14.5	2.66	2.81	1.11	1.43
	11.7					NO	
4. α-Chymotrypsin	13.2		18.0	1.72	2.02	NO	2.0
	11.2					NO	
5. Chymotrypsinogen A	14.5	17.2	18.1	1.94	2.12	NO	2.0
	12.3					NO	
6. Pepsin	27.8	18.6	20.5	4.25	4.76	3.81	2.0
	23.6					2.74	
8. RBP, holo and apo at pH 7.0[f]	24.7		19.8	3.78	3.62	3.14	1.76
	20.9					2.14	
9. β-Lactoglobulin dimer	23.8		21.6	3.31	2.93	2.61	2.13
	20.2					1.69	
10. Bovine serum albumin[g]	35.2		30.6	4.27	3.88	3.88	2.90
	29.8					2.81	

B. Globular proteins, $M > 100{,}000$[h]

Protein	$(a/b)^k$	V_{K-P}, Å³	V_{exp}, Å³	R_{K-P}, Å	R_{as}, Å	R_{exp}, Å	$(a/b)_{exp}$
10a. Lactate dehydrogenase, M_4	NO	0.514	0.409	29.6		34.7	0.409
11. β-Lactoglobulin A octamer	0.650	0.438	0.255	33.1	*33.8*	34.4	0.347
12. Glyceraldehyde-3-phosphate dehydrogenase, apo	0.536	0.406	0.389	33.8		32.1	0.636
14. Malate synthase	0.280[j]	0.314	0.363	42.4	*41.0*	39.6	0.363
15. Pyruvate kinase, apo[i]	0.195[j]	0.270	0.298	49.2	*44.4*	43.5	0.321
17. Catalase	1.72	2.39	2.24	40.4		39.8	1.91
18. Glutamate dehydrogenase	3.37	3.43	2.30	57.6		47.0	1.98

C. Spherical viruses

Protein	$(a/b)^k$	V_{K-P}, Å³	V_{exp}, Å³	R_{K-P}, Å	R_{exp}, Å
19. Turnip yellow mosaic virus	1.0	10.91×10^6	11.49×10^6	106.5	108
20. Southern bean mosaic virus	1.0	15.30×10^6	12.25×10^6	119	111

[a] Subscript "K-P" refers to calculations as described under Applications.

[b] Subscript "as" and corresponding figures refer to calculations based on accessible surface.

[c] Subscript "sm" refers to calculations based on smooth-surface model.

[d] Subscript "exp" refers to experimental values from Table I.

[e] NO = "not obtainable," because corresponding $r_1 > 2.15$ (see discussion under Monomers).

[f] The two forms have been found to have the same value of $s^0_{20,w}$ at pH 7.0 (Ref. 14). The data under No. 7 of Table I refer to the apo form at pH 3.0, where the riboflavin is released.

[g] Hydration of 0.280 assumed since BSA is monomeric.

[h] For proteins in Section C of this table, which are all oligomers, an average hydration is taken as 0.444 (see section on Calculations from Sedimentation Coefficients).

[i] Native, as opposed to fructose diphosphate liganded.

[j] Value not usable because less than $(a/b)_2$ (see discussion under Oligomers).

[k] From electron microscopy.

prolate or oblate ellipsoid of revolution or a cylinder cannot be determined without SAXS experimental data, and therefore the radius of gyration R_{K-P} and smooth-surface axial ratio $(a/b)_{1,K-P}$ must be estimated for both prolate and oblate ellipsoids of revolution. The volume again can be calculated from the anhydrous molecular weight, now assuming the appropriate average hydration of 0.444 g H_2O/g protein. As before, from the sedimentation coefficient, with a knowledge of the molecular weight, volume, and partial specific volume, the frictional coefficient and rugose axial ratio can be obtained by Eqs. (1) and (2). The solution surface areas derived in this chapter can be used in the expressions relating surface areas, volume, and radius of gyration to rugose axial ratio, Eqs. (3b) or (3c), to calculate the radius of gyration for either prolate or oblate models. Smooth-surface axial ratios can then be calculated from Eq. (3a). If the parameter r_1 is greater than 2.15 (geometrically meaningless) the rugose-surface axial ratio is assumed equal to the smooth-surface axial ratio, and the radius of gyration R_{sm} is now calculated from $(a/b)_{2,K-P}$ values with the expression $3V/(4\pi R_G^3)$ [Eq. (3a)]. (This is the case for No. 10a.)

Another constraint arises from the relationship between r_1 and r_2, when it is remembered that the point in utilizing r_2 is that it, in contrast to r_1, takes into account the excess surface area due to rugosity by translating it, for hydrodynamic purposes, into an axial ratio that is enhanced in the sense of indicating increased anisotropy.[14] Therefore, $(a/b)_2$ for a prolate ellipsoid can never be smaller than $(a/b)_1$, and vice versa for an oblate ellipsoid. If the calculated values violate one of these constraints (as they do for Nos. 14 and 15), they are not usable (so indicated in the table), nor are the related R_{K-P} and R_{as}. Instead, R_{sm}, calculated as under Monomers, is utilized.

Table IIIB shows the results of such calculations performed on Nos. 10a–18 of Table I, using only their shape, determined by SAXS, as listed in that table. Dogfish lactate dehydrogenase, β-lactoglobulin A octamer, and bakers' yeast glyceraldehyde-3-phosphate dehydrogenase have values of $3V/(4\pi R_G^3)$ greater than 2.15, based on the calculated R_{K-P}. Hence, the smooth-surface axial ratio $(a/b)_{1,K-P}$ must be assumed equal to the rugose-surface axial ratio, from which the italicized R_G is calculated. These derived radii of gyration and axial ratios, shown in Table IIIB, are in fair agreement with the experimental R_G and (a/b) from SAXS in all cases except glutamate dehydrogenase. There, neither the calculated smooth-surface and rugose-surface axial ratios nor the R_G agree with the experimentally derived SAXS values. The sedimentation coefficient predicted from SAXS in Table I for glutamate dehydrogenase, however, is quite close to the experimental value. A value of hydration larger than 0.444 for this protein[21] is a likely cause for the disagreement between calculated and experimental values of the parameters. It is of interest that this is the only apoprotein in the data set of Tables I and III for which such a large disagreement exists.

For Ligand-Induced Shape Changes. Three proteins listed in Table I (not counting the two virus particles, Nos. 19 and 20) have not been used in Table III. They are apo-RBP at pH 3.0 (No. 7), the holo form of glyceraldehyde-3-phosphate dehydrogenase (No. 13), and the liganded form of pyruvate kinase (No. 16). These proteins in each case have undergone some structural (either conformational or configurational) change from their respective native form, as may be seen from the structural parameters listed in Table I. Three types of structural change are conceivable: (1) a change in volume, with essentially constant axial ratio, as in the holo form of glyceraldehyde-3-phosphate dehydrogenase, (2) a change in axial ratio, with constant volume, as in the holo (liganded) form of pyruvate kinase, and (3) changes in both volume and axial ratio, as in the apo form of RBP at pH 3.0.

To determine from sedimentation velocity data in the absence of SAXS information which of these three categories would be appropriate for an unknown protein undergoing a structural rearrangement, some other accessory information is required. For this purpose, changes in the preferential hydration of the particle could be measured under both conditions (for example by density gradient ultracentrifugation in high salt solutions, by pycnometry, or by another suitable method), bearing in mind that changes in such a quantity can usually be determined with higher precision than the absolute value of the quantity itself. Assuming unchanged salt binding, changes in preferential hydration are, by the definition of this quantity, equal to changes in total hydration[26]; the latter, in turn, are reflected in changes in hydrated particle volume by Eq. (7). Thus, in the case of glyceraldehyde-3-phosphate dehydrogenase, Sloan and Velick[27] found a decrease in preferential hydration of 0.075 g/g concurrent with the binding of the coenzyme nicotinamide adenine dinucleotide (NAD), as determined from sedimentation velocity and relative viscosity data. (The ready availability since then of advanced density-measuring instrumentation[28,29] has considerably simplified the precise determination of preferential hydrations.[26]) From Eq. (7), with $\bar{v}_2 = 0.737$ ml/g and $A_1 = 0.444$ g/g (the average value), a decrease in ($\bar{v}_2 + A_1$) by 0.075 corresponds to a decrease in V of 6.35%. This compares with

[26] J. C. Lee, K. Gekko, and S. N. Timasheff, this series, Vol. 61, p. 26.

[27] D. L. Sloan and S. F. Velick, *J. Biol. Chem.* **248**, 5419 (1973).

[28] D. W. Kupke, *in* "Physical Principles and Techniques of Protein Chemistry" (S. L. Leach, ed.), Part C, p. 1. Academic Press, New York, 1973.

[29] O. Kratky, H. Leopold, and H. Stabinger, this series, Vol. 27, p. 98.

a corresponding volume decrease of 5.37% reported by Durchschlag *et al.*[30] for this same enzyme, although under slightly different conditions (at pH 8.5, $40°$[27] vs pH 7.4, $25°$[26]). This was accompanied by a 0.4 Å decrease in R_G, with almost no change in axial ratio. As can be seen from Eqs. (1a) and (1b) and Table I after allowing for the change in molecular weight, the amount of volume contraction or decrease in hydration can account to within about 1% for the change in sedimentation coefficient upon binding of NAD. This treatment, therefore, may be useful in determining the contribution of volume contraction (or decrease in hydration), as well as of the change in axial ratio, to changes in sedimentation coefficients for proteins under altered environmental conditions.

If, as in the case of binding of pyruvate kinase, the hydrated volume remains constant (indicated, for instance, by unchanged preferential hydration), any change in sedimentation coefficient must be attributed to a change in axial ratio alone. (Malate synthase, No. 14, behaves similarly on binding to substrate or to an analog,[31] but it does not furnish a good quantitative example because changes here are exceedingly subtle and literature data do not appear to be sufficiently explicit to permit the present kind of analysis.) For pyruvate kinase, evaluating V and S as indicated before, the increase in $s_{20,w}^0$ from 8.70 to 8.81 corresponds to a decrease in f/f_0 from 1.148 to 1.139 and an increase in axial ratio from 0.270 to 0.281. The corresponding dimensionless ratio r_2 ($\equiv R_G S/V$) would be required to go from 4.11 to 4.00 (r_2 being appropriate because axial ratios derived from frictional coefficients relate to the rugose surface). Assuming the surface area remains essentially constant, and with the volume also constant, the change in sedimentation coefficient therefore translates into an unambiguous change in radius of gyration as well. (The assumption of approximate surface constancy for proteins in excess of M_r 100,000, mentioned earlier, is verified by inspection of the pertinent data of Table I, from which it may be seen that changes in S are of the order of only 3% for both pyruvate kinase, $M \simeq 190,000$, and glyceraldehyde-3-phosphate dehydrogenase, $M \simeq 140,000$.) It follows that R_G should change in proportion, i.e., at a ratio of $4.11/4.00 = 1.028$. This prediction, made without SAXS, agrees well with the SAXS radii of gyration of Table I, according to which this ratio is $43.5/42.25 = 1.024$. To estimate R_G values individually, following the procedure described under Oligomers gives, from the above $(a/b)_{2,K-P}$ of 0.0270 and 0.281: R_{K-P}, 49.2 and 47.7; r_1, 0.749 and 0.825; $(a/b)_{1,K-P}$, 0.194 and 0.216. Inasmuch as for oblate ellipsoids $(a/b)_1$ may not be smaller than $(a/b)_2$ (see remarks under Oligo-

[30] H. Durchschlag, G. Puchwein, O. Kratky, I. Schuster, and K. Kirschner, *Eur. J. Biochem.* **19**, 9 (1971).

[31] D. Zipper and H. Durchschlag, *Eur. J. Biochem.* **87**, 85 (1978).

mers), these values of R_{K-P} are not usable. The use of 0.270 and 0.281 for p in Eq. (3a), however, gives r_1 values of 1.012 and 1.049, from which R_{sm} of 44.4 and 44.1 are obtained as best estimates, in error by less than 4% when compared to the SAXS values of 43.5 and 42.5, respectively.

By contrast, for the apo form of riboflavin-binding protein, $M \simeq$ 33,000, in going from pH 7 to pH 3 the change in S (see Table I) amounts to a decrease of over 12%. This is so substantial that only the product of R_G and S can be evaluated from the rugose-surface axial ratio and the estimated hydrated volume, but not R_G by itself. SAXS measurements would need to be made to find the separate contributions of radius of gyration and surface area.

Calculations from Difference Sedimentation and SAXS Data and Electron Microscopy

The surface areas of globular proteins may be calculated from sedimentation coefficients in conjunction with SAXS volumes and R_G values. Since the Soulé-Porod plots from which SAXS values for S/V are derived tend to be imprecise (requiring, preferably the use of a symmetrically scanning apparatus to determine the true zero angle), whereas R_G and V can be much more readily determined precisely, surface areas are easier to calculate through use of Eqs. (1)–(3). Small changes in protein surface areas induced by biological processes or environmental conditions can thus be detected by means of difference sedimentation analysis in conjunction with R_G and V values from SAXS. For the most accurate results, the molecular weight used in Eq. (1) should be obtained from sequence data or, at least, from sedimentation equilibrium, rather than from SAXS.

Section C of Table III shows the predicted radius of gyration and volume from sedimentation coefficients of two virus particles, on the basis of spherical shape as revealed by electron microscopy. These calculated values of the radius of gyration are in fair agreement with the ones determined by SAXS. Hence, it is reasonable to determine the shape of a large particle with smooth surface by electron microscopy, assuming that the fixation technique used has not distorted the sample significantly. The radius of gyration and volume can then be determined accurately from the sedimentation coefficient and the axial ratio from electron microscopy.

Note Added

Since the preparation of this manuscript, the authors have been made aware of results of work not previously available to them which bears directly on the subject of this chapter (Prof. G. Damaschun, East-Berlin, personal communication). Parameters of 10 additional globular proteins and 2 small RNA molecules have been examined with respect to the relationships of Ref. 21 and, although differences between predicted and experimental sedimentation coefficients were in several instances larger than those above, the findings in general represent excellent confirmation of the semiempirical procedure described here.[32] The reader may further find it of interest that Damaschun and co-workers have developed methods also of calculating hydrodynamic parameters from atomic coordinates or from SAXS many-body models[33,34] and of estimating the thickness of solvation layers from combined SAXS and quasi-elastic light scattering.[35]

Acknowledgment

The authors have benefitted greatly from discussions with H. M. Farrell, Jr. and from the valuable advice of J. C. Lee.

[32] J. J. Müller, H. Damaschun, G. Damaschun, K. Gast, P. Plietz, and D. Zirwer, *Stud. Biophys.* **102**, 171 (1984).
[33] J. J. Müller, O. Glatter, D. Zirwer, and G. Damaschun, *Stud. Biophys.* **93**, 39 (1983).
[34] J. J. Müller, D. Zirwer, G. Damaschun, H. Welfle, K. Gast, and P. Plietz, *Stud. Biophys.* **96**, 103 (1983).
[35] K. Gast, D. Zirwer, P. Plietz, J. J. Müller, G. Damaschun, and H. Welfle, in ''Physical Optics of Dynamic Phenomena and Processes in Macromolecular Systems.'' de Gruyter, Berlin (in press).

[12] The Use of Covolume in the Estimation of Protein Axial Ratios

By LAWRENCE W. NICHOL and DONALD J. WINZOR

For several decades attempts have been made to assess the overall geometry of protein molecules in solution by visualizing them as ellipsoids of revolution.[1–5] With the convention that a denotes the length of the semimajor axis of the ellipse and b that of the semiminor axis, rotation of the ellipse about these axes results, respectively, in prolate and oblate ellipsoids both with axial ratio $a/b \geq 1$, the limiting case where a equals b being a sphere. The ultimate aim in this context is to view the hydrated

[1] T. Svedberg and K. O. Pedersen, ''The Ultracentrifuge.'' Oxford Univ. Press, London and New York, 1940.
[2] J. L. Oncley, *Ann. N.Y. Acad. Sci.* **41**, 121 (1941).
[3] H. K. Schachman, ''Ultracentrifugation in Biochemistry.'' Academic Press, New York, 1959.
[4] J. T. Yang, *Adv. Protein Chem.* **16**, 323 (1961).
[5] C. Tanford, ''Physical Chemitry of Macromolecules.'' Wiley, New York, 1961.

molecule as an effective hard, impenetrable ellipsoid and to determine both the individual magnitudes of a and b; and the type of ellipsoid (prolate or oblate) which is appropriate. The advent of the application of X-ray crystallographic techniques, particularly with the recent ability to probe the nature of bound water,[6] may at first sight appear to render the geometric interpretation both simplistic and unnecessary. Indeed, despite some lingering doubts that the protein conformation may differ between crystal and solution states,[7] it must be acknowledged at the outset that this is likely to be the case: provided that the sole aim is the determination of the detailed structural conformation of the molecule, intrinsically laborious X-ray crystallographic studies, and to a lesser extent those involving electron microscopy, provide the more detailed picture.

However, despite this assertion, several areas exist in the study of the behavior of proteins in solution where simple geometrical representation of shape retains usefulness and may indeed be necessary for quantitative elucidation of complicated interactive phenomena. Three examples will suffice. First, recent calculations of the translational frictional coefficients and hence sedimentation coefficients of protein oligomers comprising various arrangements of constitutive subunits (monomers) are based on an assignment of the ellipsoidal shape of the monomers.[8,9] In these terms, it was found, for example, that the sedimentation coefficient of zinc–insulin hexamer in solution is consistent with that calculated for a closed structure of six monomer units (each resembling a prolate ellipsoid) with the overall shape of an oblate ellipsoid.[8] Second, in the consideration of more elaborate protein self-association patterns, such as nucleated-condensation polymerization, the formation of rod-like structures (microscopically helical) is necessarily viewed in one light as a particular type of geometrical assembly dependent on overall monomer shape.[10] Third, in the thermodynamic characterization of all types of self and mixed associations of proteins, particularly if this is conducted with reasonably concentrated solutions, due allowance must be made for composition dependence of the activity coefficients of the species.[11–14] Such allowance involves as

[6] J. T. Edsall and H. A. McKenzie, *Adv. Biophys.* **16**, 53 (1982).
[7] S. E. Harding and A. J. Rowe, *Int. J. Biol. Macromol.* **4**, 160 (1982).
[8] P. R. Andrews and P. D. Jeffrey, *Biophys. Chem.* **4**, 93 (1976).
[9] D. C. Teller, E. Swanson, and C. de Haën, this series, Vol. 61, p. 103.
[10] S. N. Timasheff, *in* "Protein-Protein Interactions" (C. Frieden and L. W. Nichol, eds.), Chapter 8. Wiley, New York, 1981.
[11] A. G. Ogston and D. J. Winzor, *J. Phys. Chem.* **79**, 2496 (1975).
[12] L. W. Nichol and D. J. Winzor, *J. Phys. Chem.* **80**, 1980 (1976).
[13] P. R. Wills, L. W. Nichol, and R. J. Siezen, *Biophys. Chem.* **11**, 71 (1980).
[14] L. W. Nichol, *in* "Protein-Protein Interactions" (C. Frieden and L. W. Nichol, eds.), Chapter 1. Wiley, New York, 1981.

one step the assessment of covolume combinations of spherical, prolate, and oblate shapes,[15] important assessments also in the elucidation of the effects of space-filling polymers on the extents and kinetics of interacting protein systems.[16–18] In at least these areas, then, one is not primarily concerned with the details of irregular surface topography nor with the possibility of more elaborate triaxial ellipsoidal representation (intrinsically interesting though they may be), but rather with making realistic *estimates* of interaction parameters and overall polymer dimensions. It is argued then, as an integral part of this endeavor, that the classical problem of defining the effective ellipsoidal geometry of a protein under noninteractive conditions retains relevance.

Desired Properties of the Shape Function

In what follows, we shall consider a single protein solute, which does not chemically react, and which has been dialyzed against buffer so that the protein component (termed 2) is that defined by Casassa and Eisenberg,[19] the macroion plus counterions required for electrical neutrality. A common approach[1] is to determine from density measurements the apparent partial specific volume, $\bar{v}_2^U$, on the basis of an anhydrous concentration scale, and to use it in conjunction with (say) sedimentation equilibrium results to determine the anhydrous molecular weight, M_2^U, at infinite dilution. This permits determination of the anhydrous volume of a molecule, V_2^U, as $M_2^U \bar{v}_2^U/N$, where N is Avogadro's number, and the translational frictional coefficient of a sphere of the same volume,

$$f_2^0 = 6\pi\eta(3M_2^U\bar{v}_2^U/4\pi N)^{1/3} \tag{1}$$

where η is the viscosity of the medium. A series of sedimentation velocity experiments is now performed at different solute concentrations to find s_2^0, the sedimentation coefficient at infinite dilution, from which the actual translational coefficient, f_2^H, may be found using,

$$f_2^H = M_2^U(1 - \bar{v}_2^U\rho)/Ns_2^0 \tag{2}$$

where it is noted that $M_2^U(1 - \bar{v}_2^U\rho) = M_2^H(1 - \bar{v}_2^H\rho)$ provided that ρ, the solution density, approximates the reciprocal of the solvent partial spe-

[15] L. W. Nichol, P. D. Jeffrey, and D. J. Winzor, *J. Phys. Chem.* **80**, 648 (1976).

[16] L. W. Nichol, A. G. Ogston, and P. R. Wills, *FEBS Lett.* **126**, 18 (1981).

[17] A. P. Minton, *Biopolymers* **20**, 2093 (1981).

[18] L. W. Nichol, M. J. Sculley, L. D. Ward, and D. J. Winzor, *Arch. Biochem. Biophys.* **222**, 574 (1983).

[19] E. F. Casassa and H. Eisenberg, *Adv. Protein Chem.* **19**, 287 (1964).

cific volume, v_1, at infinite dilution.[20] The relations between unhydrated and hydrated quantities are,[5]

$$M_2^H = M_2^U(1 + w) \tag{3a}$$
$$\bar{v}_2^H = (\bar{v}_2^U + wv_1)/(1 + w) \tag{3b}$$

where w is the number of grams of solvent bound per gram of dry solute. If a value of the degree of hydration is assumed, it is possible to interpolate at this value in the contour diagram presented by Oncley[2] for the appropriate value of the frictional ratio, f_2^H/f_2^0, and read from the ordinate axis corresponding values of axial ratios for a prolate and an oblate ellipsoid of revolution. Sets of values of a and b readily follow since $V_2^H = M_2^H \bar{v}_2^H/N$ and

$$V_2^H = 4\pi ab^2/3 \quad \text{(prolate)} \tag{4a}$$
$$V_2^H = 4\pi a^2b/3 \quad \text{(oblate)} \tag{4b}$$

The difficulties with this approach, however, are readily evident. It requires assuming a value for w and, even when this is done, leaves ambiguous the type of ellipsoid which is appropriate and hence the specific values of a and b which pertain.

For these reasons a more definitive shape function than the frictional ratio has long been sought.[21-25] We are now in the position to list the properties that such a shape function should desirably possess, in addition to the obvious requirement that constitutive parameters must be capable of experimental measurement.

1. Each parameter should be expressable as an explicit function of axial ratio.

2. Expressions for parameters should be capable of combination to yield a function independent of molecular volume and degree of hydration.

3. Values of the function when calculated theoretically for assigned axial ratios for both prolate and oblate ellipsoids of revolution should diverge so that a distinction between shapes is possible.

4. The function should be sufficiently sensitive to change in axial ratio to allow, within available experimental precision for the determination of

[20] P. D. Jeffrey, L. W. Nichol, D. R. Turner, and D. J. Winzor, *J. Phys. Chem.* **81**, 776 (1977).
[21] H. A. Scheraga and L. Mandelkern, *J. Am. Chem. Soc.* **75**, 179 (1953).
[22] C. Sadron, *Prog. Biophys. Biophys. Chem.* **3**, 237 (1953).
[23] A. G. Ogston, *J. Phys. Chem.* **74**, 668 (1970).
[24] M. Wales and K. E. Van Holde, *J. Polym. Sci.* **14**, 81 (1954).
[25] J. M. Creeth and C. G. Knight, *Biochim. Biophys. Acta* **102**, 549 (1965).

a value of the function, a reasonably accurate definition of the axial ratio by interpolation in the theoretical curve.

It is evident that the translational frictional coefficient meets the requirement of a constitutive parameter. It may be found experimentally by use of Eq. (2) or by measurement of the diffusion coefficient at infinite dilution,

$$f_2^H = RT/ND_2^0 \tag{5}$$

where R is the gas constant and T the temperature. It assumes a value, for example, of $5.60 \pm 0.05 \times 10^{-8}$ g sec^{-1} for ovalbumin,[20] the standard error being assessed in relation to the use of Eq. (2). Moreover, the frictional ratio of both a prolate and an oblate ellipsoid may be expressed as a function of axial ratio.[1,5,26,27] To do this we note that the frictional coefficient of a sphere of the same volume as V_2^H defined by Eqs. (4a) and (4b) is[1]

$$(f_2^0)^H = 6\pi\eta(ab^2)^{1/3} \qquad \text{(prolate)} \tag{6a}$$
$$(f_2^0)^H = 6\pi\eta(a^2b)^{1/3} \qquad \text{(oblate)} \tag{6b}$$

and we introduce ε, the eccentricity,[28] a dimensionless quantity lying between zero and unity and directly related to the axial ratio, a/b, by

$$\varepsilon^2 = 1 - (b^2/a^2) \tag{7}$$

In these terms, the familiar equations of Perrin[26] become,[20]

$$f_2^H/(f_2^0)^H = 2\varepsilon/\{(1 - \varepsilon^2)^{1/3} \ln[(1 + \varepsilon)/(1 - \varepsilon)]\} \qquad \text{(prolate)} \tag{8a}$$
$$f_2^H/(f_2^0)^H = \varepsilon/[(1 - \varepsilon^2)^{1/6} \sin^{-1}\varepsilon] \qquad \text{(oblate)} \tag{8b}$$

We are, of course, no further advanced at this stage than the Oncley treatment[2]; but the stage has been more firmly set for the introduction of a different type of parameter which in combination will lead to a function with several of the desired characteristics.

The Covolume

Experimental Determination. The Ω analysis of sedimentation equilibrium results[29] provides one direct means of determining the activity coefficient of a single protein solute, y_2, as a function of its weight concentra-

[26] F. Perrin, *J. Phys. Radium* **7**, 1 (1936).
[27] R. O. Herzog, R. Illig, and H. Kudar, *Z. Phys. Chem., Abt. A* **167**, 329 (1933).
[28] A. Isihara, *J. Chem. Phys.* **18**, 1446 (1950).
[29] B. K. Milthorpe, P. D. Jeffrey, and L. W. Nichol, *Biophys. Chem.* **3**, 169 (1975).

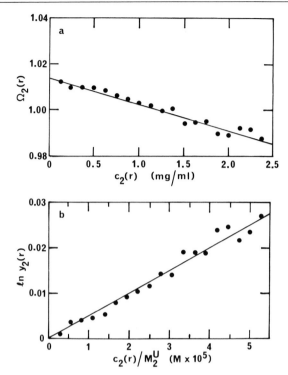

FIG. 1. Determination of the second virial coefficient, α_{22}, from a sedimentation equilibrium experiment at 20,000 rpm on ovalbumin in acetate–chloride buffer, pH 4.59, $I = 0.16$ M. (a) Evaluation of the activity coefficient for $\bar{c}_2(r_F) = 1.25$ g/liter by means of the $\Omega_2(r)$ function; (b) the plot from which α_{22} is obtained as the slope. (Adapted with permission from Jeffrey et al.[20] Copyright 1977 American Chemical Society.)

tion, c_2 (g/liter) on the anhydrous scale.[20] This is done by selecting a reference point $[r_F, c_2(r_F)]$ within the experimental distribution of $c_2(r)$ versus radial distance, r, and with a desk calculator evaluating $\Omega_2(r)$ operationally defined as

$$\Omega_2(r) = \{c_2(r) \exp[M_2^U(1 - \bar{v}_2^U\rho)\omega^2(r_F^2 - r^2)/2RT]\}/c_2(r_F) \qquad (9)$$

A plot is constructed of $\Omega_2(r)$ versus $c_2(r)$, whereupon the extrapolated value at infinite dilution equals $y_2(r_F)$. Thus, in Fig. 1a, which presents such a plot from a sedimentation equilibrium experiment on ovalbumin, a value of 1.014 is obtained for the activity coefficient $y_2(r_F)$ pertaining to the selected reference concentration, $c_2(r_F)$, of 1.25 g/liter. It has also been shown that $y_2(r) = y_2(r_F)/\Omega_2(r)$ and hence that y_2 as a function of c_2 is then readily obtained.[20] In turn, a plot of $\ln y_2$ versus c_2/M_2^U (see Fig. 1b) yields a value of α_{22}, the apparent second virial coefficient[11,30,31]: from the slope of the line in Fig. 1b, $\alpha_{22} = 500$ liters/mol. A statistical mechanical interpretation of this quantity for like effective hard-sphere interactions has shown that[13]

$$\alpha_{22} = \frac{32\pi N r_2^3}{3} + \frac{Z_2^2(1 + 2\kappa r_2)}{2I(1 + \kappa r_2)^2} - M_2^U \bar{v}_2^U \qquad (10)$$

where the first term denotes the covolume (liters/mol) of the sphere of radius r_2; the second term accounts for the electrostatic interaction written in terms of the net charge borne by the protein, Z_2, and κ, the inverse screening length of the supporting electrolyte of ionic strength I; and the third term is the molar volume. On this basis it has been suggested that subtraction from α_{22}, determined experimentally as described above, or, indeed, from osmotic pressure or light scattering results, of estimates of the second and third terms would yield a value for the covolume referring to the actual hydrated species in solution regardless of its shape.[20] Indeed, this view has been supported in the case of ovalbumin by assessing the self-covolume, U_{22}^H, as 500 liters/mol both from results obtained at the isoelectric point (pH 4.59) where $Z_2 = 0$ (Fig. 1) and at pH 7.5, where $Z_2 = -14$.[20] We see, therefore, that the covolume is a parameter which meets at least the basic requirement that it may be experimentally measured and, hinging as it does on the apparent second virial coefficient, by a variety of means.

Theoretical Expressions. A particular value of the covolume concept is that being geometrical itself, rigorous expressions may be formulated for a variety of combinations; rod–rod,[23] prolate spherocylinder–prolate spherocylinder,[32] sphere–prolate ellipsoid,[11,15] sphere–oblate ellipsoid,[11,15] and oblate ellipsoid–prolate ellipsoid.[15] In the present context, however, we are interested in the simpler self-covolume combinations, sphere–sphere, prolate ellipsoid–prolate ellipsoid and oblate ellipsoid–oblate ellipsoid. For this set of three, a single expression suffices,[15,28,32]

$$U_{22}^H = N V_2^H \left\{ 2 + \frac{3}{2} \left[1 + \frac{\sin^{-1}\varepsilon}{\varepsilon(1 - \varepsilon^2)^{1/2}} \right] \left[1 + \frac{1 - \varepsilon^2}{2\varepsilon} \ln \frac{1 + \varepsilon}{1 - \varepsilon} \right] \right\} \qquad (11)$$

[30] H. Fujita, "Mathematical Theory of Sedimentation Analysis." Academic Press, New York, 1962.

[31] J. W. Williams, K. E. Van Holde, R. L. Baldwin, and H. Fujita, *Chem. Rev.* **58,** 715 (1958).

[32] J. O. Hirschfelder, C. F. Curtis, and R. B. Bird, "Molecular Theory of Gases and Liquids." Wiley, New York, 1954.

where ε is the eccentricity defined as before in terms of the axial ratio by Eq. (7); and V_2^H is given by Eq. (4a) for a prolate ellipsoid and by Eq. (4b) for an oblate ellipsoid. The derivation of Eq. (11) by the method used by Isihara[28] is undeniably complicated, but it has been confirmed by different means in two places.[11,15] Moreover, it is reassuring that when $a = b = r$ and $\varepsilon = 0$ (a sphere), the limits of $(\sin^{-1}\varepsilon)/\varepsilon$ and of $(1/2\varepsilon) \ln[(1 + \varepsilon)/ (1 - \varepsilon)]$ are both unity, whereupon with $V_2^H = 4\pi r^3/3$, Eq. (11) simplifies to $U_{22}^H = 32N\pi r^3/3$.[20] This is in accord with the molecular covolume,[5]

$$U_{22}^H/N = 4\pi(r + r)^3/3 \quad \text{(sphere)} \quad (12)$$

visualized as the spherical space in which two identical spheres of radius r could just be accommodated. We have presumed to identify U_{22}^H given rigorously by Eq. (11) for hard ellipsoid of revolution self-covolumes with the effective covolume of the hydrated protein molecule found experimentally. In so doing we neglect any distortion of, for example, the hydration shell around the protein on near approach of two molecules (an inelastic collision); but in this connection we would warn against the visualization of self-covolume as an actual rotation of molecules around each other in solution with a squeezing of hydration shells.[33]

It is now noted, as with Eq. (8), that Eq. (11) does not permit explicit specification of protein geometry. Thus, even if a value of w were assumed and Eqs. (3a) and (3b) used to evaluate V_2^H, Eq. (11) would merely permit ε (and hence the axial ratio) to be determined without specification of the nature of the ellipsoid or specification of the values of a and b.

The Combined Shape Function, ψ

Elimination of $(f_2^0)^H$ between Eqs. (6a) and (8a), and combination of the result with Eq. (4a) permits an explicit expression for V_2^H(prolate) to be written in terms of f_2^H and ε.[20] Thus,

$$V_2^H = \frac{4\pi(f_2^H)^3(1 - \varepsilon^2)\{\ln[(1 + \varepsilon)/(1 - \varepsilon)]\}^3}{3(6\pi\eta)^3(2\varepsilon)^3} \quad \text{(prolate)} \quad (13)$$

Therefore, in keeping with the desire to write a shape function independent of molecular volume and degree of hydration, we operationally define the function ψ as,[20]

$$\psi = U_{22}^H \eta^3/N(f_2^H)^3 \quad (14)$$

---

[33] A. P. Minton, *Biophys. Chem.* **12**, 271 (1980).

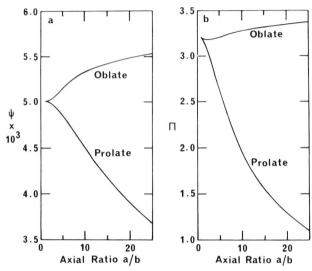

FIG. 2. A comparison of the dependence on axial ratio of two shape functions, both involving covolume as a parameter and both potentially capable of distinguishing between prolate and oblate ellipsoids of revolution: (a) the ψ function defined by Eq. (14) in terms of covolume and translational frictional coefficient, and calculated using Eqs. (7), (15), and (16); (b) the Π function defined by Eq. (20) in terms of covolume and intrinsic viscosity, and calculated using Eq. (11).

Not only may this function be experimentally evaluated as we have seen, but also its theoretical formulation with the use of Eqs. (11) and (13) shows it to be a sole function of ε, and hence axial ratio, referring unequivocally to a prolate ellipsoid of revolution.[20] Specifically,

$$\psi(\text{prolate}) = \frac{(2 + \tfrac{3}{2}AB)(1 - \varepsilon^2)\{\ln[(1 + \varepsilon)/(1 - \varepsilon)]\}^3}{1296\pi^2\varepsilon^3} \quad (15a)$$

$$A = 1 + \{(\sin^{-1}\varepsilon)/[\varepsilon(1 - \varepsilon^2)^{1/2}]\} \quad (15b)$$

$$B = 1 + [[(1 - \varepsilon^2)/2\varepsilon]\{\ln[(1 + \varepsilon)/(1 - \varepsilon)]\}] \quad (15c)$$

By entirely similar reasoning we may write using Eqs. (6b), (8b), and (11),

$$\psi(\text{oblate}) = \frac{\{2 + \tfrac{3}{2}AB\}(\sin^{-1}\varepsilon)^3(1 - \varepsilon^2)^{1/2}}{162\pi^2\varepsilon^3} \quad (16)$$

with A and B defined as before in Eqs. (15b) and (15c).

It is now possible to test the third requirement of the shape function, which has been done in Fig. 2a by calculating theoretical values of ψ(prolate) and ψ(oblate) for assigned values of ε transposed to values of axial ratio employing Eq. (7). For the first time, we see in Fig. 2a diverging plots which are potentially capable of distinguishing between the prolate and oblate models for a particular protein for which a single value of ψ is determined experimentally. It is noted that the plots converge at the value $8/162\pi^2$, that of both Eqs. (15) and (16) as $\varepsilon \rightarrow 0$. This value has no particular significance, but the demonstration of convergence as the ellipsoids approach a sphere is required.[20]

Experimental Application of the ψ Function

From the purely experimental viewpoint only Eq. (14), which defines ψ, and Fig. 2a are relevant. Let us illustrate this with mean values of parameters found for ovalbumin[20]: U_{22}^H found from sedimentation equilibrium results was 5.0×10^5 ml mol^{-1}, and f_2^H, from sedimentation velocity studies, was 5.60×10^{-8} g sec^{-1}; both sets of experiments were conducted in a medium with a viscosity, η, of 0.01016 P. These values, when substituted into Eq. (14), yield $\psi = 4.9_7 \times 10^{-3}$, which on interpolation in Fig. 2a suggests that ovalbumin behaves as a prolate ellipsoid of revolution with axial ratio, a/b, of 2.5. The corresponding value of ε is 0.92, from Eq. (7); V_2^H is 8.3×10^{-23} liters, from Eq. (13); and $a = 5.0$ nm, $b = 2.0$ nm, from Eq. (4a). Moreover, it follows from Eqs. (3a) and (3b) that,

$$w = (NV_2^H - M_2^U \bar{v}_2^U)/M_2^U v_1 \tag{17}$$

which gives an apparent value of 0.37 g/g for the degree of hydration of ovalbumin, a value within the range generally accepted for globular proteins.[5] This representation of the hydrated ovalbumin molecule obtained by using mean values of the experimental parameters in the ψ function compares quite favorably with the recent conclusion drawn by Harding[34] that ovalbumin is spheroidal (possibly prolate) with an axial ratio of 1.5 ± 0.3.

This illustration of the use of the ψ function stresses its exact fulfillment of the first three listed desirable properties of a shape function, in particular that it is potentially capable of distinguishing between prolate and oblate models. Before discussing experimental uncertainty in the function,[7,20] relevant to the fourth property listed, it is timely to comment on the consequences of uncertainty in ellipsoidal representation. Put directly, is it really necessary to define with great precision the exact geometry of an effective hydrodynamic entity? The calculated values presented in Table I address this question. Each row refers to a particular entity for which the hydrated molecular volume and the translational frictional coef-

[34] S. E. Harding, *Int. J. Biol. Macromol.* **3**, 398 (1981).

TABLE I

COMPARISON OF THE AXIAL RATIOS AND COVOLUMES OF PROLATE AND OBLATE
ELLIPSOIDS OF REVOLUTION WITH THE SAME MOLECULAR VOLUMES AND
TRANSLATIONAL FRICTIONAL COEFFICIENTS

Oblate			Prolate		
Axial ratio (a/b)	Eccentricity (E)	Y from Eq. (18)	Eccentricity (ε)	Axial ratio (a/b)	$\dfrac{U_{22}^{H} \text{ (oblate)}}{U_{22}^{H} \text{ (prolate)}}$
2	0.8860	0.8840	0.8616	1.97	1.01
3	0.9428	0.7419	0.9383	2.89	1.02
5	0.9798	0.5461	0.9761	4.60	1.06
7	0.9897	0.4286	0.9869	6.20	1.10
10	0.9950	0.3229	0.9929	8.41	1.16
13	0.9970	0.2587	0.9955	10.55	1.20
15	0.9978	0.2284	0.9964	11.80	1.25
17	0.9983	0.2044	0.9971	13.14	1.28
20	0.9987	0.1765	0.9978	15.08	1.28
23	0.9991	0.1553	0.9982	16.67	1.38
25	0.9992	0.1438	0.9984	17.68	1.39

ficient must be fixed. Accordingly, we may write using Eq. (13) and the corresponding expression for an oblate ellipsoid, Eq. (8b) of Jeffrey *et al.*,[20]

$$\frac{(\sin^{-1}E)^3(1 - E^2)^{1/2}}{E^3} = Y = \frac{(1 - \varepsilon^2)\{\ln[(1 + \varepsilon)/(1 - \varepsilon)]\}^3}{8\varepsilon^3} \tag{18}$$

where, in this context, E denotes the eccentricity (oblate) and ε the eccentricity (prolate) of ellipsoids with the same molecular volume and translational frictional coefficient. Columns 1 and 2 of Table I give values of axial ratios and corresponding values of E for the oblate form, leading via Eq. (18) to the reported values of Y in column 3. As is evident from Eq. (8), Y is the reciprocal cube of the frictional ratio. Columns 4 and 5 give the corresponding numerical solutions of Eq. (18) in terms of ε and axial ratio for the prolate form. Clearly the values of the axial ratios (columns 1 and 5) diverge, and markedly so, as the entity becomes more asymmetric. The consequence of the divergence on covolume is seen in the last column, where Eq. (11) has been used to evaluate U_{22}^{H}(oblate)/U_{22}^{H}(prolate) for each pair of E and ε and with permitted cancellation of NV_2^{H}.

At axial ratios below 5, less than 6% uncertainty is introduced in consideration of covolume regardless of whether the entity is considered to be a prolate ellipsoid or an oblate ellipsoid, each with axial ratio corresponding to the molecular volume and the translational frictional coefficient of the entity. It is in this region that assessed standard error in $\psi(4.97 \pm 0.24 \times 10^{-3}$ for ovalbumin[20]) renders difficult the distinction between the ellipsoidal shapes, a difficulty, as we now see, of little concern to the researcher involved in nonideality calculations. On the other hand, at larger values of axial ratio column 6 of Table I indicates that a distinction between shapes assumes increasing importance; it is in this region that interpolation in Fig. 2a is likely to permit the distinction. In this connection, it is noted that the magnitude of the standard error in ψ is unlikely to increase with increasing particle asymmetry and size.

While the foregoing comments attempt to place the problem in perspective, they are not intended to mask the fact that the ψ function is subject to considerable uncertainty, involving as it does the cubing of the experimentally determined value of f_2^H, and that it is relatively insensitive to change in axial ratio.[7] Thus, while the ψ function meets the first three of the desired properties listed earlier for a shape function, it falters on the fourth, if exact ellipsoidal representation is required, especially when low values of axial ratio are encountered. While this may be of little concern in the assessment of second-order nonideality effects in (for example) a self-associating globular protein system,[11,13,14] it may well prove a stumbling block to those interested in oligomer assembly based on a precise knowledge of constituent monomer shape.[8-10] Accordingly, it is timely to comment on other available shape functions.

The Π Function

Following the combination of covolume with translational frictional coefficient in the ψ function,[20] Harding[35] explored a similar combination of covolume with intrinsic viscosity, $[\eta]$. Provided that viscosity experiments are conducted with sufficiently low velocity gradients and random orientations are assumed, we may write the following expression [Eq. (23-3) of Tanford[5]],

$$[\eta] = NV_2^H \nu / M_2^U \tag{19}$$

where $V_2^H = M_2^U(\bar{v}_2^U + wv_1)/N$ in accord with Eqs. (3a) and (3b), and ν is a function of axial ratio defined by the Simha–Einstein equations.[5,36,37] The origin of Harding's Π function now becomes evident, because

$$\Pi = U_{22}^H / [\eta] M_2^U = U_{22}^H / N V_2^H \nu \tag{20}$$

[35] S. E. Harding, *Int. J. Biol. Macromol.* **3**, 340 (1981).
[36] R. Simha, *J. Phys. Chem.* **44**, 25 (1940).
[37] J. W. Mehl, J. L. Oncley, and R. Simha, *Science* **92**, 132 (1940).

This is a particularly valuable formulation for two reasons. First, it is clear from Eq. (20) that an experimenter may find a numerical value of Π for a given system, and that such an evaluation does not require cubing an experimental parameter, as did evaluation of the ψ function. Second, it is also evident from Eq. (20) that theoretical plots of Π versus axial ratio may be constructed, since U_{22}^{H}/NV_{2}^{H}, the reduced covolume, is seen from Eq. (11) to be a sole function of ε (and hence axial ratio) for both prolate and oblate ellipsoids. The corresponding values of Simha's factor, ν(prolate) and ν(oblate) are best obtained from Table I of Mehl *et al.*[37] or the graphical representation given in Figs. 19-7 and 19-8 of Tanford,[5] because these values have been calculated from the full equation given by Simha[36] rather than from approximate relations which apply only at large values of the axial ratio.[5,36] Plots of Π versus axial ratio are shown in Fig. 2b to permit direct comparison with those pertaining to the ψ function. Figure 2b is in accord with that presented by Harding,[35] but covers a greater range of axial ratios. It is noted that values of Π converge to 3.2 when $\varepsilon = 0$ (a sphere), for which $\nu = 2.5$ and the reduced covolume is 8, according to Eq. (11) with limiting value of unity for $(\sin^{-1}\varepsilon)/\varepsilon$ and $(1/2\varepsilon)\ln[(1 + \varepsilon)/(1 - \varepsilon)]$.[20]

Three points merit comment in relation to a comparison of Fig. 2a and b. First, it is clear that both ψ and Π functions are potentially capable of distinguishing between oblate and prolate ellipsoids except as the limiting shape of a sphere is approached. Second, with the calculated standard error in Π of ± 0.045 reported by Harding[35] and that of $\pm 0.24 \times 10^{-3}$ for ψ in mind, it does appear that the Π function would be the more sensitive in determining the axial ratio of a prolate ellipsoid. Third, the difficulty is noted of using either shape function to determine the precise geometry of a protein which is oblatoid.

Concluding Remarks

Whereas the ψ function intrinsically requires the conduct of sedimentation velocity and equilibrium experiments for its evaluation, the Π function necessitates viscosity measurements at low shear in addition to a means of determining covolume. Both (preferably used in conjunction) serve the fundamental purpose of defining the appropriate type of ellipsoid except at low axial ratio, where we may well query the need for fine distinction (Table I). Once the ellipsoidal nature of a reasonably asymmetric macromolecule has been thus determined, it is appropriate to enquire whether yet another shape function exists which would give a more precise definition of the axial ratio than is provided by either the ψ or Π functions. Certainly, this attribute cannot be ascribed to the β function of

Scheraga and Mandelkern,[21] which is almost completely insensitive to variation in axial ratio for oblate ellipsoids and, moreover, with experimental uncertainty can be used to distinguish between prolate and oblate ellipsoids only for extremely asymmetric molecules. Indeed, the major use of the β function has been in the determination of molecular weight.[4,24] Of the remaining shape functions recently reviewed by Harding and Rowe,[7] the ratio $k_s/[\eta]$, the R function, merits comment in this context, since it couples two hydrodynamic parameters, the sedimentation coefficient-concentration regression coefficient, k_s, and the intrinsic viscosity, $[\eta]$.

If it is assumed that linearization of sedimentation results may be effected by

$$s_2 = s_2^0(1 - k_s c_2) \quad \text{or} \quad (1/s_2) = (1 + K_s c_2)/s_2^0 \tag{21}$$

K_s may be taken as k_s in evaluating the R function since it is the limiting value as $c_2 \to 0$ that is required.[25,38] The units of k_s and of $[\eta]$ are the same and thus the R function is dimensionless, as are the ψ and Π functions. Creeth and Knight[25] have provided an excellent review of the use of the R function as an index of molecular asymmetry and of its theoretical basis in terms primarily of the work of Burgers,[39,40] Wales and Van Holde,[24] Ogston,[41] Cheng and Schachman,[42] and Yamakawa.[43] In brief, while some success was achieved in predicting the value of the ratio for random-coil polymers, the theoretical basis was judged wanting. More recently, Rowe,[44] by assuming *inter alia* that particle migration relative to solvent at low limiting concentrations is independent of concentration and that charge–charge interactions may be neglected, has suggested that

$$R = 2\{1 + [f_2^H/(f_2^0)^H]^3\}/\nu \tag{22}$$

where $f_2^H/(f_2^0)^H$ is given by Eqs. (8a) and (8b) for prolate and oblate ellipsoids, respectively, and ν is the Simha factor tabulated for ellipsoids in Table I of Mehl *et al.*[37] On this basis it is a simple matter to calculate R as a function of axial ratio, and such results are shown in Fig. 3, which is in accord with the more extensive set of calculations provided by Rowe,[44] and which shows that the R function is incapable of distinguishing between prolate and oblate ellipsoids below an axial ratio of approximately

[38] G. Kegeles and F. J. Gutter, *J. Am. Chem. Soc.* **73**, 3770 (1951).
[39] J. M. Burgers, *Proc. Acad. Sci. Amsterdam* **44**, 1045 (1941).
[40] J. M. Burgers, *Proc. Acad. Sci. Amsterdam* **45**, 9 (1942).
[41] A. G. Ogston, *Trans. Faraday Soc.* **49**, 1481 (1953).
[42] P. Y. Cheng and H. K. Schachman, *J. Polym. Sci.* **16**, 19 (1955).
[43] H. Yamakawa, *J. Chem. Phys.* **36**, 2995 (1962).
[44] A. J. Rowe, *Biopolymers* **16**, 2595 (1977).

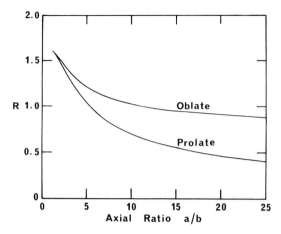

FIG. 3. The dependence on axial ratio, for both prolate and oblate ellipsoids of revolution, of the shape function R calculated using Eqs. (8a), (8b), and (22).

50. This serves to highlight one major conclusion of this chapter that this important distinction for relatively asymmetric molecules (Table I) appears at present possible only when covolume is combined with a hydrodynamic parameter as in the ψ and Π functions. Does, then, the R function, as has been suggested,[7] provide a more sensitive means of defining the axial ratio and hence values of a and b once this distinction has been made? This is likely to be a controversial point because while one group assesses the likely uncertainty in R as 1.4%,[7] Creeth and Knight[25] conclude more cautiously that probable errors of less than 5% might be obtainable: indeed, they point out that presently available values of R have probable errors of up to 12–13% when systematic errors, such as neglect to correct for radial dilution, have been avoided; and of up to 20% when they have not. Let us take the more optimistic view that at least a 5% precision in R is obtainable and that Fig. 3 has a sufficiently firm theoretical basis for use in relation to studies where charge–charge interactions have been minimized. The conclusion follows that the conjoint use of ψ, Π, and R functions may lead to reasonably precise delineation of the geometry of effective ellipsoidal representation and degree of hydration in regions of axial ratio where realistic estimates of such information are needed for other purposes.[8–14]

Two final comments should be made. First, we would emphasize that any attempt to define the effective ellipsoidal geometry of a macromolecule must ensure that, under the conditions of study, the solute is homogeneous with respect to molecular weight. Moreover, while the crystallo-

TABLE II

SHAPE FUNCTIONS OF OBLATE AND PROLATE ELLIPSOIDS OF REVOLUTION: DEPENDENCE
OF ν, ψ, Π, AND R ON AXIAL RATIO

	Simha factor ν		$\psi \times 10^3$		Π		R	
a/b	Oblate	Prolate	Oblate	Prolate	Oblate	Prolate	Oblate	Prolate
1	2.50	2.50	5.00	5.00	3.20	3.20	1.60	1.60
2	2.85	2.91	5.02	4.99	3.18	3.12	1.50	1.47
3	3.43	3.68	5.06	4.95	3.18	2.96	1.37	1.29
4	4.06	4.66	5.11	4.90	3.19	2.78	1.27	1.14
5	4.71	5.81	5.16	4.84	3.21	2.60	1.20	1.02
6	5.36	7.10	5.20	4.77	3.23	2.44	1.15	0.92
8	6.70	10.10	5.27	4.63	3.26	2.16	1.07	0.78
10	8.04	13.63	5.33	4.49	3.28	1.94	1.02	0.69
12	9.39	17.76	5.37	4.36	3.30	1.75	0.98	0.61
15	11.42	24.8	5.42	4.17	3.32	1.53	0.94	0.54
20	14.80	38.6	5.48	3.91	3.35	1.29	0.90	0.46
25	18.19	55.2	5.52	3.68	3.37	1.11	0.87	0.41
30	21.6	74.5	5.54	3.49	3.38	0.98	0.86	0.38
40	28.3	120.8	5.58	3.18	3.41	0.80	0.84	0.33
50	35.0	176.5	5.60	2.93	3.43	0.68	0.82	0.30
60	41.7	242.0	5.62	2.74	3.44	0.59	0.81	0.28
80	55.1	400.0	5.64	2.44	3.46	0.48	0.80	0.25
100	68.6	593.0	5.65	2.21	3.47	0.40	0.80	0.23
150	102.3	1222.0	5.67	1.83	3.47	0.29	0.79	0.20
200	136.2	2051.0	5.68	1.59	3.48	0.23	0.78	0.18
300	204.1	4278.0	5.69	1.29	3.48	0.17	0.77	0.16

graphic dimensions of some globular proteins, for example, cytochrome c (2.5 × 2.5 × 3.5 nm),[45] and lysozyme (4.5 × 3.0 × 3.0 nm),[46] provide reassurance that ellipsoidal representation (in these cases prolate) is highly appropriate, this is less true for other proteins.[7] Nevertheless, even in the latter cases, provided that it is understood that an *effective* representation of the entity in solution has been made, a reasonable quantitative basis has been laid for accounting for second-order effects in the interpretation of other interactions involving the protein. Second, another major aim of this work has been to present a coherent set of equations from which the various shape functions may be calculated as functions of axial ratio, the captions to Figs. 2 and 3 indicating the relevant final

[45] R. E. Dickerson and I. Geiss, "The Structure and Action of Proteins." Harper & Row, New York, 1969.
[46] C. C. F. Blake, D. F. Loenig, G. A. Mair, A. C. T. North, D. C. Phillips, and V. R. Sarma, *Nature (London)* **206,** 757 (1965).

expressions. For convenience, we have presented graphically calculations up to an axial ratio of 25, but even higher ratios have been deemed appropriate for certain proteins.[7] For completeness, we present in Table II numerical values of the ψ, Π, and R functions, discussed in this work, over the range of axial ratio originally considered by Simha.[36,37] It is believed that this type of information, based on the use of more traditional methodology, will continue to provide estimates of fundamental macromolecular parameters which will suffice in their approximate nature for several purposes, and will complement more sophisticated probes, such as X-ray crystallography, small-angle X-ray scattering, and fluorescence polarization, in refining the definition of overall shape when this is required.

Acknowledgment

The authors are indebted to Elisabeth A. Owen for assistance with numerical computations.

[13] Application of Transient Electric Birefringence to the Study of Biopolymer Structure

By Paul J. Hagerman

A medium is said to be birefringent if the velocity of light propagating through the medium is dependent upon the plane of polarization of the incident beam. For light propagating in the Z direction, the birefringence of the medium is defined as the difference in refractive index (n) between two orthogonally polarized beams,

$$\Delta n = n_x - n_y$$

A corresponding difference in absorbance

$$\Delta A = A_x - A_y$$

is referred to as dichroism. Many crystals exhibit birefringence as a consequence of the nonrandom distribution of various scattering elements within the crystal. While homogeneous solutions are normally optically isotropic, in many instances birefringence can be induced, either by applying external magnetic or electric fields, or through the establishment of a gradient of shear in the solution. The present chapter deals exclusively with electrically induced birefringence.

When the electric field is removed, the solution will again become optically isotropic, the rate of decay of birefringence being a function of the size and shape of the optically active species in the solution. Therefore, through measurement of the rate of decay of birefringence, one can obtain useful information regarding molecular structure. This technique is known as transient electric birefringence (TEB). An excellent monograph on electric birefringence and electric dichroism has been published by Fredericq and Houssier[1] and should be referred to for extensive discussions of the theory of electric birefringence.

This chapter does not consider contributions to the sign and magnitude of electrically induced birefringence, a formidable theoretical problem. Rather, it is concerned exclusively with the analysis of the rates of field-free decay of birefringence, quantities which are directly related to the principal diffusion coefficients of the macromolecule under study.[2,3]

In principle, TEB and TED (transient electric dichroism) yield the same information regarding macromolecular structure. Both should yield an identical set of field-free decay times and the signs and amplitudes of the birefringence and dichroism signals should provide information regarding the orientation of chromophoric groups within the molecule. In practice, however, TEB represents a more precise means of determining decay constants and therefore is the method of choice for hydrodynamic investigations. On the other hand, TED amplitudes, which in many instances involve a single electric transition within the molecule, are more easily interpretable in terms of absolute chromophoric orientation. Therefore, TED measurements are more suitable for studies involving a determination of internal structure.

Instrumental Design

A schematic representation of the TEB system developed by the author is presented in Fig. 1. The implementation of a dual-channel detection system has resulted in a substantial improvement in signal-to-noise (S/N) characteristics, thus allowing the determination of rates of decay of birefringence for solutions having an initial birefringence in the range of $2–5 \times 10^{-8}$. Meaningful measurements can thus be performed on DNA fragments using less than 1 μg of material.

---

[1] E. Fredericq and C. Houssier, "Electric Dichroism and Electric Birefringence." Oxford Univ. Press (Clarendon), London and New York, 1973.

[2] W. A. Wegener, R. M. Dowben, and V. J. Koester, *J. Chem. Phys.* **70**, 622 (1979).

[3] P. J. Hagerman and B. H. Zimm, *Biopolymers* **20**, 1481 (1981).

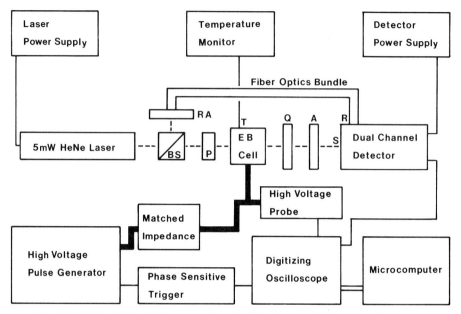

Fig. 1. Outline of the TEB system described in the text. BS, beam splitter; P, prism polarizer; T, temperature probe; EB, electric birefringence; Q, quarter-wave plate; A, analyzer (film polarizer); S, sample beam entry port; R, reference beam entry port; RA, reference analyzer (film polarizer).

He–Ne Laser

The system utilizes a 5-mW helium–neon (He–Ne) laser (Aerotech) with a remote head. The emission wavelength in 6328 Å. The output beam is polarized (500 : 1) with the plane of polarization set at 45° with respect to the vertical axis of the sample cell. During the course of development of this TEB system, it was noted that every He–Ne laser tested (eight different models were examined) demonstrated periodic oscillations in light intensity at the level of ~0.1–0.6% of the average total intensity. These oscillations appear in packets, interspersed by oscillation-free performance lasting up to an hour. The frequency of oscillation in any given packet varies from 100 to 10 kHz (approximate limits) over time. The frequency with which the packets appear is dependent upon the temperature of the laser cavity. Several manufacturers have attributed this behavior to properties of the lasing plasma. This phenomenon is of practical importance for high-precision measurements in that it falls within the frequency domain of many decay processes of interest. A second type of fluctuation in laser output intensity has also been noted and is secondary to radio frequency (RF) pickup from the high-voltage pulse generator (discussed below). These two types of fluctuation in output light intensity are effectively eliminated through the use of a dual-channel detection system.

High-Voltage Pulse Generator

Several high-voltage pulse generators suitable for TEB studies are available commercially. The one used in the author's system is a Velonex Model 360 high power (31,000 W) pulse generator with a direct output module (0–2.5 kV). Various plug-in modules are available for use with the Velonex pulse generator, depending upon the required voltage range; however, none has the performance characteristics of the direct output module with respect to the fidelity of the square-wave pulse and rate of field-off decay (40 nsec, $1/e$ value; 260 Ω load). In order to reduce RF transmission during the pulse, the direct output module was modified by the author to be fully coaxial at the point of exit of the pulse from the module.

The performance of the Velonex is optimal when the load impedance matches the output impedence of the particular plug-in module in use. Since the impedance of the birefringence cell will vary widely, depending upon the ionic strengths of solutions being studied, a noninductive load was placed in parallel with the cell. The effective load impedance is essentially constant (within a factor of two) until the impedance of the cell drops below that of the parallel load. However, the performance of the pulse generator (with direct output module) is not markedly degraded when the effective load impedance drops below the optimal value.

High-Voltage Probe

The configuration of the HV pulse is measured across the two electrode mounts on the birefringence cell holder by means of a 1000 : 1 step down probe (Tektronix P6013A).

Optical Configuration

The current system utilizes a quarter-wave plate configuration with the slow axis of the quarter wave plate at 135° with respect to the direction of the electric field (for a complete discussion of this and other optical configurations used in conjunction with TEB measurements, see Ref. 1). For this optical arrangement, the relationship between the fractional change in light intensity during the field pulse, $\Delta I(t)/I_\alpha$, and the optical retardation (δ) is given by

$$\Delta I(t)/I_\alpha = \sin^2[\alpha + \delta(t)/2]/\sin^2 - 1 \qquad (1)$$

where α represents the angle through which the analyzer is turned from the crossed position toward the polarizer, and I_α represents the steady-state (field-off) photocurrent for an analyzer setting α degrees from the crossed position. Rearranging,

$$\delta(t) = 2[\sin^{-1}[(\Delta I(t)/I_\alpha + 1)^{1/2} \sin \alpha] - \alpha] \qquad (2)$$

which is related to the time-dependent birefringence, $\Delta n(t)$, by the relation

$$\Delta n(t) = \lambda \delta(t)/360l \qquad (3)$$

where λ is the wavelength in Å, l is the path length in cm, and δ is expressed in degrees. A detailed discussion of the factors involved in setting α has been presented by Fredericq and Houssier[1] (also, see below).

The linearly polarized light from the He–Ne laser is split into a reference beam and a sample beam using a prism-type beam splitter. The reference beam then passes through a film analyzer in an adjustable mount (one minute-of-arc resolution) and impinges on a fiber optics bundle which leads to the reference port of the dual-channel detector. The reference analyzer is used to adjust the intensity of the reference beam for the purpose of balancing the two channels. After leaving the beam splitter, the sample beam passes through a prism polarizer ($10^5 : 1$), then through the birefringence cell assembly, the quarter wave plate, an adjustable analyzer, and finally through the sample port of the dual-channel detector.

Electric Birefringence Cell

The birefringence cell employed in the present system is displayed in Fig. 2. The electrode gap is 2 mm and the optical path length is 1.0 cm. Sample volumes can vary from 100 to 200 μl. The cell was designed to allow easy access for sample changes or for mixing within the cell, and to allow easy dismantling for cleaning.

Cell Housing Assembly

One of the most important functions of the cell housing is to provide a constant temperature environment. The assembly displayed in Fig. 3 is capable of providing long-term temperature stability to within 0.1 degree at 2°. The brass block upon which the cell rests contains three internal chambers: a central chamber through which coolant (water–ethanol) at the desired temperature is circulated, and two outer chambers (heat exchangers) through which predried air is circulated prior to being blown against the cell windows. The cell itself, along with the upper portion of

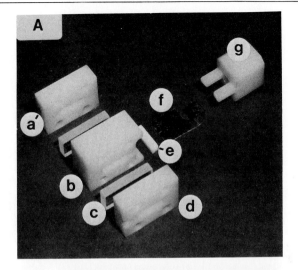

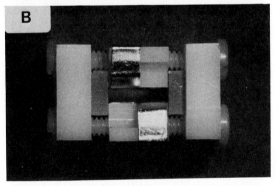

FIG. 2. Electric birefringence sample cell. (A) Exploded view of the various components comprising the cell. (a) Short line points to diagonal channel which directs a jet of predried, precooled air against the cell window; (b) main cell block; (c) glass windows; (d) window compression end-plate; (e) lower spacer for electrodes; (f) platinum electrode; (g) cell cap and upper spacer for electrodes. Lower spacer is made of Teflon, all other polymer components are made of Delrin. (B) Assembled cell, top view.

the brass block, is covered by a Plexiglas housing. The cooled air from the brass block maintains a slight positive pressure within the Plexiglas housing and eventually escapes through two small holes in the faces of the housing (through which passes the sample light beam) as well as through the electrode ports. The jets of air impinging on the two windows of the cell prevent condensation when experiments are performed at temperatures below ambient temperature. Moreover, they reduce the response time of the cell following temperature adjustments.

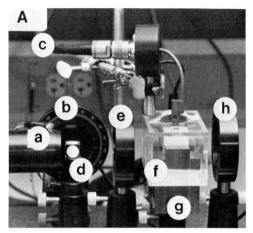

FIG. 3. Cell housing assembly. (A) Portion of the TEB instrument including the cell housing assembly. (a) Laser head; (b) reference analyzer; (c) HV coaxial cable; (d) beam-splitter; (e) prism polarizer; (f) Plexiglas cover; (g) brass block; (h) quarter-wave plate. (B) Superior diagonal view of the cell housing assembly, with the cell and Plexiglas cover removed. (a) Prism polarizer; (b) coolant and air lines; (c) electrode contact jacks; (d) electrode contacts (spring brass); (e) exit ports for air; (f) quarter-wave plate; (g) portion of brass block housing heat-exchangers for air; (h) central portion of brass block within which coolant circulates.

Temperature Monitor

A low-heat-capacity temperature probe consisting of a GB41J1 thermistor (Fenwal) encased in a thin-walled glass capillary is used to measure the temperature of the solution within the cell. The thermistor is connected to a digital monitor where the impedance of the thermistor is displayed. The temperature response of the thermistor is nonlinear and a calibration curve is required for each probe. These are developed using NBS traceable high-precision thermometers over the range 0–100°.

Dual Channel Detector

An outline of the elements of the dual channel detector is shown in Fig. 4. Light from each of the two incoming beams is detected using a PIN5 photodiode (United Detector Technology) and the output of each of the photodiodes is fed into an LH0032 operational amplifier (National Semiconductor) which acts as a differential amplifier. The output of the LH0032 is fed to an LH0002 current amplifier (National Semiconductor) which provides the signal to the oscilloscope. The gain of the amplifier

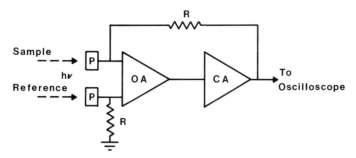

FIG. 4. Outline of the circuit of the dual-channel detector. Configuration is that of a standard differentiator utilizing two PIN photodiodes (P). There are many different fast operational amplifiers (OA) and current amplifiers (CA) currently on the market. The gain and bandwidth of the detector are determined by the value of R, which can be varied according to experimental requirements. The photodiodes are operated in the photoconductive mode.

circuit is determined by the feedback resistor, R. The bandwidth of the detector is also determined by the value of R. Therefore, R should be kept as small as possible subject to gain requirements. Improved performance of the detector is obtained if the two channels are balanced optically rather than electronically, that is, if the prepulse output of the detector is zeroed by means of adjustment of the reference analyzer.

Digitizing Oscilloscope

A Nicolet Explorer IIIA digitizing oscilloscope is used in conjunction with a Model 204 plug-in differential amplifier (8 bit, 50 ns/pt acquisition rate) as a fast analog-to-digital converter and storage buffer.

Microcomputer

An Altos/8000, Z-80-based microcomputer is linked in a bidirectional mode to the Nicolet oscilloscope, allowing data transfer from the oscilloscope to the computer or transfer of a processed curve back to the oscilloscope for display. Most curve-averaging and baseline subtraction procedures are performed with the Altos/8000 computer.

System Trigger

Upon actuation by the operator, the system trigger delivers a pulse to the HV pulse generator as well as the digitizing oscilloscope to begin a measurement.

Experimental Procedures

Calibration of the Instrument Response Time

Nitrobenzene has often been used as a means of determining the overall instrument response time since it displays a substantial electric birefringence signal. One major disadvantage of nitrobenzene, however, is its high toxicity. An alternative currently employed by the author consists of a 1:1 mixture of dimethyl sulfoxide (DMSO) and water. As currently configured, the overall response time of the author's system is <100 nsec.

Sample Preparation

The following discussion will be limited to the preparation of DNA fragments; however, the same principals of characterization and purification apply to the use of any biological macromolecule. In order to fully exploit the sensitivity of the TEB method, it is important that the DNA sample under investigation be well characterized as well as being highly purified. In many physical studies of DNA, insufficient attention has been paid to this point; a brief discussion of the purification and characterization of DNA molecules intended for physical studies will therefore be presented. Due to the high precision possible with the TEB approach, unsequenced DNA fragments produced by shearing or by nuclease digestion (e.g., production of sized DNA molecules by digestion of nucleosomal particles) should not in general be regarded as acceptable materials for use in TEB measurements, since the methods used for sizing such fragments (e.g., column chromatography, gel electrophoresis) are less precise than the TEB measurements themselves.

Techniques for the isolation and purification of plasmid DNA are now standard, and are described in detail elsewhere (e.g., this series, Vols. 65 and 68). I shall simply point out several features of the purification procedures that are not commonly recognized in the hope that several common pitfalls can be avoided.

CsCl–EtBr Density Gradient Centrifugation

This procedure is in common use as a means of separating covalently closed plasmid DNA from chromosomal DNA as well as proteins and RNA. It should be understood, however, that significant amounts of all three contaminants can be withdrawn with the supposedly purified plasmid. Double-banding will improve the purity of plasmids; however, small RNA or protein molecules whose association with DNA is primarily nonelectrostatic may still persist in the plasmid fraction. For studies of small restriction fragments, where subsequent purification steps are required, such RNA and protein contamination does not usually present a problem. If, on the other hand, the plasmid itself (including linear form) is used, a significant amount of RNA may be present. Such RNA would give a falsely high estimate of plasmid concentration based on A_{260} readings, the net effect being a low estimate of the specific Kerr constant. It should be noted that much of the contaminating RNA, consisting of small single-stranded molecules, stains extremely poorly with EtBr and may be missed entirely if EtBr-stained gels are used as a criterion of DNA purity (see below). It is therefore prudent to introduce at least one additional fractionation step following CsCl–EtBr banding in order to resolve contaminating species from the DNA molecule of interest. It should also be noted that commercial preparations of DNA (e.g., calf thymus) contain variable amounts of both proteins and RNA, depending on the supplier, and should not be considered as suitable material for TEB measurements until additional purification steps have been completed (see below). Finally, passage of DNA solutions containing EtBr over a cation exchange resin such as Bio-Rad AG-50-X8 removes EtBr more efficiently than does butanol extraction.

Isolation of DNA Restriction Fragments

A number of methods currently exist for use in the fractionation of DNA molecules of various sizes. Wells and co-workers[4] have extensively exploited high-pressure liquid chromatography (HPLC) using RPC-5 as a means of fractionating the various products of restriction digests of plasmid DNAs. RPC-5 HPLC is capable of excellent resolution of DNA molecules less than 500-bp in length, resolution decreasing as the molecular weight increases. DNA so obtained is of high purity in most cases; the exceptions generally involve DNA molecules of similar molecular weight, where resolution can become a problem, and improper use of the column material (e.g., presence of phenol and/or chloroform in the sample), in which case "bleeding" of the column material may result.

Another general method for the isolation of DNA molecules involves excising DNA bands from either acrylamide or agarose gels.[5,6] This method is generally the easiest to perform, costs the least, and usually gives the highest resolution of any of the current methods. Moreover, its use is not restricted to a particular molecular weight range. The major

---

[4] R. D. Wells, S. C. Hardies, G. T. Horn, B. Klein, J. E. Larson, S. K. Neuendorf, N. Panayatatos, R. K. Patient, and E. Selsing, this series, Vol. 65, p. 327.

[5] H. O. Smith, this series, Vol. 65, p. 371.

[6] R. C.-A. Yang, J. Lis, and R. Wu, this series, Vol. 68, p. 176.

caveat associated with this approach is the presence of gel materials in the DNA following its extraction from the gel slice. Elias and Eden[7] have reported that DNA contaminated with materials from acrylamide gels results in spurious birefringence signals. This problem can be overcome by step-gradient elution of DNA from an anion exchange resin such as DEAE-cellulose.[5] The major drawbacks of the gel-slice approach are reduced capacity and variable recoveries. However, many apparently low recoveries simply reflect the fact that substantial RNA and/or chromosomal DNA impurities were present in the original DNA mixture.

A third approach to the fractionation of DNA molecules on the basis of size involves the use of various preparative gel electrophoresis devices.[8] We have developed several continuous-collection electrophoresis columns in our own laboratory which are capable of loading in excess of 50 mg of DNA, thus approaching the capacity of RPC-5 columns. Moreover, there is no DNA size limitation with this method. Since the gels are not disrupted for DNA isolation, the possibility of contamination by gel materials is substantially reduced and, following concentration and step-gradient elution of the pooled column fractions over DEAE-cellulose, the resulting DNA is quite pure. One advantage of the use of column methods (preparative gel columns as well as RPC-5 columns) is the ability to identify UV-absorbing impurities such as RNA which are present in the original loading material (Fig. 5).

We have found it convenient to store purified DNA samples as isopropanol or ethanol slurries until use, at which time the slurries are pelleted, DNA resuspended as a concentrated stock solution in a low ionic strength buffer or water (dialyzed if necessary), and then diluted into the final solution for measurement.

Characterization of DNA Molecules under Study

Most of the following discussion is based on the assumption that the DNA molecules of interest have been cloned and properly sequenced. Such molecules are, in principle, perfectly monodisperse and their sizes (in number of base pairs) are known exactly. The major defects associated with such molecules include sites of depurination and/or depyrimidination, single-stranded nicks or gaps, and degradation of the ends of the molecules (as through the action of exonucleases). It is, of course, important to determine the extent to which such defects exist before beginning TEB measurement; however, DNA molecules may become partially degraded during the course of certain types of measurements (extremes of

[7] J. G. Elias and D. Eden, *Macromolecules* **14,** 410 (1981).
[8] E. Southern, this series, Vol. 68, p. 152.

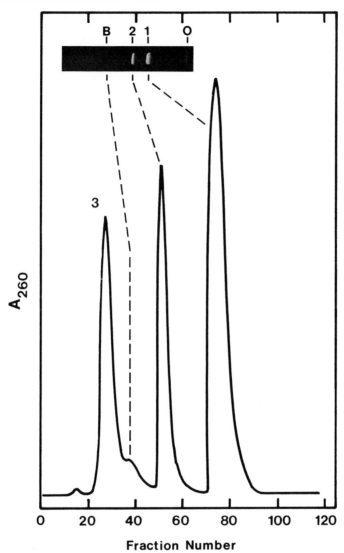

Fraction Number

FIG. 5. Demonstration of the unacceptability of ethidium bromide (EtBr)-stained gels as an indicator of DNA purity: 1.5 mg of a DNA restriction digest (100-bp + 266-bp fragments) containing approximately 25% low-molecular-weight RNA (previously subjected to RNase A digestion) was run on a 5-cm, 9% polyacrylamide electrophoretic column. The column effluent was monitored continuously. Inset: 9% polyacrylamide gel onto which 5 μg of the mixture described above had been loaded. (O), origin; (1), 266-bp fragment; (2) 100-bp fragment; (B) bromphenol blue. No EtBr staining material was observed to the left of (B). No RNA was detected by EtBr staining when 15 μg of the mixture was run on the same analytical gel.

pH, high temperatures, etc.), and such molecules should also be characterized following the experiment.

Depurination/Depyrimidation. It is well known that low pH and/or high temperature can result in significant depurination. Such adverse conditions may arise during the course of measurements, particularly if no buffer is used or if excessive power per pulse is delivered to the cell. One way to readily detect the presence of apurinic/apyrimidinic (AP) sites in the DNA molecules under study is to run a portion of the sample on alkaline (pH 12.5) agarose gels. Since AP sites are alkalai labile, the presence of such sites can be identified as single-stranded species smaller than full length.

Single-Stranded Nicks and/or Gaps. The presence of such defects can be identified in the same fashion used for the detection of AP sites, namely, through the use of alkaline agarose gels. Agarose gels up to several percent may be run in order to examine molecules less than 100 base pairs in length. Single-stranded DNA fragments may be demonstrated using autoradiography (for fragments previously end-labeled with ^{32}P), or following neutralization of the gel in a neutral buffer solution by EtBr staining.

Degradation of the Ends of DNA Molecules. One advantage of using DNA fragments generated by the action of restriction endonucleases is that the ends of the molecule are precisely defined. Ligation of the ends of a fragment should regenerate the original restriction site. Tests of the integrity of the ends of a molecule therefore consist of ligation of the molecule to higher order multimers followed by cleavage with the same restriction enzyme used to generate the original fragment. Loss of even a single base pair will result in the failure of the enzyme to cut.

Factors Determining the Configuration of the Electric Field Pulse

The major limitation of the use of TEB to study biological macromolecules has been the restriction of such measurements to very low ionic strengths (1 mM). This is due to the fact that under conditions generally used for TEB (and TED) measurements (nearly saturating fields and pulse widths sufficient for the establishment of an equilibrium distribution of orientated molecules), substantial joule heating takes place. This effect explains why early investigators were plagued with apparent sample degradation. More recently it has been observed that, in the case of DNA, field-free birefringence decay is independent of the configuration of the pulse for $E < 15$ kV/cm and for pulse widths less than 10 μsec (0.2 mM ionic strength).[9] Whereas this range of pulse amplitudes and widths can-

[9] P. J. Hagerman, *Biopolymers* **20**, 1503 (1981).

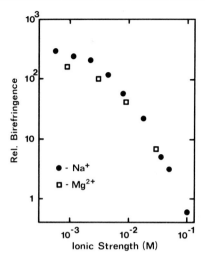

FIG. 6. Dependence of the relative birefringence intensity for a 126-bp restriction fragment as a function of ionic strength. This figure is intended only to illustrate the strong dependence of the birefringence signal on ionic strength. No corrections have been made at the higher ionic strengths for reductions in field strength due to reduced cell impedance. Moreover, the short field pulses used (although of constant length) do not produce steady-state orientation.

not be applied at much higher ionic strengths due to limitations associated with joule heating, it is generally true that field-free decay is independent of the orienting pulse under conditions where heating of the sample is not significant. It should be noted that with DNA molecules displaying complex decay curves, only the terminal portion of the curve decays at a rate which is independent of the orienting pulse.[9] Furthermore, field-induced rise times (orientation in the presence of the field) are strongly dependent on the field strength under nearly all conditions, and therefore should not be used to extract hydrodynamic information.

Another potential limitation of TEB is the reduction in signal strength at constant E with increasing ionic strength (Fig. 6). One must therefore attempt to compensate by increasing E and/or sample concentration. Our basic approach has been to establish a value of E for any given pulse width and ionic strength which will keep the associated temperature rise well below 1° at the particular ionic strength of interest. Sample concentration is then increased until a reasonable signal is obtained. Under these conditions, DNA samples can be subjected to several hundred pulses without any degradation as judged by the criteria set forth above. For relatively short DNA fragments, one has broad latitude in adjusting DNA concentration. For example, the field-free decay time associated with a 211-bp

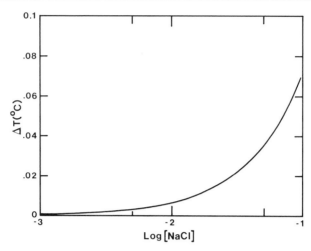

DNA fragment is constant (within 5%) over a range of fragment concentrations extending from 10 to over 500 μg/ml.[9a]

Quantitation of Joule Heating during the Electric Field Pulse

For a square-wave pulse (E = constant during pulse), the expression presented by Eigen and de Maeyer[10] for the associated temperature rise can be simplified to the form,

$$\Delta T = (E^2/R_c)(t/v\rho c_p) \tag{4}$$

where E is the voltage applied to the cell, R_c is the resistance of the cell (a function of cell dimensions and solution composition), v is the volume of the space between the electrodes, ρ is the solution density, and c_p is the specific heat at constant pressure. For a cell having a path length of 1.0 cm, an electrode gap of 0.2 cm, and an electrode height of 0.5 cm (v = 0.1 cm³), ΔT per μsec pulse-width is given as (assuming ρ, c_p =1)

$$\Delta T(^\circ C) = 2.4 \times 10^{-6}E^2/R_c$$

where E is expressed in volts and R_c is expressed in ohms. Figure 7 depicts the temperature rise expected for a 1-μsec pulse of 1 kV (5 kV/cm applied to the above cell) as a function of concentration of sodium chlo-

[9a] P. J. Hagerman, in preparation.
[10] M. Eigen and L. de Maeyer, *in* "Technique of Organic Chemistry" (S. L. Friess, E. S. Lewis, and A. Weissberger, eds.), p. 895. Wiley (Interscience), New York, 1963.

ride. Equation (4) should be used only as a rough guide when designing experiments; for more precise estimates of ΔT, we have found a combination of temperature-dependent changes in refractive index and the use of low heat capacity temperature probes to be quite useful.

Under conditions where signal averaging is required, it is important to establish that the overall temperature rise during the series of field pulses remains small. The aggregate temperature rise will be a function of the pulse repetition rate as well as the thermal conductivity of the cell. For solutions of moderate ionic strength (10–50 mM), using the cell described above, pulse repetition rates of 0.02 to 0.05 Hz do not produce any aggregate temperature rise.

Data Acquisition

Field-free birefringence decay curves are digitized, utilizing 1024 equally spaced time points, with a time range chosen such that essentially all of the useful decay information is contained within the first 20% of the record (204 points). Upon transfer of the data from a single measurement to the microcomputer, the baseline signal is determined from the last 40 time points of the sample record, and the resultant value back-subtracted from all 1024 points comprising the record. This baseline-corrected curve is then combined with previously obtained sample records in an arithmetic fashion to yield the final averaged curve. Current software routines allow the accumulation of any number of samples; however, most curves are comprised of from 1 to 40 samples, depending upon the nature of the initial signal.

Analysis of Data

The current chapter is concerned with the use of TEB to obtain information regarding the hydrodynamics of macromolecules (in particular, DNA); therefore, this section will be devoted to the extraction of decay times from birefringence decay curves. An extensive discussion of corrections for stray light and residual birefringence, along with comparisons among various means of analyzing complex decay curves, has been presented by Fredericq and Houssier[1] and will not be reproduced here.

Birefringence decay curves are displayed on a semilogarithmic scale as indicated in Fig. 8. Decay curves are usually normalized with respect to the value of the birefringent signal [$\Delta n(0)$] at the time the field pulse is turned off. Although $n(t)$ is not strictly linearly related to $\Delta I(t)$, as can be seen by inspection of Eqs. (1)–(3), the relation

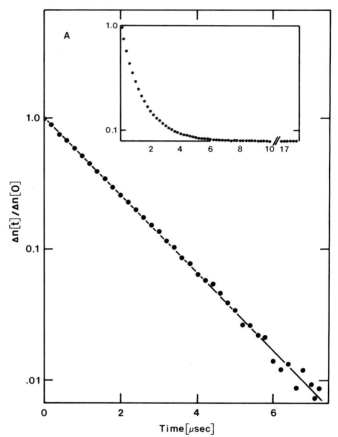

FIG. 8. Examples of birefringence decay curves. (A) Semilogarithmic plot of the decay curve for a 98-bp DNA restriction fragment displaying single-exponential decay over two orders of magnitude of birefringence signal intensity. Measurement condition: 2.0 mM Tris–HCl, pH 8.0, 4.0°; 2-μsec pulse width; 5 kV/cm pulse amplitude. Inset: same curve plotted on a linear ordinate scale. (B) Semilogarithmic plot of the decay curve for a 203-bp restriction fragment displaying an initial fast-relaxation component. Measurement conditions: 1.0 mM Tris–HCl, 0.1 mM Na$_2$-EDTA, 5 mM NaCl, pH 8.0, 4.0°; 2-μsec pulse width; 5 kV/cm pulse amplitude.

$$\frac{\Delta n(t)}{\Delta n(0)} = \frac{\Delta I(t)}{\Delta I(0)}$$

will, under conditions where $\Delta I(0)/I_\alpha < 0.1$, yield relaxation times that are accurate to within 1% when $\Delta I(t)/\Delta I(0)$ curves are analyzed directly.

For curves demonstrating single exponential decay patterns (Fig. 8A), relaxation times are determined from the slope of the semilogarithmic

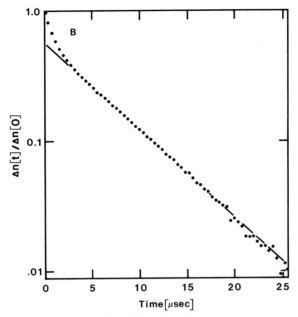

FIG. 8. (*continued*)

decay curve using least-squares curve-fitting routines. For curves demonstrating complex decay patterns (Fig. 8B) relaxation times are determined
as above, using the terminal portion of the curve. As a simple check of the
validity of this last approach, it should be demonstrated that there is no
systematic dependence of the apparent slope, and hence, τ, on the position of the first point chosen for inclusion in the analysis. It should be
noted that while the birefringence decay curves of rigid structures may
comprise up to five exponential components,[2] there is no theoretical justi-
fication for analyzing the fast component of complex decay curves for
flexible polymers (e.g., DNA) in terms of one or more exponential terms.
Therefore, a multiexponential analysis of such curves should be avoided.

*Comparison of Two Birefringence Decay Curves by Analyzing Their
Differential Decay (Differential Decay of Birefringence, DDB)*

Using the instrumental configuration described above, S/N ratios exceeding 1000 : 1 (single shot) are obtained routinely, thus allowing curve
analysis to be carried out over two decades in amplitude reduction ($t >
4\tau$). Under these conditions, τ can be determined with an attendant un-
certainty of less than 5%. However, under conditions where S/N is re-

duced (e.g., high ionic strength) or where very subtle differences in structure are being sought (i.e., small differences between two τ's), the difference between two decay curves can be analyzed directly, resulting in the ability to quantitate the difference between two decay times differing by as little as one percent.

For two normalized birefringence decay curves having associated decay times, τ_1 and τ_2, their difference decay curve is given by

$$\Delta n_2(t) - \Delta n_1(t) = e^{-t/\tau_1}[e^{-tR/\tau_1} - 1] \qquad (5)$$

where $R = \tau_1/\tau_2 - 1$.

In performing a difference analysis, the two normalized birefringence decay curves to be analyzed are subtracted in a point-by-point fashion. Using the value of τ_1 determined from the designated reference curve, a curve-fitting routine is then utilized in order to find the value of R which provides the best agreement between Eq. (5) and the experimentally derived difference curve. Alternatively, a two-parameter analysis can be performed without recourse to a separate determination of τ for either of the two decay curves.

For birefringence decay curves demonstrating a fast component, only the terminal portion of the curve should be used in conjunction with Eq. (5). However, it should be noted that the DDB method can be utilized for decay curves of arbitrary complexity when one is interested only in determining whether differences exist.

An example of the application of the DDB approach is presented in Figs. 9[10a] and 10. In this example, the uncertainty in R (± 0.01) represents the combined standard error obtained from the single-parameter (R) fit displayed in Fig. 10 and the initial determination of τ_{XHR}. The value of R determined using the single parameter analysis (τ_{XHR} determined independently) is not particularly sensitive to the choice of τ_{XHR}. A 5% uncertainty in the value of τ_{XHR} would lead to a 1.7% uncertainty in R. The overall standard error corresponds to a difference in length between two straight rods of 0.3%, thus underscoring the extreme precision of the DDB approach.

Interpretation of Results

As mentioned above, for rigid particles of arbitrary shape, the decay of birefringence can be described in terms of up to five decay times depending upon the symmetry of the particle with respect to both its charge distribution (i.e., how it orients in the field) and the directions and magnitudes of the various optical transitions within the particle. These decay

[10a] P. J. Hagerman, *Proc. Natl. Acad. Sci. U.S.A.* **81**, 4632 (1985).

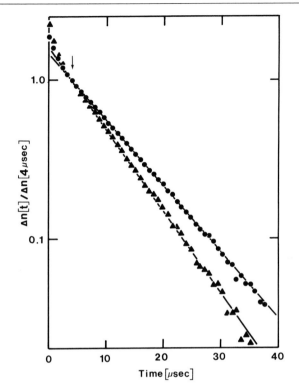

FIG. 9. Semilogarithmic plots of the decay of birefringence for the two 242-bp DNA fragments, KP242 and XHR242. Buffer composition: 1.0 mM Tris–HCl (pH 8.0), 0.1 mM MgCl$_2$. KP242 has been shown to possess stable curvature in solution.[10a] The birefringence decay curves have each been normalized with respect to their respective birefringence signals 4 μsec after removal of the electric field (descending arrow). The portion of each curve corresponding to $t < 4$ μsec displays a small fast component and is therefore excluded from the subsequent analysis of differential decay of birefringence (DDB).

times can be expressed in terms of the three principal rotational diffusion coefficients of the particle.[2] Iterative methods for the computation of rotational diffusion coefficients for rigid structures have been presented by Garcia de la Torre and Bloomfield.[11,12] Various limiting expressions for rigid particles of high symmetry (e.g., spheres, rods, ellipsoids of revolution) can be found in standard textbooks on physical biochemistry.

For flexible linear polymers such as DNA, an expression has been derived[3] which relates the largest decay time for a polymer of given con-

[11] J. Garcia de la Torre and V. A. Bloomfield, *Biopolymers* **16,** 1747 (1977).
[12] J. Garcia de la Torre and V. A. Bloomfield, *Biopolymers* **16,** 1765 (1977).

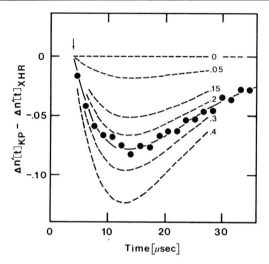

FIG. 10. Plot of DDB using the prenormalized curves from Fig. 9 [$\Delta n'(t) = \Delta n(t)/\Delta n$ (4 μsec)]. The filled circles have been obtained by subjecting the primary difference curve to a 3-point averaging procedure. The dashed lines represent expected difference curves for various values of R (= $\tau_{XHR}/\tau_{KP}-1$) as indicated. The solid line represents a least-squares best fit to the experimental difference curve ($R = 0.24 \pm 0.01$).

tour length and persistence length, P, to the decay time that same polymer would have if it were a perfectly straight, rigid rod. The ratio, $\tau(P)/\tau(\text{Rod})$, is given by

$$R_c = \tau(P)/\tau(\text{Rod}) = R_a X(1 - Y) \tag{6}$$

where

$$R_a = 1.012 - 0.248X + 0.0337X^2 - 0.00198X^3$$

$$Y = 0.0647X - 0.0115X^2 + 0.000989X^3$$

and where X is the contour length of the polymer in units of P. The range of validity of Eq. (6) is given by

$$0.1 < X < 5.0, \qquad L/d > 20.$$

For DNA, this corresponds to molecules ranging in size from 150 bp to about 730 bp (for $P = 500$ Å).[9] [For further discussions regarding the application of Eq. (6), see Refs. 3,9,13.]

Equation (6) has been used to determine the value of P for DNA directly.[9] However, it can also be used in association with a DDB analysis

[13] P. J. Hagerman, *Biopolymers* **22**, 811 (1983).

to provide a correction for residual flexibility. In this latter instance, $R_{c,1}/R_{c,2}$ can be used even if $L/d < 20$, since attendant errors will be quite small as that ratio approaches unity.

It should be noted that the assumption of "rod-like" character (i.e., structural rigidity) for flexible polymers shorter than one persistence length is rendered invalid by the extreme sensitivity of TEB. For example, a DNA molecule 100 bp long ($X = 0.68$ for $P = 500$ Å) would demonstrate a birefringence decay time approximately 18% below that expected for the rigid-rod limit. Therefore, in determining various structural features for DNA, as well as for other flexible polymers, one needs to make appropriate corrections for residual flexibility. For the 100-bp DNA molecule referred to above, failure to account for residual flexibility would lead to an apparent rise/bp of only 3.2 Å (assuming that the true rise/bp is 3.4 Å).

[14] Measurements of Protein Hydration by Various Techniques

By HELMUT PESSEN and THOMAS F. KUMOSINSKI

Introduction

The problems of measuring the interaction of water with biological macromolecules have long plagued biochemists. The importance of this interaction with respect to protein primary, secondary, and tertiary structure is well documented.[1] Nevertheless, several basic aspects of this topic remain unresolved. Among these are a clear definition of the hydration of a protein, an understanding of the relationships between different experimental values of total hydration and the methods of measurement from which they are obtained, and whether the interactions are strong or weak.

In this chapter we will deal with hydration measurements made on globular proteins in solution[2] by three techniques which have been found in recent years to afford certain useful insights: nuclear magnetic resonance (NMR) relaxation, small-angle X-ray scattering (SAXS), and hydrodynamics (velocity sedimentation). The latter two will be treated together, because sedimentation, which cannot by itself give definitive

---

[1] I. D. Kuntz and W. Kauzmann, *Adv. Protein Chem.* **28**, 239 (1974).

[2] The study of hydration by NMR measurements of nonfreezable water does not fall within the purview of this article. It has been reviewed elsewhere.[1]

METHODS IN ENZYMOLOGY, VOL. 117

values of hydration, is very useful in corroborating SAXS data. Conversely, sedimentation coefficients can be calculated independently from SAXS. NMR methodology will be treated in particular detail, both because of its potential utility for the present purpose, and because previous treatment in the literature has been somewhat limited. Specific discussions of these methods will be followed by some general remarks, including an evaluation of the different approaches.

Approaches

NMR Relaxation

Theory. Proteins, by their very nature as polyelectrolytes with large charge-to-mass ratios, are susceptible to severe interaction effects in solution. Although the consequent deviations from ideality are among the most important effects in solution theory, they have not received sufficient consideration in NMR studies of proteins. A prime example is the nonideality caused by repulsion due to net positive or negative charges, which has been demonstrated to have serious effects on measurement other than NMR (light scattering, osmotic pressure, sedimentation equilibrium and velocity, translational diffusion, and even pH titration).[1,3-7] Thus Pedersen,[5] on the basis of the theory of Tiselius,[6] as early as 1940 showed that a potential gradient is created in a sedimenting solution during both equilibrium and velocity runs as a result of high net charge on a macromolecule; Booth[7] has quantitated such behavior by a general theory for charged spheres. Sedimentation coefficients of bovine serum albumin at acid pH are lower by at least half when compared with the same protein in the presence of 0.1 M salt[8]; the salt thus minimizes the high repulsive effect of the molecular charges. Light scattering and osmotic pressure measurements of aqueous proteins carrying high net charge indicate decreased apparent weight-average molecular weights with increasing concentration, in consequence of intermolecular repulsion,[3,9-11] according to the equation:

[3] C. Tanford, "Physical Chemistry of Macromolecules," pp. 227, 293, 352, 563. Wiley, New York, 1961.

[4] H. K. Schachman, "Ultracentrifugation in Biochemistry," pp. 226ff. Academic Press, New York, 1959.

[5] K. O. Pedersen, *in* "The Ultracentrifuge" (T. Svedberg and K. O. Pedersen, eds.), p. 16. Oxford Univ. Press (Clarendon), London and New York, 1940.

[6] A. Tiselius, *Kolloid-Z.* **59**, 306 (1932).

[7] F. Booth, *J. Chem. Phys.* **22**, 1956 (1954).

[8] K. O. Pedersen, *J. Phys. Chem.* **62**, 1282 (1958).

[9] S. N. Timasheff and M. J. Kronman, *Arch. Biochem. Biophys.* **83**, 60 (1959).

$$M_{w,app}^{-1} = M_{w,0}^{-1} (1 + cd \ln \gamma/dc)$$ (1)

where γ is the activity coefficient of the protein at a concentration c, $M_{w,app}$ is the experimentally derived apparent weight-average molecular weight, and $M_{w,0}$ is the true molecular weight at infinite dilution of the protein. From the virial expansion for osmotic pressure it follows that

$$d \ln \gamma/dc = 2B_0 + 3Bc + \cdots$$ (2)

where the B quantities are the second and higher virial coefficients; for a repulsive effect usually $d \ln \gamma/dc = 2B_0 > 0$ (i.e., $\gamma > 1$), in accord with general electrolyte solution theory.

Beyond simple electrostatic repulsion, one needs to be concerned with a less obvious phenomenon, namely the charge fluctuations treated in the theory of Kirkwood and Shumaker.[12] Extensive experimental evidence for this theory has been furnished by many investigations using light scattering; for proteins in water under isoionic conditions the apparent weight-average molecular weight increases with increasing concentration.[13-16] Ordinarily, this phenomenon might be interpreted simply as an aggregation of the molecule. However, Kirkwood and Shumaker have shown that, since proteins are polyampholytes rather than simple polyelectrolytes, attractive forces arise in isoionic protein solutions from statistical fluctuations, both in charge and in charge distribution; these in turn are associated with fluctuations in the number and configurations of protons bound to the protein molecule. Briefly stated, one or more virial coefficients in $c^{n/2}$ powers must be added to Eq. (2) and these coefficients, due to progressive ionization of the macromolecule, have negative values. Moreover, the virial coefficients of the c^n terms should also usually be negative. Experimental results show the virtual elimination of such charge-fluctuation effects in the presence of a moderate amount of salt. Since virial effects are found for other solution parameters, such as sedimentation coefficients, intrinsic viscosities, and linear as well as rotatory diffusion coefficients, charge fluctuations can be expected to affect these hydrodynamic parameters also.

[10] M. J. Kronman and S. N. Timasheff, *J. Phys. Chem.* **63**, 629 (1959).
[11] S. N. Timasheff, *in* "Electromagnetic Scattering-I.C.E.S." (M. Kerker, ed.), p. 337. Macmillan, New York, 1963.
[12] J. G. Kirkwood and J. B. Shumaker, *Proc. Natl. Acad. Sci. U.S.A.* **38**, 863 (1952).
[13] J. G. Kirkwood and S. N. Timasheff, *Arch. Biochem. Biophys.* **65**, 50 (1956).
[14] S. N. Timasheff, H. M. Dintzis, J. G. Kirkwood, and B. D. Coleman, *J. Am. Chem. Soc.* **79**, 782 (1957).
[15] S. N. Timasheff and I. Tinoco, *Arch. Biochem. Biophys.* **66**, 427 (1957).
[16] S. N. Timasheff and B. D. Coleman, *Arch. Biochem. Biophys.* **87**, 63 (1960).

Specifically, the translational diffusion coefficient D_t and the rotary diffusion coefficient D_r exhibit the influence of the activity coefficient, γ, which expresses the combined result of all such effects, according to

$$D_t = (8/6)D_r = [kT/(6\pi\eta r_s)](1 + d \ln \gamma/dc) \tag{3}$$

where k is Boltzmann's constant, 1.3806×10^{-16} erg $^\circ K^{-1}$, T is the temperature in K, η is the viscosity of the solvent, and r_s is the Stokes radius of the macromolecule. Accordingly, electrostatic repulsion and charge fluctuation will affect these, as well as other hydrodynamic parameters.

Measurements of protein hydration by NMR relaxation techniques using the two-state model of Zimmerman and Brittin[17] have yielded surprisingly large values for the correlation time τ_c of bound water when cross-relaxation effects were avoided by experiments in D_2O.[18,19] When cross-relaxation effects were not taken into account (e.g., in proton NMR of water), correlation times obtained from dispersion data also were high,[18-21] even though Andree[22] had shown that apparent τ_c values from frequency dependence of spin–lattice relaxation should be smaller as a result of cross relaxation. Also, recent ^{17}O NMR T_1 and T_2 data of Halle *et al.*[23] give τ_c values larger than would be predicted by the Debye–Stokes–Einstein equation, a result attributed by the authors to long-range coulombic protein–protein interactions. However, these measurements as well as many others were made, again, on isoionic protein solutions with no added salt. But since rotary diffusion is basic to these measurements (i.e., the bound water rotates with substantially the same τ_c as the protein) and is related to τ_c by $\tau_c = r_s^2/6D_r$, it can be seen from Eq. (3) that

$$\tau_c = [4\pi\eta r_s^3/(3kT)](1 + cd \ln \gamma/dc)^{-1} \tag{4}$$

and consequently, not only electrostatic effects due to the high charge-to-mass ratio of the protein but also charge-fluctuation effects should be encountered, both as a result of the methodology of these measurements. The absence of salt and the consequently large negative virial coefficients thus account for the excessive apparent τ_c values observed. The addition of salt should then result in correlation times more in accord with those

[17] J. R. Zimmerman and W. E. Brittin, *J. Phys. Chem.* 1328 (1957).
[18] K. Hallenga and S. H. Koenig, *Biochemistry* **15**, 4255 (1976).
[19] S. H. Koenig, K. Hallenga, and M. Shporer, *Proc. Natl. Acad. Sci. U.S.A.* **72**, 2667 (1975).
[20] S. H. Koenig and W. E. Schillinger, *J. Biol. Chem.* **244**, 3283 (1968).
[21] T. R. Lindstrom and S. H. Koenig, *J. Magn. Reson.* **15**, 344 (1974).
[22] P. J. Andree, *J. Magn. Reson.* **29**, 419 (1978).
[23] B. Halle, T. Andersson, S. Forsén, and B. Lindman, *J. Am. Chem. Soc.* **103**, 500 (1981).

calculated for the protein directly from structural considerations. In fact, two recent studies,[24,25] using different means to deal with the complicating effects of cross relaxation but both using salt, obtained τ_c values in good agreement with the known structure.

Since, to address this problem, one clearly should work with solutions containing salt, the theory now needs to be expanded to accommodate a three-component system. Expressions for NMR relaxation rates are rederived here in terms of the multicomponent theory of Casassa and Eisenberg,[26] since now the salt components can compete with water for binding sites on the protein molecule (e.g., charged side chains). The expressions embodying this theory are then tested against concentration-dependent spin–lattice and spin–spin relaxation data on β-lactoglobulin A (β-Lg A) in D_2O and H_2O under associated and unassociated conditions.[24]

The model used as a point of departure for this derivation is a fast-exchange two-state, three-component system. The initial assumptions are (1) only one correlation time exists for the bound-water state, (2) there is competition between bound water and salt for the interaction sites on the surface of the protein, but (3) there is no competition between salt and protein for water since experiments will be performed only on protein solutions with added salt, not on insoluble or powdered samples. Hence, a general equilibrium expression for the binding is

$$P + iW + jX \rightleftharpoons PW_iX_j \tag{5}$$

where P, W, and X represent protein, water, and salt at free concentrations p, w, and x, bound concentrations p_b, w_b, and x_b, and total concentrations

$$P = p + p_b, \quad W = w + w_b, \quad \text{and} \quad X = x + x_b \tag{5a–c}$$

respectively. Associated with the formation of each species PW_iX_j is an apparent macroscopic association constant K_{ij}. Then

$$p_b = p\sum_{i=0}^{q} \sum_{j=0}^{r} K_{ij}w^i x^j \tag{6a}$$

where q and r are any positive integers such that $q + r = n$, the total number of binding sites per molecule. At the same time,

$$w_b = p \sum \sum iK_{ij}w^i x^j \quad \text{and} \quad x_b = p \sum \sum jK_{ij}w^i x^j \tag{6b,c}$$

[24] T. F. Kumosinski and H. Pessen, *Arch. Biochem. Biophys.* **218,** 286 (1982).
[25] H. Pessen, J. M. Purcell, and H. M. Farrell, Jr., *Biochim. Biophys. Acta* **828,** 1 (1985).
[26] E. F. Casassa and H. Eisenberg, *Adv. Protein Chem.* **19,** 287 (1964).

The average number of molecules of water or salt bound to a molecule of protein, i.e., the Scatchard hydration, $\bar{\nu}_W$, and salt binding $\bar{\nu}_W$, respectively, are then given by

$$\bar{\nu}_W = w_b/P \quad \text{and} \quad \bar{\nu}_X = x_b/P \quad (6d,e)$$

which may be also written in terms of the Eqs. (6a–c). (We designate here as Scatchard hydration all protein–water interactions characterized by multiple equilibria. Another type of water binding, preferential hydration, will be encountered further on. For a particularly clear exposition of the relationships between these concepts, see Ref. 27.)

The concentrations are commonly expressed in moles/liter but may, when convenient, equally well be expressed in moles/1000 g water, as will be the case in the following. While this will change the numerical values and dimensions of association constants, $\bar{\nu}_W$ and $\bar{\nu}_X$ will remain unaffected. Strictly speaking, all terms should be expressed as the corresponding activities rather than as molar or molal concentrations. (However, in view of the generally large molecular weights of proteins, even fairly concentrated solutions are usually no more than 10^{-13} M and, under the conditions of many experiments, virial coefficients are negligible; i.e., the protein has a small net charge. Deviations from this last assumption will be discussed specifically later.)

The experimentally derived quantity to be considered is the slope of the plot of either the spin–lattice (R_1) or the spin–spin (R_2) relaxation rate of the water vs the concentration of protein present. In analogy to other quantities encountered in solution theory, the slope $(dR/dP)_\mu$ may be termed the relaxation increment; the subscript μ indicates that solutions are at constant chemical potential and, in terms of experimental procedure, that exhaustive dialysis against buffer was performed prior to the proton NMR relaxation measurements.

Considering now the relaxation increment of an aqueous solution of protein in the presence of salt as a concentration-dependent function, and expressing the total relaxation rate of the water in the three-component system in terms of the solution components, one may write, on the basis of fast exchange of water between two fractions of rates R_b (for bound water) and R_f (for free water), and in view of Eq. (6b),

$$R = \frac{w_b}{W} R_b + \frac{w}{W} R_f = \frac{(p\Sigma\Sigma i K_{ij} w^i x^j)R_b + wR_f}{p\Sigma\Sigma i K_{ij} w^i x^j + w} \quad (7)$$

[27] G. C. Na and S. N. Timasheff, *J. Mol. Biol.* **151**, 165 (1981).

From R as a function of P, W, and X, one has the total derivative

$$\left(\frac{dR}{dP}\right)_{\mu} = \left(\frac{\partial R}{\partial P}\right)_{W,X} + \left(\frac{\partial R}{\partial W}\right)_{P,X}\left(\frac{dW}{dP}\right)_{\mu} + \left(\frac{\partial R}{\partial X}\right)_{P,W}\left(\frac{dX}{dP}\right)_{\mu} \qquad (8)$$

The coefficients of the three terms can be evaluated by partial differentiation of Eq. (7) with the use of Eqs. (6a–c). More directly, from Eqs. (7) and (5b) (where w, because of the exhaustive dialysis, is a constant),

$$R = \frac{(W - w)R_{b} + wR_{f}}{W} = R_{b} - (R_{b} - R_{f})\frac{w}{W} \qquad (9)$$

and from this,

$$\left(\frac{\partial R}{\partial P}\right)_{W,X} = 0, \qquad \left(\frac{\partial R}{\partial W}\right)_{P,X} = \frac{(R_{b} - R_{f})w}{W^{2}}, \qquad \left(\frac{\partial R}{\partial X}\right)_{P,W} = 0 \qquad (9a)$$

and therefore

$$\left(\frac{dR}{dP}\right)_{\mu} = \left(\frac{\partial R}{\partial W}\right)_{P,X}\left(\frac{dW}{dP}\right)_{\mu} = \frac{(R_{b} - R_{f})w}{W^{2}}\left(\frac{dW}{dP}\right)_{\mu} \qquad (10)$$

From Eqs. (5b) and (6d), $W = w + w_{b} = w + \bar{\nu}_{W}P$ and, with $dw/dP = 0$,

$$\left(\frac{dW}{dP}\right)_{\mu} = \bar{\nu}_{W} \qquad (11)$$

Because $w_{b} \ll W$, $w \simeq W$ ($\simeq 55.6$), and Eq. (10) becomes

$$\left(\frac{dR}{dP}\right)_{\mu} = \frac{R_{b} - R_{f}}{W}\bar{\nu}_{W} \qquad (12)$$

Since protein molecular weights are not always precisely known, it is convenient to replace the molality P by the concentration c (in g protein/g water) and the hydration $\bar{\nu}_{W}$ (in moles/mole) by $\bar{\nu}'_{W}$ (in the conventional units of g bound water/g protein). Then $c = PM_{p}/1000$ and $\bar{\nu}'_{W} = \bar{\nu}_{W}M_{W}/M_{P}$, where M_{P} and M_{W} are the molecular weights of protein and water, respectively. Also, $dP/dc = 1000/M_{p}$ and ($1000/WM_{W}$ being unity),

$$\left(\frac{dR}{dc}\right)_{\mu} = \left(\frac{dR}{dP}\right)_{\mu}\left(\frac{dP}{dc}\right) = (R_{b} - R_{f})\bar{\nu}'_{W} \qquad (13)$$

In the present work the relationship of R to c (as shown in Fig. 1 for both R_{1} and R_{2} results) was linear. One may write Eq. (9) alternatively as

$$R = R_f + (R_b - R_f)\frac{w_b}{W} = R_f + (R_b - R_f)\frac{P\bar{\nu}_W}{W}$$
$$= R_f + (R_b - R_f)c\bar{\nu}'_W \tag{13a}$$

and from Eqs. (13) or (13a), with a constant relaxation increment,

$$(dR/dc)_\mu = (R_b - R_f)\bar{\nu}'_W = k \tag{13b}$$

or

$$R = R_f + (R_b - R_f)c\bar{\nu}'_W = R_f + kc \tag{13c}$$

Absence of such linearity might be due to polydisperity and consequent changes in the $(R_b - R_f)$ factor, which would have to be allowed for. The other factor in the relaxation increment, $\bar{\nu}_W$, is generally taken to be independent of protein concentration (cf. Ref. 28). If it is not actually constant, an additional term accounting for its concentration dependence would have to be included in Eqs. (13b,c), unless a change from concentration units to activities will remove the nonlinearity, as discussed under Analysis of Data.

If it is desired to express protein concentrations as c_2 in g protein/ml solution, the solution density ρ and the concentration of the third component, c_3, need to be known, whereupon

$$c = \frac{c_2}{\rho - c_2 - c_3} \tag{14}$$

and

$$\left(\frac{dR}{dc_2}\right)_\mu = k\frac{\rho - c_3}{(\rho - c_2 - c_3)^2} \tag{14a}$$

The need to know solution densities may be eliminated for all except very high protein concentrations with a knowledge of the partial specific volume $\bar{\nu}$ and the solvent density ρ_0. From the definition of $\bar{\nu}$ it can then be shown that

$$c \simeq \frac{c_2}{\rho_0(1 - \bar{\nu}c_2) - c_3} \tag{14'}$$

and

$$\frac{dR}{dc_2} \simeq k\frac{\rho_0 - c_3}{[\rho_0(1 - \bar{\nu}c_2) - c_3]^2} \tag{14'a}$$

In cases of high protein concentration the more general Eqs. (14), (14a) may have to be used. At low protein concentration, where $\rho_0 \gtrsim 1.0$ while

[28] C. Tanford, *J. Mol. Biol.* **39**, 539 (1969).

c_2, $c_3 \ll \rho$, Eqs. (14a) and (14'a) reduce to expressions formally identical to Eq. (13b) with c_2 in place of c; a corresponding remark is true for Eqs. (14) and (14') with respect to Eq. (13c).

From Eq. (13b), $R_b = (k/\bar{\nu}'_W) + R_f$ for either R_1 or R_2, and therefore

$$R_{2b}/R_{1b} = (k_2 + R_{2f}\bar{\nu}'_W)/(k_1 + R_{1f}\bar{\nu}'_W) \tag{15}$$

But generally $R_f\bar{\nu}'_W \ll R_b$, hence, from the definition of k, one obtains the useful approximation

$$R_{2b}/R_{1b} \simeq k_2/k_1 \tag{15a}$$

The parameters k_1, R_{1f}, k_2, and R_{2f} are experimentally accessible as the slope k and intercept R_f from R_1 or R_2 vs c_2 plots, respectively, such as those of Fig. 1.

Use of these relationships in conjunction with the Kubo–Tomita–Solomon equations,[29,30] which relate R_{1b} and R_{2b} to the correlation time of the bound water, will give the requisite number of equations to permit simultaneous solution for $\bar{\nu}'_W$ and τ_c. These equations may be written as

$$R_{1b} = 2K\tau_c[(1 + \omega_0^2\tau_c^2)^{-1} + 4(1 + 4\omega_0^2\tau_c^2)^{-1}] \tag{16a}$$

and

$$R_{2b} = K\tau_c[3 + 5(1 + \omega_0^2\tau_c^2)^{-1} + 2(1 + 4\omega_0^2\tau_c^2)^{-1}] \tag{16b}$$

where $\omega_0 = 2\pi\nu_0$ is the nuclear angular procession frequency for the nuclide observed, in radians/sec, and K is a measure of the strength of the nuclear interaction, viz.

$$K_{\text{deuterons}} = (3/80)(e^2qQ/\hbar)^2(3\eta^2 + 1)^2S_{\text{deut}}^2 \tag{16c}$$

and

$$K_{\text{protons}} = (3/20)\hbar^2\gamma^4r^{-6}S_{\text{prot}}^2 \tag{16d}$$

Here e is the electronic charge, 1.6022×10^{-19} C, q is the electric field gradient, Q is the nuclear electric quadrupole moment, $\hbar$ is Planck's constant divided by 2π, 1.0546×1^{-27} erg · sec; η is a dimensionless parameter measuring the deviation from axial symmetry[31]; γ is the gyromagnetic ratio for the proton, 2.6752×10^4 rad · G^{-1} · sec^{-1}; r is the internuclear proton distance for water, 1.526 Å; and S_{deut} and S_{prot} are respective order parameters.[23] It may be noted that these expressions assume no particular model for the relaxation mechanism accompanying NMR hydration. The

[29] R. Kubo and K. Tomita, *J. Phys. Soc. Jpn.* **9**, 888 (1954).
[30] I. Solomon, *Phys. Rev.* **99**, 559 (1955).
[31] A. Abragam, "The Principles of Nuclear Magnetism," Chapter 8. Oxford Univ. Press, London and New York, 1961.

thermodynamic theory can be used whether isotropic relaxation ($S = 1$) or anisotropy of the bound water ($S < 1$) is hypothesized, where in the latter case the "bound" should be understood in the sense of "hydrodynamically influenced layers" or "surface-induced probability distribution of water molecules." [23,32,33] The above treatment can easily be extended to a model postulating three or more states.

Experimental Procedures. Preparation of solutions. The following procedures are described to illustrate methods that have led to satisfactory results in a study of the whey protein β-lactoglobulin (β-Lg) in solution.[24] Protein solutions to be used for proton resonance measurements, prepared 1 day before use, were exhaustively dialyzed overnight against buffer at 0–5°; dilutions for the concentration series to be studied were made with the appropriate dialyzate. Solutions to be used for deuteron resonance measurements were made up from a stock solution prepared by partial deuterium exchange. A suitable amount of crystalline protein in a stoppered vial was allowed to equilibrate repeatedly for 24-hr periods at 4° as a slurry with a small quantity of D_2O, followed by high-speed centrifugation and addition of fresh D_2O, for a total of five times. The solutions were buffered by direct addition of solid potassium phosphate, and the pH was adjusted by addition of 0.1 N NaOD in D_2O. Concentrations of β-Lg were determined spectrophotometrically from an absorption coefficient of 0.96 ml mg^{-1} cm^{-1} at 278 nm.[34]

Relaxation measurements. Resonance relaxation spectra were obtained by Pulse Fourier Transform spectroscopy with a JEOL FX60Q spectrometer[35] operating at a nominal frequency of 60 MHz. The frequency of observation for protons was 59.75 MHz; for deuterons, 9.17 MHz. Raw data were in the form of relative intensities as calculated by the JEOL 980B computer.

Since the high concentration of water in a dilute solution produces an intense signal, a single accumulation at the particular sample temperature (2, 10, or 30 ± 1°) was sufficient for each spectrum. Even then, care was necessary to avoid exceeding the dynamic range of the computer with consequent truncation. To this end, as well as to economize on the limited amount of the A variant of the protein available, small sample volumes were employed by use of a microcell assembly with an expendable 35-μl sample bulb, available from Wilmad Glass Company, Inc. The protein

[32] R. H. Walmsley and M. Shporer, *J. Chem. Phys.* **68,** 2584 (1978).
[33] H. J. C. Berendsen and H. T. Edzes, *Ann. N.Y. Acad. Sci.* **204,** 459 (1973).
[34] R. Townend, R. J. Winterbottom, and S. N. Timasheff, *J. Am. Chem. Soc.* **82,** 3161 (1960).
[35] Reference to brand or firm name does not constitute endorsement by the U.S. Department of Agriculture over others of a similar nature not mentioned.

solution was introduced very slowly into the spherical bulb by means of a fine-gauge syringe needle inserted through its capillary neck, to avoid the inclusion of any air bubbles which, if trapped below the neck, could lead to vortex formation in the spinning sample bulb and vitiate the necessary assumption of spherical sample geometry. The bulb, suspended by its neck from a chuck attached to a plastic cap, was positioned snugly inside a precision 5-mm-o.d. sample tube which, initially, contained also the lock-signal solvent. The small amount of this solvent in the residual annular space outside the bulk was not always sufficient to assure maintenance of the lock; occasional failure of the lock during a lengthy series of automatic measurements resulted in loss of usable data. A second arrangement was then used in which the 5-mm tube, containing the sample bulb but no solvent, was positioned by means of fluorocarbon plastic spacers concentrically within a precision 10-mm-o.d. sample tube accommodating a much larger quantity of lock-signal solvent. Incidental advantages of this arrangement were that the outside of the sample bulb was thus kept dry, and that the solvent could be sealed within the annular space between the two tubes and so kept from contamination for a greatly extended time. Except for these advantages, either arrangement resulted in the same measurements. The cell assembly, in either case, was positioned in the JEOL FX60Q 10-mm ^{1}H/^{13}C dual probe insert.

Longitudinal (spin–lattice) relaxation rates R_1 were measured by the inversion–recovery method,[36] where the repetition time T in the pulse sequence [...$T...\pi...\tau...\pi/2...$] was chosen to be at least five times T_1 ($\equiv R_1^{-1}$) and the values of the variable delay time τ ranged from 10 msec to 3 sec, for a total of between 5 and 20 τ-values, depending on the detail desired. From the Bloch equations[37] under the conditions of this method, the relation of the peak intensity A_τ to the pulse delay time τ becomes

$$A_\tau = A_\infty[1 - 2\exp(-R_1\tau)] \qquad (17)$$

where A_∞ is the limiting peak intensity for $\tau \to \infty$. Independent measurement of A_∞, a source of irreducible error, can be dispensed with, and the problem of weighting the data points in the conventional linear plot (logarithm of a function of relative peak heights vs τ) can be eliminated, by fitting directly to the data points (τ, A_τ) by least-squares an exponential of the form of Eq. (17), from which the two parameters A_∞ and R_1 can be obtained.

The factor 2 preceding the exponential in Eq. (17) is based on theory which predicts $A_\infty = -A_0$, where A_0 is the peak intensity for $\tau = 0$. It is

[36] R. L. Vold, J. S. Waugh, M. P. Klein, and D. E. Phelps, *J. Chem. Phys.* **48**, 3831 (1968).
[37] F. Bloch, *Phys. Rev.* **70**, 460 (1946).

recognized that in point of fact this equality rarely holds exactly because of slightly imperfect adjustment of the flip angle π. When Eq. (17) is used in its logarithmic form, A_∞ must be determined by explicit measurement, and A_0 then is usually seen, from the ordinate intercept of the straight-line plot of $\ln(A_\infty - A_\tau)$ vs τ, to differ somewhat from its theoretical value. This, however, is essentially without relevance with respect to the only parameter of intrinsic interest, namely the R_1 obtained from the slope, except possibly insofar as agreement of the intercept with $\ln 2A_\infty$ would be some measure of the quality of the data as a whole. When the equation is used, as it is here, in the exponential form with A_∞ not explicitly determined, the assumption $A_\infty = -A_0$ implies a two-parameter exponential fit, whereas $A_\infty \neq -A_0$ would imply a three-parameter exponential fit. The latter has been advocated or practiced in the past by various authors.[38-41] Such a practice, however, has ignored the definitive treatment of this matter by Leipert and Marquardt,[42] concurred in by Becker et al.,[43] which has shown conclusively that introduction of a third adjustable parameter, so far from improving the statistics, leads to a significant loss of precision in the estimate of R_1 because of an undesirable correlation between what should be statistically independent parameters, R_1 and A_∞. The effect, on the other hand, of even a considerably misadjusted flip angle on the value of R_1 obtained was shown to be insignificant for values of $R_1 > 0.1$ sec, a condition generally satisfied.

The fitting of the two-parameter exponential was carried out by computer by means of an iterative program. For each sample R_1 was determined at least four times, and the results were averaged. This procedure was repeated at each concentration; a minimum of six concentrations were used under each set of conditions of temperature and pH at which the resonance relaxation of each nuclide was examined.

Transverse (spin–spin) relaxation rates R_2 were determined by spin-locking measurement[44] of $R_{1\rho}$, the longitudinal relaxation rate in the rotating frame. $R_{1\rho}$ equals R_2 in dilute solutions of low viscosity whenever the magnitude of $R_{1\rho}$ is independent of $H_{1\rho}$, the spin-locking radio-frequency field in the rotating frame; this was the case, within the limits of experi-

[38] R. Gerhards and W. Dietrich, *J. Magn. Reson.* **23**, 21 (1976).
[39] M. Sass and D. Ziessow, *J. Magn. Reson.* **25**, 263 (1977).
[40] J. Kowalewski, G. C. Levy, L. F. Johnson, and L. Palmer, *J. Magn. Reson.* **26**, 533 (1977).
[41] A. Brunetti, *J. Magn. Reson.* **28**, 289 (1977).
[42] T. K. Leipert and D. W. Marquardt, *J. Magn. Reson.* **24**, 181 (1976).
[43] E. D. Becker, G. A. Ferretti, R. K. Gupta, and G. H. Weiss, *J. Magn. Reson.* **37**, 381 (1980).
[44] T. C. Farrar and E. D. Becker, "Pulse and Fourier Transform NMR: Introduction to Theory and Methods," p. 92. Academic Press, New York, 1971.

mental error, in the present work. R_2 was evaluated as described above for R_1, except that the relation between peak intensity A_τ and decay time τ derived from the Bloch equations in this case becomes

$$A_\tau = A_0 \exp(-R_2\tau) \tag{18}$$

where the initial intensity A_0 replaces A_∞ as the maximum peak intensity. Again, a least-squares two-parameter exponential fit to the data points was performed by an iterative computer program, from which A_0 and R_2 were obtained.

For each sample, R_2 was determined with the same number of replications as R_1. Measurements of one mode of relaxation were made on the identical samples and immediately following the completion of measurements of the other mode, or at latest the next day. In this manner, measurements for proton relaxation at 59.75 MHz were made at pH 6.2 at 2 and 30°, at pH 4.65 at 2 and 30°, and at pH 2.7 at 10°. Measurements for deuteron relaxation at 9.17 MHz were made at pH 6.2 and 4.65, at both 2 and 30°.

Analysis of Data. Activities and the multicomponent expression. An observation regarding predictions resulting from the derived multicomponent expressions may immediately be in order. Since these are based ultimately on equilibrium constants, the mass terms should be properly expressed as activities instead of concentrations. Consequently, relaxation rate vs concentration curves should be expected to be nonlinear whenever there are appreciable protein activity coefficients. Figure 1 shows concentration dependences of R_1 and R_2 for β-Lg A at pH 6.2 and 30°, and of R_2 at pH 4.65 and 2°. Under these conditions the charge-to-mass ratio and the second virial coefficient are relatively small ($B_0 = 0.9$ ml/g).[34] In fact, all these data exhibit linear relationships over a concentration range from 0 to 0.08 g protein/g water, where γ does not differ greatly from unity. Figure 2, on the other hand, shows corresponding plots for pH 2.7 and 10°, where the net charge is approximately 40 and the second virial coefficient is 8.5 ml/g.[45] The plots in terms of concentration here are clearly nonlinear. A change in the concentration scale to g protein/g water had no significant effect on the nonlinearity of the plots. (Low-temperature data were used at this pH for experimental reasons: the protein under these conditions undergoes a dissociation from dimer to monomer, but the amount of monomer at low temperature at concentrations in excess of 0.01 g/ml is negligible.[45])

A polynomial curve-fitting program was used with all data on Figs. 1 and 2 as well as all other concentration-dependent data in order to deter-

[45] S. N. Timasheff and R. Townend, *J. Am. Chem. Soc.* **83**, 470 (1961).

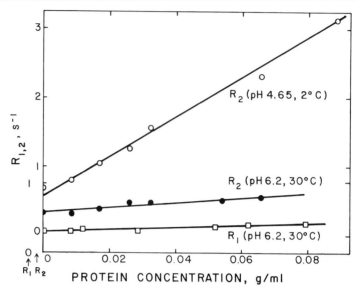

FIG. 1. Dependence of proton relaxation rates on β-lactoglobulin A concentrations (g protein/g water) in H_2O. Transverse relaxation rates R_2 at pH 4.65 and 2° (○) and at pH 6.2 and 30° (●). Longitudinal relaxation rates R_1 (□) at pH 6.2 and 30° (shown for comparison only; not used in calculations). Points represent experimental values; lines represent least-squares fits. All points show linear relationship of relaxation rates to concentration at pH 4.65 and 6.2. (Taken from Ref. 24.)

mine linearity or nonlinearity, the degree of the polynomial being determined by goodness of fit as judged by F test. (The regression program used selects the lowest degree expression for which the sum of squares due to the addition of one higher degree is statistically insignificant.) The two concentration plots of Fig. 2 give polynomials of degree two. Protein concentrations, c, were then transformed into activities, a, by means of the relationship $a = \gamma c$, where the activity coefficient $\gamma = \exp(2B_0c)$ was obtained [see Eq. (2)] from the second virial coefficient $B_0 = 8.5$ ml/g, as cited above. Activity plots corresponding to the concentration plots are also shown in Fig. 2; it is evident that these are linear by the same criteria.

Under the conditions of Fig. 2, γ is sensibly larger than unity and, even in the presence of salt, failure to treat R as a function of activity rather than concentration will evidently lead to excessively high values of hydration [cf. Eq. (13c)]. The opposite would be true under conditions when, in the absence of salt, charge fluctuations and consequent intermolecular attraction exist. With γ less than unity, a concentration plot in place of the correct activity plot for R must lead to inordinately low values

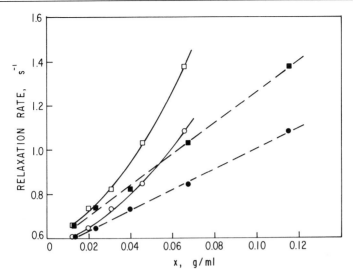

FIG. 2. Dependence of water proton relaxation rates of β-lactoglobulin A on both concentration and activity, at pH 2.7 and 10°. Transverse relaxation rates (squares) as function of X ($\square$, X = concentration; $\blacksquare$, X = activity). Longitudinal relaxation rates (circles) as function of X ($\bigcirc$, X = concentration; $\bullet$, X = activity). All points represent experimental values; lines represent least-squares polynominal fits of highest degree to make a statistically significant contribution to goodness of fit (F test). Dependence on concentration is found to be of second degree as consequence of charge effects at pH 2.7. (Polynominal coefficients for R_1, a_0 = 0.562, a_1 = 2.926, a_2 = 75.86; for R_2, a_0 = 0.606, a_1 = 3.468, a_2 = 125.9.) Activities take these effects into account, and dependence on activities is linear. (Coefficients of straight line: for R_1, a_0 = 0.541, a_1 = 4.691; for R_2, a_0 = 0.560, a_1 = 7.061.) (Taken from Ref. 24.)

of hydration. Alternatively, protein concentrations could continue to be used, provided the right-hand side of Eq. (13) is multiplied by $(da/dc)_\mu$.

These results may be taken to demonstrate the validity of the multicomponent expression. This expression will be used in the following in analyzing the data to describe the hydration of the genetic A variant of β-lactoglobulin, β-Lg A, under nonaggregation conditions, as well as under the well-characterized dimer-to-octamer association.[46-50] (This phenomenon, because it involves a 4-fold association of dimer under conditions where dissociation to monomer is negligible, will be referred to in the following as "tetramerization.") It has been demonstrated that here the

[46] J. Witz, S. N. Timasheff, and V. Luzzati, *J. Am. Chem. Soc.* **86**, 168 (1964).
[47] R. Townend and S. N. Timasheff, *J. Am. Chem. Soc.* **82**, 3168 (1960).
[48] T. F. Kumosinski and S. N. Timasheff, *J. Am. Chem. Soc.* **88**, 5635 (1966).
[49] S. N. Timasheff and R. Townend, *J. Am. Chem. Soc.* **83**, 464 (1961).
[50] S. N. Timasheff and R. Townend, *Protides Biol. Fluids* **16**, 33 (1969).

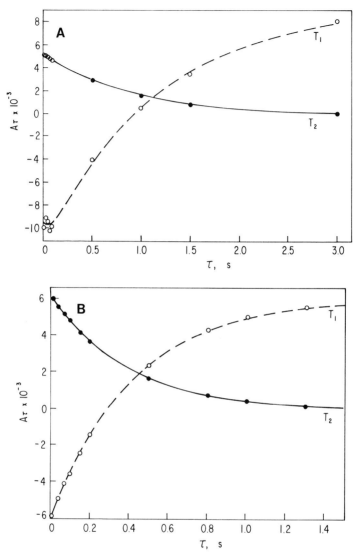

FIG. 3. (A) Proton resonance peak intensities A as functions of time τ, for solutions of β-Lg A in H_2O (6.09×10^{-2} g protein/g H_2O) at pH 6.2, 30°. Intensities for transverse relaxation (T_2, ●) as function of decay time, from spin-locking measurements of $T_{1\rho}$; intensities for longitudinal relaxation (T_1, ○) as function of delay time, from inversion-recovery measurements. Points represent experimental values. Line for T_2 represents two-parameter exponential fit. Line for T_1 represents four-parameter double-exponential fit; need for this fit shown by initial course of this line, indicative of cross-relaxation. (B) Deuteron resonance peak intensities A as function of time τ, for solutions of β-Lg A in D_2O (2.86×10^{-2} g protein/g D_2O) at pH 6.2, 30°. In this case, the indication of cross-relaxation in T_1 is absent and a two-parameter exponential shows excellent fit to the experimental points even at shortest times. (A and B, taken from Ref. 24).

virial coefficients are small,[34,45] so that it is permissible to simplify the treatment by using protein concentrations.

Isotropic binding, two-state model. Determination of correlation times. Proton spin–spin relaxation measurements of water in solutions of β-Lg A gave results indicating a single relaxation rate, whereas spin–lattice results could be fitted only by the sum of two exponential functions (Fig. 3A). This behavior is consistent with the work of Edzes and Samulski[51] and others,[22,52] who found a cross-relaxation mechanism between the water-bound protons and the protein protons to make a significant contribution to the spin–lattice relaxation rate in water–collagen systems. Subsequently, Koenig *et al.*[53] showed that cross-relaxation also exists in spin–lattice relaxation (T_1 processes) in globular protein solutions, i.e., the cross-relaxation rate disperses as a T_1 process and not as a T_2 process. With the notation of Edzes and Samulski, for the solution illustrated in Fig. 3A parameters in the equation $m(t) = c^+ \exp(-R_1^+ t) + c^- \exp(-R_1^- t)$ are $R_1^+ = 15.0$, $R_1^- = 0.45$, $c^+ = 0.01$, and $c^- = 1.01$, where the reduced magnetization $m(t) \equiv (A_\infty - A_\tau)/2A_\infty$. The statistics here were poor because of instrument limitations.[54] (Cross relaxation, calculated for β-Lg in a novel way,[25] was found to contribute to the observed apparent spin–lattice relaxation to the extent of about 90%.) As an alternative approach, protein solutions were made up in D_2O, and the dependence of R_1 and R_2 on protein concentration was measured by deuteron NMR at 9.17 MHz which, in effect, eliminated cross relaxation from the T_1 measurements (Fig. 3B). The concentration dependence of R_2 of protons in solutions of β-Lg A in H_2O under the same environmental conditions was also measured (Fig. 1).

Concentration plots of R_1 and R_2 for D_2O (Fig. 4) showed no evidence of nonlinearity, at either pH and either temperature, over the concentration range studied. This agrees with the low virial coefficient of β-Lg A under these conditions.[34] The relaxation increments k_1 and k_2, together with the corresponding intercepts R_{1f} and R_{2f} (Fig. 4), were used in Eqs. (13) and (16) to determine the bound-water correlation times τ_c presented in Table I. As can be seen, τ_c increased as the temperature decreased; this is in quantitative agreement with the requirement of Stokes' equation that τ_c increase with both increasing viscosity and decreasing temperature. Furthermore, τ_c also increased when the pH was lowered from 6.2 to 4.65,

[51] H. T. Edzes and E. T. Samulski, *J. Magn. Reson.* **31**, 207 (1978).
[52] A. Kalk and H. J. C. Berendsen, *J. Magn. Reson.* **24**, 343 (1976).
[53] S. H. Koenig, R. G. Bryant, K. Hallenga, and G. S. Jacob, *Biochemistry* **17**, 4348 (1978).
[54] T. F. Kumosinski, H. Pessen, and J. M. Purcell, *Fed. Proc., Fed. Am. Soc. Exp. Biol.* **39**, 1673 (1980).

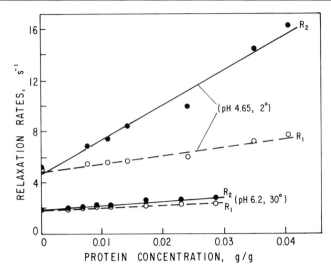

FIG. 4. Dependence of deuteron relaxation rates on β-Lg A concentrations (g protein/g water) in D_2O. Transverse relaxation rates R_2 (●) at pH 4.65, 2° and at pH 6.2, 30°. Longitudinal relaxation rates R_1 (○) under the same two sets of conditions. Points represent experimental values; lines represent least-square fits. Points at all concentrations show linear relationship of relaxation rates to concentration, at both sets of pH and temperature conditions, and for both modes of relaxation. (Taken from Ref. 24.)

as would be predicted from the work of Timasheff and Townend,[55] which showed that β-Lg A associates at the lower, but not at the higher pH, and that the association increases with decreasing temperature.

Determination of hydration parameters. Since the extent of a possible intermolecular contribution, R'_{2b}, to the spin–spin relaxation of bound-water protons and the number of protein protons so contributing are unknown, determination of Scatchard hydrations was attempted by three different methods and the results were compared.

Method I. Only the deuterium NMR relaxation increments were used, with a value of $e^2qQ/\hbar = 215.6$ kHz[56] and with the asymmetry parameter η assumed to be zero. This type of calculation gives low values of $\bar{v}'_W$, as shown in Table II; however, low values could be expected because the relaxation increment probably samples only a percentage of the total hydration of a protein, since at 9.17 MHz any bound water with τ_c values less than 6 nsec would have a T_1/T_2 ratio of unity. At pH 4.65, where tetramerization occurs, the hydration markedly increases with decreasing

[55] S. N. Timasheff and R. Townend, *J. Am. Chem. Soc.* **82**, 3157 (1960).
[56] P. Waldstein, S. W. Rabideau, and J. A. Jackson, *J. Chem. Phys.* **41**, 3407 (1964).

TABLE I

DEUTERON NMR RELAXATION AND HYDRODYNAMIC RESULTS FOR β-LACTOGLOBULIN IN SOLUTION[a]

pH	T, °C	k_1^a	k_2	R_{1f}, sec^{-1}	R_{2f}, sec^{-1}	τ_c, nsec	r_{NMR}[b], Å	r_{sed}, Å	$(\tau_c)_{calc}$, nsec
6.2	30	20.7 ±0.9	31.9 ±1.9	1.93 ±0.01	1.93 ±0.04	10.0 ±2.7	23.2 ±1.9	27.0	10.2
	2	17.9 ±2.1	58.9 ±5.2	5.46 ±0.24	5.25 ±0.28	25.6 ±4.1			22.5
4.65	30	25.4 ±1.9	73.2 ±5.0	2.01 ±0.04	1.79 ±0.12	22.5 ±3.3	30.4 ±1.5	43.5	65.9
	2	63.9 ±7.1	274.2 ±19.0	4.90 ±0.16	4.67 ±0.44	32.2 ±4.6			145.0

[a] Adapted from Ref. 24.
[b] Error terms in this and subsequent tables represent the standard error of the parameter.
[c] Spherical model assumption.

TABLE II

Hydration and Thermodynamics for β-Lactoglobulin from Deuteron NMR

Using $(e^2qQ/\hbar) = 215.6$ kHZ[a,b]

pH	T, °C	$\bar{v}'_W$, g H$_2$O/g protein	ΔG, kcal	$-\Delta H$, kcal	$-\Delta S$, eu
6.2	30	0.0063 ±0.0008	0.90 ±0.08		
				0.8 ±3.0	6 ±10
	2	0.0072 ±0.0020	0.74 ±0.15		
4.65	30	0.0095 ±0.0002	0.65 ±0.01		
				6.9 ±1.6	24.8 ±5.8
	2	0.0301 ±0.0003	−0.044 ±0.006		

[a] Adapted from Ref. 24.
[b] Method I, as described under Analysis of Data.

temperature, whereas at pH 6.2, where none occurs, the hydration is lower and independent of temperature. This is consistent with the findings of Timasheff and co-workers[46,57] from small-angle X-ray scattering that the geometry of the octamer must include a large central cavity in which trapped water could reside.

Method II. A combination of the τ_c values found by deuteron NMR at 9.17 MHz (Table I) and the k_2 values found from proton NMR at 59.75 MHz was used (Fig. 1). The reason for this procedure is that the quadrupole coupling constant for the bound water should actually decrease as hydrogen bonding increases.[58] Here R_{2b} can be calculated from Eq. (16b) at 59.75 MHz, and the $\bar{v}'_W$ values can be easily obtained from the simple relationship $\bar{v}'_W = k_2/(R_{2b} - R_{2f})$, derived from Eq. (13b). Such hydration values (Table III) are slightly higher than those from deuteron relaxation measurements only (Table II), but show the same temperature and pH dependence. (Combination experiments of this kind would be best performed at the same Larmor frequency; however, the availability of only a single spectrometer with no variable frequency capability would preclude this possibility. These experiments could shed light on such problems as the constancy of the hydrogen-bond distance under various conditions

[57] S. N. Timasheff and R. Townend, *Nature (London)* **203**, 517 (1964).
[58] M. J. Hunt and A. L. Mackay, *J. Magn. Reson.* **15**, 402 (1974).

TABLE III

HYDRATION AND THERMODYNAMICS FOR β-LACTOGLOBULIN DERIVED FROM τ_c VALUES OF TABLE I BY METHODS II AND III[a]

pH	T, °C	k_2	R_{2f}, sec^{-1}	R_{2b}, sec^{-1}		$\bar{\nu}'_w$, g H$_2$O/g protein		ΔG, kcal		$-\Delta H$, kcal		$-\Delta S$, eu	
				II[b]	III[c]	II[b]	III[c]	II[b]	III[c]	II[b]	III[c]	II[b]	III[c]
6.2	30	3.5 ±0.5	0.34 ±0.02	228	337	0.0152 ±0.0021	0.0103 ±0.0007	0.35 ±0.07	0.59 ±0.04				
	2	8.1 0.4	0.70 ±0.02	530	784	0.0152 ±0.0008	0.0103 ±0.0014	0.32 ±0.03	0.54 ±0.07	0	0	1.2 ±0.2	1.9 ±0.1
4.65	30	9.2 ±0.8	0.27 ±0.04	468	694	0.0197 ±0.0016	0.0133 ±0.0011	0.20 ±0.05	0.43 ±0.04				
	2	28.2 ±1.1	0.56 ±0.05	660	978	0.0428 ±0.0017	0.0289 ±0.0011	-0.25 ±0.02	-0.032 ±0.021	4.62 ±0.73	4.62 0.70	15.9 ±2.6	16.7 ±2.4

[a] Adapted from Ref. 24.

[b] Method II uses τ_c from Table I and proton k_2.

[c] Method III uses, in addition to the procedure of Method II, the assumption $\bar{R}_{2b} = R_{2b} + 12\,R'_{2b}$, with the intermolecular proton distance for β-Lg A calculated from the partial specific volume as 2.61 Å, as described under Analysis of Data.

and the existence of a distribution of correlation times in the total hydration shell.)

Method III. This is a combination procedure also, with an extra intermolecular interaction term R'_{2b} added to the proton spin–spin relaxation rate R_{2b}. Based on the small-angle X-ray scattering results of Witz *et al.* for β-Lg A[46] in conjunction with the known molecular weight and amino acid composition, a simple consideration of the molecular geometry shows that, on the average, each proton in the protein will have six neighboring protons at a distance of 2.61 Å. From this average intermolecular distance, together with the τ_c values of Table I and the relationship $\bar{R}_{2b} = R_{2b} + 12\, R'_{2b}$ (where $\bar{R}_{2b}$ is the total spin–spin relaxation rate), the hydration can be calculated as $\bar{\nu}'_W = k_2/(\bar{R}_{2b} - R_{2b})$. These values (Table III) are in close agreement with those in Table I. Altogether, no great difference exists in the $\bar{\nu}'_W$ values from all three methods.

Comparison of results with other structural information. Dynamics of β-Lg dimer. With the τ_c values calculated from k_2/k_1 from deuteron NMR spin–spin and spin–lattice relaxation increments, dR_2/dc and dR_1/dc, a Stokes radius r_{NMR} for the bound water can be calculated from the Stokes–Einstein relation[59] on the basis of a spherical model (Table I). At pH 6.2, where β-Lg A exists as the unassociated dimer, r_{NMR} is slightly lower than that for the protein itself derived from hydrodynamic data, r_{sed}.[60] This discrepancy could be due to the spherical approximation inherent in the use of the Stokes–Einstein equation, since the β-Lg dimer has an axial ratio of approximately 2 : 1.[34] Moreover, the Stokes radius of the protein obtained from sedimentation includes the water of hydration and should therefore be larger than the Stokes radius of the bound water calculated from the ^{2}H NMR relaxation data. What should be compared with the τ_c of the bound water is the τ_c of the protein without any contribution from hydration. For the latter, values of 10.2 nsec at 30° and 22.5 nsec at 2° (Table I) can be calculated for the protein with the use of its partial specific volume, 0.751 ml/g, and an asymmetry factor of 1.168[61] to account for the dimer axial ratio of 2 : 1. These values are in excellent agreement with the experimental τ_c of the bound water at pH 6.2 at 30 and 2° (Table I).

Hydration and dynamics of β-Lg octamer. At pH 4.65, where the protein exists to a large extent as the octamer even at 30° at concentrations above 0.01 g/ml,[47] the Stokes radius of the bound water is about 30% less than the Stokes radius of the octamer itself. However, this value is

[59] C. P. Poole and H. A. Farach, "Relaxation in Magnetic Resonance," p. 228. Academic Press, New York, 1971.
[60] H. Pessen, T. F. Kumosinski, and S. N. Timasheff, this series, Vol. 27, p. 151.
[61] P. Wahl and S. N. Timasheff, *Biochemistry* **8,** 2945 (1969).

still much closer to the theoretical value than those obtained by other investigators for other proteins.[18-22] Furthermore, the 422-symmetry model for the octamer according to Timasheff and Townend[57] possesses a large central cavity which could accommodate trapped water; if the NMR experiment observed this trapped water, the τ_c value found would be less than that of the protein. Also, if the assumption is made that the NMR hydration of the octamer itself at 2° equals $(\bar{\nu}_W')_{pH\ 4.65} - (\bar{\nu}_W')_{pH\ 6.2}$, values from 0.019 to 0.028 g H_2O/g protein can be calculated by the three methods described here. The total volume of the cavity, approximated by an internal sphere tangent to the subunits on the basis of known structural parameters,[57] amounts to about 6500 $\mathring{A}^3$. Taking the specific volume of water as unity and thus its molecular volume as 30 $\mathring{A}^3$/molecule, this would correspond to about 220 mol H_2O/mol of octamer, or 0.027 g H_2O/g protein, which is within range of the NMR-derived hydration values for the octamer.

Since the derived NMR correlation times are number-average values, the hypothesis that the increase in hydration accompanying octamer formation is largely due to trapped water may be tested by calculating a number-average correlation time from the relationship $(\bar{\nu}_W)_{4.65}\tau_c = (\bar{\nu}_W)_{6.2}(\tau_c)_0 + [(\bar{\nu}_W)_{4.65} - (\bar{\nu}_W)_{6.2}](\tau_c)_{cc}$, where $(\tau_c)_0$ is the correlation time of the octamer at 2° (i.e., 145 nsec, see Table I), $(\tau_c)_{cc}$ is the correlation time of the central cavity of volume 6500 $\mathring{A}^3$ (i.e., 1.4 nsec), and $(\bar{\nu}_W)_{6.2}$ and $(\bar{\nu}_W)_{4.65}$ are the NMR hydration values at pH 6.2 and 4.65, respectively. Calculation of τ_c from ^2H NMR hydration values by Method I at 2° gives 36 nsec, in fair agreement with the ^2H NMR experimental value of 32.2 ± 4.6 nsec at pH 4.65 and 2°. However, the results of this calculation furnish an indication only of the reasonableness of the approach and not of any exact mechanism of increased hydration accompanying octamer formation.

In contrast to Method I, Method II assumes no constant quadrupole coupling constant and therefore may serve, incidentially, to calculate values of $e^2qQ/\hbar$ (assuming $\eta = 0$) from the $\bar{\nu}_W'$ values and the relaxation increments obtained by deuteron NMR, together with the experimentally derived τ_c. The quadrupole coupling constants calculated for the respective methods range from 120 to 160 kHz. Hunt and MacKay[58] have correlated O...D...O and N...D...O hydrogen bond distances with values of quadrupole coupling constants. From their relationships and the above $e^2qQ/\hbar$ values, one obtains distances for O...D...O from 1.5 to 1.7 $\mathring{A}$, and for N...D...O from 1.6 to 2.0 $\mathring{A}$. These are in agreement with linear hydrogen bond lengths of 1.81 to 1.87 $\mathring{A}$ recently reported by Ceccarelli et al.[62] in an extensive review of neutron diffraction data.

[62] C. Ceccarelli, G. A. Jeffrey, and R. Taylor, J. Mol. Struct. 70, 255 (1981).

Contrast of NMR hydration with preferential hydration. To contrast these NMR hydration results with results from another physical method which measures water–protein interactions, preferential hydrations were obtained by Wyman's theory of linked functions,[28,63] from which it follows that for a tetramerization reaction (dimer → octamer, in the case of β-Lg A),

$$d \ln k_T/d \ln a_{X,T} = (\bar{\nu}_{X,T})_{pref} - 4(\bar{\nu}_{X,M})_{pref}$$
$$= -(W_T/X_T)[(\bar{\nu}_{W,T})_{pref} - 4(\bar{\nu}_{W,M})_{pref}] \qquad (19)$$

where the preferential interactions are defined by

$$(\bar{\nu}_X)_{pref} = \bar{\nu}_X - (X/W)\bar{\nu}_W \text{ and } (\bar{\nu}_W)_{pref} = \bar{\nu}_W - (W/X)\bar{\nu}_X \qquad (19a)$$

Here K_T is the association constant, a_X is the activity of salt, $(\bar{\nu}_{X,T})_{pref}$ and $(\bar{\nu}_{X,M})_{pref}$ are the preferential salt binding of octamer (subscript T) and dimer (subscript M), and $(\bar{\nu}_{W,T})_{pref}$ and $(\bar{\nu}_{W,M})_{pref}$ are the preferential hydration, respectively. Preferential interaction parameters can thus be readily obtained from the slope of a plot of association constants at various salt concentrations vs the activity of the salt at the corresponding concentrations.

Association constants can be calculated by the use of Gilbert's theory for rapidly reequilibrating association in a sedimenting boundary.[64,65] Gilbert has shown that for a reversible association there exists a minimum concentration c_{min} above which bimodality of a schlieren ultracentrifuge pattern appears. Furthermore, the area of the slow peak remains constant as the loading concentration is increased well above c_{min}, which is related to the equilibrium constant of the association. Thus, for a tetramerization (in this case, dimer → octamer),[46–48]

$$K_T = M_m^3 \delta[1 + \delta/4(1 - \delta)]^3/[16(1 - \delta)c^3] \qquad (20)$$

where $\delta \equiv (s - s_1)/(s_4 - s_1)$, ($s$, s_1, and s_4 being the sedimentation coefficients for the leading edge of the boundary, for the dimer, and for the octamer, respectively), K_T is the association constant, c is the total loading concentration, and M_m is the dimer molecular weight. Since $\delta_{min} = (n - 2)/3(m - 1)$ and $\delta = 2/9$ for $n = 4$ (where n is the degree of association), $K_t = 1.087 \times 10^{12}/c_{min}^3$. The product of the percentage of the slow peak and the loading concentration equals c_{min}, and K_T can thus be evaluated.

[63] J. Wyman, Jr., *Adv. Protein Chem.* **19**, 223 (1964).
[64] G. A. Gilbert, *Discuss. Faraday Soc.* **20**, 68 (1955).
[65] G. A. Gilbert, *Proc. R. Soc. London, Ser. A* **250**, 377 (1959).

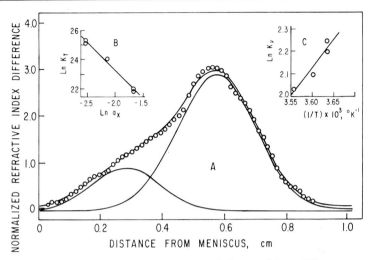

FIG. 5. (A) Sedimentation of β-Lg A under association conditions. Gilbert pattern (open circles) and decomposition into two gaussian peaks (solid lines). (B) Linked-function plot. Plot of logarithm of tetramerization equilibrium constant vs logarithm of activity of salt. The slope is a measure of preferential hydration. (C) van't Hoff plot. Plot of logarithm of equilibrium constant of water binding vs reciprocal of absolute temperature. The slope gives a value of -5.5 ± 1.2 kcal for the enthalpy of hydration. (A, B, and C, taken from Ref. 24.)

In this way, sedimentation velocity data for β-Lg A at pH 4.65 give a linked-function plot (Fig. 5B), which clearly shows a negative slope. From the least-squares value of this slope, a preferential solvation of -3.78 mol salt/mol octamer is calculated. With the assumption that the tetramerization does not release any salt, a value of 0.258 g H_2O/g protein for the preferential hydration is obtained, presumed to be equal to $(\bar{v}'_W)_{pH\ 4.65} - (\bar{v}'_W)_{pH\ 6.2}$ from NMR at 2°. The linked-function preferential hydration thus appears to differ significantly from the NMR hydration difference.

However, as previously noted, NMR probably samples only a certain percentage of the total hydration of a protein. For example, if a major portion of the bound water has a τ_c value of less than 1 nsec, its contributions to the spin–lattice and spin–spin relaxations of the bound state would be equal since at 9.17 MHz the T_1/T_2 ratio for this fast-tumbling water would be unity. Therefore, the correlation time and hydration values calculated from the NMR results by use of a two-state model approximation would yield number-average values weighted toward the slow-tumbling component. It would probably be more realistic to compare the enthalpies of hydration derived by NMR with those from linked functions by the van't Hoff relationship (Fig. 5C). This is possible by assuming a thermodynamic model involving the transfer of free water to bound water as protein is added to the solution, i.e., $\Delta G = KT \ln(\bar{v}_W/5.56)$. Values of ΔG, ΔH, and ΔS of hydration for each of the three NMR methods are presented in Tables II and III.

From the slope of the van't Hoff plot of Fig. 5C, a ΔH of -5.5 ± 1.2 kcal/mol dimer is obtained. ΔS values calculated at each temperature average -14.1 ± 0.03 eu. The small standard deviation for ΔS indicates constancy of the entropy with respect to temperature and lends support to the model. Furthermore, ΔH of hydration values from NMR range from approximately -6 to -9 kcal/mol dimer, seemingly in agreement, within experimental error, with the value of -5.5 derived from linked functions. Nevertheless, it needs to be remembered that the linked-function method measures the difference between the total hydrations of octamer and dimer, whereas the NMR method yields a quantity proportional to the total hydration of the octamer. Direct comparison of the temperature dependence of these two methods, by assuming that the linked-function hydration is proportional to the difference in the NMR hydration at pH 4.65 and 6.2, is not feasible since the proportionality constant itself should change as a function of temperature. However, the total hydration of the dimer appears to be temperature independent, as indicated by essentially constant NMR hydration values at pH 6.2 for 30 and 2°. These considerations lead to an alternative evaluation of the NMR data in terms of a three-state model.

Isotropic binding, three-state model. For the reasons indicated, an attempt was made to evaluate the ^{2}H NMR relaxation increment on the basis of a three-state model (i.e., free water, fast-, and slow-tumbling bound water), as used by Cooke and Kuntz in treating lysozyme data,[66] with the following assumptions. First, the fast-tumbling hydration component of the dimer was assumed to be 0.1 g water/g protein on the following grounds. The lysozyme crystal has been determined to contain 80 molecules of water per protein molecule, or 0.1 g H_2O/g protein,[67] and 100 to 300 water molecules per crystalline protein molecule distributed over the molecular surface have generally been reported.[66] For β-Lg, 200 mol water/mol protein is equivalent to 0.1 g water/g protein. Also, Teller *et al.*[68] have shown that experimentally derived frictional coefficients from sedimentation results are in agreement with those calculated from known X-ray crystallographic structures provided water molecules are added to each charged side chain on the protein surface. The β-Lg dimer has 90

[66] R. Cooke and I. D. Kuntz, *Annu. Rev. Biophys. Bioeng.* **3**, 95 (1974).
[67] A. T. Hagler and J. Moult, *Nature (London)* **272**, 222 (1978).
[68] D. Teller, E. Swanson, and C. De Haën, this series, Vol. 61, p. 103.

charged amino acids[69]; a value of 0.1 g water/g protein corresponds, therefore, to about two water molecules bound to each charged side-chain amino acid.

Second, since from Teller's work this water would be bound to charged side chains, and since Brown and Pfeffer[70] have recently shown that deuterium-modified lysine groups tumble at 48 psec, a τ_c value of 48 psec can be assumed. Any water bound to lysine should then have a τ_c value no lower than that of the side chain itself. It was assumed also that, because of the fast segmental motion of the side chain, any increase in association of the protein would not affect this τ_c unless the side chain was directly involved in the interaction site. However, the viscosity and temperature effect on the Stokes–Einstein relationship was taken into account when the temperature changes from 30 to 2°. Finally, the fast-tumbling hydration values at pH 4.65 were increased, by 0.101 at 30° and by 0.258 g H_2O/g protein at 2°, in line with the linked-function results, which indicate such preferential hydrations at these temperatures.

Subtraction of the fast-tumbling contribution from the ^2H NMR spin–lattice and spin–spin relaxation increments yields new values from which the correlation time of the slow component $(\tau_c)_s$ and its corresponding hydration $(\bar{\nu}'_W)_s$ can be calculated. Table IV shows that the $(\tau_c)_s$ are slightly larger than the τ_c of Table I, and the $(\bar{\nu}'_W)_s$ are slightly larger than the $\bar{\nu}'_W$ of Table II for the two-state model. However, the corresponding values are probably within experimental error, as are the derived enthalpy of hydration of the slow component (Table IV) and the enthalpy of hydration derived from the two-state model (Table II). Calculation of a number-average correlation time of the slow component of the octamer, by assuming that the increase in hydration upon octamer formation is due to water trapped in the cavity of the octamer, yields 37 nsec. This is in reasonable agreement with the experimentally derived value of $(\tau_c)_s$ at pH 4.65 and 2° of 42.3 ± 4.6 nsec (Table IV).

Anisotropic binding mechanism. The preceding calculations assume an isotropic relaxation mechanism, as detailed under Theory. In the presence of salt, all the relaxation at pH 6.2 can be accounted for by a slow-tumbling and a fast-tumbling component, amounting to 13 and 204 mol H_2O/mol dimer, respectively, and increasing at pH 4.65 and 2° to 61 and 730 mol H_2O/mol dimer; these may be considered reasonable values for the hydration of a protein.[1,71]

[69] R. McL. Whitney, J. R. Brunner, K. E. Ebner, H. M. Farrell, Jr., R. V. Josephson, C. V. Morr, and H. E. Swaisgood, *J. Dairy Sci.* **59**, 785 (1976).

[70] E. M. Brown and P. E. Pfeffer, *Fed. Proc., Fed. Am. Soc. Exp. Biol.* **40**, 1615 (1981).

[71] B. P. Schoenborn and J. C. Hanson, *in* "Water in Polymers" (S. P. Rowland, ed.), p. 217. Am. Chem. Soc., Washington, D. C., 1980.

TABLE IV
TOTAL HYDRATION FOR β-LACTOGLOBULIN DERIVED FROM A THREE-STATE
MODEL (^{2}H NMR)[a,b]

pH	T, °C	$(\bar{v}'_W)_f$	$(\bar{v}'_W)_s$	$(\tau_c)_s$, nsec	$-(\Delta H)_s$, kcal
6.2	30	0.1	0.0054 ±0.0008	11.0 ±2.7	
					0
	2	0.1	0.0056 ±0.0020	33.9 ±4.1	
4.65	30	0.201	0.0085 ±0.0002	28.8 ±3.3	
					6.0 ±1.6
	2	0.358	0.0235 ±0.0003	42.3 ±4.6	

[a] Assumptions: $(\tau_c)_f$ = 48 psec; $(R_{1b})_f = (R_{2b})_f$ = 31.94 sec^{-1} at 30° for pH 6.2 and 4.65; $(R_{1b})_f = (R_{2b})_f$ = 70.27 sec^{-1} at 2° for pH 6.2 and 4.65. Subscripts f and s refer to fast-tumbling and slow-tumbling fractions, respectively.
[b] Adapted from Ref. 24.

This does not, however, eliminate the possibility of an anisotropic relaxation mechanism for hydrodynamically bound water. The present results may be interpreted equally well on the basis of the three-component derivation in conjunction with either a two- or three-state model and an appropriate order parameter $S < 1$ [cf. Eqs. (16c,d)]. Here a three-state model is defined, as for the isotropic mechanism (Table IV), as comprising free-motion water, a slow-motion component (i.e., $\tau_c > 5$ nsec), and a fast-motion component (i.e., $\tau_c \cong 48$ psec, as assumed in Table IV). For the latter, under extreme-narrowing conditions the factor S^2 attached to Eqs. (16c,d) is changed to $(1 - S^2)$.[23] The slow motion, in either the two- or three-state anisotropic mechanism, may be due to such processes as protein reorientation, internal motion of the protein, or translational diffusion of water along the protein surface.[23,32]

Reported values of $S = 0.06$ from ^{17}O relaxation[23] have been obtained by application of line-splitting data for a liquid crystal to a protein, on the assumption that 3 to 6 water molecules are bound to carboxyl groups and 1 to 3 to hydroxyl groups. Theoretical results of Walmsley and Shporer[32] give relationships for S (termed the scaling factor by these authors) based on ^{1}H, ^{2}H, and ^{17}O relaxation. From these it follows that a value of $S = 0.06$ for ^{17}O would imply $S = 0.12$ for ^{2}H. At pH 6.2 (30 and 2°) and pH

4.65 (30 and 2°) one obtains [from Eqs. (13a–c) and the ^2H NMR data of Table I for the two-state, and those of Table IV for the three-state model] hydrations of 0.483, 0.500, 0.660, and 2.090 for the two-state, and 0.298, 0.295, 0.509, and 1.250 for the three-state model, respectively. A more reasonable estimate is obtained from the theoretical relationships of Walmsley and Shporer, together with the experimental results of Koenig et al.,[18–20] which give $S = 0.23$ and corresponding hydrations of 0.119, 0.136, 0.180, 0.569, and 0.102, 0.105, 0.163, 0.435. The ΔH of hydration at pH 4.65 is found to be −6.8 for the two-state and −5.9 kcal for the three-state model. These results agree with the isotropic mechanism (Tables II and IV), since S enters simply as a factor in the Kubo–Tomita–Solomon equations.

Alternatively, equating the preferential hydration from linked functions with the difference between the 2° ^2H NMR hydrations at pH 4.65 and pH 6.2, one obtains $S = 0.30$ for the two-state and $S = 0.26$ for the three-state model, both not far from the 0.23 predicted from the theory of Walmsley and Shporer. Furthermore, with the above values of S the pH 4.65 enthalpies of hydration are −6.8 and −5.9 kcal for the two- and three-state models, respectively. Thus, the increase in hydration as well as the corresponding enthalpy change attendant on octamer formation are the same for either assumption of relaxation mechanism.

The foregoing considerations add up to considerable agreement between certain theoretically and experimentally derived quantities. However, none of the above arguments should be interpreted as proof of any particular NMR mechanism or model, nor of the identity of the particular groups on the protein surface that interact with water. Even without such conclusions, and in place of the quest for absolute values of hydration, it can be useful to scrutinize relative changes in hydration, when these can be taken as functions of changes in secondary, tertiary, or quaternary structure of a protein.

Small-Angle X-Ray Scattering and Velocity Sedimentation

Theory. The theory of SAXS has been presented in some detail in this series by Pessen et al.[60] and by Pilz et al.[72] Elsewhere, with special relevance to the present context, Kumosinski and Pessen[73,74] have used SAXS results, which implicitly contain the hydration, to calculate sedimentation coefficients without any assumptions concerning the mechanism of hydration. In order to deal with problems of hydration, a multi-

---

[72] I. Pilz, O. Glatter, and O. Kratky, this series, Vol. 61, p. 148.
[73] T. F. Kumosinski and H. Pessen, *Arch. Biochem. Biophys.* **219,** 89 (1982).
[74] T. F. Kumosinski and H. Pessen, *Arch. Biochem. Biophys.* **214,** 714 (1982).

component expression for the chemical potential adapted from Schachman[4] can be used:

$$\mu_{123} = \mu_2 + (km_1 + \alpha)\mu_1 + km_3\mu_3 \qquad (21)$$

Here μ_{123} is the total chemical potential of the sedimenting unit (identical with the fluctuating unit in SAXS), containing component 1 (water), 2 (macromolecule), and 3 (salt); μ_i and m_i ($i = 1, 2, 3$) are the chemical potential and the molality of the respective component; k is a proportionality constant, equal to the ratio of the fraction of salt proportionally bound (proportionally, that is, to the molality of salt in the bulk of the solution) to the molality of the protein; and α is the preferential hydration of the protein, i.e., if positive, the hydration beyond that corresponding to the bulk ratio of water to salt, in moles water bound preferentially per mole protein. It is readily seen that there are $(km_1 + \alpha)$ moles of water and km_3 moles of salt per mole of protein bound to the macromolecule. The term α is related to the preferential salt binding $(\partial m_3/\partial m_2)_\mu$, used in investigations with other experimental techniques, by the expression

$$\alpha = (m_1/m_3)(\partial m_3/\partial m_2)_\mu \qquad (21a)$$

Differentiating Eq. (21) with respect to pressure at constant temperature, rearranging, replacing the molal units by concentrations in grams per gram of water, and noting that the partial specific volume of the sedimenting or fluctuating unit, $\bar{v}_{123}$, can be expressed as $V_{123}N/M_{123}$, yields

$$V_{123}N/M_2 = \bar{v}_2 + (k'g_1 + \xi_1)\bar{v}_1 + k'g_3\bar{v}_3 \qquad (22a)$$

Here N is Avogadro's number; M_i, $\bar{v}_i$, g_i ($i = 1, 2, 3$) are the molecular weight, partial specific volume, and concentration in grams per gram of water of the respective component; $k' = 1000\, k/M_2$ equals the amount of salt bound in proportion to its concentration in the bulk of the solution, in grams of component so bound per gram of protein; and $\xi_1 = \alpha M_1/M_2$ is the preferential binding, in grams of water so bound per gram of protein. In $0.1\ M$ salt solution, $g_3 \cong 0.006 \ll 1$, and Eq. (22a) reduces to

$$V_{123}N/M_2 \cong \bar{v}_2 + (k'g_1 + \xi_1)\bar{v}_1 \qquad (22b)$$

where g_1 is unity, by definition, and $\bar{v}_1 \cong 1$, since sedimentation coefficients are routinely extrapolated to zero protein concentration.

But the sum of the hydration proportional to bulk concentration, $k'g_1$, and the preferential hydration, ξ_1, by the definition of the latter, equals the total hydration A_1 [i.e., $\xi_1 \equiv A_1 - k'g_1$, see Ref. 27, where A_3/g_3 evidently is identical to our k'; A_1 is conceptually, though not necessarily experimentally, identical to the $\bar{v}'_w$ of Eq. (13), while V_{123} is the hydrated volume, V, obtained by SAXS]. Thus

$$V_{123}N/M_2 \cong \bar{v}_2 + A_1\bar{v}_1 \tag{22c}$$

and the total hydration is measured by the difference between the hydrated molecular volume and the partial specific volume, in compatible units. The effect of salt binding in this respect is negligible as long as salt concentrations are of the order indicated above. In solutions of high salt concentration, or even moderate concentration when salt binding is strong [i.e., when the preferential salt binding is positive and the preferential hydration in consequence negative; cf. Eq. (21a)], salt will contribute to the solvated volume by way of the third term in Eq. (22a).

As mentioned, it has been shown that, in addition to Stokes radii and frictional coefficients, sedimentation coefficients can be calculated from SAXS data.[73,75] Differentiating Eq. (21) as before and combining the result with the Svedberg transport equation,[5] with due regard for the makeup of the sedimenting unit, gives

$$s_{20,w}^0 = \frac{M_2(1 - \bar{v}_2\rho) + \alpha M_1(1 - \bar{v}_1\rho) + km_1M_1(1 - \bar{v}_1\rho) + km_3M_3(1 - \bar{v}_3\rho)}{f_{123}N} \tag{23}$$

where f_{123} is the frictional coefficient of the sedimenting unit, and the other symbols have been defined above. Again, $\bar{v}_1$, $\rho \cong 1$ and $m_3 \ll 1$, so that all terms beyond the first become negligible and

$$s_{20,2}^0 \cong \frac{M_2(1 - \bar{v}_2\rho)}{f_{123}N} \tag{23a}$$

From this, it may appear at first that the solvation of the protein should have no effect on the sedimentation coefficient. This, however, would be losing sight of the variability of the term f_{123}, which indeed has been commonly neglected in the context of sedimentation coefficients, even though the variability of the analogous term f_{12} has long been acknowledged.[75,76] As a matter of principle it would, therefore, be an error not to consider this effect here.

The frictional coefficient f_{123} is obtained from

$$f_{123} = (f/f_0)_{123} \, 6\pi\eta N(r_s)_{123} \tag{24}$$

where $(r_s)_{123}$, the Stokes radius of the sedimenting unit (solvated protein), has traditionally been evaluated from the protein partial specific volume, $\bar{v}_2$. The present approach more appropriately uses the hydrated volume,

[75] J. J. Müller, H. Damaschun, G. Damaschun, K. Gast, P. Plietz, and D. Zirwer, *Stud. Biophys.* **102**, 171 (1984).
[76] V. Bloomfield, W. C. Dalton, and K. E. Van Holde, *Biopolymers* **5**, 135 (1967).

V, from SAXS, so that $(r_s)_{123}$ equals $(3V/4\pi)^{1/3}$. There are then two terms that depend upon the binding of salt and water to the protein, namely, V_{123} and $(f/f_0)_{123}$. V_{123} is readily related to the hydration by Eq. (22c). The binding contribution to the term $(f/f_0)_{123}$ is not obtained by this method, but it may become possible to compare (f/f_0) evaluated from X-ray diffraction data with the $(f/f_0)_{123}$ from SAXS to find out where the water molecules are located.

Experimental Procedures. Experimental SAXS procedures have been described and discussed in detail in this series[60,72] and elsewhere.[73,74]

The technique of sedimentation analysis has been extensively treated by Svedberg and Pederson,[5] Schachman,[4] and others. Use of ultraviolet absorption with a photometric scanner interfaced to a computer, besides greatly speeding data collection, facilitates the requisite extrapolation to infinite dilution because it permits sensitive measurements at very low concentrations.[74]

Analysis of Data. In the last column of Table V are listed the values of A_1 calculated from the SAXS volume for 19 globular proteins and 2 spherical viruses.[73] Here, the first nine proteins have an average value of 0.280 g H_2O/g protein, in fair accord with generally assumed value of about 0.25 g H_2O/g protein.[1,3,77] The last 10, which have higher molecular weights ($>100,000$) and are actually oligomeric structures, average near 0.444 g per g of protein. This higher value might be expected since the phenomenon of trapped solvent (internal solvation) has been observed in such multisubunit structures as casein micelles, viruses, and aspartate transcarbamylase.[3]

There may also be a dynamic contribution to the hydration, since it is measured by the difference between two volumes. The concept of protein "breathing" has been emphasized before,[78] and the entire topic of protein dynamics has been reviewed recently.[79,80] The effects of dynamic changes such as fluctuations (e.g., ring flipping and domain hinge bending) on packing volumes and accessible surface areas remain unclear. Progress on these questions may come from dynamic modeling by computer simulation.

Unless a particle is known to be spherical (e.g., a spherical virus, from electron microscopic evidence), sedimentation analysis by itself can only determine a frictional coefficient which is a combination of a structural and a hydration contribution,[3,77] and from which, therefore, the structural contribution must be detached in order to get at the hydration. SAXS,

[77] J. L. Oncley, *Ann. N.Y. Acad. Sci.* **41**, 121 (1941).
[78] T. E. Creighton, *Prog. Biophys. Mol. Biol.* **33**, 231 (1978).
[79] J. A. McCammon and M. Karplus, *Acc. Chem. Res.* **16**, 187 (1983).
[80] A. Petsko and D. Ringe, *Annu. Rev. Biophys. Bioeng.* **13**, 331 (1984).

STRUCTURAL AND HYDRODYNAMIC PARAMETERS FROM SAXS[a,l]

Macromolecule	Model[b]	M^c	$\bar{v}$, ml/g[d]	V, Å³	From V and $\bar{v}$ A_1, g/g[e]
1. Ribonuclease (bovine pancreas)[l,2]	PE	13,690[f]	0.6964[4]	22,000	0.272
2. Lysozyme (chicken egg white)[2]	PE	14,310[g]	0.702[4]	24,200	0.317
3. α-Lactalbumin (bovine milk)[2]	PE	14,180[h]	0.704[4]	25,100	0.362
4. α-Chymotrypsin (bovine pancreas)[5]	PE	22,000[i]	0.736	37,170[j]	0.282
5. Chymotrypsinogen A (bovine pancreas)[5]	PE	25,000[i]	0.736	37,790[j]	0.175
6. Pepsin[6,k]	PE	34,160[7]	0.725[8]	54,870[j]	0.243
7. Riboflavin-binding protein, apo (pH 3.0) (chicken egg white)[9]	PE	32,500[i]	0.720	66,500	0.513
8. Riboflavin-binding protein, holo (pH 7.0) (chicken egg white)[9]	PE	32,500[i]	0.720[m]	55,600	0.311
9. β-Lactoglobulin A dimer (bovine milk)[10]	PE	36,730[n]	0.751[4]	60,250[o]	0.237
10. Bovine serum albumin[11]	PE	66,300[p]	0.735[4]	115,940[q]	0.318
10a. Lactate dehydrogenase, M₄ (dogfish)[r]	OE	138,320	0.741	253,300[r]	0.362
11. β-Lactoglobulin A octamer (bovine milk)[10]	OE	146,940	0.751[s]	215,000[o]	0.130
12. Glyceraldehyde-3-phosphate dehydrogenase, apo (bakers' yeast)[12]	HOC	142,870[t]	0.737[13]	264,200	0.377
13. Glyceraldehyde-3-phosphate dehydrogenase, holo (bakers' yeast)[12]	HOC	145,520[u]	0.737[m]	250,000	0.298
14. Malate synthase (bakers' yeast)[14]	OE	170,000[i]	0.735	338,000	0.463

(continued)

TABLE V (continued)

Macromolecule	Model[b]	Parameters			From V and v̄
		M[c]	v̄, ml/g[d]	V, Å³	A₁, g/g[e]
15. Pyruvate kinase, apo[v] (brewer's yeast)[15]	OEC	190,800[n]	0.734[16]	406,000	0.548
16. Pyruvate kinase, holo[v] (brewers' yeast)[15]	OEC	192,160[n]	0.734[m]	406,000	0.539
17. Catalase (bovine liver)[17]	PC	248,000[18]	0.730[4]	420,000	0.290
18. Glutamate dehydrogenase (bovine liver)[19]	PC	312,000[20]	0.749[20]	668,000	0.541
19. Turnip yellow mosaic virus[21]	S	4.97×10^6 [22]	0.666[22]	11.49×10^6 [j]	0.727
20. Southern bean mosaic virus[23]	S	6.63×10^6 [24]	0.696[24]	12.25×10^6 [j]	0.417

[a] Excerpted from amended compilation of Ref. 1 of this table. Superscript numbers following entries indicate original references, listed in l below. Tabulated data were taken from the references thus designated in this column, unless noted otherwise for a particular parameter.

[b] Geometric model used to describe scattering particle: PE, prolate ellipsoid; OE, oblate ellipsoid; PC, prolate (elongated) cylinder; OEC, oblate (flattened) elliptical cylinder; HOC, hollow oblate cylinder; S, sphere.

[c] Molecular weights, by preference, were based on amino acid compositions and sequences wherever available, except in some cases where the cited authors' values appeared more reliable or consistent with other parameters under the conditions of measurement.

[d] Partial specific volumes were the cited authors' values or, in some cases, more accurate values found in the literature. Corrections for temperature differences between 25 and 20°C were not in general made for v̄ because resulting differences were minimal.

[e] From Eq. (9c).

[f] From Dayhoff (Ref. 3, this table), p. D-130.

[g] From Dayhoff (Ref. 3, this table), p. D-138.

[h] From Dayhoff (Ref. 3, this table), p. D-136.

[i] Value reported by cited authors (see Note c). For 15, molecular weight calculated from value for subunits by same authors.

[j] Secondary parameter, calculated with use of indicated model from values of primary parameters of cited authors.

[k] Origin of preparation not stated.

[l] Key to References:

¹ T. F. Kumosinski and H. Pessen, Arch. Biochem. Biophys. 219, 89 (1982).

² H. Pessen, T. F. Kumosinski, and S. N. Timasheff, J. Agric. Food Chem. 19, 698 (1971).

³ M. O. Dayhoff, ed., "Atlas of Protein Sequence and Structure," Vol. 5. Natl. Biomed. Res. Found., Washington, D.C., 1972.

[3] J. C. Lee and S. N. Timasheff, *Biochemistry* **13**, 257 (1974).

[4] W. R. Krigbaum and R. W. Godwin, *Biochemistry* **7**, 3126 (1968).

[5] A. A. Vazina, V. V. Lednev, and B. K. Lemazhikin, *Biokhimiya (Moscow)* **31**, 629 (1966).

[6] T. G. Rajagopalan, S. Moore, and W. H. Stein, *J. Biol. Chem.* **241**, 4940 (1966).

[7] T. L. McMeekin, M. Wilensky, and M. L. Groves, *Biochem. Biophys. Res. Commun.* **7**, 151 (1962).

[8] T. F. Kumosinski, H. Pessen, and H. M. Farrell, Jr., *Arch. Biochem. Biophys.* **214**, 714 (1982).

[9] J. Witz, S. N. Timasheff, and V. Luzzati, *J. Am. Chem. Soc.* **86**, 168 (1964).

[10] V. Luzzati, J. Witz, and A. Nicolaieff, *J. Mol. Biol.* **3**, 379 (1961).

[11] H. Durchschlag, G. Puchwein, O. Kratky, I. Schuster, and K. Kirschner, *Eur. J. Biochem.* **19**, 9 (1971).

[12] R. Jaenicke, D. Schmid, and S. Knof, *Biochemistry* **7**, 919 (1968).

[13] D. Zipper and H. Durchschlag, *Eur. J. Biochem.* **87**, 85 (1978).

[14] K. Müller, O. Kratky, P. Röschlau, and B. Hess, *Hoppe-Seyler's Z. Physiol. Chem.* **353**, 803 (1972).

[15] H. Bischofberger, B. Hess, and P. Röschlau, *Hoppe-Seyler's Z. Physiol. Chem.* **352**, 1139 (1971).

[16] A. G. Malmon, *Biochim. Biophys. Acta* **26**, 233 (1957).

[17] J. B. Summer and N. Gralén, *J. Biol. Chem.* **125**, 33 (1938).

[18] I. Pilz and H. Sund, *Eur. J. Biochem.* **20**, 561 (1971).

[19] E. Reisler, J. Pouyet, and H. Eisenberg, *Biochemistry* **9**, 3095 (1970).

[20] P. W. Schmidt, P. Kaesberg, and W. W. Beeman, *Biochim. Biophys. Acta* **14**, 1 (1954).

[21] R. Markham, *Discuss. Faraday Soc.* **11**, 221 (1951).

[22] B. R. Leonard, Jr., J. W. Anderegg, S. Shulman, P. Kaesberg, and W. W. Beeman, *Biochim. Biophys. Acta* **12**, 499 (1953).

[23] G. L. Miller and W. C. Price, *Arch. Biochem. Biophys.* **10**, 467 (1946).

[m] Value for apoenzyme used, since $\bar{v}$ for holoenzyme not available.

[n] From M. O. Dayhoff, ed., "Atlas of Protein Sequence and Structure," Suppl. 1, p. S-83. Natl. Biomed. Res. Found., Washington, D.C. 1973.

[o] Unpublished data of authors of Ref. 10, this table (S. N. Timasheff, personal communication).

[p] From M. O. Dayhoff, ed., "Atlas of Protein Sequence and Structure," Suppl. 2, p. 267. Natl. Biomed. Res. Found., Washington, D.C., 1976.

[q] Molecular volume as well as molecular weight reported by the cited authors appeared to be high, pointing to the possible presence of aggregation products. Values listed are the amino acids sequence molecular weight and a proportionally adjusted molecular volume.

[r] Unpublished data of H. Pessen, T. F. Kumosinski, G. S. Fosmire, and S. N. Timasheff.

[s] Value for 9 (dimer) used, since $\bar{v}$ for octamer not available.

[t] From Dayhoff (Ref. 3, this table), pp. D-147, D-148.

[u] Calculated from value for apoenzyme.

[v] The designations "apo" and "holo," not strictly correct here, are used for brevity to refer to "native" and "fructose diphosphate liganded," respectively.

[w] Molecular weight calculated from value for subunits reported by Bischofberger et al. (Ref. 16, this table).

which permits shape analysis, is a means to obtain this structural contribution, making use of known relationships between shape and structural frictional coefficient.[3,68,73,74,80–82]

This combination of SAXS with sedimentation analysis has been validated for its predictive value by the work of several authors.[73,75] (See also the chapter "Structural Interpretation of Hydrodynamic Measurements of Proteins in Solution through Correlation with X-Ray Data" in this volume.[83]) It may be noted that correlations between SAXS and hydrodynamic methods other than sedimentation (viscosity, diffusion), and between various combinations of hydrodynamic data with each other, have been less successful.[3,68,73]

Discussion of Approaches

The hydrodynamic approach, as pointed out in that section, cannot stand alone. At best, it requires supplementation by a structural method, preferably SAXS, to give unequivocal information on hydration. Furthermore, it has been demonstrated that sedimentation and SAXS information can be correlated to the extent that sedimentation parameters are fairly predictable[73,75] and, in effect, do not constitute an independent method. It would appear, then, that for the purpose of investigating hydration, SAXS could be considered a more useful approach than sedimentation, and that it might replace the latter, were it not for the circumstance that, in contemporary laboratories, SAXS instruments are even rarer than analytical ultracentrifuges.

The SAXS method, for practical purposes, suffers from the lack of a substantial published data base. It would be helpful if investigators reporting SAXS results on a protein undertook the labor of calculating and disseminating a full set of the parameters their data are capable of generating[74] instead of merely those engaging their immediate attention. Further, it is to be hoped that larger numbers of those interested in hydration and structural information will be enabled to use this powerful and versatile method which, with the commercial availability in recent years of position-sensitive detectors, has been a very productive method as well.

In lieu of SAXS data, NMR is capable of giving information on hydration. Absolute values of NMR hydration are not completely independent of assumptions regarding mechanism, whereas relative values are largely independent of such assumptions and can serve to assess changes in structure or biochemical behavior as a function of varying environmental

[81] F. Perrin, *J. Phys. Radium* **7**, 1 (1936).
[82] J. García de la Torre and V. A. Bloomfield, *Q. Rev. Biophys.* **14**, 81 (1981).
[83] T. F. Kumosinski and H. Pessen, this volume [11].

conditions. The interpretation of such changes of course will depend on the system studied.

One way of using NMR relaxation to obtain this kind of information has been by means of frequency dependence.[20] There are, however, few instruments available that allow dispersive measurements. Attention must be paid to the frequencies chosen: if measurements are made at two different frequencies to eliminate one of the modes of relaxation used here [cf. Eqs. (16a) and (16b)], care must be taken that the frequencies are sufficiently different to exhibit a substantial enough dispersion for adequate relaxation statistics. In any case, the frequencies must be high enough (and so must the molecular weight, on which the correlation time depends) to escape the line-narrowing region, i.e., $\omega_0^2 \tau_c^2$ must be distinctly larger than 1. Otherwise, the appropriate simultaneous equations will either have no solution, or the solution will be very imprecise.

In place of frequency dependence, concentration dependence (accessible with any NMR instrument) has been emphasized here. Two modes of relaxation, spin–lattice and spin–spin, were used with deuteron resonance to avoid the cross-relaxation encountered with spin–lattice relaxation of protons. ^{17}O relaxation has also been used, for similar reasons.[19,23] In a different method, proton spin–lattice relaxation has been employed, but cross-relaxation effects were determined and allowed for by the joint examination of two genetic variants of the same protein, which differed only in the known extent of an association reaction.[25] Again, the combination of protein molecular weight and resonant frequency must satisfy the condition for avoiding the line-narrowing mentioned.

In addition to hydration, NMR relaxation gives other valuable information that can be correlated with structural and biochemical changes. Correlation times, obtained from Eqs. (16a) and (16b), are related to molecular size and shape and are relevant to hydrodynamics. Virial coefficients, for which concentration dependence data are indispensable, can be evaluated directly from a concentration plot. From Eq. (13c), in light of the discussion of Fig. 2, it follows for either mode of relaxation that $R = \bar{v}'_w(R_b - R_f)c \exp(2B_0 c) + R_f$. It can be seen that nonlinear regression applied to this relationship gives the virial coefficient B_0 directly. The data of Fig. 2 gave $B_0 = 8.22$ ml/g, which is in close agreement with the literature value of 8.5 ml/g, obtained from light scattering.[45] Virial coefficients, related to the average net charge carried by molecules, provide a particularly useful measure of molecular interaction.

[15] Correlation Function Profile Analysis of Polydisperse Macromolecular Solutions and Colloidal Suspensions

By B. CHU, J. R. FORD, and H. S. DHADWAL

Introduction

Recent advances in laser light scattering instrumentation have made it possible to measure extremely precise time correlation function (or power spectrum) profiles using photon correlation spectroscopy (PCS) [or fast Fourier transform (FFT) spectral analysis]. This development has permitted us, at least in principle, to obtain an accurate description of the distribution of characteristic decay times, which can be related to molecular parameters of interest. For example, in polydispersity analysis of macromolecules in solution at infinite dilution or independent colloidal particles in suspension, the characteristic linewidth distribution can be related to a distribution of equivalent hydrodynamic radii or of molecular weights when rotational and internal motions of the macromolecules or suspended particles are insignificant.[1-3] Alternatively, these other modes of motion can be analyzed, usually in the absence of polydispersity, to provide information about the shape[4] and internal motions[5,6] of the scatterers, including filament flexibility[7] and active cross-bridge motions in isolated *Limulus* thick myofilaments in suspension.[8] Laser Doppler velocimetry can be used to measure the motions of suspended particles in the gas as well as the liquid phase, providing a valuable means of studying turbulent flow[9]; in the presence of an external electric field, the electrophoretic mobility of suspended particles can be determined.[10] Indeed there is a wealth of information present in the characteristic linewidths. It is, therefore, crucial to gain an understanding of the process whereby we can try

---

[1] Es. Gulari, Er. Gulari, Y. Tsunashima, and B. Chu, *J. Chem. Phys.* **70**, 3965 (1979).

[2] A. DiNapoli, B. Chu, and C. Cha, *Macromolecules* **15**, 1174 (1982).

[3] B. Chu, J. R. Ford, and J. Pope, *Soc. Plast. Eng., ANTEC 83*, **26**, 547 (1983).

[4] A. Patkowski, S. Jen, and B. Chu, *Biopolymers* **17**, 2643 (1978).

[5] B. Chu and T. Nose, *Macromolecules* **12**, 599 (1979); **13**, 122 (1980).

[6] A. Z. Akcasu, M. Benmouna, and C. C. Han, *Polymer* **21**, 866 (1980).

[7] K. Kubota and B. Chu, *Macromolecules* **16**, 105 (1983).

[8] K. Kubota, B. Chu, S.-F. Fan, M. M. Dewey, P. Brink, and D. E. Colflesh, *J. Mol. Biol.* **166**, 329 (1983).

[9] See, for example, *in* "Proceedings of the 5th International Conference on Photon Correlation Techniques in Fluid Mechanics" (E. O. Schulz-DuBois, ed.), Springer Ser. Opt. Sci., pp. 30–225. Springer-Verlag, Berlin and New York, 1983.

[10] B. A. Smith and B. R. Ware, *Contemp. Top. Anal. Clin. Chem.* **2**, 29 (1978).

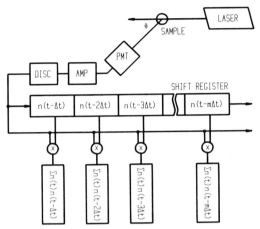

FIG. 1. Schematic of the photon correlation spectroscopy (PCS) experiment. θ = scattering angle, $n(t)$ = number of photocounts measured during the sample time Δt at time t. For clipped correlation, a clipping circuit would precede the shift register, the contents of the shift register would become the binary signal $n_k(t)$, and the multiplications indicated by $\times$ would be replaced by AND gates. It should be noted that the correlation function obtained associated with the photocounts $n(t)$, $n(t - \Delta t)$, ... is the same as that predicted by Eq. (1) corresponding to the photocounts $n(t)$, $n(t + \Delta t)$....

to determine the characteristic linewidth distribution function from the measured time correlation function. In addition, we want to be able to estimate the uncertainties in our results so as to provide a solid foundation on which to build further interpretations, such as molecular weight, size, shape, mobility, based on our data analysis procedure.

The first barrier to understanding the PCS technique lies with the hidden intricacies of the time correlation function itself. Considering the experimental setup shown in Fig. 1, the sample is illuminated by a coherent laser source, and the scattered light is detected at a scattering angle θ. Motions of the scatterers modulate the laser light, giving rise to the intensity time correlation function. For example, thermal Brownian motion of particles will cause laser light to be scattered quasielastically with a Lorentzian spectral profile. By mixing the scattered light intensity either with a portion of the unshifted (or shifted) incident radiation, referred to as a homodyne (or heterodyne) experiment, or with itself, referred to as the self-beating experiment, the intensity spectrum can be measured using either an FFT spectrum analyzer or a digital correlator.

A digital correlator samples the incoming signal from a photomultiplier tube after appropriate standardization of the photoelectron pulses, by counting the number of photon events occuring in a sample time Δt, giving rise to a train of numbers $n(t)$, $n(t + \Delta t)$, $n(t + 2\Delta t)$, ... and computes the average products

$$\langle n(t)n(t + \Delta t)\rangle = G^{(2)}(\tau_1 = \Delta t)$$
$$\langle n(t)n(t + 2\Delta t)\rangle = G^{(2)}(\tau_2 = 2\Delta t) \qquad (1)$$
$$\langle n(t)n(t + 3\Delta t)\rangle = G^{(2)}(\tau_3 = 3\Delta t)$$
$$\cdots$$

where τ is the delay time which is an integral multiple of Δt, and the average is performed over all sample times. This photoelectron count autocorrelation function can be related to the Fourier transform of the power spectrum according to the Wiener–Khinchine theorem.[11] Hence the two techniques of correlation and spectrum analysis in principle yield the same information. The input signal [the $n(t_i)$'s] is split in the digital correlator into two paths and a delay introduced in one of these (see Fig. 1) allowing the multiplications to be performed in parallel. In practice, these multiplications are often approximated by "clipping" the signal in one path such that if $n(t)$ exceeds a particular value (the clipping level), a "one" is propagated, otherwise a "zero" results. The products of Eq. (1) then become

$$\langle n(t)n_k(t + \Delta t)\rangle = G_k^{(2)}(\tau_1 = \Delta t)$$
$$\langle n(t)n_k(t + 2\Delta t)\rangle = G_k^{(2)}(\tau_2 = 2\Delta t) \qquad (2)$$
$$\cdots$$

with

$$n_k(t_i) = \begin{cases} 1 \text{ if } n(t_i) > k \\ 0 \text{ otherwise} \end{cases}$$

For certain signal characteristics[12] this can be shown to approach the full correlation function of Eq. (1), and has the major advantage that the process can be performed in real time at smaller values of Δt, typically down to around 50 nsec, while 4-bit $\times$ n "full" correlators have sample times typically down to around 100 nsec. The relation between $G^{(2)}(\tau)$ and $G_k^{(2)}(\tau)$ has been described elsewhere.[13,14] For Gaussian signals, the single-clipped photoelectron count autocorrelation function can be related to the

[11] E. O. Schulz-DuBois, in "Proceedings of the 5th International Conference on Photon Correlation Techniques in Fluid Mechanics" (E. O. Schultz-DuBois, ed.), Springer Ser. Opt. Sci., p. 6. Springer Verlag, Berlin and New York, 1983.

[12] E. Jakeman and E. R. Pike, J. Phys. A: Gen. Phys. [1] 2, 411 (1969).

[13] B. Chu, "Laser Light Scattering." Academic Press, New York, 1974.

[14] R. J. Adrian, in "Proceedings of the 5th International Conference on Photon Correlation Techniques in Fluid Mechanics" (E. O. Schulz-DuBois, ed.), Springer Ser. Opt. Sci., p. 242. Springer-Verlag, Berlin and New York, 1983.

first-order normalized electric field correlation function $g^{(1)}(\tau)$ by the Siegert relation

$$G_k^{(2)}(\tau) = N_s\langle n_k\rangle\langle n\rangle[1 + \beta|g^{(1)}(\tau)|^2]$$ (3)

where N_s is the total number of samples, $\langle n_k\rangle$ and $\langle n\rangle$ are the mean clipped and unclipped counts per sample time, and k is the clipping level. β is related to a spatial coherence factor and is usually taken as an unknown parameter to be determined by the analysis routines. Note that Eq. (3) is valid only for the self-beating experiment. $G_k^{(2)}(\tau)$ decays from an initial value of $N_s\langle n \cdot n_k\rangle$ to a baseline of $N_s\langle n\rangle\langle n_k\rangle$ which we will subsequently refer to as A. The function $g^{(1)}(\tau)$ which has an initial value of 1 and decays to zero at $\tau = \infty$ is a function that is related to the characteristic linewidth distribution $G(\Gamma)$. In the absence of directed particle motions,

$$g^{(1)}(\tau) = \int_0^\infty G(\Gamma)e^{-\Gamma\tau}\,d\Gamma$$ (4)

which shows that the first-order electric field autocorrelation function is the Laplace transform of the characteristic linewidth distribution $G(\Gamma)$. The problem of data analysis therefore consists of inverting the Laplace transform and forms the main subject of this chapter.

Experimental Considerations

Importance of the Baseline

From Eq. (3) we see that the *net* unnormalized electric field autocorrelation function

$$(A\beta)^{1/2}g^{(1)}(\tau_i) = [G_k^{(2)}(\tau_i) - A]^{1/2} \equiv b_i$$ (5)

is the experimental quantity of interest, assuming that the baseline A is known to sufficient precision. We have found that errors of more than a few tenths of a percent in the quantity A will preclude meaningful correlation function profile analysis of the data, and that in order to obtain even an approximate form for $G(\Gamma)$, this quantity must be known to within about 0.2% in order to avoid significant distortion of the resultant linewidth distribution function from the Laplace inversion. Usually the correlator will provide both a "measured" and a "calculated" baseline. The measured baseline is computed just as are the other channels in the correlation function, except that the signal is delayed by some additional blank channels, typically on the order of $400\Delta t$, so that, for $\beta = 1$, $[G_k^{(2)}(\tau = 400\Delta t)/A] - 1 < 0.002$. The calculated baseline is based on the fact that $G_k^{(2)}(\tau \to \infty)$ approaches the value $N_s\langle n_k\rangle\langle n\rangle$, and the single-clipped correlator uses additional separate counters to determine the quantities N_s (total samples), $N_s\langle n_k \rangle$ (total clipped counts), and $N_s\langle n \rangle$ (total unclipped counts). The measured baseline will be affected by signals having characteristic linewidths that extend out through the baseline delay channels where significant contributions, say at $400\Delta t$ or whatever the baseline delay channels are set for, to the net electric field correlation function remain. These contributions could include signals from dust and from statistical number fluctuations in the concentration of particles in the scattering volume. The calculated baseline will also be affected by long-term fluctuations, although the effects are harder to evaluate quantitatively.

Recent work by Oliver[15] indicates that the best statistical accuracy in measuring the net intensity correlation function can be achieved by taking short batches of data and normalizing by the calculated baseline for each batch separately, then averaging the net intensity correlation functions all together. This approach may provide a biased estimate of the correlation function depending upon the correlator design. However, the bias usually decreases with the duration of each batch. Many commercial correlators begin an experiment by first clearing the signal paths, and then start accumulating the correlation function. Thus, after Δt, the first channel of the correlator, $G_k^{(2)}(\tau_1)$ [or $G^{(2)}(\tau_1)$], has a value corresponding to the first point in the correlation function; all other channels have zeros. After the next sampling, the first channel now has two estimates of $G_k^{(2)}(\tau_1)$ while the second channel is getting its first estimate of $G_k^{(2)}(\tau_2)$; the third through the end are still zeros. Repeating this process m times with m being the number of delayed channels in the correlator not including the measured baseline, we have an m-channel correlation function that is severely distorted toward higher values in the low channels. It should be noted that the measured baseline is still zero after $m\Delta t$ delays. However, as the duration of the experiment goes to ∞, the bias goes to zero. When setting the duration of a ''short'' experiment one must choose it long enough so that the bias is negligible ($<0.1\%$) and yet short enough to still get some advantage from taking data in short bursts. This quantity can be estimated by using N_s such that m/N_s is less than 0.1%. However, most of the new correlators coming on the market are free of the above bias introduced in the short experiment.

Range of Delay Times

For a single exponential decay corresponding to a single Lorentzian curve in the power spectrum, Oliver and co-workers[16,17] have shown that

[15] C. J. Oliver, *J. Phys. A: Math. Gen.* **12**, 591 (1979).
[16] A. J. Hughes, E. Jakeman, C. J. Oliver, and E. R. Pike, *J. Phys. A: Math., Nucl. Gen.* **6**, 1327 (1973).
[17] C. J. Oliver, *Adv. Phys.* **27**, 387 (1978).

the delay time increment Δt should be chosen so that $\Gamma\tau_{max}$ is between 2.5 and 3, where Γ is the linewidth and τ_{max} ($\equiv m\Delta t$) is the maximum delay of the correlation function, not including the baseline delay. For nonsingle exponential decays, we often use the criterion $2.5 < \bar{\Gamma}\tau_{max} < 3$ where $\bar{\Gamma}[\equiv \int G(\Gamma)\Gamma \, d\Gamma]$ is the average linewidth. However, this generalization can sometimes be misleading—in fact, it is desired that the minimum characteristic linewidth or the slowest decay be sampled long enough so that the decay can be evaluated (i.e., τ_{max} must be sufficiently large) while simultaneously one wishes to see the fastest decay show up in at least two or three channels (i.e., $\tau_{min} = \Delta t$ must be small enough). These are demanding requirements for a correlator consisting of a finite number of delay channels with equally spaced delay time increments. One alternative is to space the channels nonlinearly, thereby covering a larger dynamic range using the same limited number of channels. Commercial instruments with this capability are appearing on the market. An expedient approach of taking several correllograms at different ranges of delay time increments and patching them together is not recommended due to normalization difficulties, unless the correllograms are accumulated simultaneously.

Finally, it must be noted that the samples must be relatively free of dust particles as their presence often contributes a slow decay to the measured correlation function. Correction after the measurement has been attempted, e.g., by floating the value of A in the fitting routines, is occasionally possible, but the interpretation of the results must be considered with great care.

Nature of the Problem

Equation (4) is a Fredholm integral equation of the first kind; the problem associated with recovering $G(\Gamma)$ is ill conditioned when the data are less than perfect. In recent years much attention has been focused on this inversion problem.[18–29] In particular, the reader is advised to refer to

[18] B. Chu, Es. Gulari, and Er. Gulari, *Phys. Scr.* **19**, 476 (1979).

[19] J. G. McWhirter and E. R. Pike, *J. Phys. A: Math. Gen.* **11**, 1729 (1978).

[20] N. Ostrowski, D. Sornette, P. Parker, and E. R. Pike, *Opt. Acta* **28**, 1059 (1981).

[21] E. R. Pike, *in* "Scattering Techniques Applied to Supramolecular and Non-Equilibrium Systems" (S. H. Chen, B. Chu, and R. Nossal, eds.), p. 179. Plenum, New York, 1981.

[22] G. Wahba, *SIAM J. Numer. Anal.* **14**, 651 (1977).

[23] G. B. Stock, *Biophys. J.* **16**, 535 (1976); **18**, 79 (1978).

[24] S. W. Provencher, *Biophys. J.* **16**, 27 (1976); *J. Chem. Phys.* **64**, 2772 (1976); *Makromol. Chem.* **180**, 201 (1979).

[25] S. W. Provencher, J. Hendrix, L. DeMaeyer, and N. Paulussen, *J. Chem. Phys.* **69**, 4273 (1978).

[26] J. R. Ford and B. Chu, *in* "Proceedings of the 5th International Conference on Photon Correlation Techniques in Fluid Mechanics" (E. O. Schulz-DuBois, ed.), Springer Ser. Opt. Sci., p. 303. Springer-Verlag, Berlin and New York, 1983.

specific topics in chapters of two recent books[27,28] dealing with linewidth polydispersity analysis for details. Due to the ill conditioning, the amount of information about $G(\Gamma)$ that can be recovered from the data is quite limited. Attempts to extract too much information will result in physically unreasonable solutions. Correct inversion of Eq. (4) therefore must involve some means of limiting the amount of information requested from the problem; this can be done in the following ways.

1. Limiting the statement of the problem in the beginning so that only the correct amount of information is resolved[19,29];
2. Applying regularization criteria on the solution, so the unreasonable solutions are penalized[24];
3. Stating the problem in terms of a large number of information elements and applying a rank reduction technique midway through the solution.[26]

By examining the inversion of Eq. (4) in terms of information theory, and by emphasizing the eigenvalue analysis of the Laplace kernel, McWhirter and Pike[19] have shown how a knowledge of the eigenvalue spectrum of the Laplace kernel coupled with reliable estimates of the noise in the data can be used to determine the maximum amount of information that can be recovered about $G(\Gamma)$. The eigenvalues are shown to decrease rapidly below the noise level of the data, at which point it becomes impossible to extract further knowledge about the characteristic linewidth distribution. Unfortunately, it is difficult to formulate a quantitative description of the noise in the data and therefore to determine the number of terms that should be retained in the eigenfunction expansion. Bertero *et al.*,[29] for example, have shown that the obtainable resolution is increased as the support ratio γ ($\equiv \Gamma_{max}/\Gamma_{min}$, where Γ_{max} and Γ_{min} are the upper and lower frequencies of the characteristic linewidth distribution) of $G(\Gamma)$ is decreased. It should be noted that here Γ_{min} and Γ_{max} refer to the bounds of the *true* distribution. In addition, it is often possible by the introduction of physical constraints to recover more detailed information than would be indicated by the eigenvalue spectrum of the general Laplace kernel. The singular value decomposition technique[30-32] provides a

[27] See "Essentials of size distribution measurement," *in* "Measurement of Suspended Particles by Quasi-Elastic Light Scattering" (B. E. Dahneke, ed.), p. 81–252. Wiley, New York, 1983.

[28] See "Photon correlation spectroscopy of Brownian motion: Polydispersity analysis and studies of particle dynamics," *in* "Proceedings of the 5th International Conference on Photon Correlation Techniques in Fluid Mechanics" (E. O. Schulz-DuBois, ed.), Springer Ser. Opt. Sci., pp. 286–315. Springer-Verlag, Berlin and New York, 1983.

[29] M. Bertero, P. Boccacci, and E. R. Pike, *Proc. R. Soc. London, Ser. A* **383**, 15 (1982).

[30] G. H. Golub and C. Reinsch, *Numer. Math.* **14**, 403 (1970).

means of determining the information elements of the actual problem (including constraints) and, by ordering these elements in decreasing importance, is a useful method which can estimate the amount of information recoverable for a given problem.

Regularization methods aim at seeking solutions to Eq. (4) from a class of functions that make the inversion of the integral equation well behaved. These pseudo-solutions generally correspond to solutions of minimum energy, nonnegativity, or some other reasonable physical constraints. However, it is important that the regularization method be chosen such that one can determine the relationship between the regularized solution $\tilde{G}(\Gamma)$ and the actual solution $G(\Gamma)$. Provencher[33] has made available an automated procedure that combines regularization with the eigenfunction analysis to determine the "smoothest" solution to Eq. (4).

Mathematical Background

Least-Squares Minimization Techniques

In general we seek solutions to Eq. (4) that minimize some measure of the discrepancies between the discretely sampled data points b_i [defined by Eq. (5)] and the corresponding values that would be calculated from our solution, subject to the requirement that the solution be well behaved. We shall minimize the L^2 norm of the residuals (i.e., least squares minimization); a different sensitivity to data errors may be achieved by using some other norm as a metric. Discussion of the use of these alternate norms may be found in Rice[34] and will not be considered further here. Each of the approaches we describe below, with the exception of the regularization method, will involve the solution, in a least-squares sense, of an overdetermined set of simultaneous equations arising from a discrete analog to Eq. (4). For example, the cumulants approach[35] takes advantage of the formal equivalence between $g^{(1)}(\tau)$ and the moment generating function $M(-\tau;\Gamma) \equiv \langle \exp(-\Gamma\tau) \rangle_{av} = \int G(\Gamma)e^{-\Gamma\tau}d\Gamma$ to define a cumulants generating function $K(-\tau;\Gamma) \equiv \ln M(-\tau;\Gamma)$. The least-squares minimization problem then amounts to determining the coefficients of the

[31] R. J. Hanson, *SIAM J. Numer. Anal.* **8**, 616 (1971).
[32] C. L. Lawson and R. J. Hanson, "Solving Least Squares Problems." Prentice-Hall, Englewood Cliffs, New Jersey, 1974.
[33] S. W. Provencher, *in* "Proceedings of the 5th International Conference on Photon Correlation Techniques in Fluid Mechanics" (E. O. Schultz-DuBois, ed.), Springer Ser. Opt. Sci., p. 322. Springer-Verlag, Berlin and New York, 1983.
[34] J. R. Rice, "The Approximation of Functions," Vol. I. Addison-Wesley, Reading, Massachusetts, 1964.
[35] D. E. Koppel, *J. Chem. Phys.* **57**, 4814 (1972).

(truncated) McLaurin expansion of K, i.e., the cumulants of the distribution. Details are presented below; the procedure is that a few parameters (the first few cumulants) are to be determined so as to achieve the best fit to a large number of data points. The multiexponential and histogram methods choose a mathematically convenient form to approximate $G(\Gamma)$ that again results in an overdetermined set of equations. The cumulants method is nonlinear, whereas the multiexponential and histogram models lead to either linear or nonlinear problems depending on whether or not the model for $G(\Gamma)$ is allowed to float in Γ-space. The cumulants method and the nonlinear double exponential model where the two characteristic linewidths as well as their corresponding amplitudes are unknown parameters can often be solved by application of a standard nonlinear least-squares algorithm. As the number of adjustable parameters increases, however, the problem rapidly becomes more susceptible to the ill conditioning of Eq. (4), and one must be careful to include some form of rank reduction step in the analysis procedure. We have found the singular value decomposition technique quite helpful for the linear multiexponential and histogram approximations.

Nonlinear Methods

Methods of solving the nonlinear least-squares problem generally involve an iteration scheme that aims at descending the χ^2 surface to the minimum. The χ^2 surface is defined by the value of the sum of the squares of the residuals

$$\sum_{i=1}^{m} [b_i - f(\tau_i;\mathbf{P})]^2$$

at the point located by the vector $\mathbf{P} = (\mathbf{P}_1,\mathbf{P}_2,...\mathbf{P}_n)$, with b_i defined by Eq. (5) and $f(\tau_i;\mathbf{P})$ being the value of b_i calculated from the model at the point τ_i where the parameters of the model have the values given by the $\mathbf{P}$ vector. Therefore, an initial estimate of the $\mathbf{P}$ vector is supplied to the routine (i.e., a starting point at which to evaluate the χ^2 surface), and an appropriate algorithm is used to choose the direction of descent (i.e., to find some suitable modification to the $\mathbf{P}$ vector so as to reduce χ^2). Three methods are in common use and will be described further.

The gradient method assumes that the gradient vector points out the direction of the minimum. Therefore, at the current point $\mathbf{P}^{(N)}$, the $(N + 1)^{st}$ estimate of the coordinates of the minimum $\mathbf{P}^{(N+1)}$ is given by

$$\mathbf{P}^{(N+1)} = \mathbf{P}^{(N)} - aJ^{\mathrm{T}}F^{(N)} \tag{6}$$

where a is a scalar quantity defining the size of the step to take along the gradient, $F^{(N)}$ is the residual vector at the Nth estimate with $(F^{(N)})_i = b_i - f(\tau_i;\mathbf{P}^{(N)})$, and J is the Jacobian matrix whose elements are given by $J_{ij} = \partial(F^{(N)})_i/\partial(\mathbf{P}^{(N)})_j$; the superscript (N) on the Jacobian has been omitted for clarity but it should be kept in mind that this matrix must be reevaluated at each point $\mathbf{P}$. The notation J^{T} indicates the transpose of the matrix J, i.e., $(J^{\text{T}})_{ij} = J_{ji}$. The quantity $J^{\text{T}}F^{(N)}$ represents a displacement along the gradient vector, as can be verified by expanding out the terms of the matrix multiplication; the reason for the somewhat abstruse notation should become clear shortly. The derivatives involved can be evaluated numerically using central or forward difference algorithms,[36,37] or can be calculated explicitly if the functional form of the derivatives are known, as will be the case for the models described in this chapter. Note that the gradient vector points in the direction of the steepest *increase* in χ^2; hence the subtraction in Eq. (6). Bevington[38] gives a discussion of the gradient method as well as the other nonlinear methods we will describe below. It suffices to say here that the gradient method is quite good at approaching the general vicinity of the solution when it is far away from the solution, but when its gets close, the slope of the χ^2 surface tends to zero (as it should near the minimum) and the search often becomes erratic.

An alternative is to use the Gauss–Newton method, an extension of Newton's method to many variables. In this approach, a Taylor's series expansion of the χ^2 surface is used to obtain an estimate of the position of the minimum. Only the first-order terms in the expansion are used in order to keep the computation time finite. The $(N+1)^{\text{st}}$ estimate is obtained from

$$\mathbf{P}^{(N+1)} = \mathbf{P}^{(N)} - J^{-1}F^{(N)} \tag{7}$$

where the J and $F^{(N)}$ are the same as in Eq. (6) and J^{-1} refers to the inverse of J. Equation (7) can be compared to the one-dimensional expression describing Newton's method

$$x^{(N+1)} = x^{(N)} + f(x^{(N)})/f'(x^{(N)}) \tag{8}$$

with f' representing the first derivative of f with respect to x. Again, expansion of the terms in Eq. (7) should convince the persistent reader that the expressions are correct. The obvious problem with this technique is that unless one is quite close to the correct minimum, the truncation after the linear terms may lead to poor estimates of the minimum, and

[36] K. M. Brown and J. E. Dennis, Jr., *Numer. Math.* **18**, 289 (1972).
[37] M. Abramowitz and I. A. Stegun, "Handbook of Mathematical Functions," pp. 877ff. Dover, New York, 1972.
[38] P. R. Bevington, "Data Reduction and Error Analysis for the Physical Sciences." McGraw-Hill, New York, 1969.

hence slow convergence. Indeed, if the curvature of the χ^2 surface is negative, the algorithm proceeds in the wrong direction.

It became obvious to Levenberg[39] in 1944 and independently to Marquardt[40] in 1963 that what was needed was some combination of the two approaches. Taking the $(N+1)^{st}$ estimate to be found from

$$\mathbf{P}^{(N+1)} = \mathbf{P}^{(N)} - [\lambda^{(N)}I + J^T J]^{-1} J^T F^{(N)} \tag{9}$$

where I is the identity matrix with $I_{ij} = \delta_{ij}$ and $\lambda^{(N)}$ is a scalar quantity that may change with each iteration, it is apparent that this algorithm combines the virtues of the two previous methods. For $\lambda^{(N)} = 0$, the algorithm reduces to the Gauss–Newton method; for $\lambda^{(N)} \to \infty$, the algorithm approaches the gradient method. The quantity $\lambda^{(N)}$ is often referred to as the "Marquardt" parameter. Initially it should be chosen large enough to force a gradient search; at each subsequent step it can be decreased if the search direction was correct (i.e., the value of χ^2 decreased) or increased if not. Actual values (i.e., what is "large enough") of the Marquardt parameter may depend on implementation of the algorithm; in our case we generally start with $\lambda^{(N)} = 10$, decreasing by a factor of 2 when the algorithm is improving, and increasing by 4 when χ^2 increases. Convergence to a solution can be determined by one of several criteria, among them (1) the norm of the gradient decreases below a certain value, indicating that the χ^2 surface is getting quite flat; (2) the change in χ^2 is less than a fixed (small) amount, again indicating a minimum in χ^2, and (3) the parameters of the model are changing by less than a certain (small relative) amount from one iteration to the next.

Although the above algorithms are quite straightforward from a conceptual point of view, the reader should be aware of possible problems arising from the finite precision arithmetic of digital computers. Several texts are available[41]; suffice it to say that in the inversion of the Laplace transform, consideration *must* be given to the numerical accuracy of the results. Although Bevington[38] presents FORTRAN IV code that will implement the Marquardt routine, we have found instances when the results differ by more than 10% from the IMSL[42] routine ZXSSQ, and there is quite general disagreement on the order of a few percent in all cases. Commercial routines such as those provided by IMSL are *highly* recommended for the user who has not had extensive training in numerical methods.

[39] K. Levenberg, *Q. Appl. Mech.* **2**, 164 (1944).
[40] D. W. Marquardt, *J. Soc. Ind. Math.* **11**, 431 (1963).
[41] See, for example, R. F. Churchhouse, ed., "Handbook of Applicable Mathematics," Vol. III, Wiley, New York, 1981.
[42] IMSL, 7500 Bellaire Blvd., Houston, TX 77036.

Linear Methods

The linear least squares minimization problem can be stated as

$$CP \simeq b \qquad (10)$$

where C is the $(m \times n)$ curvature matrix, $\mathbf{P}$ is the parameter vector of length n, and b is again the data vector of length m defined in Eq. (5). m and n are, respectively, the number of data points in the net correlation function and the number of adjustable parameters of the model. For example, under the linear multiexponential approximation, the elements of C can be determined from comparison of Eqs. (4) and (54) with Eq. (10) to be

$$c_{ij} = e^{-\Gamma_j \tau_i} \qquad (11)$$

where i and j are the row and column indices, respectively, and the parameters to be determined are the $\mathbf{P}_j$, i.e., the weighting factors of the δ functions [see Eq. (54)]. The symbol $\simeq$ in Eq. (10) is intended to imply solution of the overdetermined set of equations subject to the least squares criterion (minimization of the Euclidean norm of the residual vector $\|b - CP\|$, where the Euclidean norm of a vector $\mathbf{h}$ can be computed from $\|\mathbf{h}\| = (\Sigma_i \mathbf{h}_i^2)^{1/2}$, a special case of Eq. (30) below. Prior to entering the singular value decomposition routine, we generally scale the columns of C to unit norm in order to improve the numerical stability of the inversion (see for example, Lawson and Hanson[32] pp. 185–188). The scaling transforms Eq. (10) to

$$Ax \simeq b \qquad (12)$$

where

$$A = CH \qquad (13)$$

and

$$x = H^{-1}\mathbf{P} \qquad (14)$$

with H being a diagonal matrix whose nonzero elements are the reciprocals of the norms of the corresponding column of C,

$$\mathbf{h}_{ij} = (\sum_{j=1}^{n} c_{ij}^2)^{-1/2} \delta_{ij} \qquad (15)$$

The resulting problem [Eq. (12)] is passed to a singular value decomposition algorithm (for example, subroutine SVDRS from appendix C of Lawson and Hanson[32]) which transforms Eq. (12) to

$$USV^{-1}x = b \qquad (16)$$

where U and V^{-1} are orthogonal matrices and S is a diagonal matrix whose nonzero elements are the (monotonically decreasing) singular values of the stated problem. The explicit procedure of finding this singular value decomposition (i.e., determining the matrices U, S, and V^{-1}) involves the application of properly chosen orthogonal transformations to the matrix A such that Eq. (16) holds; details of the procedure may be found in Lawson and Hanson[32] and will not be discussed further. It is precisely this decomposition that subroutine SVDRS of Lawson and Hanson's book accomplishes. Defining the new vectors $\mathbf{y}$ and $\mathbf{g}$ by

$$x = V\mathbf{y} \qquad (17)$$

and

$$\mathbf{g} = U^T\mathbf{b} \qquad (18)$$

we have

$$S\mathbf{y} = \mathbf{g} \qquad (19)$$

The singular value routine will typically return the matrices V and S and the vector $U^T\mathbf{b}$. Since S is diagonal, we have immediately.

$$\mathbf{y}_i = \mathbf{g}_i/s_i \qquad (20)$$

with s_i being i^{th} singular value, S_{ii}. Equation (20) represents the full-rank solution to problem (12) and assumes that all singular values are nonzero, i.e., the problem was nonsingular. This is generally not the case, since Eq. (12) is often pseudo-rank-deficient in that not all n elements of x are recoverable as independent resolution elements, due to the data noise. A rank-reduction step is employed to limit the amount of information necessary to express the solution. If we know the problem to be of pseudorank k, we can retain the first k diagonal elements of S and ignore the rest, i.e., we will compute the pseudo-inverse of S, S^+, and place zeros in the locations of S^+ that correspond to near-zero singular values s_i. This is consistent with the viewpoint based on information theory, since the square of the singular values are the eigenvalues of the matrix A^*A (A^* is the adjoint of A) and thus are closely related to the fundamental elements of resolution transmitted through the integral operator of Eq. (4). Generally, k is unknown; we therefore define a set of "candidate solutions" $[x^{(k)}]$ from Eq. (17) as

$$x^{(k)} = V\mathbf{y}^{(k)} \qquad (21)$$

where

$$\mathbf{y}^{(k)} = \begin{cases} \mathbf{g}_i/s_i \text{ for } 1 < i \le k \\ 0 \text{ for } k < i < n \end{cases} \tag{22}$$

In terms of the original problem [Eq. (10)] we have a set of candidate solutions for the $\mathbf{P}$ vector defined by

$$\mathbf{P}^{(k)} = Hx^{(k)} \tag{23}$$

Furthermore, this procedure defines a set of residual vectors $[\mathbf{r}^{(k)}]$ such that

$$\mathbf{r}^{(k)} = C\mathbf{P}^{(k)} - b \tag{24}$$

There are several criteria for the selection of the particular value of k that can be applied (see chapters 25 and 26 of Lawson and Hanson[32]). One that we have found useful is to examine a plot of $\ln\|\mathbf{r}^{(k)}\|$ vs $\ln\|\mathbf{P}^{(k)}\|$. We endeavor to select k such that $\|\mathbf{r}^{(k)}\|$ is "sufficiently small" without $\|\mathbf{P}^{(k)}\|$ getting "too large." This is directly analogous to the "energy" constraint of some regularization methods.

Subroutine SVDRS of Lawson and Hanson would seem to deal satisfactorily with possible numerical problems that might arise during computation. Alternatively the IMSL routine LSVDF or equivalent may be used; again we caution the amateur computer enthusiast against attempting to "improve" on these routines.

Methods of Regularization

In this section we describe a regularized scheme for recovering the characteristic linewidth distribution function $G(\Gamma)$ and show the technique to possess some desirable features regarding the stability of the solution in the presence of noise. In addition, such a priori constraints as positivity are incorporated into the solution. The analysis below is presented in terms of continuous functions and discretization is introduced to show how the technique can be implemented. This leads to unavoidable complications in notation; in general, we have tried to use unsubscripted characters for continuous functions, introducing subscripts upon discretization in such a way as to be as consistent as possible with the notation of the previous sections.

In the absence of noise the inverse problem is described by the integral equation (4), which belongs to the general class of linear Fredholm integral equations of the first kind

$$\bar{b}(\tau) = \int_{\Gamma_{min}}^{\Gamma_{max}} K(\tau,\Gamma)G(\Gamma)d\Gamma, \qquad 0 < \tau < \infty \tag{25}$$

where $\bar{b}(\tau)$ is the continuous noise-free data function from which the distribution $G(\Gamma)$ has to be determined, and $K(\tau,\Gamma)$ is a general kernel given by $e^{-\Gamma\tau}$ for the special case of the Laplace transform. We have already remarked on the ill-posed nature of the problem associated with the inversion of Eq. (25) when the data are less than perfect. Methods of imposing stability on the solution by the use of known or plausible constraints (the so-called regularization techniques) have been the subject of many studies.[24,43–47] Regularization involves essentially a modification of the original problem, which is transformed into one possessing the desired properties. This is by no means the only way of restoring stability to the ill-posed problem. Indeed, such a problem may be regularized by, for example, changing the definition of what is meant by ill posed. A discussion of this is given by Nashed.[47]

Abbiss et al.[48] used Miller's regularization procedure[43] to derive, iterative and noniterative, solutions for the inversion of a similar equation which arises in Fourier optics and in the extrapolation of band-limited signals. Following the steps outlined in Abbiss et al.,[48] let us denote the perturbation to the data (for example, the effects of additive noise) by the function $r(\tau)$ and rewrite Eq. (25)

$$\bar{b}(\tau) = \int_{\Gamma_{min}}^{\Gamma_{max}} K(\tau,\Gamma)G(\Gamma)d\Gamma + r(\tau) \tag{26}$$

briefly in operator form,

$$b = KG + r \tag{27}$$

Since the noise term r [not to be confused with $\mathbf{r}^{(k)}$ in Eq. (24)] is unknown, Eq. (27) cannot be solved directly for G. One might, as an alternative, attempt to determine the unique distribution ϕ which satisfies the equation

$$b = K\phi \tag{28}$$

This, however, would be a fundamentally unsatisfactory procedure because of the ill posedness of the problem: small perturbations of b due for example to noise will induce in general very large fluctuations to the solution ϕ, which may then take a completely different form from the

[43] K. Miller, SIAM J. Math. Anal. **1**, 52 (1970).
[44] A. N. Tikhonov and V. Y. Arsenin, "Solutions of Ill-Posed Problems," V. H. Winston & Sons, Washington, D.C., 1977.
[45] M. Bertero, C. DeMol, and G. A. Viano, Top. Curr. Phys. **20**, 161 (1980).
[46] M. Bertero, G. A. Viano, and C. DeMol, Opt. Acta **27**, 307 (1980).
[47] M. Z. Nashed, IEEE Trans. Antennas Propag. **AP-29**, 220 (1981).
[48] J. B. Abbiss, C. DeMol, and H. S. Dhadwal, Opt. Acta **30**, 107 (1983).

original distribution. The singular value decomposition routine described above determines a pseudo-solution to the discrete analogue of Eq. (28), with the rank reduction step acting to prohibit these large fluctuations. Regularization theory on the other hand requires that we solve a related equation whose solution is constrained to be "close to" the true distribution, that is, in some well defined sense, and to respond stably to pertubations to the data.

In Abbiss et al.[48] a regularized approximation to the true distribution function was obtained by minimizing the following functional:

$$F(u) = \|b - Ku\|^2 + \alpha\|Cu\|^2 \qquad (29)$$

where u represents the possible set of functions satisfying Eq. (29), α is a small positive quantity referred to as the regularization parameter, and $\|\mathbf{h}\|$ denotes the L^2 norm of $h(x)$ on $[x_1, x_2]$

$$\|\mathbf{h}\|^2 = \int_{x_1}^{x_2} |h(x)|^2 \, dx \qquad (30)$$

It has been assumed that G and b are square integrable function on $[\Gamma_{min}, \Gamma_{max}]$ and on $[0, \infty]$ respectively. In Eq. (29) C is any linear constraint operator which allows the incorporation of additional a priori knowledge of the solution. For example, we may restrict the norm of the solution if C is taken to be the identity operator. Alternatively we may express preference for a smoother solution by choosing C to be a differential operator. Equation (29) represents the minimization of the residual norm $\|b-Ku\|$ subject to the a priori knowledge of u, $\|Cu\| < E$ with E representing an upper bound for the norm of Cu. By the application of variational calculus[49] the minimum of Eq. (29) can be shown to occur for that estimate $\tilde{G}$ which is the solution of the equation

$$[K^*K + \alpha C^*C]\tilde{G} = K^*b \qquad (31)$$

where the asterisk denotes the adjoint operator.[50] The solution to Eq. (31) is

$$\tilde{G} = HK^*b \qquad (32)$$

where $H = [K^*K + \alpha C^*C]^{-1}$. Equation (32) could be used to solve for $\tilde{G}$ directly. We note that the singular value decomposition technique could be used for solving Eq. (32) rather than Eq. (10), thereby including linear constraints represented by the operator C. We will not pursue this any

[49] C. W. Groetsch, "Generalized Inverses of Linear Operators." Dekker, New York, 1977.
[50] V. Hutson and J. S. Pym, "Applications of Functional Analysis and Operator Theory." Academic Press, New York, 1980.

further; rather we describe a means for calculating $\tilde{G}$ by iteration that allows a simple incorporation of the positivity constraint at each step. This may be derived in the following way. Equation (31) is rewritten in the form

$$\tilde{G} = \tilde{G} + K^*b - [K^*K + \alpha C^*C]\tilde{G}$$

This suggests the iterative scheme,

$$\tilde{G}^{(N)} = K^*b + [(1 - \alpha)I - K^*K]\tilde{G}^{(N-1)} \tag{33}$$

where we have taken C to be the identity operator, I. The notation $\tilde{G}^{(N)}$ represents the Nth iteration estimate of $\tilde{G}$. Equation (33) has the added advantage that for the unregularized solution ($\alpha = 0$) we can use the number of iterations to restrict the solution to a finite set of singular values, analogous to the selection of a value for k in the singular value decomposition rank reduction step described above. The convergence of Eq. (33) is guaranteed if the norm of the operator $[(1 - \alpha) I - K^*K]$ is strictly bounded by unity for all $\alpha < 1$. Using the singular values of the Laplacian kernel computed by Bertero et al.,[29] it can be shown that Eq. (33) in the limit of infinite iterations converges to the solution of Eq. (31). At this point, it should be noted that the convergence of $\tilde{G}^{(N)}$ to $\tilde{G}$ when N tends to infinity does not ensure by itself regularization. In other words, convergence of an iterative sequence to a well-behaved solution, by itself, is not a sufficient condition to establish regularization. This latter property is completely distinct since it concerns the convergence of $\tilde{G}$ to the true solution G when the noise on the data b tends to zero. We will later show this to be true by invoking a singular function decomposition of the solution, but first describe how the iterative scheme of Eq. (33) can be implemented on a digital computer in order to determine the regularized approximation to the characteristic linewidth distribution function satisfying Eq. (31).

By KG and K^*b we mean as follows

$$(KG)(\tau) = \int_{\Gamma_{min}}^{\Gamma_{max}} e^{-\Gamma\tau}G(\Gamma)d\Gamma, \qquad 0 < \tau < \infty \tag{34}$$

$$(K^*b)(\Gamma) = \int_0^\infty e^{-\Gamma\tau}b(\tau)d\tau, \qquad \Gamma_{min} < \Gamma < \Gamma_{max} \tag{35}$$

Therefore,

$$(K^*KG)(\Gamma) = \int_0^\infty e^{-\Gamma\tau}d\tau\int_{\Gamma_{min}}^{\Gamma_{max}} e^{-z\tau}G(z)dz$$

By changing the order of integration we find

$$(K*KG)(\Gamma) = \int_{\Gamma_{min}}^{\Gamma_{max}} \frac{1}{\Gamma + z} G(z)dz \tag{36}$$

In the analysis of real data, $b(\tau)$ will be known only over (τ_{min}, τ_{max}). Equation (35) becomes

$$(K*b)(\Gamma) = \int_{\tau_{min}}^{\tau_{max}} e^{-\Gamma\tau} b(\tau)d\tau, \qquad \Gamma_{min} < \Gamma < \Gamma_{max} \tag{37}$$

Therefore,

$$(K*KG)(\Gamma) = \int_{\Gamma_{min}}^{\Gamma_{max}} \frac{[e^{-\tau_{min}(\Gamma+z)} - e^{-\tau_{max}(\Gamma+z)}]}{(\Gamma + z)} G(z)dz \tag{38}$$

Up to now we have been dealing with the continuous functions $b(\tau)$ and $G(\Gamma)$ and integral operators. The extension to discrete distributions is a trivial one. Let b_i be defined as in Eq. (5) to be the discretely sampled first-order electric field autocorrelation function. If $\tilde{G}_j$ represents the estimates of $\tilde{G}(\Gamma)$ at discrete points $\Gamma_1, \Gamma_2, ..., \Gamma_n$, then the first estimate of the regularized linewidth distribution function [found by setting $\tilde{G}^{(0)} = 0$ in Eq. (33)] is

$$\tilde{G}_j^{(1)} = K*b = \sum_{i=1}^{m-1} 0.5(\tau_{i+1} - \tau_i)(b_{i+1}e^{-\Gamma_j\tau_{i+1}} + b_ie^{-\Gamma_j\tau_i})$$

$$= 0.5(\tau_2 - \tau_1)b_1 \, e^{-\Gamma_j\tau_i}$$

$$+ 0.5\sum_{i=2}^{m-1} (\tau_{i+1} - \tau_{i-1})b_i \, e^{-\Gamma_j\tau_i} \tag{39}$$

$$+ 0.5(\tau_m - \tau_{m-1})b_m \, e^{-\Gamma_j\tau_m}$$

Note that we have used the trapezoidal rule for approximating the integration of Eq. (37). This expression has been written in its most general form in order to allow the use of nonlinear sampling of both the correlation and linewidth distribution functions.

In Eq. (33) the Nth estimate of $\tilde{G}$ involves the integral operator $[(1 - \alpha) + K*K]$. We can also approximate this by an equivalent discrete matrix operator P,

$$P = (1 - \alpha)I - S \tag{40}$$

where S is the discrete matrix operator corresponding to the integral operator of Eq. (38)

$$S_{i1} = 0.5(\Gamma_2 - \Gamma_1)W_{i1}$$

$$S_{ij} = 0.5(\Gamma_{j+1} - \Gamma_{j-1})W_{ij} \qquad 2 \leq j \leq n - 1$$

$$S_{in} = 0.5(\Gamma_n - \Gamma_{n-1})W_{in}$$

$$W_{ij} = \frac{[e^{-\tau_{min}(\Gamma_i + \Gamma_j)} - e^{-\tau_{max}(\Gamma_i + \Gamma_j)}]}{(\Gamma_i + \Gamma_j)}$$

Now in matrix notation, the Nth estimate of the linewidth distribution function is

$$[\tilde{G}^{(N)}] = [\tilde{G}^{(1)}] + [P][\tilde{G}^{(N-1)}] \tag{41}$$

where $[G\text{'s}]$ are the $n \times 1$ column matrices and $[P]$ is $n \times n$.

The procedure for recovering the characteristic linewidth distribution function from the experimental data requires the computation of the P matrix for user defined values of α, Γ_{min}, and Γ_{max}. The iterative sequence in Eq. (41) is then used to estimate the Nth approximation to $\tilde{G}$. The regularization parameter is a function of the level of noise, ε, in the data, that is, $\alpha = \varepsilon^\rho$, with $\rho < 1$. Predicting the optimum choice of α, assuming one exists, requires knowledge of the characteristics of the linewidth distribution to be recovered as well as that of the noise process. However, there is no mystery involved in selecting a particular value for a given set of data. It should be remembered that α is essentially a smoothing parameter, having a value between 0 and 1, which when close to unity has the effect of completely suppressing noise but unfortunately at the expense of smoothing the characteristic linewidth distribution to be recovered. Values close to zero allow the propagation of noise through the solution. There is no one correct value. The only, but sufficient, guideline requires that α be as close to zero as possible in order to allow the maximum amount of information to be retrieved from the data and yet retaining the stability of the solution. The optimum value of α can be found by requiring the relative difference in the norms of the solution, say after 900 and 1000 iterations, to satisfy a user defined convergence criterion. The selection of Γ_{min} and Γ_{max} is achieved by searching for the zero crossings of the solution, say after 100 iterations. The actual procedures used are illustrated through the analysis of experimental data.

Error in the Regularized Procedure

In order to quantitatively assess the closeness of the regularized solution to the true distribution, let us first summarize some of the properties of the Laplacian kernel that will be required to perform the desired manipulations. The operator K admits a singular system $\{k_i, U_i, V_i\}$

$$K(\tau, \Gamma) = \sum_{i=0}^{\infty} k_i U_i V_i \qquad (42)$$

The sequences $\{U_i(\Gamma)\}$ and $\{V_i(\tau)\}$ constitute orthonormal systems with properties[51]

$$KU_i = k_i V_i \qquad K^*V_i = k_i U_i \qquad (43)$$
$$K^*KU_i = k_i^2 U_i \qquad KK^*V_i = k_i^2 V_i$$

so that U_i and V_i are the eigenfunctions of the symmetric (self-adjoint) operators K^*K and KK^*, respectively associated with the eigenvalue k_i^2. The k_i^2 are necessarily real and positive and can be arranged $k_i > 0$. The set $\{U_i\}$ is a basis in $L^2[\Gamma_{min}, \Gamma_{max}]$ and the set $\{V_i\}$ is a basis in $L^2[0, \infty]$.

In Eq. (33) if we set $\tilde{G}^{(0)} = 0$ the first estimate of the distribution will be

$$\tilde{G}^{(1)} = K^*b$$

The singular function expansion for $b(\tau)$ is

$$b(\tau) = \sum_{i=0}^{\infty} B_i V_i(\tau)$$

$$B_i = \int_0^{\infty} b(\tau)V_i(\tau)d\tau$$

the first estimate becomes

$$\tilde{G}^{(1)} = \sum_{i=0}^{\infty} k_i B_i V_i \qquad (44)$$

Note that the B_i are the coefficients in the series expansion of $b(\tau)$ and are not to be confused with the discretely sampled data b_i. Using the above properties [Eq. (43)] we can show that the Nth regularized estimate is

$$\tilde{G}^{(N)} = \sum_{i=0}^{\infty} q_i(\alpha, N)B_i U_i/k_i \qquad (45)$$

where

$$q_i(\alpha, N) = \frac{k_i^2}{k_i^2 + \alpha} [1 - (1 - \alpha - k_i^2)^N]$$

$q_i(\alpha, N)$ is seen to be a low pass filter which admits singular values up to a certain value depending on the value of α and N. When $\alpha = 0$ and

[51] G. Miller, in "Numerical Solution of Integral Equations" (L. M. Delves and J. Walsh, eds.), p. 175. Oxford Univ. Press, London and New York, 1974.

$N = \infty$, the norm of the unregularized solution is seen to increase without bound as k_i becomes larger than the level of noise in the data. For $0 < \alpha < 1$ and $N < \infty$, q_i is a filter which is flat when $k_i > \alpha$ and rolls off to zero for $k_i < \alpha$.

If we now expand $G(\Gamma)$ and $r(\tau)$ as a singular function expansion, substitute them into Eq. (27) and make use of the orthogonality of $\{U_i\}$ and $\{V_i\}$, we find

$$B_i = k_i\, G_i + R_i \tag{46}$$

where G_i is

$$G_i = \int_{\Gamma_{\min}}^{\Gamma_{\max}} G(\Gamma)U_i\,(\Gamma)d\Gamma$$

and

$$R_i = \int_0^{\infty} r(\tau)V_i\,(\tau)d\tau$$

Equation (45) becomes

$$\tilde{G}^{(N)} = \sum_{i=0}^{\infty} (G_i + R_i/k_i)q_i\,(\alpha,\,N)U_i \tag{47}$$

We note that since $|1 - \alpha - k_i| < 1$ for all $0 < \alpha,\, k_i < 1$, Eq. (47) in the limit of infinite iterations converges to the solution of Eq. (31)

$$\tilde{G} = \sum_{i=0}^{\infty} \frac{k_i^2}{k_i^2 + \alpha} \frac{B_i}{k_i}\, U_i$$

To quantitatively assess the closeness of the estimate of Eq. (47) to the true distribution we shall calculate in the manner of Rushforth and Harris[52] an ensemble average for the integrated error, e^2, over the range of support of the distribution $\gamma(=\Gamma_{\max}/\Gamma_{\min})$. Thus

$$e^2 = \left\langle \int_{\Gamma_{\min}}^{\Gamma_{\max}} [G(\Gamma) - \tilde{G}^{(N)}\,(\Gamma)]^2 d\Gamma \right\rangle \tag{48}$$

Using the expansion for $G(\Gamma)$ and Eq. (47) we find

$$e^2 = \left\langle \int_{\Gamma_{\min}}^{\Gamma_{\max}} \left\{ \sum_{i=0}^{\infty} (1 - q_i\,(\alpha,\,N))G_i\,U_i \right. \right.$$

$$\left. \left. - \sum_{i=0}^{\infty} q_i\,(\alpha,\,N)\,R_i\,U_i/k_i \right\}^2 d\Gamma \right\rangle \tag{49}$$

[52] C. K. Rushforth and R. W. Harris, *J. Opt. Soc. Am.* **58**, 539 (1968).

Assuming that the noise and the linewidth are uncorrelated, i.e., $\langle R_i \, G_i \rangle = 0$,

$$e^2 = \sum_{i=0}^{\infty} [1 - q_i \, (\alpha, \, N)]^2 \langle G_i^2 \rangle + \sum_{i=0}^{\infty} \frac{q_i^2(\alpha, \, N)}{k_i^2} \langle R_i^2 \rangle \qquad (50)$$

To proceed further, other characteristics of the linewidth distribution and the noise process must be known. In order to gain some insight into the solution let us assume that $r(\tau)$ is a zero-mean white noise process with power spectrum $\varepsilon^2 \delta_{ij}$, $\langle R_i^2 \rangle = \varepsilon^2$. Since $q_i \, (\alpha, \, N)$ decreases as k_i for large values of i, we note that the first series in Eq. (50) will only converge for classes of objects such that

$$\sum_{i=0}^{\infty} \langle G_i^2 \rangle < +\infty$$

(This excludes, for instance, "white" objects for which a modified error has to be defined.) For further details we refer the reader to Abbiss et al.[53]

Moreover, for the iterative regularized scheme of Eq. (33), it can be seen from Eq. (45) that $q_i \, (\alpha, \, N)$ tends to $k_i^2/(k_i^2 + \alpha)$ as the number of iterations tends to ∞, and hence

$$\lim_{N \to \infty} e^2 = \sum_{i=0}^{\infty} \frac{\alpha^2}{(k_i^2 + \alpha)^2} \langle G_i^2 \rangle + \varepsilon^2 \sum_{i=0}^{\infty} \frac{k_i^2}{(k_i^2 + \alpha)^2} \qquad (51)$$

We remark that for $\alpha = 0$, i.e., for the unregularized scheme, the contribution of the noise to this expression is unbounded, since the series $\sum_{i=0}^{\infty} k_i^{-2}$ diverges. When $\alpha = \varepsilon^\rho$, $\rho < 1$, however, the second series is not only convergent, but also tends to zero as $\varepsilon \to 0$. This result follows from the dominated convergence theorem, since on the one hand each term of the series tends to zero as $\varepsilon \to 0$, and on the other hand, for any fixed ε, the series is dominated by the convergent series $\sum_{i=0}^{\infty} k_i^{-2}$. Similarly, the first series in Eq. (51), which is dominated by the series $\sum_{i=0}^{\infty} \langle G_i^2 \rangle$, assumed to be convergent, also tends to zero with ε, provided that $\rho > 0$. Hence we have established that, provided the regularization parameter is correctly related to the spectral density ε of the noise, the iterative scheme of Eq. (33) is indeed regularized, in the sense that it converges to an estimate which tends to the original characteristic linewidth distribution as the noise tends to zero.

[53] J. B. Abbiss, M. Defrise, C. DeMol, and H. S. Dhadwal, *J. Opt. Soc. Am.* **73**, 1470 (1983).

Analysis of the Linewidth Distribution

In this section we describe the specific implementation of several methods of determining $G(\Gamma)$ and illustrate the application of the mathematical techniques described above for the correlation data shown in Table I. The sample used for this measurement was polymethylmethacrylate (PMMA) in methylmethacrylate undergoing thermal polymerization at 50°. The monomer was used as obtained from Aldrich without further purification. The measurements were taken 3 hr after the onset of polymerization (as determined from the change in scattered intensity) at a scattering angle of 35°. The concentration of polymer (PMMA) was determined by Raman scattering measurements to be 1.34×10^{-3} g/cm³. A Malvern Instruments K7023 96-channel single-clipped digital correlator was used to obtain the photocount autocorrelation function. Sample time (Δt) used was 12 μsec and the clip level set at 1. The average intensity from the sample was 1.2×10^5 counts/sec; the duration of the experiment was 64 sec. The data of Table I have had the measured baseline of 3429707 subtracted (baseline delay used was $412\Delta t$); the calculated baseline of $N_s\langle n\rangle\langle n_k\rangle$ is computed to be 3431632 from $N_s = 5,373,100$ (=total num-

TABLE I
NET PHOTOCOUNT AUTOCORRELATION FUNCTION[a,b]

1.78550E + 05	1.69120E + 05	1.61470E + 05	1.54000E + 05	1.45600E + 05
1.37810E + 05	1.27660E + 05	1.21880E + 05	1.15850E + 05	1.13540E + 05
1.07800E + 05	9.90700E + 04	9.12500E + 04	9.13900E + 04	8.39700E + 04
7.84200E + 04	7.74600E + 04	7.40100E + 04	7.06600E + 04	6.34500E + 04
6.13300E + 04	6.06300E + 04	5.77000E + 04	5.66500E + 04	5.39500E + 04
5.27200E + 04	4.85700E + 04	4.65200E + 04	4.89500E + 04	4.36200E + 04
4.21800E + 04	4.01800E + 04	3.90300E + 04	3.64300E + 04	3.51400E + 04
3.54600E + 04	3.49200E + 04	3.28300E + 04	3.25900E + 04	2.72100E + 04
2.58700E + 04	2.44100E + 04	2.37500E + 04	2.36600E + 04	2.16000E + 04
2.03300E + 04	2.16000E + 04	1.80600E + 04	1.69900E + 04	1.90300E + 04
1.67800E + 04	1.61000E + 04	1.41900E + 04	1.37100E + 04	1.39500E + 04
1.48700E + 04	1.57000E + 04	1.49300E + 04	1.42000E + 04	9.66010E + 03
1.26300E + 04	1.22100E + 04	1.21900E + 04	1.04600E + 04	8.74010E + 03
9.47010E + 03	1.09100E + 04	9.80010E + 03	8.77010E + 03	9.46010E + 03
9.57010E + 03	1.03000E + 04	7.58010E + 03	6.92010E + 03	7.42010E + 03
7.27010E + 03	8.81010E + 03	7.78010E + 03	9.21010E + 03	5.50010E + 03
7.78010E + 03	8.62010E + 03	7.74010E + 03	5.38010E + 03	6.85010E + 03
4.88010E + 03	6.32010E + 03	3.80010E + 03	3.27010E + 03	2.34010E + 03
3.98010E + 03	3.86010E + 03			

[a] PMMA in MMA at $\theta = 35$ and 50°. Measured baseline of 3,429,707 subtracted, $\Delta t = 12 \mu$ sec, clip level 1.
[b] $\sqrt{A\beta}\, g^{(1)}(\tau = \Delta t) = 1.78550 \times 10^5$; $\sqrt{A\beta}\, g^{(1)}(\tau = 92\Delta t) = 3.86010 \times 10^3$.

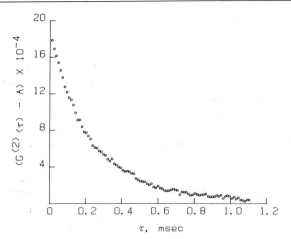

FIG. 2. Net photocount autocorrelation function of the PMMA data of Table I. Delay time $\Delta t = 12$ sec. Number of channels $m = 92$. The value of A was determined from the measured baseline of 3,429,707.

ber of samples) and the total clipped and unclipped counts $N_s\langle n_k \rangle$ and $N_s\langle n \rangle = 2{,}318{,}900$ and $7{,}951{,}400$, respectively. The net autocorrelation function is shown in Fig. 2 and listed in Table I. It should be noted that this was an unusually short measurement, necessitated by the fact that the system was undergoing polymerization but facilitated by the high intensity scattered by the polymer. The agreement to within 0.06% between the measured and calculated baseline is some indication that the baseline may be quite close to the true value. As this was a short experiment, we check that the value of m/N_s is small; the value <0.005% indicates the virtual absence of bias extending even out to the baseline channels (i.e., with $m = 412$). We used these experimental data as a basis for our demonstration of the fitting procedures because the samples can be prepared readily and represents typical molecular weight distributions with M_w/M_n ~ 2.

Cumulants

The method of cumulants[25] is well suited to reasonably narrow, well-behaved characteristic linewidth distribution function. $G(\Gamma)$ is described in terms of a moment expansion

$$\ln |g^{(1)}(\tau)| = -\bar{\Gamma}t + \frac{1}{2!}\,\mu_2\tau^2 - \frac{1}{3!}\,\mu_3\tau^3 + \frac{1}{4!}\,[\mu_4 - 3\mu_2]\tau^4 - \cdots$$

$$= K_m(\Gamma)(-\tau)^m/m! \tag{52}$$

where K_m (Γ) is the mth cumulant. $K_1 = \mu_1 = 0$, $K_2 = \mu_2$, $K_3 = \mu_3$, and $K_4 = \mu_4 - 3\mu_2$; $\bar{\Gamma} = \int \Gamma\, G(\Gamma)d\Gamma$ and $\mu_i = \int(\Gamma - \bar{\Gamma})^i G(\Gamma)d\Gamma$. One can fit this directly to the quantity b_i defined in Eq. (5) obtained from the measured data, recognizing that a constant term will be added to the right-hand side of Eq. (52) to account for the $A\beta$ term. Note that $A\beta$ (not A) is an adjustable parameter that will be determined by the algorithm as presented here. This is a straightforward weighted linear least-squares problem, the weighting factors being necessary since the operation of taking the logarithm of the data has affected the (approximately) equal weighting of the data points. In point of fact, the data points are not equally significant. Jakeman et al.[54] have derived expressions for the variance of the correlation function as a function of delay channel, but only in the limit of single exponential decays. Nevertheless, if we assume that the major source of statistical uncertainty is due to counting statistics, and that the errors in adjacent channels are uncorrelated, we conclude that the error in a particular channel ought to be approximately constant, since the first channel (with the maximum in the correlation function) rarely differs by more than 40% from the last. If we were to introduce a weighting of each data point by another factor corresponding to b_i^{-1}, additional complications may result as the net value of the data points becomes very small near $\tau_{\max}$. Alternatively we can use the net autocorrelation function directly (after subtracting the baseline) and fit to

$$[G_k^{(2)}(\tau_i) - A] \equiv \mathbf{B}_i = A\beta \exp\left[2\left(-\Gamma\tau_i + \frac{1}{2!}\mu_2\tau_i^2 - \frac{1}{3!}\mu_3\tau_i^3 + \cdots\right)\right] \quad (53)$$

We introduce the upper case $\mathbf{B}_i$ to distinguish the data vector of this problem from that of Eq. (5); there is no relation to the B_i of Eq. (44) above. The least squares techniques described in the section on mathematical background are, however, directly applicable, which motivates our choice of the letter $\mathbf{B}$ for this vector. This problem is a nonlinear one, but has the advantage that the relative statistical weighting of the data is unaffected. For the implementation of the nonlinear least-squares algorithm we have in this case the parameter vector $\mathbf{P}$ defined by

$$\mathbf{P}_1 = A\beta$$
$$\mathbf{P}_2 = \bar{\Gamma}$$
$$\mathbf{P}_{j,j>2} = K_j(\Gamma)/(j-1)!$$

Equation (53) then becomes

$$\mathbf{B}_i = \mathbf{P}_1 \exp[2(-\mathbf{P}_2\tau_i + \mathbf{P}_3\tau_i^2 - \mathbf{P}_4\tau_i^3 + \cdots)]$$

[54] E. Jakeman, E. R. Pike, and S. Swain, J. Phys. A: Gen. Phys. [1] 4, 517 (1971).

and the elements of the Jacobian matrix are

$$J_{i1} = -\exp[2(-\mathbf{P}_2\tau_i + \mathbf{P}_3\tau_i^2 - \mathbf{P}_4\tau_i^3 + \cdots)] = -\exp(Z)$$
$$J_{i2} = 2\mathbf{P}_1\,\tau_i\,\exp(Z)$$
$$J_{i3} = -2\mathbf{P}_1\,\tau_i^2\,\exp(Z)$$
$$J_{i4} = 2\mathbf{P}_1\,\tau_i^3\,\exp(Z)$$
$$\cdots$$

where the expansion in Eq. (53) can be truncated after 2, 3, or 4 terms giving rise to second, third, or fourth order cumulants fits. Starting estimates for the parameters can be provided directly from the data themselves:

$$\mathbf{P}_1 = \mathbf{B}_1$$
$$\mathbf{P}_2 = \ln[\mathbf{B}_{10}/\mathbf{B}_1]/2(\tau_1 - \tau_{10})$$
$$\mathbf{P}_{n,n>2} = 0$$

where the expression for $\mathbf{P}_2$ is derived from Eq. (52) as $\tau \to 0$. The use of the first and tenth points to determine the initial slope helps to supress the effect of statistical noise in the data, although the estimate need not be very close for the algorithm to converge. For wider or bimodal distributions, the cumulants approach does not usually describe the data well, and it is not often possible to recover accurately more than the first three cumulants. Nevertheless, the simplicity of the method and the lack of a priori assumptions about the distribution make it suitable as a first step in a more complicated analysis.

Results of the third order nonlinear cumulants analysis for the PMMA data of Table I are shown in Table II. The values of the $\mathbf{P}_j$ at each iteration are listed along with the value of the Marquardt parameter $\lambda^{(N)}$ and

$$\chi^2 = \sum_{i=1}^{m} \{\mathbf{B}_i - \mathbf{P}_1\,\exp[2(-\mathbf{P}_2\tau_i + \mathbf{P}_3\tau_i^2 - \cdots)]\}^2$$

The convergence criterion satisfied was that the parameters were stable to three significant digits on successive iterations, i.e., no parameter changed by more than 0.1%. This should not be taken to mean that the uncertainty in the parameters is 0.1%, but rather that the parameters are most likely within a few tenths of a percent of the values they will eventually converge to. The variance is defined as $\mu_2/\bar{\Gamma}^2$ and the value of 0.323 indicates a reasonably wide distribution for this sample.

Multiexponential

The multiexponential approaches model $G(\Gamma)$ as a weighted sum of Dirac delta functions

TABLE II
CUMULANTS ANALYSIS OF PMMA DATA[a,b,c]

		P_j			
χ^2	$\lambda^{(N)}$	$j = 1$	$j = 2$	$j = 3$	$j = 4$
1.595E + 09	1.000E + 01	1.786E + 05	2.096E + 03	0.000E − 01	0.000E − 01
1.122E + 09	1.000E + 01	1.792E + 05	2.085E + 03	3.135E + 04	−5.243E + 07
6.966E + 08	5.000E + 00	1.801E + 05	2.072E + 03	6.758E + 04	−1.130E + 08
4.743E + 08	2.500E + 00	1.810E + 05	2.064E + 03	9.677E + 04	−1.652E + 08
4.058E + 08	1.250E + 00	1.819E + 05	2.065E + 03	1.115E + 05	−2.020E + 08
3.741E + 08	6.250E − 01	1.829E + 05	2.078E + 03	1.166E + 05	−2.325E + 08
3.448E + 08	3.125E − 01	1.839E + 05	2.099E + 03	1.230E + 05	−2.643E + 08
3.227E + 08	1.562E − 01	1.850E + 05	2.126E + 03	1.418E + 05	−2.867E + 08
3.087E + 08	7.812E − 02	1.860E + 05	2.154E + 03	1.868E + 05	−2.767E + 08
2.951E + 08	3.906E − 02	1.868E + 05	2.186E + 03	2.731E + 05	−2.182E + 08
2.786E + 08	1.953E − 02	1.876E + 05	2.226E + 03	4.111E + 05	−1.098E + 08
2.636E + 08	9.766E − 03	1.885E + 05	2.276E + 03	5.883E + 05	3.320E + 07
2.557E + 08	4.883E − 03	1.894E + 05	2.323E + 03	7.550E + 05	1.691E + 08
2.536E + 08	2.441E − 03	1.899E + 05	2.352E + 03	8.573E + 05	2.532E + 08
2.534E + 08	1.221E − 03	1.901E + 05	2.362E + 03	8.940E + 05	2.836E + 08
2.534E + 08	6.104E − 04	1.902E + 05	2.364E + 03	9.010E + 05	2.895E + 08
2.534E + 08	3.052E − 04	1.902E + 05	2.364E + 03	9.017E + 05	2.900E + 08
2.534E + 08	1.526E − 04	1.902E + 05	2.364E + 03	9.017E + 05	2.901E + 08

[a] $[G_k^{(2)}(\tau) - A] = P_1 \exp[2(-P_2\tau + P_3\tau^2 - \cdots)]$. Third order fit to 92 points beginning with first point.
[b] $\bar{\Gamma} = 2.36 \times 10^3 \text{ sec}^{-1}; \mu_2/\bar{\Gamma}^2 = 0.323$.
[c] The iterative sequence used for computing the P_j's was terminated when none of the parameters changed by more than 0.1% between successive iterations.

$$G(\Gamma) = \sum_{j=1}^{n} P_j \delta (\Gamma - \Gamma_j) \qquad (54)$$

Linear methods fix the location of the δ functions (i.e., the Γ_j) and fit for best values of the P_j; nonlinear methods allow the Γ_j to float as well. The choice of the number of δ functions n, and hence the number of adjustable parameters (n for the linear approximations, $2n$ for the nonlinear) depends upon the method of inversion that is to be used. It must be kept in mind that the amount of information available is limited. Unless the method of inversion includes an appropriate rank-reduction step, the range in $G(\Gamma)$ determines n, as the resolution is fixed by the noise on the data. The nonlinear methods rapidly become very time consuming and are subject to convergence problems as n is increased. In practice it is found that the nonlinear model, where the location of the δ functions is an adjustable parameter, is essentially limited to the determination of a double-exponential decay, i.e., $n = 2$. The nonlinear problem can be solved by standard nonlinear least squares techniques. By writing our model as

$$G(\Gamma) = w\delta(\Gamma - \Gamma_1) + (1 - w)\delta(\Gamma - \Gamma_2) \tag{55}$$

where w is a normalized weighting factor, we have [after appropriate substitution of Eq. (55) into Eq. (4)]

$$b_i = \mathbf{P}_1\exp(-\mathbf{P}_2 \tau_i) + \mathbf{P}_3\exp(-\mathbf{P}_4 \tau_i) \tag{56}$$

with the elements of the Jacobian given by

$$\begin{aligned}
J_{i1} &= -\exp(-\mathbf{P}_2\tau_i) \\
J_{i2} &= \mathbf{P}_1\tau_i\exp(-\mathbf{P}_2\tau_i) \\
J_{i3} &= -\exp(\mathbf{P}_4\tau_i) \\
J_{i4} &= \mathbf{P}_3\tau_i\exp(-\mathbf{P}_4\tau_i)
\end{aligned} \tag{57}$$

Initial estimates of the parameters can be obtained from the results of the cumulants analysis; we have found it simpler and usually sufficient to use the starting values

$$\begin{aligned}
\mathbf{P}_1 &= \mathbf{P}_3 = \mathbf{b}_1/2 \\
\mathbf{P}_2 &= \bar{\Gamma}/2 \\
\mathbf{P}_4 &= 2\bar{\Gamma}
\end{aligned}$$

where $\bar{\Gamma}$ is estimated from the initial slope of the correlation function as in the cumulants technique. The results of the nonlinear double exponential analysis are presented in Table III in a fashion similar to Table II. Again, convergence of the Levenberg–Marquardt algorithm was based on the estimated number of significant digits in each of the parameters, i.e., none of the parameters changed by more than 0.1% in successive iterations. We note that although the value of $\bar{\Gamma}$ agrees with that of the cumulants analysis, the variance $(=\mu_2/\bar{\Gamma}^2)$ is significantly lower.

The linear problem, where the Γ_j are fixed, is often used to obtain more information about the shape of a well-behaved unimodal linewidth distribution function. In practice we have found it extremely dfficult to obtain data with sufficient accuracy to warrant the extraction of enough parameters to describe a bimodal distribution without additional constraints. Indeed, our studies on generated data have shown that with this approach resolution of a bimodal distribution composed of two narrow peaks of equal height and $\Gamma_2/\Gamma_1 = 4$ is only marginally possible with just the numerical noise of single-precision (32-bit) computations present. In the linear problem one fixes the locations of the δ functions based again on estimates made from the results of the cumulants or double exponential analysis. As a first approximation, the characteristic linewidth distribution

TABLE III
Double Exponential Analysis of PMMA Data[a,b,c]

		P_j			
χ^2	$\lambda^{(N)}$	$j = 1$	$j = 2$	$j = 3$	$j = 4$
1.519E + 04	1.000E + 01	2.113E + 02	1.048E + 03	2.113E + 02	4.192E + 03
1.085E + 04	1.000E + 01	2.125E + 02	1.042E + 03	2.137E + 02	4.122E + 03
7.091E + 03	5.000E + 00	2.140E + 02	1.036E + 03	2.169E + 02	4.036E + 03
5.097E + 03	2.500E + 00	2.153E + 02	1.037E + 03	2.202E + 02	3.954E + 03
4.339E + 03	1.250E + 00	2.160E + 02	1.048E + 03	2.227E + 02	3.885E + 03
3.965E + 03	6.250E − 01	2.161E + 02	1.067E + 03	2.240E + 02	3.814E + 03
3.719E + 03	3.125E − 01	2.159E + 02	1.088E + 03	2.238E + 02	3.721E + 03
3.558E + 03	1.562E − 01	2.155E + 02	1.107E + 03	2.225E + 02	3.613E + 03
3.484E + 03	7.812E − 02	2.152E + 02	1.124E + 03	2.211E + 02	3.520E + 03
3.466E + 03	3.906E − 02	2.150E + 02	1.134E + 03	2.202E + 02	3.464E + 03
3.464E + 03	1.953E − 02	2.150E + 02	1.138E + 03	2.198E + 02	3.444E + 03
3.464E + 03	9.766E − 03	2.151E + 02	1.139E + 03	2.195E + 02	3.441E + 03
3.464E + 03	4.883E − 03	2.154E + 02	1.140E + 03	2.192E + 02	3.443E + 03
3.463E + 03	2.441E − 03	2.160E + 02	1.142E + 03	2.187E + 02	3.449E + 03
3.463E + 03	1.221E − 03	2.171E + 02	1.146E + 03	2.176E + 02	3.459E + 03
3.463E + 03	6.104E − 04	2.190E + 02	1.152E + 03	2.158E + 02	3.476E + 03
3.463E + 03	3.052E − 04	2.218E + 02	1.162E + 03	2.130E + 02	3.502E + 03
3.463E + 03	1.526E − 04	2.251E + 02	1.173E + 03	2.098E + 02	3.534E + 03
3.462E + 03	7.629E − 05	2.278E + 02	1.181E + 03	2.072E + 02	3.561E + 03
3.462E + 03	3.815E − 05	2.291E + 02	1.186E + 03	2.059E + 02	3.575E + 03
3.462E + 03	1.907E − 05	2.295E + 02	1.187E + 03	2.055E + 02	3.579E + 03
3.462E + 03	9.537E − 06	2.296E + 02	1.187E + 03	2.054E + 02	3.580E + 03

[a] $[G_k^{(2)}(\tau) - A]^{1/2} = P_1 \exp(-P_2\tau) + P_3 \exp(-P_4\tau)$. Fit to 92 points beginning with first point.
[b] $\bar{\Gamma} = 2.32 \times 10^3 \ \text{sec}^{-1}$; $\mu_2/\bar{\Gamma}^2 = 0.266$.
[c] The iterative sequence used for computing the P_j's was terminated when none of the parameters changed by more than 0.1% between successive iterations.

can be considered to be Gaussian and the first two moments of the cumulants expansion used to estimate the linewidth range that contains some large percentage of the distribution. One then finds the relative contributions from each of the δ functions using a linear least-squares algorithm. If the range of the distribution is not estimated correctly, the inversion will often indicate this by letting the weighting function cross the axis (negative points in the distribution indicating that the range was overestimated on this end of the distribution) or by failure to achieve a distribution that approaches the axis at the endpoints (indicative of underestimation of the range). One must be careful to address the problem of the ill-conditioning properly, however, as the trends described above can be a result of over-specification of the problem rather than true indicators of the proper linewidth range.

The multiexponential problem with fixed δ functions can be further modified by the introduction of a priori constraints such as nonnegativity, smoothness, etc. In these cases the problem again becomes a nonlinear one, and the choice of minimization must be made carefully and with a full understanding of the problem of the ill posedness, since the aim of the introduction of the constraints is to reduce the ill conditioning, and one must be cognizant of how much improvement has been effected.

Substituting the model [Eq. (54)] into Eq. (4) gives

$$b_i = \sum_{j=1}^{n} P_j \exp(-\Gamma j \tau_i) \tag{58}$$

In this linear problem, the elements of C are as described in Eq. (11) above, and the $\mathbf{P}$ vector holds the weighting factors P_j.

It should be added here that the eigenvalue analysis of the Laplace transform by McWhirter and Pike showed that the Γ_j should be spaced logarithmically in Γ (i.e., linear spacing in $\ln \Gamma$) to maximize the transmission of information across Eq. (4). When the δ-functions are spaced unequally, e.g., logarithmically, one should be careful in viewing the results. We have implicitly assumed that the continuous distribution can be sufficiently well represented by the discrete model. It must be kept in mind that the model, and thus the results, are *discrete*. One cannot connect a line between the points (Γ_j, P_j) and expect this to be a picture of the continuous distribution. If it is desired to *estimate* the behavior of the (assumed to be approximately equivalent) continuous distribution, one must correct for the unequal spacing of the Γ_j. This can be done by plotting the points $(\Gamma_j, P_j/\Gamma_j)$ and drawing a continuous curve through them, as has been shown in the figures that show the multiexponential analysis results (Fig. 3a–d), with dashed vertical lines drawn from the Γ axis to the curve to represent the locations of the functions used in the model. It is imperative to keep in mind that although the output of the algorithm can be a large number of parameters (typically 20), this does not mean that 20 *independent* parameters have been recovered from the data. Indeed, the $\mathbf{P}$ vector is reconstructed from a limited number (k, typically 3 or 4) of basis functions.

Results of the singular value decomposition analysis of the multiexponential approximation applied to the PMMA data are shown in Fig. 3a–d. Initial estimates were obtained from the cumulants results and a little Kentucky windage. In Fig. 3a is shown the $k = 3$ solution to the problem where the range of $G(\Gamma)$ was overestimated. The zeros of $G(\Gamma)$ were

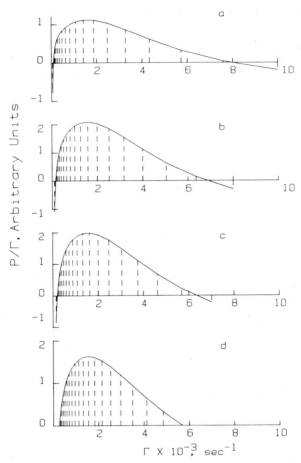

FIG. 3. Results of the linear multiexponential singular value decomposition analysis of the PMMA data of Table I. (a) $\Gamma_{min} = 50$ sec^{-1}, $\Gamma_{max} = 10^4$ sec^{-1}; (b) $\Gamma_{min} = 100$ sec^{-1}, $\Gamma_{max} = 8000$ sec^{-1}; (c) $\Gamma_{min} = 135$ sec^{-1}, $\Gamma_{max} = 7000$ sec^{-1}; (d) $\Gamma_{min} = 260$ sec^{-1}, $\Gamma_{max} = 5700$ sec^{-1}. For each range the $k = 3$ solution is shown; the $k = 4$ solution exhibited similar behavior. The values of Γ_{min} and Γ_{max} in (b) were estimated from the axis intercepts of (a) and similarly (c) from (b). (d) was the first solution satisfying positivity, and occurred several iterations after (c).

interpolated and used as improved estimates of the range; this process was repeated (Fig. 3b and c) until the resulting linewidth distribution was nonnegative (Fig. 3d). At each analysis, we required that the solution we felt to be the correct one for the range being investigated had a $k + 1$ solution which exhibited similar behavior. If the k and $k + 1$ solutions were not similar, we felt that the results did not indicate a true solution.

TABLE IV
MULTIEXPONENTIAL ANALYSIS OF PMMA DATA[a,b]

j	Γ_j	$\mathbf{P}_j$	$\mathbf{P}_j/\Gamma_j$
1	2.60000E + 02	3.19001E − 01	1.22693E − 03
2	3.05877E + 02	4.45129E + 00	1.45525E − 02
3	3.59849E + 02	9.44640E + 00	2.62510E − 02
4	4.23345E + 02	1.54904E + 01	3.65906E − 02
5	4.98045E + 02	2.28029E + 01	4.57848E − 02
6	5.85925E + 02	3.16337E + 01	5.39894E − 02
7	6.89312E + 02	4.22522E + 01	6.12961E − 02
8	8.10942E + 02	5.49178E + 01	6.77210E − 02
9	9.54033E + 02	6.98258E + 01	7.31901E − 02
10	1.12237E + 03	8.70107E + 01	7.75239E − 02
11	1.32042E + 03	1.06200E + 02	8.04292E − 02
12	1.55341E + 03	1.26616E + 02	8.15085E − 02
13	1.82750E + 03	1.46751E + 02	8.03014E − 02
14	2.14997E + 03	1.64188E + 02	7.63677E − 02
15	2.52933E + 03	1.75555E + 02	6.94078E − 02
16	2.97564E + 03	1.76722E + 02	5.93895E − 02
17	3.50069E + 03	1.63260E + 02	4.66364E − 02
18	4.11839E + 03	1.31107E + 02	3.18346E − 02
19	4.84508E + 03	7.72596E + 01	1.59460E − 02
20	5.70000E + 03	2.98877E + 01	5.24346E − 05

[a] $G(\Gamma) = \Sigma \mathbf{P}_j \delta(\Gamma - \Gamma_j)$. Fit to 92 points beginning with first. Parameters of the $k = 3$ candidate solution.
[b] $\bar{\Gamma} = 2.30 \times 10^3 \ \text{sec}^{-1}$; $\mu_2/\bar{\Gamma}^2 = 0.262$.

This requirement is overstrict, but is the only requirement we have found that enables us to distinguish the occasional misleading results. Values of $\bar{\Gamma}$ and $\mu_2/\bar{\Gamma}^2$ computed from the distribution of Fig. 3d are shown in Table IV along with the values of the $\mathbf{P}_j$. Table V shows the contents of the matrices V and S as well as the $\mathbf{U}^T\mathbf{b}$ vector. We note that $\bar{\Gamma}$ again agrees with the previous results; the variance agrees well with that determined by the double exponential method.

Histogram

A simple refinement to the linear multiexponential approximation is to approximate $G(\Gamma)$ as a histogram distribution

$$G(\Gamma) = \sum_{j=1}^{n} \mathbf{P}_j S(\Gamma) \tag{59}$$

where

$$S(\Gamma) = \begin{cases} 1 \text{ for } \Gamma_j \leq \Gamma < \Gamma_{j+1} \\ 0 \text{ otherwise} \end{cases}$$

Again, the choice of n is dependent upon the intended method of fitting the parameters. This model is a straightforward extension of the multiexponential model described above, and all the considerations of a priori constraints, etc. are directly applicable. The only implementation difference from the multiexponential model is that the elements of the curvature matrix C are now given by

$$c_{ij} = \tau_i^{-1}[\exp(-\Gamma_j\tau_i) - \exp(-\Gamma_{j+1}\tau_i)] \tag{61}$$

TABLE V
MATRICES OF THE SINGULAR VALUE DECOMPOSITION

			V matrix (first 5 columns)		
i	V_{i1}	V_{i2}	V_{i3}	V_{i4}	V_{i5}
1	−2.202E − 01	2.578E − 01	−2.610E − 01	2.694E − 01	−2.773E − 01
2	−2.212E − 01	2.485E − 01	−2.320E − 01	2.145E − 01	−1.900E − 01
3	−2.224E − 01	2.374E − 01	−1.988E − 01	1.542E − 01	−9.983E − 02
4	−2.237E − 01	2.242E − 01	−1.609E − 01	8.934E − 02	−9.991E − 03
5	−2.251E − 01	2.085E − 01	−1.181E − 01	2.112E − 02	7.478E − 02
6	−2.267E − 01	1.899E − 01	−7.031E − 02	−4.821E − 02	1.481E − 01
7	−2.282E − 01	1.678E − 01	−1.776E − 02	−1.152E − 01	2.022E − 01
8	−2.298E − 01	1.418E − 01	3.876E − 02	−1.750E − 01	2.279E − 01
9	−2.312E − 01	1.114E − 01	9.770E − 02	−2.212E − 01	2.166E − 01
10	−2.324E − 01	7.595E − 02	1.564E − 01	−2.456E − 01	1.620E − 01
11	−2.332E − 01	3.528E − 02	2.110E − 01	−2.397E − 01	6.425E − 02
12	−2.332E − 01	−1.065E − 02	2.558E − 01	−1.957E − 01	−6.505E − 02
13	−2.325E − 01	−6.143E − 02	2.836E − 01	−1.095E − 01	−1.982E − 01
14	−2.306E − 01	−1.161E − 01	2.861E − 01	1.476E − 02	−2.904E − 01
15	−2.275E − 01	−1.732E − 01	2.548E − 01	1.598E − 01	−2.870E − 01
16	−2.231E − 01	−2.305E − 01	1.823E − 01	2.913E − 01	−1.469E − 01
17	−2.175E − 01	−2.859E − 01	6.460E − 02	3.567E − 01	1.165E − 01
18	−2.108E − 01	−3.371E − 01	−9.820E − 02	2.893E − 01	3.807E − 01
19	−2.033E − 01	−3.821E − 01	−3.006E − 02	1.870E − 01	3.633E − 01
20	−1.951E − 01	−4.197E − 01	−5.324E − 01	−5.139E − 01	−3.962E − 01

			s vector (singular values)		
i	s_i	s_{i+1}	s_{i+2}	s_{i+3}	s_{i+4}
1	4.280E + 00	1.278E + 00	2.282E − 01	2.908E − 02	2.776E − 03
6	2.050E − 04	1.194E − 05	5.305E − 07	1.672E − 07	9.507E − 08
11	8.745E − 08	8.568E − 08	8.506E − 08	8.138E − 08	7.637E − 08
16	7.069E − 08	6.841E − 08	6.841E − 08	6.749E − 08	6.037E − 08

TABLE V (*continued*)

	U^{T}b vector				
i	(U^{T}b)$_i$	(U^{T}b)$_{i+1}$	(U^{T}b)$_{i+2}$	(U^{T}b)$_{i+3}$	(U^{T}b)$_{i+4}$
1	−1.931E + 03	−1.531E + 02	5.923E + 01	−2.685E + 00	−9.774E − 01
6	−1.684E + 01	2.703E + 01	−2.777E + 00	−5.826E + 00	−1.823E + 00
11	7.084E − 01	−4.147E + 00	1.430E + 00	4.244E + 00	−2.945E + 00
16	−1.545E + 00	8.734E − 01	3.928E + 00	2.416E + 00	−4.101E + 00
21	−3.017E + 00	7.295E − 01	−2.466E + 00	−1.816E − 01	1.391E + 00
26	1.788E + 00	−1.274E + 00	−2.448E + 00	8.058E + 00	−3.554E − 01
31	8.784E − 01	2.177E + 00	2.823E + 00	−1.297E + 00	−8.427E − 01
36	2.833E + 00	6.076E + 00	2.742E + 00	7.554E + 00	−4.637E + 00
41	−5.274E + 00	−6.375E + 00	−1.955E + 00	3.004E + 00	−1.843E + 00
46	−3.338E + 00	5.208E + 00	−6.836E + 00	−5.629E + 00	7.044E + 00
51	−3.416E + 00	1.702E + 00	−6.936E + 00	−3.764E + 00	−4.796E − 01
56	5.742E + 00	7.112E + 00	9.298E + 00	8.361E + 00	−1.153E + 01
61	6.649E + 00	3.782E + 00	5.748E + 00	−3.104E + 00	−9.371E + 00
66	−4.234E + 00	4.202E + 00	6.103E − 01	−6.344E + 00	−8.184E − 01
71	4.150E + 00	5.128E + 00	−7.452E + 00	−6.938E + 00	−3.806E + 00
76	−6.451E + 00	2.795E + 00	−7.466E − 01	1.189E + 01	−8.816E + 00
81	3.881E + 00	1.100E + 01	9.260E + 00	−4.996E + 00	5.680E + 00
86	−1.082E + 00	9.162E + 00	−5.225E + 00	−6.953E + 00	−1.350E + 01
91	3.618E + 00	6.266E + 00			

as can be verified by putting Eq. (59) into Eq. (4), performing the integration, and comparing the result to Eq. (10). Again, the sampling of the Γ_j benefits from logarithmic spacing. Here, however, the model represents a continuous (though irrelevantly not continuously differentiable) curve, and therefore there is no ambiguity when plotting the results directly. One might wonder, then, why the multiexponential analysis should ever be preferred to the histogram model—again we point out that determination of the linewidth distribution $G(\Gamma)$ is often only a first step in the analysis of a sample, and that subsequent processing may be considerably simplified if the distribution is discrete. This is the case when molecular weight transformations are desired.

The histogram analysis of the PMMA data is presented in Fig. 6 and Table VI. The endpoints of the $G(\Gamma)$ distribution were in this case estimated from the results of the multiexponential fit of Fig. 3d, and therefore immediately gave a suitable distribution (nonnegative, approaching zero at the end points). This is consistent with our assumption that a continuous distribution can be closely represented by a discrete distribution. We have found in general the values of $\bar{\Gamma}$ and $\mu_2/\bar{\Gamma}^2$ computed from the histogram and multiexponential models to be virtually identical, and the form

TABLE VI
HISTOGRAM ANALYSIS OF PMMA DATA[a,b]

j	Γ_j	$\mathbf{P}_j$
1	2.50000E + 02	7.21827E − 06
2	2.92458E + 02	5.78234E − 05
3	3.42128E + 02	1.02371E − 04
4	4.00233E + 02	1.41920E − 04
5	4.68206E + 02	1.77125E − 04
6	5.47723E + 02	2.08625E − 04
7	6.40744E + 02	2.36672E − 04
8	7.49564E + 02	2.61332E − 04
9	8.76865E + 02	2.82346E − 04
10	1.02579E + 03	2.99071E − 04
11	1.20000E + 03	3.10528E − 04
12	1.40380E + 03	3.15330E − 04
13	1.64221E + 03	3.11885E − 04
14	1.92112E + 03	2.98647E − 04
15	2.24739E + 03	2.74466E − 04
16	2.62907E + 03	2.39064E − 04
17	3.07557E + 03	1.93300E − 04
18	3.59791E + 03	1.39312E − 04
19	4.20895E + 03	8.02791E − 05
20	4.92378E + 03	1.99627E − 05
21[c]	5.76000E + 03	0.00000E − 01

[a] $G(\Gamma)$ as defined by Eqs. (59) and (60). Fit to 92
points beginning with first. Parameters of the
$k = 3$ candidate solution.
[b] $\bar{\Gamma} = 2.30 \times 10^3$ sec^{-1}; $\mu_2/\bar{\Gamma}^2 = 0.268$.
[c] Twenty-one values are required to specify the
endpoints of 20 histogram boxes.

of the resulting distribution essentially equivalent. Choice of one model
over the other then seems to be merely a matter of convenience, depending on any subsequent processing the investigator has in mind.

Functional Approximations

If a priori knowledge of the functional form of $G(\Gamma)$ is known, minimization techniques can be used to find the parameters of the assumed
distribution. For example, if $G(\Gamma)$ is assumed to follow a Pearsons type I
distribution

$$G(\Gamma) = C \left(\frac{\Gamma - \Gamma_{min}}{\Gamma_{max} - \Gamma_{min}} \right)^{m1} \left(\frac{\Gamma_{max} - \Gamma}{\Gamma_{max} - \Gamma_{min}} \right)^{m2} \tag{62}$$

we can solve (in a least-squares sense) for the best values of Γ_{min}, Γ_{max}, $m1$, and $m2$. Functional approximations for $G(\Gamma)$ generally lead to nonlinear least-squares problems, and involve extensive numerical integration techniques. In the light of the limited amount of information that can be extracted from Eq. (4), and the fact that the functional form of $G(\Gamma)$ is not generally known, these approximations are usually not worth the effort.

Regularized Solution

We now demonstrate the behavior of the regularized iterative solution given by Eq. (33) to determine the characteristic linewidth distribution function from the experimental data. The first estimate $\tilde{G}_j^{(1)}$ is computed from the given data b_i by using Eq. (39). If, however, a better estimate of $\tilde{G}^{(1)}$ is known we may use that instead. For any α, Γ_{min} and Γ_{max} we used the iterative sequence defined by Eq. (41) to estimate $\tilde{G}^{(N)}$. The analysis begins by estimating the range, Γ_{min} and Γ_{max} of the distribution. This is achieved by studying the solution after, say 100 iterations, (with $\alpha = 0$) for any pair of values of Γ_{min} and Γ_{max}. In addition, positivity is incorporated into the solution by replacing the negative values of $\tilde{G}^{(N-1)}$ with zero before computing $\tilde{G}^{(N)}$. The effects of having chosen Γ_{min} or Γ_{max} incorrectly are shown in Fig. 4a and b, respectively. Figure 4c shows the approximate range of the distribution and the result of omitting the above step which incorporates positivity is shown in Fig. 4d. Note that the discrete points obtained from the regularized routine fall on a continuous curve and we have shown this by drawing a line between the points of the regularized estimates.

The next step involves the determination of the optimum value for the regularization parameter, α. As we showed in Eq. (51) it is not possible to predict this parameter without knowledge of both the noise process and the characteristic linewidth distribution to be recovered. However, we have found the following procedure to give a consistent value for the "optimum" α. As can be ascertained from Eq. (45) the iterative sequence converges to a finite norm solution for any α between 0 and 1. In practice, the number of iterations required before it reaches this solution may be very large. Thus, we are forced to limit the number of iterations to some sensible value so that the time required for obtaining a "stable" solution does not exceed, say 2 min. Using 32-bit single precision arithmetic we found that 1000 iterations took 1.5 min. By a "stable" solution we mean that solution which meets the convergence test

$$\frac{\left\|\tilde{G}^{(1000)}\right\| - \left\|\tilde{G}^{(900)}\right\|}{\left\|\tilde{G}^{(1000)}\right\|} < 0.001$$

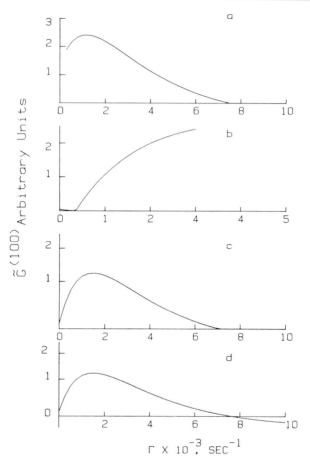

FIG. 4. Results of using the regularized iterative procedure of Eq. (41) to estimate the range of the distribution. $N = 100$, $\alpha = 0$, and positivity was incorporated at each iteration. (a) $\Gamma_{min} = 300$ sec^{-1}, $\Gamma_{max} = 10^4$ sec^{-1}; (b) $\Gamma_{min} = 1$ sec^{-1}, $\Gamma_{max} = 3000$ sec^{-1}; (c) $\Gamma_{min} = 1$ sec^{-1}, $\Gamma_{max} = 10^4$ sec^{-1}; (d) $\Gamma_{min} = 1$ sec^{-1}, $\Gamma_{max} = 10^4$ sec^{-1}. (d) shows the effect of relaxing the positivity constraint.

Figure 5a to d shows the $\tilde{G}^{(1000)}$ solutions, solid line, for $\alpha = 0$, $\alpha = 0.0001$, $\alpha = 0.0005$, and $\alpha = 0.003$. The $N = 900$ solutions lie on top of these and are not shown. The latter three solutions in Fig. 5 satisfy the above convergence test but not the first. None of the solutions looks significantly different, however, if we let $N \to \infty$; the norm of the solutions will become unbounded for values of α smaller than the "optimum" but remain bounded for values of α larger than the "optimum." We show this effect by computing the solution after 5000 iterations, shown by the

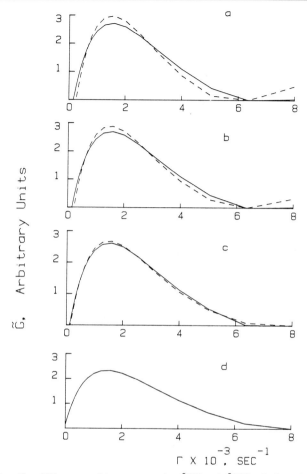

FIG. 5. Equation (41) was used to compute the $\bar{G}^{(900)}$ and $\bar{G}^{(1000)}$ solutions (solid line) for various values of α with positivity incorporated at each iteration. (a) $\alpha = 0$, (b) $\alpha = 0.0001$, (c) $\alpha = 0.0005$, (d) $\alpha = 0.003$. The relative norm $(\equiv[\|\bar{G}^{(1000)}\| - \|\bar{G}^{(900)}\|]/\|\bar{G}^{(1000)}\|)$ of the solution after 900 and 1000 iterations was (a) 0.0011, (b) 0.001, (c) 0.0003, and (d) 0.00001. The dotted line shows the effect of increasing the number of iterations to 5000.

dashed lines in Fig. 5. The onset of instability is clearly seen in Fig. 5a and b. In Fig. 5d the $N = 10000$ solution also lies on top of the $N = 1000$ solution. Thus, the "optimum" α lies between 0.0005 and 0.003. For $\alpha = 0.0005$ we find $\bar{\Gamma} = 2.38 \times 10^3 \text{ sec}^{-1}$ and $\mu_2/\bar{\Gamma}^2 = 0.299$ compared with $\bar{\Gamma} = 2.56 \times 10^3 \text{ sec}^{-1}$ and $\mu_2/\bar{\Gamma}^2 = 0.366$ for $\alpha = 0.003$. From these we note that the $\bar{\Gamma}$ values differ by 7.6% for the two values of α. As discussed earlier we choose the solution corresponding to the smaller value of α. However, a better estimate of α and N may be obtained by forcing the solution to minimize the discrepancy between $\bar{\Gamma}$ obtained from the singular value decomposition technique. To assist the reader in the debugging stage we have listed in Table VII the regularized solution at $N = 1, N = 2, N = 900$, and $N = 1000$ for $\alpha = 0.0005$. Positivity was incorporated at each iteration and the $N = 0$ solution was identically zero for all values of α.

TABLE VII

INTERMEDIATE RESULTS OF THE REGULARIZED ITERATIVE PROCEDURE OF Eq. (41)[a]

Γ	$N = 1$	$N = 2$	$N = 900$	$N = 1000$
0.10000E + 02	0.19165E + 00	0.27103E − 03	0.00000E + 00	0.00000E + 00
0.12592E + 02	0.19146E + 00	0.27081E − 03	0.00000E + 00	0.00000E + 00
0.15857E + 02	0.19121E + 00	0.27054E − 03	0.00000E + 00	0.00000E + 00
0.19967E + 02	0.19090E + 00	0.27019E − 03	0.00000E + 00	0.00000E + 00
0.25144E + 02	0.19051E + 00	0.26976E − 03	0.00000E + 00	0.00000E + 00
0.31662E + 02	0.19003E + 00	0.26921E − 03	0.00000E + 00	0.00000E + 00
0.39869E + 02	0.18942E + 00	0.26852E − 03	0.00000E + 00	0.00000E + 00
0.50205E + 02	0.18866E + 00	0.26766E − 03	0.00000E + 00	0.00000E + 00
0.63220E + 02	0.18770E + 00	0.26659E − 03	0.00000E + 00	0.00000E + 00
0.79608E + 02	0.18651E + 00	0.26524E − 03	0.00000E + 00	0.00000E + 00
0.10025E + 03	0.18503E + 00	0.26356E − 03	0.00000E + 00	0.00000E + 00
0.12623E + 03	0.18319E + 00	0.26147E − 03	0.76351E − 05	0.61180E − 05
0.15896E + 03	0.18091E + 00	0.25887E − 03	0.25787E − 04	0.24445E − 04
0.20016E + 03	0.17811E + 00	0.25565E − 03	0.47398E − 04	0.46258E − 04
0.25205E + 03	0.17467E + 00	0.25169E − 03	0.72659E − 04	0.71741E − 04
0.31739E + 03	0.17050E + 00	0.24683E − 03	0.10184E − 03	0.10117E − 03
0.39967E + 03	0.16546E + 00	0.24092E − 03	0.13465E − 03	0.13427E − 03
0.50328E + 03	0.15945E + 00	0.23379E − 03	0.17011E − 03	0.17002E − 03
0.63375E + 03	0.15238E + 00	0.22527E − 03	0.20671E − 03	0.20689E − 03
0.79804E + 03	0.14418E + 00	0.21522E − 03	0.24183E − 03	0.24226E − 03
0.10049E + 04	0.13485E + 00	0.20354E − 03	0.27170E − 03	0.27229E − 03
0.12654E + 04	0.12448E + 00	0.19023E − 03	0.29176E − 03	0.29241E − 03
0.15935E + 04	0.11324E + 00	0.17538E − 03	0.29695E − 03	0.29751E − 03
0.20066E + 04	0.10141E + 00	0.15922E − 03	0.28313E − 03	0.28347E − 03
0.25267E + 04	0.89352E − 01	0.14214E − 03	0.24825E − 03	0.24826E − 03
0.31817E + 04	0.77472E − 01	0.12465E − 03	0.19456E − 03	0.19424E − 03
0.40066E + 04	0.66142E − 01	0.10734E − 03	0.12849E − 03	0.12799E − 03
0.50452E + 04	0.55659E − 01	0.90781E − 04	0.60749E − 04	0.60409E − 04
0.63531E + 04	0.46210E − 01	0.75470E − 04	0.32086E − 05	0.35785E − 05
0.80000E + 04	0.37874E − 01	0.61746E − 04	0.10294E − 08	0.55642E − 07
		$\bar{\Gamma} = 3.02 \times 10^3$	$\bar{\Gamma} = 2.38 \times 10^3$	$\bar{\Gamma} = 2.38 \times 10^3$
		$\mu_2/\bar{\Gamma}^2 = 0.56$	$\mu_2/\bar{\Gamma}^2 = 0.30$	$\mu_2/\bar{\Gamma}^2 = 0.30$

[a] With $\alpha = 0.0005$ and positivity incorporated at each iteration. The values of $\bar{G}^{(N)}$ for $N = 1, N = 2, N = 900, N = 1000$ are tabulated with the corresponding values of $\bar{\Gamma}$ and $\mu_2/\bar{\Gamma}^2$. $\bar{G}^{(0)}$ was identically zero for all values of Γ.

Discussion

We have presented detailed descriptions of five methods of obtaining information about the characteristic linewidth distribution function $G(\Gamma)$ from measured photocount autocorrelation functions, each with its own advantages and disadvantages. The cumulants and nonlinear double exponential approaches require no a priori information about $G(\Gamma)$, but are severely limited in the form of the distribution functions they can adequately represent (unimodal, narrow, and reasonably symmetric for cumulants, bimodal with narrow peaks in the double exponential). If it is not known in advance that one of these models is correct, it can be difficult to assess the results of these fits. In addition, both methods ignore the ill conditioning of the problem and rely on the fact that only a few parameters are asked for to avoid problems. Both methods are useful in providing starting estimates for the other techniques. The linear multiexponential and histogram approaches with singular value decomposition, as well as the regularized inversion, address the ill conditioning and may therefore be capable of more detailed description of $G(\Gamma)$. The singular value decomposition methods require a value for the range of $G(\Gamma)$ in order to set up the model, are not constrained to physically reasonable distributions, and require an interactive rank reduction step to achieve a meaningful solution. It is occasionally difficult for the inexperienced user with poor starting estimates to arrive at a satisfactory linewidth distribution. Experience has indicated that these methods will not converge to a proper solution when there is a significant ($>0.2\%$) baseline error; this can be frustrating, but does provide a useful piece of information that might be missed by the regularized inversion, which may give a solution that is unacceptably biased in favor of the low frequencies when such error is present—the error in the baseline may only become evident after the data have been further processed, e.g., representing an unrealistic molecular weight distribution. On the other hand the regularized inversion requires no a priori knowledge of the distribution, is quite stable in the presence of noise, and can provide valuable insight into the errors inherent in the analysis.

The results of the histogram and multiexponential singular value decomposition and regularization techniques have been plotted in Fig. 6 for comparison. We have also tabulated the $\bar{\Gamma}$ values and $\mu_2/\bar{\Gamma}^2$ for all the approaches in Table VIII. We bring to attention the good agreement for the value of $\bar{\Gamma}$ obtained from the different methods. The larger spread in the values of $\mu_2/\bar{\Gamma}^2$ reflects in part the inability of the different methods to adequately represent the true characteristic linewidth distribution.

The investigator must be aware of the physical origins of the characteristic linewidth distribution function in order to both successfully

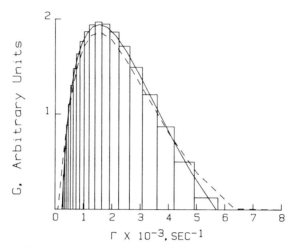

FIG. 6. Comparison of the characteristic linewidth distributions obtained by using the histogram (boxes) and multiexponential (solid line) singular value decomposition and the regularized iterative (dashed line) techniques. Values of $\bar{\Gamma}$ and $\mu_2/\bar{\Gamma}^2$ for these distributions are summarized in Table VIII.

TABLE VIII

COMPARISON OF THE VALUES OF $\bar{\Gamma}$ AND $\mu_2/\bar{\Gamma}^2$
OBTAINED FROM VARIOUS TECHNIQUES

Method	$\bar{\Gamma}$, sec^{-1}	$\mu_2/\bar{\Gamma}^2$
Cumulants	2.36×10^3	0.323
Double exponential	2.32×10^3	0.266
Multiexponential	2.30×10^3	0.262
Histogram	2.30×10^3	0.268
Regularization		
$\alpha = 0.0005$	2.38×10^3	0.299
$\alpha = 0.0030$	2.56×10^3	0.366

choose the experimental conditions (sample time, baseline delay, duration of the experiment) and correctly interpret the results of the data analysis procedures.[55] No single procedure appears sufficient for the analysis; rather the results of the above methods complement each other and taken together can provide an answer consistent with all available information.

These methods have all been implemented in FORTRAN IV code

[55] B. Chu, in "The Application of Laser Light Scattering to the Study of Biological Motion" (J. C. Earnshaw and M. W. Steer, eds.). Plenum, New York, 1983.

running on 64-kbyte microprocessors. The regularized inversion can easily determine 50 points on the continuous distribution function using 150 data points; the other programs use 92 data points and up to 20 parameters without overlay code. Run times typically are 1 or 2 min on a PDP 11/35 with floating point processor, and slightly longer on a Z80-based system.

Finally, we remark that Provencher's technique[24,25] for inverting the Laplace transform is based on the Tikhonov regularization and as such is not different from the one described here. However, implementation of the former scheme requires quadratic programming whereas the latter uses linear techniques. While Provencher's technique has the potential to resolve multimodal characteristic linewidth distributions, the multiexponential and regularization techniques described here are applicable essentially for unimodal characteristic linewidth distributions.

Acknowledgments

We wish to thank Day-chyuan Lee for providing the measurements on the PMMA system. Support of this research by the National Science Foundation, Polymers Program (DMR 8314193), the U.S. Army Research Office, and the Petroleum Research Fund administered by the American Chemical Society is gratefully acknowledged.

Section II

Interaction of Macromolecules with Ligands and Linkages

[16] Nonlinear Least-Squares Analysis

By MICHAEL L. JOHNSON and SUSAN G. FRASIER

Introduction

The recent proliferation of computer resources has made available to biological scientists a number of techniques for the analysis of experimental data which were rarely used a few years ago. One such procedure is nonlinear least-squares parameter estimation. While there are a large number of scientists who are aware of the problems and pitfalls of nonlinear least-squares methods, the majority of biological scientists acquire a computer program from their local computer center or some other scientist without understanding the possible problems which they may encounter.

It is critically important to realize that the collection of experimental data and its subsequent analysis by a technique such as least-squares is not a process of two independent sequential steps. The inherent random and systematic experimental errors in any type of experimental data have a profound influence on the method of choice for the analysis of data. Therefore, the experimental procedure needs to be carefully chosen such that the data with their concomitant error are compatible with the method of analysis. The scientist needs to be aware of the assumptions inherent within different analysis procedures as well as the problems that may arise in matching experimental data to an analysis method.

Least-squares is a numerical method for choosing an optimal set of parameters, α, of an equation, $G(\alpha, X)$, such that this equation will describe a set of data points, X_i and Y_i. This numerical method is simply an algorithm which, when given an initial guess for vector α, will find a better guess for α. The procedure is then applied in an iterative fashion until the vector α does not change within some specified tolerance. There are a number of published algorithms for performing this procedure which will be discussed later in this chapter. For most problems, any of the algorithms will yield essentially the same parameter values within realistic confidence limits. However, each algorithm has different properties, such as the rate of convergence or sensitivity to experimental error, which may dictate a choice of one over another.

The proper choice of form for the parameters to be determined α, the data points X_i and Y_i, and the function $G(\alpha, X)$ is extremely important. For example, it is better to use a free energy change than an equilibrium constant as one of the parameters to be determined. This forces the equilibrium constant to be positive at all times. In addition, since the free energy change is of greater interest than the equilibrium constant, the error statistics are evaluated in terms of free energy by using it as one of the fitting parameters. In general, the parameters to be estimated should be in the ultimate desired form. The choice of dependent variable, the Y_i's, and the independent variables, the X_i's, is also very important. For instance, if the data being analyzed are from a chemical relaxation experiment, such as temperature jump, then the optical density of the solution can, in general, be described as a series of exponential decays with time. Therefore, logical choices for the presentation of the data points (Y_i and X_i) might be optical density vs time or the logarithm of optical density vs time. The graph of the second of these presentations is a straight line for a single exponential decay. The preferred method of presentation depends on the distribution of the random experimental errors in the data, but would probably be optical density or percentage transmittance vs time. The choice of functional form, $G(\alpha,X)$, is dictated by the theoretical basis of the experimental procedure and the choice of dependent variables, independent variables, and parameters to be determined. The proper choice of each of these is ultimately a consequence of the basic assumptions of the method of least-squares.

Linear least-squares is a special case of the more general nonlinear least-squares method. Linear does not imply a straight line in this context since higher order polynomials can also satisfy the definition. A system is referred to as linear when all of the second derivatives of its function with respect to the parameters being estimated are zero:

$$\frac{\partial^2 G(\alpha,X)}{\partial\alpha_i\partial\alpha_j} = 0 \tag{1}$$

where the subscripts refer to individual parameters. For example, if G represents a quadratic equation of the form $\alpha_1 + \alpha_2 X + \alpha_3 X^2$, then the second derivatives, with respect to α_1, α_2, and α_3 will all be zero. Consequently, a least-squares fit to a quadratic polynomial is a linear least-squares fit. The important item to note is that the only difference between linear and nonlinear least-squares techniques is the number of iterations required to perform the analysis, because the assumptions and problems associated with nonlinear least-squares parameter estimation are also present in linear least-squares analysis. Consequently, we are presenting the more general nonlinear least-squares technique here.

The Assumptions of the Least-Squares Method

The least-squares parameter estimation method makes the following assumptions. (1) All of the experimental error of the data can be attributed

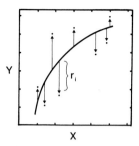

FIG. 1. Graphical depiction of the method of least-squares parameter estimation.

to the ordinate (the Y axis). (2) The random experimental errors of the data can be described by a Gaussian (bell shaped) distribution. (3) No systematic error exists in the data. All three of these assumptions must be satisfied for the least-squares procedure to be valid. Furthermore, each of the assumptions has marked implications as to the "proper" method to proceed in analyzing experimental data.

The first assumption is graphically depicted in Fig. 1. The least-squares method picks values for the various parameters of the function such that the sum of the squares of the vertical distances, r_i, is a minimum. It should be noted that it is only the vertical distances which are minimized, not the horizontal or perpendicular distances. Therefore, in order to be able to use correctly a least-squares analysis, the problem must be formulated in such a manner that the precision of the determination of the values along the abscissa (X axis) must be significantly greater than the precision of the ordinate (Y axis). In addition, the experiments must be designed such that the random experimental noise of the dependent variables is not directly correlated with the noise in the independent variables. In general, there is no method to circumvent this requirement and still utilize a least-squares procedure. It cannot be done by "appropriate weighting factors."

The second assumption means that if a given data point on Fig. 1 was experimentally measured an infinite number of times the resulting distribution of error values (or r_i's) could be described by a Gaussian distribution. This is a reasonable assumption for many, but not all, experimental procedures. The second assumption also implies that the number of data points is sufficient to ensure a good statistical sampling of the random error. The exact number of data points required to meet this assumption is difficult to specify. However, what can be stated is that there is no practical difference between experimentally determining six data points distributed within a given range in triplicate, or measuring 18 separate data points within that range. Furthermore, if 6 data points were measured in triplicate, the analysis should be performed on the 18 individual values, not the 6 average values.

The third assumption clearly states that systematic errors cannot exist in the data. The only way to circumvent this requirement is to include a series of terms in the function, G, to describe the systematic errors. For example, in hormone binding experiments, nonspecific binding is a source of systematic error for the determination of specific binding. It has been shown in previous publications that the function, G, can be written in such a manner that nonspecific binding is explicitly included, and is thus no longer a systematic error of the data.[1,2]

With the assumptions presented previously, it is relatively easy to demonstrate that the values of the parameters, α, with the highest probability of being correct are given by the method of least-squares. In order to demonstrate this, we must introduce one additional assumption: (4) The functional form, $G(\alpha, X)$, is correct. It is necessary, but not sufficient that the functional form be capable of predicting the experimental data. In order for the determined parameters to have any physical meaning $G(\alpha, X)$ must be the correct mathematical description of the phenomena being studied and any systematic errors which might be present in the data. In other words, the dependent variable, Y_i, can be expressed as a function, G, of an optimal set of parameter values and the independent variable, X_i:

$$Y_i = G(\alpha, X_i) + \text{noise}_i \qquad (2)$$

where noise_i represents the experimental uncertainty, or noise, in the measurement.

With these four assumptions, we can write the probability, P_i, for making the observed measurement, Y_i, at the independent variable, X_i, and any particular values of α:

$$P_i(\alpha) = \frac{1}{\sigma_i \sqrt{2\pi}} \exp\left\{ -\frac{1}{2} \left[\frac{Y_i - G(\alpha, X_i)}{\sigma_i} \right]^2 \right\} \qquad (3)$$

where σ_i represents the standard deviation of the random experimental error at the particular data point. The probability of making a series of these measurements at n distinct, independent data points is then

[1] M. L. Johnson and S. G. Frasier, *in* "Methods in Diabetes Research" (J. Larner and S. Pohl, eds.), Vol. 1, Part A, p. 45. Wiley, New York, 1984.

[2] P. J. Munson and D. Rodbard *Anal. Biochem.* **107**, 220 (1980).

$$P(\alpha) = \prod_{i=1}^{n} P_i(\alpha)$$

$$= \left\{ \prod_{i=1}^{n} \left(\frac{1}{\sigma_i \sqrt{2\pi}} \right) \right\} \exp \left\{ -\frac{1}{2} \sum_{i=1}^{n} \left[\frac{Y_i - G(\alpha, X_i)}{\sigma_i} \right]^2 \right\} \qquad (4)$$

where the product $\prod$ and summation Σ apply to each of the n data points with subscript i. The best estimates for the parameters, α, given the four stated assumptions are the values which maximize the probability of Eq. (4). Since the first term of Eq. (4) is constant and is independent of the values of the coefficients, maximizing the probability of $P(\alpha)$ is the same as minimizing the sum in the exponential term of Eq. (4); that is minimizing:

$$\chi^2 = \sum \left[\frac{Y_i - G(\alpha, X_i)}{\sigma_i} \right]^2 \qquad (5)$$

Thus, the method dictated by our assumptions for finding the optimal values of the parameters from the data is to minimize the weighted sum of squares of the deviations, χ^2. This is commonly referred to as the method of least-squares.

All parameter estimation procedures minimize some norm of the data. For the four assumptions which we presented earlier the appropriate norm is the sum of the squares of the deviations, the χ^2 in Eq. (5). It should be noted, however, that if the assumptions are not strictly followed then the appropriate norm to minimize will not necessarily be χ^2. Consequently, the use of a least-squares procedure when the assumptions are not satisfied will yield values for the parameters which are not the maximum likelihood estimates of their true values.

Numerical Methods

The least-squares minimization process can be performed by a number of commonly used numerical procedures: Nelder–Mead, Newton–Raphson, Gauss–Newton, Steepest Descent, and Marquardt–Levenberg.[3] For the majority of problems, all of these methods will yield equivalent results. Consequently, we will discuss the advantages and disadvantages of each of these methods briefly and then present the particular method which we prefer.

---

[3] M. E. Magar, "Data Analysis in Biochemistry and Biophysics," p. 144. Academic Press, New York, 1972.

The numerical algorithms will be presented in terms of a two-dimensional problem: Y vs X. It should be noted that these same algorithms will work for multiple dependent and/or independent variables. An example of such an expansion might be a protein which binds two different ligands at the same binding site. In this case two dependent variables would be the amount of ligand "a" bound to the protein and the amount of ligand "b" bound to the protein. This problem would then have at least three independent variables: the concentrations of protein, ligand "a" and ligand "b." In order to expand the following numerical methods to multiple dimensions, each data point (X_i, Y_i) is simply replaced with two vectors: one for the dependent variables at that data point and one for the independent variables at that data point. The summations also are expanded to include the additional terms.

Nelder–Mead

The Nelder–Mead algorithm[4] is a multidimensional search for a minimum norm, usually χ^2, performed as a series of carefully selected one-dimensional searches.

The Nelder–Mead procedure to determine two parameters requires that three points, i.e., sets of the parameters, be selected at random, with the stipulation that they do not form a straight line. For n parameters, it requires $n + 1$ parameter points in the n-dimensional parameter space. The next step is to evaluate, at each of these parameter points, a statistical norm of the data to be minimized. The parameter point corresponding to the largest value of the norm is selected, and a centroid of the remaining points is evaluated. This centroid is literally the average of all of the parameter points except the point which corresponds to the maximum of the norm. The next step is to perform a search along the straight line connecting the maximum parameter point with the centroid until a point is found which corresponds to a value of the norm less than at least one of the parameter points averaged to form the centroid. The maximum parameter point is then replaced by the new parameter point. The process is repeated by determining the new highest value parameter point, evaluating the new centroid from the remaining points, and searching along the new line for another parameter point corresponding to a lower norm. This process is continued until all $n + 1$ points are, essentially, coincident. These individual searches are, in effect, one-dimensional, since each can be expressed as a search to find a single distance along a straight line.

The only disadvantage of this method is that it sometimes converges very slowly. Its advantages are (1) that it always converges eventually, (2)

[4] J. A. Nelder and R. Mead, *Comput. J.* **7**, 308 (1965).

that it does not require derivatives of the function G with respect to the parameters, and (3) that it can be used to minimize χ^2 and/or any other norm of the data and function.

Newton–Raphson

The Newton–Raphson method[3] is simply a method of solving a system of equations of the form

$$\frac{\partial \chi^2}{\partial \alpha_j} = 0 \tag{6}$$

with a different equation for each of the parameters being estimated, α_j. The basis of this method is that a function will have a minimum, or a maximum, when its first derivatives are zero. The advantages of this method are (1) that it will minimize any norm, not only χ^2, and (2) it is extremely fast. The disadvantages are (1) that it requires the evaluation of both the first and second derivatives of the function G, and (2) it is particularly sensitive to the initial parameter values used to start the iterative process.

Gauss–Newton

The Gauss–Newton algorithm is based on a first-order series expansion of Eq. (2). That is, the data points are assumed to be approximated by the function evaluated at the current parameter values, α, after $k + 1$ iterations:

$$Y_i \simeq G(^{k+1}\alpha, X_i) \tag{7}$$

and further that this can be approximated as a series expansion about the previous iteration:

$$Y_i \simeq G(^{k+1}\alpha, X_i) \approx G(^k\alpha, X_i) + \sum_{j=1}^{n} \frac{\partial G(^k\alpha, X_i)}{\partial ^k\alpha_j} [^{k+1}\alpha_j - {}^k\alpha_j] \tag{8}$$

where the i subscripts refer to each of the data points, the j subscripts refer to each of the parameters being evaluated, and the k and $k + 1$ superscripts refer to the iteration number. It should be noted that Eq. (8) is actually a series of equations, one for each of the data points. For any initial values of the parameters, $^k\alpha$, these equations can be used to evaluate a new set of parameters, $^{k+1}\alpha$, to be used as the initial values for the next iteration. The procedure is repeated until α does not change within some specified tolerance, usually a fractional change of one part in ten thousand or more.

The method of evaluating $^{k+1}\alpha$ is easiest to understand in matrix notation. Thus, Eq. (8) can be written as

$$\mathbf{Y}^* = P\boldsymbol{\epsilon} \tag{9}$$

where $\mathbf{Y}^*$ is a vector of residuals whose individual elements are

$$\mathbf{Y}_i^* = [\mathbf{Y}_i - G(^k\boldsymbol{\alpha}, X_i)]/\sigma_i \tag{10}$$

P is a Hessian matrix whose individual elements are

$$P_{ij} = \left[\frac{\partial G(^k\boldsymbol{\alpha}, X_i)}{\partial^k\boldsymbol{\alpha}_j}\right]\Big/\sigma_i \tag{11}$$

and $\boldsymbol{\epsilon}$ is a correction vector whose elements are

$$\boldsymbol{\epsilon}_j = (^{k+1}\boldsymbol{\alpha}_j - {}^k\boldsymbol{\alpha}_j) \tag{12}$$

It should be noted that we have also introduced a weighting factor $1/\sigma_i$ into Eqs. (10) and (11). This weighting factor was included to allow for a variation of the precision of determination of the individual data points. As it is used here, this individual data point weighting factor has a value of one over the standard deviation of the experimental error of the particular data point. Equation (9) cannot easily be solved for $\boldsymbol{\epsilon}$ since the matrix, P, is not a square matrix. However, Eq. (9) can be rewritten as

$$P'\mathbf{Y}^* = (P'P)\boldsymbol{\epsilon} \tag{13}$$

where P' is the transpose of the matrix P. Equation (13) can then be solved for $\boldsymbol{\epsilon}$ as

$$\boldsymbol{\epsilon} = (P'P)^{-1}P'\mathbf{Y}^* \tag{14}$$

where $(P'P)^{-1}$ is the inverse of the matrix $P'P$. The vector $\boldsymbol{\epsilon}$, as defined in Eq. (12), is then evaluated by Eq. (14). Then $\boldsymbol{\epsilon}$ can be used with a rearranged form of Eq. (12) to find a new set of parameter values:

$$^{k+1}\boldsymbol{\alpha} = {}^k\boldsymbol{\alpha} + \boldsymbol{\epsilon} \tag{15}$$

and the process repeated until $\boldsymbol{\epsilon}$ is arbitrarily close to zero, or conversely $^k\boldsymbol{\alpha} \simeq {}^{k+1}\boldsymbol{\alpha}$.

This is obviously a cyclic, or iterative, process. Parameter values, $^k\boldsymbol{\alpha}$, for the first iteration are obtained by some outside method, usually a guess. These can then be used to evaluate each of the terms in Eq. (8) except the vaues of $^{k+1}\boldsymbol{\alpha}$. It is the values of $^{k+1}\boldsymbol{\alpha}$ which we are attempting to determine since they will then be used as the initial values in the next cycle. To reiterate, matrix notation was used in Eqs. (9)–(12) such that the vector $\boldsymbol{\epsilon}$ is the only unknown quantity. The vector $\boldsymbol{\epsilon}$ was evaluated by the process in Eqs. (13)–(14). The definition of $\boldsymbol{\epsilon}$, Eq. (12), was rewritten as Eq. (15) to enable the evaluation of the $^{k+1}\alpha$ term in Eq. (8). This process is repeated until $^{k+1}\alpha$ is approximately equal to $^{k}\alpha$.

The advantages of the Gauss–Newton procedure are (1) that when it converges it does so extremely rapidly, and (2) with reasonable initial guesses, it will converge the vast majority of the time. The disadvantages are (1) that it does not always converge, and (2) that it can only be used to minimize χ^2.

Steepest Descent

The method of Steepest Descent is a modification of the basic Gauss–Newton procedure which circumvents most of the problems of lack of convergence. This is accomplished by making all of the off-diagonal elements of the $P'P$ matrix equal to zero. Off-diagonal elements are those matrix elements whose row and column indices are not equal:

$$P'P_{lj} = 0, \quad \text{if } l \neq j \tag{16}$$

Unfortunately, the Steepest Descent modification also significantly decreases the speed of convergence as compared with the Gauss–Newton procedure.

Marquardt–Levenberg

The Marquardt–Levenberg[5] algorithm is probably the most commonly used algorithm for minimizing χ^2, even though it usually converges slower than the Gauss–Newton method. The Marquardt algorithm begins as a Steepest Descent algorithm, utilizing that method's guaranteed convergence property. As the convergence proceeds, it undergoes a smooth transition to the Gauss–Newton algorithm to take advantage of that method's rapid convergence property near the minima.

Our Method

The algorithm we prefer is another variation on the basic Gauss–Newton procedure which is designed to improve the convergence properties without adversely slowing the procedure. This modification is to introduce an arbitrary parameter, δ, into Eq. (15) as

$$^{k+1}\alpha = {}^{k}\alpha + \delta\epsilon \tag{17}$$

where δ is a positive constant chosen at each iteration such that the χ^2 for

---

[5] D. W. Marquardt, *SIAM J. Appl. Math* **14**, 1176 (1963).

that iteration is a minimum. The value of δ is initially assumed as 1. For subsequent iterations, the starting value of δ is the optimal value determined by the previous iteration. In order to evaluate δ for each iteration, a search is performed by either multiplying or dividing the magnitude of δ by two until two values for δ are found: one whose corresponding variance is less than the variance of the previous iteration and one whose corresponding variance is greater than the previous iteration. The value of the variance from the previous iteration and the variances evaluated at the new values of δ are then used to determine a quadratic polynomial which is differentiated to find the distance, δ, corresponding to the minimum variance. Occasionally, one of the points used in the search will have a lower variance than the predicted minimum, and in this case, the lowest point is taken. Such a search procedure forces the variance to decrease with each iteration. This greatly improves the convergence properties of the Gauss–Newton procedure, thus relaxing the requirements on the initial guesses. This modification is a generalization of a procedure suggested by Hartley[6] and is sometimes referred to as an Aitken's δ^2 process.[7]

Constraining the Parameter Estimation Process

Under certain circumstances it is desirable to constrain some parameter, or function of the parameters to have a particular value within some specified standard deviation. For example, it has been observed experimentally that the average number of Bohr protons released at the first oxygenation state of hemoglobin at pH 7.4 is 0.64 ± 0.07.[8] Consequently, the least-squares analysis of the oxygen binding data should be constrained such that the number of Bohr protons is required to be 0.64 ± 0.07 at the first step. The classical methods of constrained least-squares fitting by linear programming[9] allow for particular parameter constraints to be within a specified region. However, they do not allow for these particular values to be specified within a given standard deviation. Furthermore, classical methods do not easily allow for constraint of an arbitrary function of the determined parameters within a specified range. In actuality, the basic least-squares algorithm already limits a function of the

[6] H. O. Hartley, *Technometrics* **3**, 269 (1961).
[7] F. B. Hildebrand "Introduction to Numerical Analysis," p. 445. McGraw-Hill, New York, 1956.
[8] M. L. Johnson, B. L. Turner, and G. K. Ackers, *Proc. Natl. Acad. Sci. U.S.A.* **81**, 1093 (1984).
[9] M. E. Magar, "Data Analysis in Biochemistry and Biophysics," p. 126 Academic Press, New York, 1972.

parameters in this way. We can therefore achieve the desired goal by simply adding an additional data point, corresponding to the desired constraint, which is fit to its unique function. That is, if we wish to analyze n data points which correspond to oxygen bound versus free oxygen concentration and include the constraint that the average number of Bohr protons bound at the first step is 0.64 ± 0.07, then the least-squares fit is actually performed on $n + 1$ data points. The first n points are treated as described previously with a weighting factor for each data point which corresponds to the inverse of its approximate standard deviation. For the $n + 1$ data point the dependent variable is 0.64 in this case. The weighting factor is $1/0.07 = 14.29$ for this example and the function G_{n+1} [Eq. (2)] has a totally different form than G_1 to G_n. This new functional form is simply the calculation of the number of Bohr protons bound at the first step as predicted by the current best values of the experimental parameters.

A method of constraining parameters to have only positive values was alluded to in the introduction. Parameters such as equilibrium constants are physically meaningful only if they have positive values. The least-squares technique makes no assumption, requirement, or constraint that parameters have only positive values, and as a consequence will sometimes predict negative equilibrium constants. This problem can be circumvented by redefining the parameters such that negative values cannot occur. For example, if the parameter being determined is a free energy change, rather than an equilibrium constant, then the corresponding equilibrium constant will always be positive. This is a consequence of the nonlinear transformation of free energy changes into equilibrium constants. That is, any value of a free energy change, positive or negative, will correspond to a positive equilibrium constant.

Simultaneous Analysis of Multiple Sets of Data

One of the best methods to improve the precision of a set of determined parameters is to simultaneously analyze multiple experiments. These multiple experiments could be replicates of each other, in which case they are treated as a single experiment with more data points. The individual replicates should not be averaged; all of the data points should be used. It is probably more useful to simultaneously analyze multiple sets of data from different experiments.

The analysis of multiple experiments, either of a different type or variations on the same experiment pose two major problems. First, the amounts of random experimental noise will probably be different from one experiment to another or between data points. Second, there will be some parameters which are common to both sets of experiments and others which are unique to only one of the experiments. The problem of variations of experimental noise from data sets can usually be treated by a careful selection of weighting factors for each data point as described in the discussion of Eqs. (8)–(12). It is almost always possible to determine a reasonable estimate of the expected experimental noise on an individual data point by a careful inspection of the sources of random error in the experiment. Typical sources which should be considered are pipetting errors, dilution errors, and instrumental errors such as radioactive counting error.

The problem of different sets of parameters from experiment to experiment can easily be solved by introducing a different functional form, $G(\alpha, X)$, for each experiment. The computer program can easily distinguish individual experiments by introducing an additional independent variable, which would be nothing more than an identifier number for each data set. The first step in the evaluation process is to determine for each data point which functional form is appropriate. The program then proceeds to evaluate the appropriate function and its derivatives. The vector of parameters, α, would contain all of the parameters of all the functions.

Reliability of the Parameter Estimation Process

A number of statistical tests have been developed for linear least-squares procedures to measure the reliability of each process. Since most of these tests have been developed with limiting assumptions, they are not exact for a nonlinear least-squares case. They are useful as an indication but not an exact description of parameter reliability.

One such statistical test determines cross-correlation coefficients. The cross-correlation between fitting parameters, CC_{lj}, is evaluated from the elements of the inverse of the $P'P$ matrix at the solution,

$$CC_{lj} = (P'P)_{lj}^{-1}/[(P'P)_{ll}^{-1} (P'P)_{jj}^{-1}]^{1/2} \quad l \neq j \qquad (18)$$

The cross-correlation coefficient is particularly useful to test the reliability of the estimated parameters. If any of the values of CC_{lj} approach plus or minus one, then any variation in α_l can be almost totally compensated for by a variation of α_j. As a consequence, unique values of α_l and α_j cannot be evaluated without additional data. Unfortunately, the degree to which the CC_{lj} can approach plus or minus one without indicating a serious problem with the parameter estimation is not well defined. However, a critical limit of $\pm \sim 0.96$ seems to be reasonable.[10]

[10] M. L. Johnson, J. J. Correia, H. R. Halvorson, and D. A. Yphantis, *Biophys. J.* **36**, 575 (1981).

A second method of testing the reliability of the parameter estimation procedure is to determine the number of linearly independent parameters of the matrix P by matrix rank analysis.[11] The number of parameters being estimated should not exceed the rank of the matrix. However, this method also assumes a linear system of equations and thus is only approximate.

Uniqueness of the Estimated Parameters

There is no general method to guarantee that the determined parameters are unique. All of the presented methods will usually find a minimum of some norm for a given set of data, function, and parameters. The problem is that there may well be a different set of parameters of the function which will describe the data with equal or better precision.

Some measure of confidence can be obtained by performing multiple least-squares parameter estimations with a variety of different starting values for the parameters. If all of these converge to the same set of parameters, it may be reasonable to assume that it is a unique estimation of the parameters.

The only method to ensure uniqueness would be to test all physically reasonable combinations of the parameters. This would, of course, require an exorbitant amount of computer time for the general case. It should be noted, however, that with current state-of-the-art computers it is not unreasonable to perform a grid search which evaluates the norm at several million combinations of the parameters.

Estimating Confidence Intervals

Once a set of parameters has been determined, the investigator is required to address the question, "How accurately can the desired parameters be determined by least-squares fitting of a specific set of experimental data to a specific functional form?" Another way of expressing this question is to ask what the confidence interval is for each of the fitted parameters.

It has been shown previously that true confidence intervals of estimated parameters can be approximated by parameter values which yield a variance ratio predicted by an F statistic.[1,9,12-14] The variance ratio is calculated as the ratio of the variance at any point in the parameter space to the variance at the values of the parameters estimated by the minimiza-

[11] Y. Bard, "Nonlinear Parameter Estimation," p. 301 Academic Press, New York, 1974.
[12] G. D. P. Box, *Ann. N.Y. Acad. Sci.* **86,** 792 (1960).
[13] L. Endrenyi and F. H. F. Kwong, *Acta Biol. Med. Ger.* **31,** 495 (1973).
[14] M. L. Johnson, *Biophys. J.* **44,** 101 (1983).

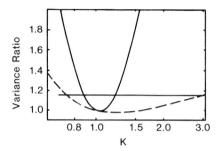

FIG. 2. Confidence intervals for a simple single-parameter least-squares analysis of a rectangular hyperbola. Redrawn from Ref. 1.

tion procedure. The F statistic is a function of the desired confidence probability and the number of degrees of freedom of the two variances. A 67% confidence region for a Gaussian distribution is the mean plus or minus approximately one standard deviation.

One example of this method of confidence interval estimation is given in Fig. 2. This example is a simulation of a simple ligand binding problem where the number of binding sites is known independently and there is no cooperative interaction of any type between the binding sites. With these assumptions, the fractional occupancy can be expressed as $K[X]/(1 + K[X])$ where K is the equilibrium binding constant of the ligand and $[X]$ is the free concentration of the ligand. This is the commonly used functional form for the evaluation of hormone–receptor interactions.

Each curve of Fig. 2 was calculated by generating 50 data points with equal logarithmic spacing in $[X]$. The data were then perturbed with Gaussian distributed pseudo-random error of standard deviation of 0.05. The variance which corresponds to any value of K can be calculated from the residuals, the difference between the data points and the function evaluated at the corresponding free ligand concentration. The variance ratio, F statistic, corresponding to a given value of K can then be calculated as the ratio of the variance for that particular value of K and the variance evaluated at the best estimated value for K. The solid curve in Fig. 2 was generated with data where the free ligand concentration varied over four orders of magnitude centered on the simulated dissociation constant: $0.01 < [X] < 100$. The critical value of the F statistic for a 67% confidence interval is 1.14 for this example, shown as a horizontal line. This predicts that the equilibrium constant from these data is 1.03 with a one standard deviation confidence interval of 0.89 to 1.21. This confidence interval is determined as the values of K at the intersection of the critical F statistic (the horizontal line) and the solid curve. It should be noted that in real experiments, data are rarely obtainable over a perfectly selected four orders of magnitude in concentration. Consequently, the dashed curve in Fig. 2 was calculated by the same procedure with the range of $[X]$ being only one order of magnitude in concentration and not centered around the dissociation constant, $10 < [X] < 100$. The same critical value of the F statistic predicts that the equilibrium constant is 1.20 with a one standard deviation confidence interval of 0.75 to 2.79.

It is interesting to note that this pseudo-realistic one-parameter curve fitting problem can yield very asymmetrical confidence intervals. Such asymmetrical confidence intervals can be determined only by searching the variance space for the critical value of the F statistic.

Most of the generally available computer programs for nonlinear least-squares employ a method for the evaluation of confidence intervals which is based on the method used for linear least-squares fitting. This method assumes that the parameters are not correlated with each other, that the variance at any set of parameters can be predicted from the minimum variance and the derivatives of the function, and that the answer can be approximated by a linear least-squares fit. The consequences of these assumptions are that the computer program will predict a symmetrical confidence interval, which is an underestimate of a true confidence interval. The previous example demonstrates that the true confidence interval for a nonlinear problem is not symmetrical and thus cannot be described as plus or minus a standard error. The investigator should be cautious of any computer program which expresses the confidence interval as plus or minus a single standard error, since in practice the confidence interval is asymmetrical.

The search method described in Fig. 2 can be expanded easily to multiple dimensions to evaluate the confidence intervals when multiple parameters are evaluated simultaneously. However, it should be noted that this mapping of the variance ratio must include both mapping along each of the parameters independently, as in Fig. 2, as well as mappings where multiple parameters can vary simultaneously. Consequently, the amount of computer time required to perform the mapping increases as the power of the number of parameters. Clearly, for a complex multiple parameter evaluation a faster approximation is required.

One method of approximating the confidence intervals of the fitted parameters in multiple dimensions is to assume that the variance space can be approximated by a multidimensional hyperellipsoid defined by the solutions, Ω, of the following matrix equation[9,12,13]:

$$(A - \Omega) \, PP' \, (A - \Omega) < n\sigma^2 \ (F \text{ statistic}) \tag{19}$$

where n is the number of parameters and σ^2 is the variance. Confidence

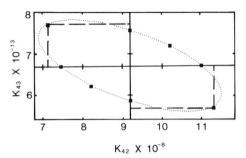

FIG. 3. Confidence intervals for the analysis of the second and third product Adair binding constants (K_{42} and K_{43}) from the data of Roughton and Lyster.[18] Data were at pH 7 in 0.6 M phosphate and approximately 2% hemoglobin. Redrawn from Ref. 14.

intervals evaluated by this procedure are by definition symmetrical about the optimal values determined as the minimum by the least-squares procedure. This procedure assumes that the shape of the variance space can be predicted from the curvature of the minimum. These last two assumptions are at best tenuous in some cases.

A better method of evaluation of the confidence intervals is to search the variance space for the desired F statistic. This method does not assume that the confidence region is elliptically shaped. Furthermore, the method does not assume that the confidence region is symmetrical or that it can be predicted by the curvature of the space at the minimum. The amount of computer time required is proportional to the number of points required to determine the confidence region contour. Consequently, we have evaluated only the confidence region contour along particular sets of directions. First, each of the parameters is searched independently in both an increasing and decreasing direction until the desired variance ratio is found. Second, each of the axes of the hyperellipse defined by Eq. (19) is searched in both directions for the desired F statistic. The confidence region for a given parameter is then determined by projecting each of the mapped points along the hyperellipse onto the axis for the given parameter and taking the maximum deflection in each direction. Figure 3 provides an example of this search method for finding the confidence limits of fitted parameters in a multidimensional space. The experimental data points in this example describe the binding of oxygen to hemoglobin. The oxygenation-linked subunit assembly of human hemoglobin can be described by seven independent parameters: four describe the oxygen binding to tetramers, two for oxygen binding to dimers, and the dimer–tetramer association constant.[15] However, if the experimental data are measured at very high concentrations, then only the oxygen binding to tetramers need be considered. These experiments were performed at such

[15] G. K. Ackers and H. R. Halvorson, *Proc. Natl. Acad. Sci. U.S.A.* **71,** 4312 (1974).

TABLE I[a]

Constant	Value	67% confidence interval	
K_{41}	2.96×10^4	1.98×10^4	4.00×10^4
K_{42}	9.21×10^8	0.16×10^8	17.63×10^8
K_{43}	6.69×10^{13}	0.35×10^{13}	10.18×10^{13}
K_{44}	1.51×10^{19}	1.44×10^{19}	1.59×10^{19}
k_{41}	2.96×10^4	1.98×10^4	4.00×10^4
k_{42}	3.11×10^4	0.04×10^4	8.89×10^4
k_{43}	7.27×10^4	1.96×10^4	627.66×10^4
k_{44}	2.26×10^5	1.50×10^5	4.35×10^5

[a] Macroscopic Adair constants as determined by a least-squares curve fit of the data of Roughton and Lyster[18] to Eq. (20). Also given are the stepwise Adair binding constants as determined by an error propagation of the product Adair constants.

a high hemoglobin concentration.[16] Consequently, these data can be used as a simple, but real, example of the nonlinear least-squares technique.

At high concentrations, the oxygen binding to tetrameric human hemoglobin can be described by four product Adair constants: K_{41}, for the average affinity of the first oxygen; K_{42}, for the average affinity for binding the first two oxygens; K_{43}, for the average affinity for binding the first three oxygens; and K_{44}, for the average affinity for binding all four oxygens.[17] The fractional saturation can then be expressed as:

$$\bar{Y} = \frac{\sum\limits_{q=1}^{4} q K_{4q} [X]^q}{4 \sum\limits_{q=0}^{4} K_{4q} [X]^q} \tag{20}$$

where $K_{40} = 1$.

Figure 3 was generated by first performing a four-parameter least-squares fit of the data of Roughton and Lyster[18] to Eq. (20). The parameter values determined by the least-squares fit are also given in Table I. Since it is difficult to visualize a four-dimensional space, a second least-squares fit was performed to generate the example in Fig. 3. In this second fit, an assumption was made that the values of K_{41} and K_{44} were known to be the values given in Table I. Consequently, this second least-squares fit is a two parameter least-squares fit to determine only K_{42} and K_{43}. The fitted parameters for this second fit are presented in Table II.

[16] M. L. Johnson and G. K. Ackers, *Biochemistry* **21**, 201 (1982).
[17] M. L. Johnson and G. K. Ackers, *Biophys. J.* **7**, 77 (1977).
[18] F. J. W. Roughton and R. L. J. Lyster, *Hvalradets Skr.* **48**, 185 (1965).

TABLE II[a]

Constant	Value	67% confidence interval	
K_{41}	2.96×10^4		
K_{42}	9.21×10^8	7.07×10^8	11.34×10^8
K_{43}	6.69×10^{13}	5.65×10^{13}	7.74×10^{13}
K_{44}	1.51×10^{19}		
k_{41}	2.96×10^4		
k_{42}	3.11×10^4	2.39×10^4	3.84×10^4
k_{43}	7.27×10^4	4.98×10^4	10.95×10^4
k_{44}	2.26×10^5	1.95×10^5	2.67×10^5

[a] Macroscopic Adair constants as determined by a least-squares parameter estimation of the data of Roughton and Lyster[18] to Eq. (20), with the assumption that K_{41} and K_{44} are known. Consequently, the least-squares fit presented in this table is to two parameters only, i.e., K_{42} and K_{43}. Also given are the stepwise Adair binding constants as determined by an error propagation of the product Adair constants.

The dotted elliptically shaped curve in Fig. 3 was calculated by exhaustively searching the two parameter space (K_{42} vs K_{43}) for the dividing line between acceptable and unacceptable parameter pairs which yield an F statistic as determined by the desired confidence probability and number of degrees of freedom ($F = 1.22$ in this case). The best method of approximating the true confidence interval is to take the extreme values determined by such a search. However, this search takes an excessively large amount of computer time when least-squares fitting to multiple parameters, and consequently some approximation is needed.

The eight solid squares in Fig. 3 show the points on this contour which were evaluated by our search procedure. The confidence region for a given parameter is then determined by projecting each of the mapped points onto the axis for the given parameter and taking the maximum deflection in each direction. The long dashed lines in Fig. 3 show an example of this procedure. The 67% confidence intervals presented in Tables I and II, corresponding to plus or minus one standard deviation, were determined by this method. The amount of computer time required is proportional to the number of parameters being determined, not a power of the number of parameters.

Error Propagation of Confidence Intervals

Once a set of parameters has been determined for a given set of data and a particular functional form, G, it is sometimes desirable to propagate these parameter values into the values of a different set of parameters which are of interest. It would be most desirable to perform this propagation without simply repeating the least-squares fit for a different function and set of parameters. The first least-squares fit to function G has imposed constraints on the allowable values of its parameters and it is of interest to see how these implied constraints will propagate into a different set of parameters. Furthermore, a simple method of propagating these parameters and confidence intervals will save considerable amounts of computer time.

When the fitted parameters have been evaluated, they can be used to evaluate other parameters of interest without repeating the least-squares process. For example, Eq. (20) can also be expressed in terms of the stepwise Adair constants: k_{41} for the affinity of binding the first oxygen; k_{42} for the affinity for adding the second oxygen to singly oxygenated hemoglobin; k_{43} for the affinity for adding the third oxygen to doubly oxygenated hemoglobin; k_{44}, etc. These stepwise and product Adair constants are related by Eqs. (21)–(24).

$$K_{41} = k_{41} \tag{21}$$
$$K_{42} = k_{41}k_{42} \tag{22}$$
$$K_{43} = k_{41}k_{42}k_{43} \tag{23}$$
$$K_{44} = k_{41}k_{42}k_{43}k_{44} \tag{24}$$

For a more complete description of the linkage between subunit assembly and oxygen binding in human hemoglobin, the reader is referred to Ackers and Halvorson.[15] It is these relationships which will be used as the example of the propagation of confidence intervals determined by least-squares curve fitting in one space to parameters of another space. The actual example will be to determine the product Adair constants and their associated confidence intervals and then determine from them the stepwise Adair constants and corresponding confidence intervals without repeating the least-squares fit.

Serious problems arise in propagating the confidence intervals to determine the confidence intervals of the stepwise Adair constants. It should be noted that standard methods of error propagation assume that the confidence interval is symmetrical about the estimated best value and most assume that the parameters are not correlated. However, the cross-correlation coefficient between the parameters used to generate Fig. 3 was -0.65 and the suboptimal case in Fig. 2 is obviously asymmetrical.

Our basic method of confidence interval propagation is actually a simple problem of mapping one parameter space into another parameter space. The optimal method is to map the entire critical F statistic contour from one space to another. This requires a large amount of computer time, as noted previously, since it requires that the entire contour be

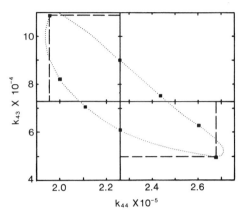

FIG. 4. Propagation of the confidence intervals shown in Fig. 3 to derive the third and fourth stepwise Adair constants (k_{43} and k_{44}). See text for details. Redrawn from Ref. 14.

evaluated again. Therefore we developed a method to approximately map the entire space.

An example of this mapping from one parameter space to another is given in Fig. 4. The elliptically shaped confidence contour in Fig. 3 is a two-dimensional contour, i.e., K_{43} vs K_{42}. When this contour in product Adair constant space is mapped to the stepwise Adair constant space, it becomes a three-dimensional contour in k_{42}, k_{43}, and k_{44}. This added dimension arises from the assumption that K_{41} and K_{44} are known previously. For ease of visualization, a two-dimensional projection, k_{44} vs k_{43}, is presented in Fig. 4. It is obvious that this contour is neither symmetrical nor elliptical.

Also shown in Fig. 4 are the eight points presented in Fig. 3 which were mapped into the new parameter space. These points are sufficient to define the extremes of the contour, and thus the projections of these points onto the individual stepwise Adair constant axes are close approximations of the true confidence intervals of the propagated parameters.

The effects of cross correlation between parameters can be seen also by a comparison between Tables I and II. If, as in Table II, the assumption is made that K_{41} and K_{44} are known, then the value of K_{42} is 9.21×10^8 with a confidence interval of 7.07 to 11.34 $\times$ 10^8. If the effects of the possible variations of all four parameters are included, as in Table I, then the confidence intervals are substantially larger.

Implementation Notes

Comments on some of the nuts and bolts of developing a computer program to perform a parameter estimation are in order before proceeding to a discussion of examples of the use of these methods.

First, most of the methods require the evaluation of the derivatives of some function with respect to the parameters being determined. This can, of course, be implemented by the evaluation of an explicit analytical derivative. This process requires the user of the computer program to find the analytical form of the derivatives, which is at best a nuisance and in some cases may be extremely difficult. An alternative approach is to evaluate these derivatives numerically, using for example, the derivative of a five-point Lagrangian interpolation function.[19] These formulas can be applied with any desired precision and need not be reprogrammed if the form of the function G is changed. The important point is that the value of the derivative is required, not the method by which it is evaluated.

The second comment is in reference to the method of inverting the matrix in Eq. (14). For parameter estimation problems of the type being presented here the matrix $P'P$ will be difficult to invert, i.e., it is almost singular. In the process of developing our computer program, we tested most of the standard methods for inverting matrices and have found by empirical observation that the Square Root method[20] of matrix inversion seems to work for problems which could not be solved by other methods. The difference is worth coding the Square Root method rather than using a matrix inverter from your computer library.

The third comment concerns the method of deciding that convergence has been reached. The most commonly used criterion is that convergence is complete when the value of χ^2 does not change within some specified value, commonly as low a fractional change as one part in a thousand. This is an extremely weak standard since sometimes the parameter values can change by large amounts without altering the value of χ^2. A far better convergence criterion is when the *parameter values* do not change within some specified limit, preferably as small as one part in 100,000.

The last point is in reference to the actual computer language in which the algorithms are coded. Some of the examples which will be presented at the end of this chapter are significantly easier to code if the language allows recursion. Recursion means that a subroutine can call itself either directly or indirectly. Languages such as C allow this technique, but most implementations of FORTRAN do not.

Examples

In this section we will present several examples of the analysis of typical biochemical data. These examples will be chosen specifically to

[19] F. B. Hildebrand, "Introduction to Numerical Analysis," p. 82. McGraw-Hill, New York, 1956.
[20] V. N. Faddeeva, "Computational Methods of Linear Algebra," p. 81. Dover, New York, 1959.

emphasize the need to select carefully which variables are to be dependent or independent, as well as the parameters for the least-squares process. The reader should be aware that the formulation of the problem for computer analysis, that is, the choice of dependent and independent variables, is the same as choosing a way to represent the data graphically. In other words, irrespective of whether a graph is actually drawn the computer program must generate internally a graph of the data. This pseudograph is subject to all of the problems associated with any particular graphical procedure and these problems will alter the results of a nonlinear least-squares analysis. Thus, it is critically important to pick carefully the computer equivalent of the graphical procedure, the proper choice being dictated by the distribution of experimental error, or noise, in the data.

Hormone Receptor Interactions and the Choice of Independent Variables

In a typical insulin-binding experiment, a series of receptor preparations are incubated in the presence of a small amount of ^{125}I-labeled insulin plus varying concentrations of unlabeled insulin. After incubation, the preparations are fractionated by filtration or sedimentation, washed, and the amounts of radioactivity are measured by counting the filters or pel-

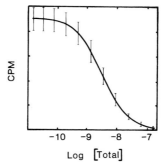

Log [Total]

FIG. 5. Synthetic example of a typical insulin-binding experiment. Data are simulated assuming two independent noninteracting classes of binding sites with K_a's of 2.0×10^9 and $2.1 \times 10^8 \, M^{-1}$, and respective binding capacities of 5×10^{-10} and $1.5 \times 10^{-9} \, M$.[21] It should be noted that in the hormone receptor field, the binding capacities are usually expressed as per some amount of total protein or tissue preparation. For simplicity of simulation, we have expressed all concentrations in molarity. The error bars correspond to a radioactive counting error of 10%, when the total concentration is approximately $10^{-11} \, M$, and increases to 20% as the number of counts decreases at higher unlabeled insulin concentrations. It is assumed that nonspecific binding did not occur in these data. It was further assumed that the concentration of added unlabeled insulin was known to significantly better precision than the counting error. Redrawn from Ref. 1.

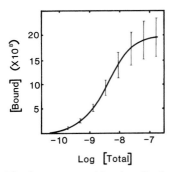

FIG. 6. The data used in Fig. 5 are corrected for the effective specific activity of labeled insulin and presented as moles bound versus the log of total insulin concentration in moles. The error bars in this figure, as well as Figs. 7–8, are evaluated by simulating a data point at each end of the error bars in Fig. 5 and then performing the appropriate transformations. Redrawn from Ref. 1.

lets, respectively. Figure 5 shows a synthetic example of such an experiment plotted as radioactivity bound versus the logarithm of the total molar concentration of insulin. The numerical values of the various constants are approximately those reported for the binding of insulin to liver plasma membranes[21] and are presented in the figure legend. These data can then be perturbed by known amounts of "random" error varying with the total number of radioactive counts to demonstrate how this type of error affects the graphical presentation.

Data of this type are usually transformed from radioactivity bound into concentration bound, as is shown in Fig. 6. This transformation is accomplished by dividing the bound counts of radioactivity by the effective specific activity of the combination of labeled and unlabeled insulin. It is interesting to observe the way that the radioactive counting error is propagated from Fig. 5 to Fig. 6. This increasing experimental error at higher concentrations of unlabeled hormone arises from the combination of two phenomena. First, at higher concentrations of unlabeled hormone, the number of radioactive counts is quite low. Under these conditions most investigators do not count to a constant number of counts above background, i.e., a constant counting error. Most investigators will let the counting error increase in order to decrease the amount of counting time required. Second, at high concentrations of unlabeled ligand the effective specific activity approaches zero. Since the bound concentration is calculated by dividing this poorly determined low number of counts by a spe-

[21] C. R. Kahn, P. Freychet, J. Roth, and D. M. Neville, Jr., *J. Biol. Chem.* **249,** 2249 (1974).

cific activity which approaches zero the absolute error in the bound concentration increases markedly.

It should be noted that most of the mathematical formulations for ligand-binding problems are presented in terms of the free (or unbound) concentration of ligand (or hormone in this case), rather than the total concentration. Accordingly, the free concentration is calculated as the total concentration less the bound concentration. Typically, the data are graphed showing bound versus either the free or the logarithm of the free concentration as in Fig. 7. It is quite interesting that when the concentration of free hormone falls below 10^{-8} M in this particular example (Fig. 7) the error bars are no longer vertical. This happens as a result of the method used to calculate the free concentration from the bound concentration. At these low total concentrations the bound concentration can, in some cases, be a significant fraction of the total concentration. When this occurs, as in this example, any small experimental error in the bound concentration will generate an error in the free concentration, since the free concentration is calculated as total concentration minus the bound concentration. Thus, since both the bound and free concentrations have the same associated errors, the error bars will no longer be vertical. The same nonvertical error bars occur when plotted against the free concentration but are less obvious due to the format of the data presentation.

Probably the most commonly used and abused graphical procedure for the analysis of hormone binding data is the Scatchard plot. If the chemical mechanism of binding the hormone to the receptor can be described as a single class of binding sites, then the Scatchard plot will be a straight line. The same simulated data set is again presented in Fig. 8 as the bound divided by the free concentrations versus the bound concentration (the

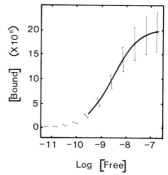

FIG. 7. The data of Fig. 5 are presented as amount bound versus the log of free molar concentration. The calculated line has been omitted at low concentrations so that the angle of the error bars is easily observable. See Figs. 5 and 6 for details.

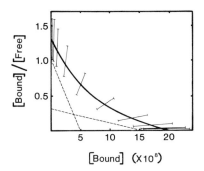

FIG. 8. Scatchard plot (bound divided by free molar concentrations versus bound molar concentration) for the data in Fig. 5. The dashed lines correspond to the Scatchard plots of the individual classes of binding sites. See Figs. 5 and 6 for details.

Scatchard transformation). Also shown in Fig. 8 are the Scatchard plots for the two individual classes of binding sites, represented by the dashed lines. It is interesting to note that the error bars in this formulation seem to radiate from the origin. On the right side of this graph the total concentration of unlabeled hormone is very high and well determined while the bound concentration is significantly lower and poorly determined. Under these conditions, the bound to free concentration ratio (Y axis) will be very small with a small absolute error, while the bound concentration (X axis) will be large and the error bars will be almost horizontal. At extremely low total concentrations (the left side of this graph) the free concentration will be very small. Thus the bound to free concentration ratio will be very large and have a large absolute error compared to the bound alone, and the error bars will appear to be almost vertical.

The first assumption listed for the least-squares method was that all of the error of the data can be attributed to the ordinate (Y axis). An examination of Figs. 5–8 indicates that this assumption is violated by the Scatchard plot (Fig. 8), and by the plots of bound versus the free (not shown) and bound versus the logarithm of the free concentration (Fig. 7). This is a consequence of the method of calculating the free concentration from the total and bound concentrations. Thus it appears that the only valid formulations are plots of amount bound or radioactivity versus the total concentration or the logarithm of the total concentration. This is the formulation used by the computer program of Munson and Rodbard.[14,22]

At first it might seem that this problem could be circumvented by a direct measurement of the free concentration. However, the usefulness of

[22] P. J. Munson and D. Rodbard, *Science* **220**, 979 (1983).

this approach is doubtful because the free concentration will generally correspond to a very low level of radioactivity and thus will be subject to large errors in measurement. In this case, it is necessary for the investigator to allow the number of counts at the low end of the CPM scale to remain high, even though this requires a large amount of counting time.

The data of Figs. 5–8 were simulated assuming an increased radioactive counting error when proportionately greater amounts of unlabeled hormone are added. This is a standard approach to such experiments. It is important to be aware that the predicted distribution of experimental error due to counting uncertainties is not Gaussian; it is a Poisson distribution. At large numbers of counts, the Poisson distribution can be approximated by a Gaussian curve, but not as the number of counts approaches zero. This means that if an investigator wants to apply least-squares with some measure of validity at low specific activities, i.e., high concentrations of unlabeled hormone, it is essential that the investigator count all samples to a predetermined number of counts (for example, 3000) rather than to a predetermined time.

There is currently an ongoing discussion in the literature[22–24] as to which of these plots (Figs. 5–8) is the proper method to analyze and present hormone binding data. Munson and Rodboard[2] have pointed out that "the statistical information content of the data is not altered by presentation in one or the other coordinate system: a simple algebraic manipulation will convert one format into the other." This is obviously true. However, as pointed out by Klotz,[23,24] what is altered is our perception of the data. Klotz presented several examples from the literature[23,24] in which various investigators have not collected data at high enough concentrations to reach the inflection point of the plot of bound versus the logarithm of the free concentration. These investigators presented their data as a Scatchard plot without noting that the amount bound had not reached saturation. Klotz points out that under these conditions it is impossible to determine either the binding capacity or the equilibrium constant from the data.

Neither Klotz nor Munson and Rodbard noted what is perhaps the most serious abuse of the Scatchard plot. If the Scatchard plot is not a straight line, then either the underlying molecular mechanism is not a single class of noninteracting binding sites, or the nonspecific binding has been treated incorrectly. In the case of nonlinear Scatchard plots, some investigators will draw two straight lines through the limiting slopes of the Scatchard plot and assume that these slopes reflect the binding affinities

[23] I. M. Klotz, *Science* **217**, 1247 (1982).
[24] I. M. Klotz, *Science* **220**, 981 (1983).

of the high and low affinity classes of sites. First, these investigators are assuming that two classes of sites exist, although the evidence may be insufficient to support this idea. Second, it is nearly impossible to perform such an operation with reasonable precision or statistical validity. Third, and perhaps most important, when the two classes of sites have affinities which are reasonably close, such as within a factor of 10, as shown in our simulated Scatchard plot (Fig. 8), the limiting slopes do not correspond to the individual binding affinities. This is obvious in Fig. 8 by a comparison of the dashed lines and the limiting slopes.

After choosing a representation of the experimental data which is consistent with the assumptions of the method of least-squares, we are ready to generate a functional form, G, which describes the data as presented in our formulation. The fourth assumption of the nonlinear least-squares techniques requires a knowledge of the particular molecular mechanism, in this example hormone binding. This assumption is necessary because the nonlinear least-squares technique requires an exact explicit function, G, which relates the dependent variables with the independent variables: that is the bound concentration versus the logarithm of the total concentration, for this particular case.

Typically the assumption is made that the molecular mechanism of interest is an unknown number of independent noninteracting binding sites. For this assumption, the correct functional form to describe the bound concentration as a function of the free concentration will be

$$G(\text{free}) = [R] \sum_{j=1}^{m} \frac{n_j[\text{free}]}{K_{d_j} + [\text{free}]} \qquad (25)$$

where $G(\text{free})$ describes the bound concentration of hormone, R is the concentration of receptor molecules, j is an index which refers to each of the m different classes of binding sites, n_j is the number of binding sites in each of the m classes of sites, and K_{d_j} is the corresponding dissociation constant for each of the classes of binding sites. As shown earlier, a graph of bound versus free concentration violates the basic assumptions of the least-squares method of parameter estimation. In order to utilize the least-squares technique, the data must be represented as a function of the total concentration. This can be accomplished by noting that the total concentration is a function of the free, i.e.,

$$[\text{total}] = [\text{free}] + G(\text{free}) \qquad (26)$$

For any particular set of values of m, n_j's, K_{d_j}'s, and total concentrations, Eq. (26) can be solved numerically to find the free concentration of hormone. This free concentration can then be used to evaluate the bound concentration, G(free), at any value of the total concentration. There are a few available computer programs which use this strategy.[14,22]

The most common definition for non-specific binding is the unsaturable low affinity specific binding which the investigator is not currently interested in studying. The key words in this definition are unsaturable and specific binding. By acknowledging that the nonspecific binding is actually a low affinity specific binding, it can be incorporated into Eq. (25) by simply adding an additional class of binding sites. A word of caution is in order regarding the key term, unsaturable. Unsaturable implies that the K_{d_j} for the nonspecific binding is always significantly greater than the free concentration. When this occurs the corresponding term for nonspecific binding in Eq. (25) degenerates to the form

$$\frac{n_j}{K_{d_j}} \, [\text{free}] \tag{27}$$

It is important to note that this form of the binding equation implies that n_j and K_{d_j} are always correlated with each other. Unique values of n_j and K_{d_j} cannot be found, but their ratio is determinable and can be substituted into Eq. (25) to yield the form

$$G(\text{free}) = K_{ns}[\text{free}] + [R] \sum_{j=1}^{m} \frac{n_j[\text{free}]}{K_{d_j} + [\text{free}]} \tag{28}$$

where K_{ns} is an arbitrary constant which describes the nonspecific binding: i.e., $K_{ns} = n_j/K_{d_j}$. This is basically the method which is employed by Munson and Rodbard.[22]

To recapitulate, the least-squares method for computer analysis generates internally a graph of the bound concentration versus the logarithm of the total concentration. This graph is then analyzed by utilizing a combination of Eq. (26) and (28). By this method, the nonspecific binding is determined simultaneously with the specific binding.

Relaxation Phenomena and the Choice of Dependent Variables

One of the classic problems for which the nonlinear least-squares technique has been employed is the evaluation of rate constants and amplitudes from chemical relaxation data (cf. Ref. 25). Typical experiments which yield this type of data are temperature-jump, pressure-jump, stopped-flow, and flash-photolysis. In experiments of this type the position of a chemical equilibrium is altered by an external perturbation, and

[25] L. Endrenyi, "Kinetic Data Analysis." Plenum, New York, 1981.

the concentration of one or more chemical constituents is monitored as a function of time, as the chemical equilibrium relaxes to its new position. The mathematical form which describes the relaxation is usually assumed to be

$$G(\alpha, X_i) = \sum_j \text{Amp}_j \exp(-t_i/\tau_j) \qquad (29)$$

or

$$= \sum_j \text{Amp}_j \exp(-K_j t_i) \qquad (30)$$

where the individual amplitudes (Amp_j) and rate constants (K_j) or decay times (τ_j) correspond to the parameters being determined (α's), and the time after the perturbation (t_i) corresponds to the independent variable (X_i).

Figure 9 shows a synthetic example of such a decay. This example was generated by assuming a single characteristic decay time of 1.0 and an amplitude of 10.0 in arbitrary units. The data are then presented as a decay to a new equilibrium position of 0.5, again in arbitrary units. Gaussian-distributed pseudo-random noise with a standard deviation of 0.26 has been superimposed upon these data to simulate a real experiment. The solid curve in Fig. 9 corresponds to the simulated decay without the noise.

One of the most critical problems encountered in the analysis of data of this type is small uncertainties in the final equilibrium value. To illustrate this point, the data shown in Fig. 9 have been terminated at a time before the decay process reached equilibrium. The analysis of these data with several assumed values for the equilibrium position are given in

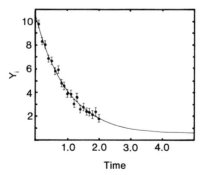

FIG. 9. Simulated single exponential decay data: amplitude = 10.0; decay rate = 1.0; infinite time value = 0.5; Gaussian distributed pseudo-random noise with SD = 0.26. The solid line corresponds to the "correct" answers; Case 1 of Table III.

TABLE III[a]

Case	Amp	K	Equilibrium value
1	10.0 (9.6, 10.4)[b]	1.00 (0.94, 1.06)[b]	0.50[c]
2	9.7 (9.3, 10.2)[b]	1.14 (1.07, 1.22)[b]	1.00[c]
3	10.3 (9.9, 10.7)[b]	0.89 (0.84, 0.95)[b]	0.00[c]
4	9.9 (9.4, 10.3)[b]	1.08 (0.90, 1.25)[b]	0.78 (0.10, 1.42)

[a] Least-squares analysis of the data in Fig. 9 with different assumed equilibrium values.
[b] Confidence interval corresponding to ± 1 SD.
[c] Assumed value.

Table III. A comparison of cases 1, 2, and 3 illustrates the sensitivity of the determined value of the rate constant to the assumed equilibrium value. A change in the equilibrium value by only 5% in amplitude will, as in case 2, generate a 14% systematic error in the determination of the rate constant. The calculated lines which correspond to cases 2 and 3 are shown in Fig. 10 to illustrate how both of these lines can easily describe the experimental data.

Case 4 in Table III shows the values when the equilibrium value is determined as one of the parameters in the fitting process. This is accomplished by assuming that one of the decay rate constants is zero, and determining its corresponding amplitude. It is interesting to note how the inclusion of this third parameter in the least-squares process increases the range of the confidence interval for the apparent value of the rate constant.

One of the commonly used methods for the analysis of multiple exponentials is a process which is sometimes referred to as curve peeling. In this method the amplitude and time constant of the slowest relaxation are

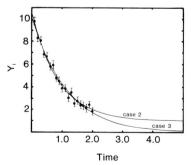

FIG. 10. The synthetic data with calculated lines which correspond to Cases 2 and 3 of Table III.

determined. These are then extrapolated back to zero time and subtracted from the original data. This modified data set is then used to determine the next slowest time constant and amplitude. The effect is then subtracted from the data and the process repeated as many times as is required to describe the data totally. This method has at least two severe problems and should never be used. One of these problems is that small errors in the determination of the slowest relaxations will have a compounded effect on the determination of the faster relaxations, as in Table III, case 2. The second problem is that it is impossible to determine reasonable confidence intervals for the parameters if they are determined one at a time.

The method to circumvent these two problems is to determine simultaneously all of the parameters and their associated confidence intervals. If the data are not of sufficient precision to allow the simultaneous determination of all of the desired parameters, then the curve peeling process will yield erroneous answers.

The standard graphical method for the analysis of exponential data is to plot the logarithm of the difference between a current value (Y_i) and the value at infinite time vs time. If the data can be described by a single exponential decay then the graph will be a straight line, the slope of this line being proportional to the decay rate, (K).

This log plot has several problems. First, the infinite time value must be determined independently and uncertainties in its determination cannot be treated. An example of this problem is shown in Fig. 11. This graph is a log plot of the two curves in Fig. 10. It is obvious that the two assumed infinite values used in Fig. 10 yield two different slopes in Fig. 11 with no easy method to determine which, if either, is correct.

A second problem is that the size of the error bars in the logarithmic

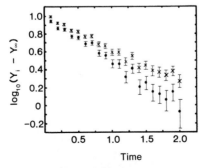

FIG. 11. Logarithmic plots of the data in Fig. 9 with the same assumed infinite time values of Fig. 10 and Cases 2 and 3 of Table III.

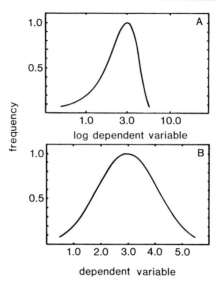

FIG. 12. Illustration of the effect of a logarithmic transformation on a Gaussian distribution: (A) transformed distribution; (B) original Gaussian distribution.

plot is not constant. This can, in theory, be compensated for by an appropriate choice of the weighting factors to be used for the least-squares analysis.

A more serious problem is also indicated in Fig. 11. A careful inspection of the data points in the lower right of Fig. 11 shows that as the decay approaches the infinite time value, the distribution of errors around the data points is no longer Gaussian. A better illustration of this point is made in Fig. 12 where the effect of the logarithmic transformation on the noise distribution is shown. Figure 12B shows a Gaussian distribution with a mean of three and a standard deviation of one. This is typical of the noise distribution which might be expected in Fig. 10 as the decay approaches the equilibrium value. The same noise distribution is shown in Fig. 12A except that the distribution has been altered by a logarithmic transformation as in Fig. 11.

The purpose of graphing the data logarithmically is to generate a straight-line graph from which a simple linear least-squares analysis technique may be used to evaluate the rate constant. However, this may not be a valid approach. The data in Fig. 10 were generated with a Gaussian distribution of error for which it is valid to apply a nonlinear least-squares method. However, after performing the logarithmic transformation as in Fig. 11, the noise distribution is no longer Gaussian, which violates one of the basic assumptions for the use of the least-squares technique. Consequently, it is not statistically valid to apply a least-squares analysis to determine the slopes of the lines in Fig. 11. This problem cannot be circumvented by an "appropriate choice of weighting factors."

It is interesting to note that a similar problem might arise if the output of a hypothetical kinetic instrument records percentage transmittance instead of optical density. In this case the error distribution superimposed on the transmittance data can probably be assumed to be a Gaussian. If so, it would not be valid to transform the data to optical densities before performing the least-squares analysis. The general rule to follow is that no transformations should be made on the dependent variables which will alter the shape of the noise distribution. The only generally used transformations which do not violate this rule are simple addition of and multiplication by a constant. An example of an allowed dependent variable transformation was presented as the correction for specific activity shown in Fig. 6.

Consequences of Truncating the Range of Independent Variables

Klotz[23,24] presented several examples from the literature in which various investigators have not collected ligand binding data at high enough concentrations to reach the inflection point of the plot of bound versus the logarithm of the free concentration. These investigators presented their data as a Scatchard plot without noting that the amount bound had not reached saturation. Klotz points out that under these conditions it is impossible to determine either the binding capacity or the equilibrium constant from the data. Figure 13 is an example of this point. In this figure, data were simulated for a single class of binding sites with a total concentration range of 10^{-11} M to 3×10^{-9} M. Since the K_d for these simulated data was 5×10^{-9} M, with a corresponding binding capacity of 1.5×10^{-9} M, these simulated data only reach approximately 30% of saturation. The solid line in Fig. 13 corresponds to the correct answers. The dotted and dashed lines correspond to binding capacities of 1.0×10^{-9} and 5×10^{-9} M. Note that all of these appear to describe the data over their limited range. This is a common problem when the Scatchard plot is used.

It is obvious that under circumstances such as limited solubility or availability of ligands or binding components, an investigator may be forced to truncate a set of data at less than an optimal range of concentration. But, this should be avoided if at all possible.

A second example of a truncated set of data is the simulated exponential decay presented in Fig. 10. This example was very sensitive to small

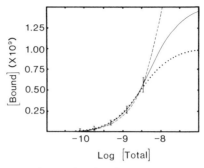

FIG. 13. A simulated experiment showing the consequences of failing to approach a saturation level. Data were simulated as in Fig. 1 except we assumed a single class of binding sites with K_d of 5×10^{-9} M and a binding capacity of 1.5×10^{-9} M. The solid line corresponds to the correct answers, the dashed line corresponds to a K_d of 2.15×10^{-8} M and a binding capacity of 5×10^{-9} M, and the dotted line corresponds to a K_d of 2.47×10^{-9} M and a binding capacity of 1×10^{-9} M. Redrawn from Ref. 1.

errors in the determination of the infinite time value. This sensitivity could have been reduced if the decay was allowed to proceed for four or five half-lives instead of only two. The data collected at longer time or closer to completion of the experimental phenomena are very important for a correct determination of the infinite time value and should not be eliminated arbitrarily.

Circumventing the Assumptions

We previously presented the four basic assumptions of the technique of least-squares. These assumptions are absolutely required to perform a least-square analysis with any measure of statistical validity and to have confidence in the value and meaning of the determined parameters. However, experiments in the laboratory cannot always be tailored to meet rigorously all of these requirements. The purpose of this section is to give the reader a feel for some of the possible methods to bend these assumptions if it is absolutely required. In this section the assumptions will be discussed in reverse order.

If the assumption of the molecular reaction mechanism is wrong, then the determined parameters and their associated confidence intervals will not correspond to what the investigator purports. Consequently, in a sense, all reports utilizing least-squares analysis should be required to have a section which begins, "These are the opinions upon which I shall base my facts."[26] This does not mean that the determined parameters (apparent affinity constants and binding capacities) are completely mean-

[26] J. Faulkner, *Nature (London)* **253**, 231 (1975).

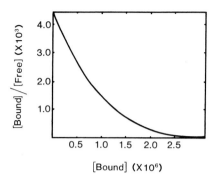

[Bound] (×10⁶)

Fig. 14. Simulated Scatchard plot for the binding of rhodamine 6G to 10 μM glucagon at pH 10.6, 0.6 M phosphate.[31] Redrawn from Ref. 1.

ingless. An investigator can legitimately use these parameters, under defined conditions, as an experimental probe in the same way that a biologist might use muscle contraction, uterine weight, or behavioral changes as a measure of the effect of an agonist or antagonist on a system.[22] It means simply that the investigator should clearly state all of the assumptions, conditions and methods employed.

The best method to explain this point is to present an example. Several groups of workers have suggested that receptor aggregation, clustering, or binding of a second protein are important for the functioning of the insulin receptor.[27-29] If such a phenomena is present it should be included in the mathematical formulation of the binding. The Scatchard plot for the binding of a ligand to a binding protein which is simultaneously undergoing a self-association is shown in Fig. 14. The particular example was calculated to simulate the binding of rhodamine 6G (the ligand) to glucagon (the binding protein). At pH 10.6 in 0.6 M phosphate, glucagon exists in an equilibrium between a monomer and a trimer with an equilibrium constant of $4.8 \times 10^7 \ M^{-1}$. Rhodamine 6G binds to the trimer with an affinity of $9.4 \times 10^4 \ M^{-1}$ but does not bind to the monomeric form.[30] Glucagon is thus a model for the binding of hormones, in this case, rhodamine 6G, to a receptor which polymerizes, in this case glucagon. It

[27] C. R. Kahn and K. L. Baird, in "Physical Chemical Aspects of Cell Surface Events in Cellular Regulation" (C. DeLisi and R. Blumenthal, eds.), p. 119. Elsevier/North-Holland, New York, 1979.
[28] C. DeLisi, in "Physical Chemical Aspects of Cell Surface Events in Cellular Regulation" (C. DeLisi and R. Blumenthal, eds.), p. 261. Elsevier/North-Holland, New York, 1979.
[29] J. T. Harmon, C. R. Kahn, E. S. Kempner, and M. L. Johnson, in "Current Views on Insulin Receptors" (D. Andreani, R. DePirro, R. Lauro, J. M. Olefsky, and J. Roth, eds.), p. 37. Academic Press, New York, 1981.
[30] M. L. Johnson, S. Formisano, and H. Edelhoch, J. Biol. Chem. 253, 1353 (1978).

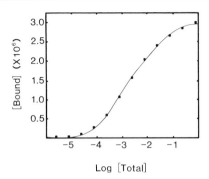

Log [Total]

Fig. 15. The simulated data for rhodamine 6G binding to glucagon (as in Fig. 14) presented as the bound versus the logarithm of the total concentration. The solid line was calculated for a model corresponding to two classes of binding sites. Redrawn from Ref. 1.

should be noted that the shape of the Scatchard plot in Fig. 14 would seem to indicate the existence of multiple classes of binding sites, or some form of heterogeneity or negative cooperativity. But remember also that these data were simulated assuming a single binding site with no heterogeneity or cooperativity other than the concomitant self-association of the receptor.

Figure 15 presents an analysis of the same data as in Fig. 14, but assumes that the molecular mechanism is two classes of binding sites. The apparent binding affinities from this analysis are 1.79×10^3 and $40.7\ M^{-1}$ with corresponding binding capacities of 1.8 and 1.2 μM. It is apparent that these constants are significantly different from the actual values. However, observe that the calculated solid line in Fig. 15 is easily capable of describing the data.

This example was chosen specifically to demonstrate a single point: the analysis of ligand binding data can yield multiple sets of answers depending upon what assumptions are made for the analysis. These multiple sets of answers can, in some cases, still be utilized as an arbitrary mathematical description of the data. Munson and Rodbard[22] have pointed out that a large amount of information can be obtained without actually determining the true mechanism of hormone binding by simply considering the determined parameters as "apparent" values. These apparent values can still be used as a probe of the hormone receptor interaction without the assumption that they are actually correct.

The third assumption was that no systematic error exists in the experimental data. There is no parameter estimation method which is totally insensitive to systematic errors in the data. Therefore, all systematic error should be eliminated by either of two methods. The most reliable method

TABLE IV[a]

Model	N_1	K_1	N_2	K_2	K_{ns}
1	1.50×10^{-9}	2.10×10^8	0.50×10^{-9}	2.0×10^9	0
2	1.55×10^{-9}	7.75×10^8	0	0	0.0032

[a] Values of equilibrium constants employed for the calculated lines in Fig. 16. K's are presented as association constants. Model 1 was used to simulate the data points for Figs. 5–8 and 16, as described in the legend for Fig. 5. The dotted line in Fig. 16 was calculated from Model 2.

is to redesign the experiment to eliminate the systematic error. If elimination of the systematic error is impossible, it might be possible to include a mathematic description of it in the formulation of the problem, i.e., the function, G.

An example of such a mathematical elimination of systematic errors was presented in Eqs. (27)–(28). In this example, the nonspecific binding which is usually present in experimental studies of hormone binding to membrane bound receptors was explicitly incorporated into the function which describes the data. An example which illustrates the importance of the correct inclusion of nonspecific binding is presented in Table IV and Fig. 16.

Figure 16 gives the same simulated data as were presented in Figs. 5–8. These data correspond to two classes of binding sites with no nonspecific binding. These two classes differ by an order of magnitude in affinity, with the lower affinity class having three times the binding capacity. The exact constants are presented in Table IV as Model 1. The dotted line in Fig. 16 (Model 2 in Table IV) corresponds to the analysis of these data as

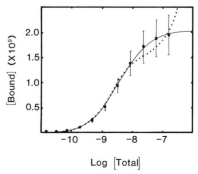

Fig. 16. The data from Fig. 6 analyzed by two different models. The actual values of the constants are given in Table IV.

a single class of binding sites and 0.32% nonspecific binding. Both of these models lie within the expected error bars of the data points. An inspection of Table IV shows that the affinity constants determined for Model 2 are quite different from those of Model 1. A visual inspection of Fig. 16 would indicate that the only way to distinguish these models would be to collect data at even higher concentrations and with greater precision.

The second assumption is that the random systematic errors of the data can be described as a Gaussian distribution. There are currently a number of "robust" mathematical methods being developed in an attempt to circumvent this assumption.[31–39] In general these robust methods consist of substituting a variety of different norms, other than least-squares, to be minimized. Either the Nelder–Mead or the Newton–Raphson procedure can be utilized to minimize any arbitrary norm of the data. It is highly recommended that a review of robust statistics be undertaken if the second assumption of least-squares, Gaussian noise distribution cannot be met. This may also be a required alternative if the number of data points is so severely limited that they do not constitute a good statistical sampling of the noise, even if it can be represented by a Gaussian distribution.

The first assumption which we presented for least-squares methods is that the independent variables are known to significantly greater precision than the dependent variables. This assumption cannot be circumvented by "appropriate weighting factors." It can be treated, however, by substituting a different norm to be minimized.[40] If we assume that both the ordinate and the abscissa have independent Gaussian distributed random experimental errors superimposed on them and that our assumptions 3 and 4 hold, we can derive a new norm to be minimized by the following procedure which is analogous to Eqs. (3)–(5).

[31] D. F. Andrews, P. J. Bickel, F. R. Hampel, P. H. Huber, W. H. Rogers, and J. W. Tukey, "Robust Estimates of Location." Princeton Univ. Press, Princeton, New Jersey, 1972.
[32] D. F. Andrews, in "Alternative Calculations for Regression and Analysis of Variance Problems in Applied Statistics" (R. F. Gupta, ed.), p. 1. North-Holland Publ., Amsterdam, 1975.
[33] S. M. Stigler, J. Am. Stat. Assoc. 68, 872 (1973).
[34] L. Breiman, "Statistics." Houghton Mifflin, Boston, Massachusetts, 1973.
[35] F. Mosteller and J. W. Tukey, "Data Analysis and Regression." Addison-Wesley, Reading, Massachusetts, 1977.
[36] J. W. Tukey, in "A Survey of Sampling from Contaminated Distributions in Contributions to Probability and Statistics" (I. Olkin, ed.), p. 448. Stanford Univ. Press, Standford, California, 1960.
[37] W. J. Kennedy, Jr., and J. E. Gentle, "Statistical Computing." Dekker, New York, 1980.
[38] P. J. Huber, "Robust Statistics." Wiley, New York, 1981.
[39] I. Isenberg, Biophys. J. 43, 141 (1983).
[40] M. L. Johnson, Anal. Biochem. (1985), in press.

The probablity, P_i, for observing a given data point (Y_i and X_i) and any particular values of α is given by

$$P_i(\alpha) = \frac{1}{2\pi\sigma_{X_i}\sigma_{Y_i}} \exp\left\{-1/2\left[\frac{Y_i - G(\alpha,\bar{X}_i)}{\sigma_{Y_i}}\right]^2\right\}$$

$$\exp\left\{-1/2\left[\frac{\bar{X}_i - X_i}{\sigma_{X_i}}\right]^2\right\} \tag{31}$$

where σ_{X_i} and σ_{Y_i} represent the independent standard deviations of the random experimental error at the particular data point and $\bar{X}_i$ is the "optimal" value of the independent variable. The probability of making a series of measurements at n independent data points is

$$P(\alpha) = \Pi \, P_i(\alpha)$$

$$= \left\{\Pi\left(\frac{1}{2\pi\sigma_{X_i}\sigma_{Y_i}}\right)\right\} \exp\left\{-1/2\sum\left[\frac{Y_i - G(\alpha,\bar{X}_i)}{\sigma_{Y_i}}\right]^2\right\}$$

$$\exp\left\{-1/2\sum\left[\frac{\bar{X}_i - X_i}{\sigma_{X_i}}\right]^2\right\} \tag{32}$$

where the product Π and summation Σ are taken over each of the n data points with subscript i. This equation can be reorganized to the form

$$P(\alpha) = \left\{\Pi\frac{1}{2\pi\sigma_{X_i}\sigma_{Y_i}}\right\}$$

$$\exp\left\{-1/2\sum\left[\left(\frac{Y_i - G(\alpha,\bar{X}_i)}{\sigma_{Y_i}}\right)^2 + \left(\frac{\bar{X}_i - X_i}{\sigma_{X_i}}\right)^2\right]\right\} \tag{33}$$

The best estimates for the parameters, α, with the current assumptions will be those values of α which maximize the probability given by Eq. (33). As described previously, this is accomplished by minimizing the summation in the exponential term in Eq. (33). The norm to minimize would then be

$$\text{NORM} = \sum\left[\left(\frac{Y_i - G(\alpha,\bar{X}_i)}{\sigma_{Y_i}}\right)^2 + \left(\frac{\bar{X}_i - X_i}{\sigma_{X_i}}\right)^2\right] \tag{34}$$

This is simply an extension of our previous derivation of the least-squares norm to include the possibilities of experimental uncertainties in the independent variables.

This new statistical norm, Eq. (34), can be minimized by a recursive application of a curve fitting algorithm such as Nelder–Mead or a Newton–Raphson. Equation (34) can be written as

$$\text{NORM} = \sum_{i=1}^{n} D_i^2 \tag{35}$$

where D_i is the weighted distance between the given data point (X_i, Y_i) and the point where an elliptically shaped confidence region centered at the ith data point is tangent to the fitted line $[\bar{X}_i, G(\alpha, \bar{X}_i)]$. The relative sizes of the axes of this ellipse are given by the corresponding standard deviations of the dependent and independent variables. The functional form of this ellipse is

$$D_i^2 = \left(\frac{Y_i - G(\alpha, \bar{X}_i)}{\sigma_{Y_i}}\right)^2 + \left(\frac{X_i - \bar{X}_i}{\sigma_{X_i}}\right)^2 \tag{36}$$

The minimization of this norm is performed by a nested minimization procedure. The parameter estimation procedure to evaluate the values of α is essentially the same as that described earlier. An initial estimate of α is found by some method. This estimate is employed to calculate the D_i^2 at each data point. From these, the NORM, given by Eq. (35) is evaluated as well as any required derivatives. These are used to predict new values for the parameters being estimated, α. This cyclic process is repeated until the values do not change within some specified limit. An inconvenience arises in the evaluation of D_i^2 at each data point and iteration because it requires the value of $\bar{X}_i$. This value of $\bar{X}_i$ is actually the value of the independent variable at the point where the ellipse given by Eq. (36) is tangent to the function evaluated at the current estimate of the parameters, α. This implies that the values of $\bar{X}_i$ will be different for each iteration. As a consequence of this these values of $\bar{X}_i$ must be reevaluated for each iteration.

If the original parameter estimation procedure is carefully developed, it can be used recursively to evaluate $\bar{X}_i$. That is, at each iteration and data point of the parameter estimation procedure the routine calls itself to evaluate D_i^2. For this recursive application of the algorithm, the parameters being estimated are now the $\bar{X}_i$, and the function being minimized is D_i^2. Use the values of X_i or the previous values of $\bar{X}_i$ as starting values.

It should be noted that this method is not restricted to a two-dimensional problem, Y vs X. As with the least-squares norm described previously all that is required to expand this method to include multiple dependent and multiple independent variables is to consider each of the Y_i, X_i, and $\bar{X}_i$'s as vectors instead of scalars, and add the appropriate additional terms in the summations.

This method will accomplish two goals. It will allow for experimental uncertainties of the dependent variables as described above. It also will allow for the analysis of ligand-binding data as a function of the total concentration of ligand (as in Fig. 6). This is accomplished by considering the total concentration as a dependent variable with its corresponding standard error. The functional form which corresponds to this dependent variable would simply be a conservation of mass equation for that particular ligand in terms of the free concentration of ligand. Any reasonable value of the free concentration can then be used as the dependent variable as long as its corresponding standard deviation is taken to be ±infinity.

It should be noted that this method might require an excessive amount of computer time. It would be better to use a classical least-squares method if at all possible.

Conclusion

The primary motivation for this chapter has been to acquaint the reader with one of the methods for the analysis of experimental data, as well as the assumptions, advantages, and disadvantages of the method. We would like to close by emphasizing some of the major points which we have covered.

The most important experimental detail to understand for any parameter estimation procedure is the sources and magnitudes of the random and nonrandom experimental errors superimposed on the data. It is a knowledge of these which dictates the actual procedure to be employed for the analysis of the data.

It is also important to understand the method of analysis which you plan to employ because it is the assumptions of this method which dictate the details of how the experimental data are to be collected.

Computers are not oracles. The investigator needs to be continually aware that the output of any computer program is no better than what goes into it. It is of critical importance that the person who develops computer programs for the analysis of experimental data has a thorough understanding of both the experimental data being analyzed and the numerical methods to be employed. We have found a number of computer programs supplied by computer manufacturers, computer centers, software houses, and other scientists which produce wrong answers. The investigator should be careful to test and possibly calibrate any computer programs with synthetic data which have superimposed pseudo-random noise. These synthetic data should be representative of actual experiments which will be performed and the pseudo-random noise should be representative of the actual experimental uncertainties.

It is important to realize that a computer is in essence the same as any other instrument in a laboratory. In order to obtain optimal results it must be used correctly. In order to use it correctly, the user must understand the assumptions which went into the development of the computer programs and their limitations.

For further readings in the general question of parameter estimation the reader is referred to a number of the books and articles which we have referenced.[1,2,11–14,25,41] An excellent entry level textbook which is highly recommended is by P. R. Bevington.[41]

Upon written request, we will provide information about how to obtain a copy of our program.

[41] P. R. Bevington, "Data Reduction and Error Analysis for the Physical Sciences." McGraw-Hill, New York, 1969.

[17] Ligand Binding to Proteins by Equilibrium Gel Penetration

By HIROSHI KIDO, ALBERTO VITA, and B. L. HORECKER

Gel filtration methods have been widely used for studies of the binding of small ligands to macromolecules. In the procedure described by Hummel and Dryer[1] as modified by Fairclough and Fruton[2] and Kemp and Krebs[3] a small volume of solution containing the macromolecules is passed through a column containing a gel of appropriate porosity previously equilibrated with the ligand to be tested. The column is developed with the same concentration of ligand and fractions are collected and analyzed for the total concentration of ligand present. The amount of ligand bound is calculated from the excess ligand eluted with the macromolecule, or from the decrease in ligand concentration in the trough that usually follows the fractions containing the macromolecule. A complete chromatographic analysis is required for each concentration of ligand to be tested and many fractions, usually 25–35 per column, must be collected and analyzed.

A more rapid and convenient technique, based on equilibrium solute partitioning,[4,5] has been employed by Fahien and Smith[6] and by Mac-

[1] J. P. Hummel and W. J. Dreyer, Biochim. Biophys. Acta 63, 530 (1962).
[2] C. F. Fairclough, Jr. and J. S. Fruton, Biochemistry 5, 673 (1966).
[3] R. G. Kemp and E. G. Krebs, Biochemistry 6, 423 (1967).
[4] G. K. Ackers, Biochemistry 3, 723 (1964).
[5] R. Valdos, Jr. and G. K. Ackers, this series, Vol. 61, p. 125.
[6] L. A. Fahien and S. E. Smith, J. Biol. Chem. 249, 2696 (1974).

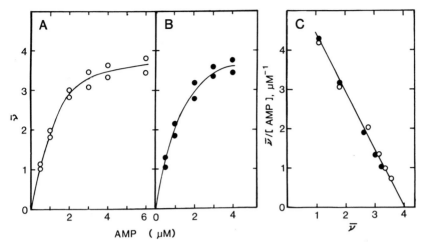

AMP (μM)

FIG. 1. Binding of AMP to rabbit liver Fru-P$_2$ase. The binding experiments were carried out at 4° as described in the text. (A) The total aqueous volume was 1.0 ml, containing 32 μg of Fru-P$_2$ase, 0.1 mM Fru-P$_2$, 80 mg of Sephadex G-50 (coarse), and concentrations of [U-^{14}C]AMP as indicated. (B) The total aqueous volume was 5.0 ml, containing 160 μg of Fru-P$_2$ase, 0.1 mM Fru-P$_2$, 400 mg of Sephadex G-500 (coarse), and concentrations of [U-^{14}C]AMP as indicated. The samples were analyzed and the amount of AMP bound calculated as described in the text. Each experiment was carried out in duplicate and duplicate tubes lacking Fru-P$_2$ase were included for each concentration of AMP. The results shown represent the maximum and minimum values calculated from the radioactivity measured in the duplicate tubes with and without protein. The concentration of protein was estimated assuming the value of M_r = 143,000.[13] (C) Scatchard plots[14] of the data in A and B. Open circles represent the data from A and closed circles the data from B.

Gregor et al.[7] to evaluate complex formation between pairs of proteins. A similar gel equilibration method was applied by Tso and Burris[8] to studies of the binding of ATP to *Clostridium pasteurianum* nitrogenase under anaerobic conditions. In these procedures only a single reaction vessel and one analysis is required for each solute concentration tested.

Binding Assay

The method is illustrated by the measurement of binding of AMP to rabbit liver fructose-1,6-bisphosphate (Fig. 1). The reagents employed were [U-^{14}C]AMP (New England Nuclear), AMP, and fructose 1,6-bis-phosphate, Na salt (Sigma), Sephadex G-50 (coarse), and Blue Dextran

[7] J. S. MacGregor, V. N. Singh, S. Davoust, E. Melloni, S. Pontremoli, and B. L. Horecker, *Proc. Natl. Acad. Sci. U.S.A.* **77**, 3889 (1980).
[8] M.-Y. Tso and R. H. Burris, *Biochim. Biophys. Acta* **309**, 263 (1973).

2000 (Pharmacia). Fructose-1,6-bisphosphatase was purified from rabbit liver[9] and stored at 4° as a suspension in 80% saturated $(NH_4)_2SO_4$. Before use, the protein was collected by centrifugation, dissolved in 0.02 M triethanolamine/0.02 M diethanolamine (TEA/DEA) buffer, pH 7.5, containing 1 mM EDTA, and dialyzed overnight at 4° against the same buffer–EDTA mixture. Protein concentration was calculated from the absorption at 280 nm and a value of 0.63 for the absorbance of a solution containing 1 mg/ml.[10]

The reaction mixtures for the binding assay (1.0 ml) contained 20 mM TEA/DEA buffer, pH 7.5, 0.1 μM fructose 1,6-bisphosphate, 32 μg of fructose-1,6-bisphosphatase, and concentrations of [^{14}C]AMP (5700 cpm/nmol) ranging from 0.5 to 6.0 μM. After addition of protein, the reaction mixtures were equilibrated for 5 min and transferred to 75 × 10-mm plastic vials (Walter Starstedt) containing precisely 80 mg of dry Sephadex G-50 (coarse) powder and several small glass beads. The vials were capped and rocked at 4° for 5 hr to allow for swelling of the gel and for equilibration of the free ligand into the solvent inside the gel. After equilibration, the vials were centrifuged and 50-μl aliquots of the clear supernatant solution were taken for measurement of radioactivity.

For each concentration of AMP tested an equivalent reaction mixture lacking the protein was equilibrated with the gel in the same way. To determine the excluded volume, a solution of Blue Dextran in the same buffer ($A_{615 \text{ nm}}$ = 2.0) was equilibrated with the gel and the absorbance of an aliquot of the supernatant solution measured at 615 nm.

To test for adequacy of equilibration, similar analyses were carried out in 5-ml volumes, with all of the quantities increased proportionately. The samples were equilibrated at 4° in 10-ml plastic scintillation vials using a reciprocal shaker.

Calculations

The quantity (B) of ligand bound to protein was calculated from the equation:

$$B = \frac{\Delta[L][V_E V_T]}{V_T - V_E}$$

where: V_T = total volume of aqueous phase
          V_E = volume of excluded phase

[9] S. Pontremoli, E. Melloni, F. Salamino, B. Sparatore, and B. L. Horecker, *Arch. Biochem. Biophys.* **188**, 90 (1978).

[10] P. S. Lazo, O. Tsolas, S. C. Sun, S. Pontremoli, and B. L. Horecker, *Arch. Biochem. Biophy.* **188**, 308 (1978).

$V_T - V_E$ = volume of included phase

L_T = total quantity of ligand

$\Delta[L] = [L]^{E'} - [L]^{E°}$

$[L]^{E°}$ = concentration of ligand in excluded phase in the absence of protein

$[L]^{E'}$ = concentration of free plus bound ligand in excluded phase in the presence of protein

In the absence of protein $[L]^{E°} = L_T/V_T$

In the presence of protein $[L]^{E'} = [(L_T - B)/V_T] + B/V_E$

Example of Binding of AMP to Rabbit Liver Fructose-1,6-bisphosphatase

Binding of AMP to rabbit liver fructose-1,6-bisphosphatase has been shown to depend on the presence of the substrate, fructose 1,6-bisphosphate.[11,12] For each concentration of AMP the analysis was carried out in duplicate and duplicate analyses were also carried out in the absence of enzyme protein. Two vials contained Blue Dextran. Thus a total of 22 vials and 160 μg of enzyme protein were required for determination of binding at 10 concentrations of AMP (Fig. 1A). Similar results were obtained in the experiments carried out in 5 ml volumes (Fig. 1B).[13] In each case, the results confirmed the presence of 4 binding sites for AMP (one per subunit) and the Scatchard plot (Fig. 1C)[14] yielded a value of 0.7 × 10^{-6} M for the dissociation constant.

Evaluation of the Method

The equilibrium solute partitioning method appears to offer several advantages over column chromatography methods. (1) All of the analyses can be carried out at the same time with the same protein sample, and duplicate analyses require only 40 measurements of radioactivity. (2) A very small volume (total = 8.0 ml) of radioactive solution is required.

Sephadex G-50 (coarse) was selected as the permeant gel because of its relatively rapid rate of swelling and equilibration to the buffer employed.

The major source of error arises from the relatively small difference in concentration of free ligand in the vials with and without protein in com-

[11] S. Pontremoli, E. Grazi, and A. Accorsi, *Biochemistry* **7**, 3628 (1968).
[12] M. G. Sarngadharan, A. Watanabe, and B. M. Pogell, *Biochemistry* **8**, 1411 (1969).
[13] S. Traniello, S. Pontremoli, Y. Tashima, and B. L. Horecker, *Arch. Biochem. Biophys.* **146**, 161 (1971).
[14] G. Scatchard, *Ann. N.Y. Acad. Sci.* **51**, 660 (1949).

parison to the total ligand concentration. This source of error can be minimized by employing conditions that yield a relatively small excluded volume, namely a high ratio of resin volume to total volume. In the experiment illustrated in Fig. 1A, the total volume was 1.0 ml and the excluded volume was 0.45 ml. With these conditions, values of $\Delta[L]$ ranged from 153% of $[L]^{E°}$ at the lowest concentration of AMP to 115% of $[L]^{E°}$ at the highest concentration. At higher concentrations of ligand the values of $\Delta[L]$ became unreliable. Sephadex G-75 would provide a lower proportion of excluded volume, but would require at least 24 hr for complete swelling and equilibration.

The value for K_D ($0.7 \times 10^{-6} M$) is smaller than the values ($\sim 10^5$–10^{-5} M) reported by earlier workers.[11] However, the earlier experiments were carried out with a proteolytically modified form of the enzyme, which Traniello et al.[13] have reported to be considerably less sensitive to inhibition by AMP. More recent studies in which the equilibrium gel penetration method was employed with preparations of native, neutral fructose-1,6-bisphosphatase[15] have confirmed the value of K_D reported here.

[15] A. Vita, H. Kido, S. Pontremoli, and B. L. Horecker, Arch. Biochem. Biophys. 209, 598 (1981).

[18] Measurement of Protein–Ligand Interactions by Gel Chromatography

By José Manuel Andreu

The measurement of ligand binding to protein by application of the gel chromatography technique of Hummel and Dreyer[1] will be discussed in this chapter. Particular attention will be given to the criteria of application and the measurement of weak binding effects.

Principle of Measurement and Possible Sources of Error

The method consists in the chromatography of protein through a size exclusion gel column equilibrated with a known ligand concentration. The perturbations in ligand concentration in the effluent due to binding are observed. This technique[1–3] is based on three requirements: (1) the chro-

[1] J. P. Hummel and W. J. Dreyer, Biochim. Biophys. Acta 63, 530 (1962).
[2] G. F. Fairclough and J. S. Fruton, Biochemistry 5, 673 (1966).
[3] G. K. Ackers, this series, Vol. 27, p. 441.

matographic gel efficiently excludes the protein while including the ligand, (2) attainment of binding equilibrium is faster than the separation process, and (3) there are convenient methods to measure accurately both protein and ligand in the effluent. Under the first and second conditions the procedure can be regarded as an equilibrium thermodynamic technique, even though it involves a transport separation process. In this sense the gel technique is formally analogous to equilibrium dialysis, where the inside of the dialysis bag is now the excluded volume of the column (outer volume V_o, accessible to both the macromolecule P and the ligand A) and the outer dialysis compartment is equivalent to the included volume (inner volume V_i, accessible only to ligand A). Binding equilibrium is established in the vicinity of the macromolecule as it passes through the column. The total macromolecule concentration, $[P] + [PA]$, and the total ligand concentration, $[A] + [PA]$, of the outer volume can be measured in the effluent. The unbound ligand concentration $[A]$ is measured in the vicinity of the macromolecule fractions and is essentially the same as the ligand concentration with which the column has been equilibrated. Therefore, we can adjust the free ligand concentration, and the bound ligand concentration in the protein fractions is calculated by its difference to the total ligand concentration measured. The sample applied to the column usually includes a total ligand concentration exactly equal to the ligand concentration with which the column is equilibrated and therefore a trough is observed in the effluent at $V_o + V_i$ due to the depletion of ligand in the sample by binding to the protein. The decrease of ligand in the trough is equal to the increase of ligand in the peak. A related approach is employed in binding measurements by batch gel partition procedures,[4,5] with the important difference that the free ligand concentration is not adjusted, but calculated from the difference of bound to total ligand concentration. These methods are valuable for small samples and estimate binding from the ligand concentration of the outer phase of gel suspensions with and without macromolecule, with the appropriate volumetric corrections. However, they involve more measurements, calculations, and corrections than the Hummel and Dreyer[1] technique and will not be discussed here.

The gel chromatography equilibrium technique is of great value when the binding assay required has to be simple and considerably faster than equilibrium dialysis, provided enough ligand is available to run the columns. This technique can be automated[2] and also performed with high-performance liquid chromatography systems.[6] Provided equilibrium is at-

[4] P. Fasella, G. G. Hammes, and P. R. Schimmel, *Biochim. Biophys. Acta* **103,** 708 (1965).
[5] M. Hirose and Y. Kano, *Biochim. Biophys. Acta* **251,** 376 (1971).
[6] B. Sebille, N. Thuaud, and J. P. Tillement, *J. Chromatogr.* **167,** 159 (1978).

tained the technique is still subject to several potential pitfalls, most of which can be adequately circumvented. (1) The technique is not applicable to ligands that adsorb strongly to the gel; however, this can be avoided by changing, for example, from dextran to polyacrylamide. A small adsorption of ligand to the columns (that is, a retardation of its elution beyond the position of small molecules) should not cause problems once the column is well equilibrated with ligand. (2) When the ligand under study is charged, the Donnan effect must be taken into account, as in the case of equilibrium dialysis. Provided the charge of the macromolecule is known under the conditions of the experiment, the uneven distribution of ions in both compartments can be calculated[7] and corrected for. This effect can be practically abolished by working with dilute macromolecule solutions in moderately high concentrations of neutral salt.[7] (3) The volume occupied by the macromolecule can be a substantial portion of the outer gel compartment and therefore the small solute concentration can be reduced. This effect may cause an artifactual decrease in ligand concentration associated to protein elution, so that in cases of no interaction an apparent negative binding would be measured. This effect can be corrected if the partial specific volume of the macromolecule is known, or made very small by the use of low macromolecule concentrations ($<1\%$). (4) Large binding effects can lead to substantial depletion of ligand from the inner gel compartment so that the actual free ligand concentration of the protein solution is smaller than the concentration used to equilibrate the gel. This may be noticed by a relatively large peak surrounded by ligand concentrations smaller than in the rest of the effluent. This effect is easily avoided using a larger column or, as above, low protein concentrations.

Experimental Procedures

The procedures and examples discussed below are those with which the author is most familiar and proceed from work with the protein tubulin. However, they are easily applied to other systems.

Sample Preparation and Chromatography

Samples containing 1 to 4 mg of protein and a chosen total ligand concentration in the desired buffer (final volume < 1 ml) were prepared by careful dilution of concentrated protein and ligand solutions (this can be made gravimetrically if great precision is required in order to measure binding from the trough) were taken to temperature and applied to $0.9 \times$ 25-cm Sephadex G-25 columns previously equilibrated with the same

---

[7] C. Tanford, "Physical Chemistry of Macromolecules," p. 221. Wiley, New York, 1961.

buffer of identical ligand concentration. Column temperature was controlled to $\pm 0.2°$ by means of water jackets and a circulating bath. The column flow was not interrupted during sample application and was kept constant during the whole experiment by means of adjustable LKB peristaltic pumps. The binding time (taken as the mean chromatographic elution time of the protein) could be varied among different experiments, in order to ensure attainment of equilibrium, between 5 and 100 min, with an accuracy of $\pm 5\%$ by simply changing the pump setting. Fractions of 1.05 $\pm$ 0.05 ml were collected and the protein was determined spectrophotometrically, subtracting any small contributions of bound ligand when necessary. Ligand concentrations in the effluent were measured as described below. Experiments were performed at different free ligand concentrations in order to obtain a binding isotherm, which was then analyzed according to the pertinent binding scheme. This was done at different temperatures and the apparent thermodynamic parameters of the binding reaction measured by van't Hoff analysis of the data.[8,9]

Measurement of Labeled Ligand Concentration

The radioactive ligand concentration was measured throughout the column effluent by means of carefully taken aliquots (using a 0.5-ml delivery pipet that afforded a reproducibility of $\pm 0.25\%$, as determined by weighing buffer and protein solutions) that were added to 10 ml of aqueous counting scintillant and counted twice to a statistical counting error smaller than 0.3% (95% confidence) in a liquid scintillation spectrometer. Duplicate aliquots were taken in the peak region. The baseline counts per minute were determined from the regions outside the peak and trough typically to a standard deviation <0.5% of the absolute value. The radioactive ligand concentrations in the effluent had to be very close to that applied to the column. Experiments without stable baseline in the vicinity of the peak or not giving a good separation of peak and trough were discarded. The amount of bound ligand was calculated from the measured increment in eluate radioactivity coupled to protein elution; this had to be coincident, within experimental error, with the absolute value of the decrement in the trough. The standard deviation of the baseline was taken as an estimate of the standard deviation of measurements of bound ligand. As an example of the sensitivity of this procedure, Fig. 1 shows an equilibrium binding gel chromatography experiment in which the weak binding of [³H]tropolone methyl ether to tubulin was measured [under these conditions a binding equilibrium constant of $(7.5 \pm 2) \times 10^2\ M^{-1}$ was

[8] J. M. Andreu and S. N. Timasheff, *Biochemistry* **21,** 534 (1982).
[9] J. M. Andreu, M. J. Gorbunoff, J. C. Lee, and S. N. Timasheff, *Biochemistry* **23,** 1742 (1984).

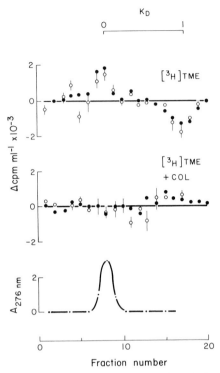

FIG. 1. Interaction of tropolone methyl ether (TME) with tubulin. The results of a typical equilibrium binding Sephadex G-25 chromatography experiment with 1.05×10^{-4} M [^{3}H]TME (1.4×10^5 cpm ml^{-1}) in 10 mM sodium phosphate buffer 0.1 mM GTP, pH 7.0 at 18° are shown by the open circles. The changes in ligand concentration of the upper profile were produced by chromatography of 3.2 mg of tubulin, while the middle profile was obtained under identical conditions after the addition of 10^{-4} M colchicine (COL) to column and sample; the vertical bars show the standard deviation of the measurements. The filled circles are the average of three (upper profile) and two (middle profile) runs, which reduce noise and show unequivocally the position of the peak and trough. Reproduced from Ref. 8.

obtained[8]]. These binding measurements involved ligand concentration increments in the order of only 1% of the absolute value and the specificity of these effects was verified by their inhibition by colchicine (see Fig. 1), a high affinity ligand that binds to the same site.[8] The purity of the ligands used was carefully checked since it is well known that the presence of radiochemical impurities in the labeled ligand preparation employed may lead to large errors that have been discussed by others.[10]

[10] S. E. Builder and I. H. Segel, *Anal. Biochem.* **85**, 413 (1978).

Measurement of Ligand Concentration by Light Absorption

The ligand concentration in the column effluent is often measured spectrophotometrically. This procedure is subject to the same general considerations as radioactive counting. In particular, it relies frequently on the measurement of small absorbance increments due to bound ligand and is therefore subjected to several potential sources of error: interference of light absorption by the protein, absorption increments due to protein effects on bound ligand, interfering light scattering by the protein and instrumental noise. In our case, the ligand concentration measurements were made at a wavelength where the protein does not absorb light. The second effect was not large and could be easily overcome if needed by measuring the ligand absorbance at an isosbestic point of the protein–ligand interaction difference spectrum.[9] The third source of error is a systematic one and is important in proteins which, like tubulin, have a tendency to aggregate. The second and third effects were easily avoided by careful addition of a small volume of concentrated sodium dodecyl sulfate to each fraction, giving a final concentration of 0.4% detergent; this displaced the protein–ligand interaction and solubilized any aggregated protein (giving typically $A_{350} \leq 0.001$ in the absence of ligand). Since the detergent may also interact with the ligand, the extinction coefficients of the protein and ligand in detergent solution were determined independently. It was also ascertained that the detergent was in excess in the protein-containing fractions of the effluent (that is, above its effective critical micelle concentration in the presence of protein) since otherwise artifactual absorption increments could be generated. Finally, noise was reduced by carefully repeated measurements through the whole column effluent, leading to a good statistical estimation of the baseline concentration. The absorption increments due to bound ligand were estimated by subtraction of the baseline values as in the case of radioactive counting. When small absorbance increments had to be measured this was done from recordings of the difference spectra of the peak fractions versus those from the baseline.[9] Figure 2 shows the elution profile generated by the interaction of 1.8 mg of tubulin chromatographed in a column equilibrated with $1.63 \times 10^{-5}\ M$ 2-methoxy-5-(2,3,4-trimethoxyphenyl)tropone at 25° and its inhibition by $2 \times 10^{-4}\ M$ podophyllotoxin, a ligand that does not absorb light at the wavelength employed and binds to a partially overlapping site.[8,9] Typically, these measurements were useful for ligands binding to protein with equilibrium constants $\geq 10^3\ M^{-1}$. Any other procedure to measure ligand concentration that is found to give good accuracy can, in principle, be employed.

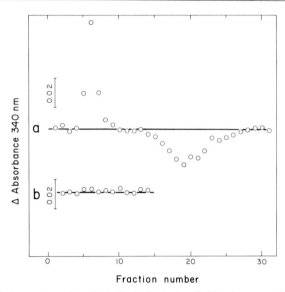

Fraction number

FIG. 2. Gel chromatography of tubulin in Sephadex G-25 columns equilibrated with 2-methoxy-5-(2′,3′,4′-trimethoxyphenyl)tropone. (a) Tubulin (1.8 mg) was chromatographed in a column equilibrated with 1.63×10^{-5} M ligand in 10 mM sodium phosphate buffer 0.1 mM GTP pH 7.0 at 25°. (b) An identical experiment in which the elution buffer contained 1.63×10^{-5} M ligand and 2×10^{-4} M podophyllotoxin. The flow rate was 40 ml/hr and fractions were approximately 1.5 ml. Taken from Ref. 9.

Experimental Criteria of Applicability; Abnormal Profiles; Protein Self-Association

There are several easily verifiable characteristics to ensure attainment of equilibrium in the Hummel and Dreyer technique.[1-3] First, and most important, the ligand concentration immediately before and after the protein peak must return to baseline, so that the peak and the trough are separated by a region of nil concentration increment. Any other type of profile may lead to incorrect measurements. For example, a tailing peak that does not separate from the trough indicates imperfect equilibrium. When the ligand concentration used in the sample is increased or decreased, the rest of the experiment remaining the same, this will be reflected by a corresponding change in the trough, but the ligand increment of the peak must remain constant. Second, the number of moles of ligand bound per mole protein must be independent of small changes of the chromatographic flow, otherwise the separation time is too close to the time needed to attain equilibrium. Third, the binding measured must be independent of protein concentration. This is easily checked by loading

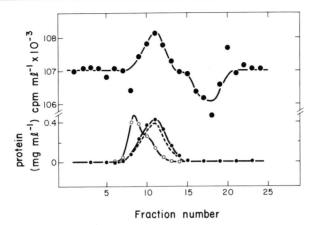

Fraction number

Fig. 3. Interaction of octylglucoside with tubulin. Tubulin (1.5 mg) was chromato-graphed in a 0.9×26-cm Sephacryl S-300 column equilibrated with 2.1×10^{-2} M [^{14}C]oc-tylglucoside in 10 mM sodium phosphate 0.1 mM GTP pH 7.0 at $25°$. Three identical experiments were averaged to obtain the upper radioactivity elution profile. In the lower part (•) is the protein elution profile, (–––) is the protein elution profile without detergent, and (○) the protein elution profile with 4.0×10^{-2} M octylglucoside. From Ref. 13.

different protein amounts on the columns and also, within the same run, by determining the binding separately in each protein containing fraction and looking for any trend with protein concentration. A dependence of ligand binding on protein concentration indicates protein self-association linked to ligand binding[11] and therefore the measurements no longer reflect intrinsic ligand binding to monomeric protein, but a combination of the bindings to all macromolecular species. Ignoring such linkage effects can lead to large errors, as has been shown, for example, with the tubu-lin–vinblastine system.[12] On the other hand, ligand-induced protein self-association can also give abnormal Hummel and Dreyer profiles. Besides performing the pertinent self-association studies, gross self-association effects induced by ligand may be detected during the binding measurements. For this purpose a chromatographic support that partially includes the protein is used. Figure 3 shows the ligand concentration profiles obtained by interaction of the detergent [^{14}C]octylglucoside with tubulin in Sephacryl S-300 at $25°$.[13] The amphiphile concentration employed was 2.1×10^{-2} M, just below its critical micelle concentration and the protein elution profile (filled circles) was not significantly different from one ob-

[11] J. Wyman, *Adv. Protein Chem.* **19**, 223 (1964).
[12] G. C. Na and S. N. Timasheff, *Biochemistry* **19**, 1355 (1980).
[13] J. M. Andreu, *EMBO. J.* **1**, 1105 (1982).

tained without detergent. Increasing the detergent concentration led to an abnormal ligand elution profile in which the peak and trough were not well separated (not shown) together with the bimodal protein profile shown by the open circles in Fig. 3, where most of the protein was no longer included but eluted in the void volume of the column, indicating the formation of large aggregates.

Acknowledgment

I wish to thank Dr. A. Hargreaves for his critical reading of the manuscript.

[19] Ultrafiltration in Ligand-Binding Studies

By ALKIS J. SOPHIANOPOULOS and JUDITH A. SOPHIANOPOULOS

Studies of the reversible interaction of small molecules with macromolecules have played a major role in the unraveling of biological mechanisms. Equilibrium dialysis has been virtually the only method for measuring free ligand directly. The introduction of permselective membranes and related apparatus has facilitated various operations with biological materials such as desalting and concentrating of macromolecular solutions; their use in binding studies however has been limited.

There are two major approaches to binding studies using semipermeable membranes. In ultrafiltration, a portion of the contents of the macromolecular solution is filtered through a membrane impermeable to the macromolecule. No components are added to the macromolecular solution in this process. In diafiltration, simultaneous to the filtering process, components are added to the macromolecular solution to maintain its volume nearly constant.

It was thought that ultrafiltration would give only approximate values of free ligand in binding studies and thus was not used in highly precise binding studies. It was shown recently that ultrafiltration is theoretically equivalent to equilibrium dialysis.[1] The conclusion[1] was that during ultrafiltration, the concentration of free ligand remains constant, although the concentration of the other components increases. During ultrafiltration of a system at equilibrium, although the concentration of the macromole-

---

[1] J. A. Sophianopoulos, S. J. Durham, A. J. Sophianopoulos, H. L. Ragsdale, and W. P. Cropper, *Arch. Biochem. Biophys.* **187**, 132 (1978).

cule–ligand complex increases, the total amount of it in the cell remains constant. Since no net uptake or removal of ligand takes place, the rate of removal of the filtrate does not disturb the equilibrium.

The Gibbs–Donnan equilibrium affects ultrafiltration as it does equilibrium dialysis. The results will also be affected by concentration-dependent processes such as self-association that may depend on ligand binding. To the extent that two or more ligands bind independently, both can be measured simultaneously. Conversely, if there are reciprocal effects, ultrafiltration could be used to study such effects.

In ultrafiltration the quantity determined experimentally is the concentration of free ligand. Thus, ultrafiltration is suitable for systems in which an appreciable amount of the total ligand in solution is bound. If the free ligand is more than 90% of the total ligand, the noise of the data will be too high. In such cases, it would be advisable to use a method that would determine directly the complex, or ligand bound. The experimental conditions for the use of ultrafiltration in titrations with ligand have been investigated recently.[2] We shall describe the method and we shall discuss briefly some of its variations used by others.

Quantities To Be Determined in Binding Studies

A few comments are necessary concerning the system to be studied and the experimental quantities necessary to describe it. A general scheme involves the binding of one or more molecules of a small ligand species L to a macromolecular component P.

$$P + L \rightleftharpoons PL$$
$$PL_{m-1} + L \rightleftharpoons PL_m \tag{1}$$

where m is the number of ligand molecules bound per macromolecule. The binding constant at any step may be written as

$$K_m = \frac{[PL_m]}{[PL_{m-1}][L]} \tag{2}$$

A general scheme would be one where n_j ligand molecules bind to sites of equal affinity and there are i sites of different affinity. The experimental quantity ν, the number of moles of ligand bound per mole of macromolecule, is given by

$$\nu = \frac{[L_T] - [L]}{[P_T]} = \sum_{i,j} \frac{n_j[L]}{1/K_i + [L]} \tag{3}$$

[2] J. A. Sophianopoulos, A. J. Sophianopoulos, and W. C. MacMahon, *Arch. Biochem. Biophys.* **223**, 350 (1983).

where $[L_T]$, $[L]$, and $[P_T]$ are the concentrations of total ligand, free ligand, and total macromolecule, respectively. Equation (3) contains two experimental quantities, ν and $[L]$. Thus, in principle, the various constants involved could be determined by measuring ν as a function of free ligand concentration. If all molecules bind with equal affinity, Eq. (3) may be reduced to

$$\nu/[L] = nk - K\nu \qquad (4)$$
$$[L]/\nu = 1/[nK] + [L]/n \qquad (5)$$

In the above equations, K is the association constant and n is the maximum number of moles of ligand L bound per mole of protein.

Equation (4) is known as the Scatchard equation and Eq. (5) is known as the Eadie–Hofstee equation. Our primary interest here is to describe how to obtain the two quantities ν and $[L]$. Some remarks will be made about analyzing the data by using equations such as Eqs. (4) and (5). There are several excellent reviews about analyzing binding data.[3-5] Such reviews may be consulted for more detailed information.

Experimental Procedures

Equipment and Experimental Set-Up

Ultrafiltration cells are available from several companies, and the kind of cell to use may depend on some specific requirement of the system being studied. Manufacturers supply detailed instructions for the use and care of the cells and such information will not be repeated here. A simple cell has an inlet for applying pressure, usually by means of compressed nitrogen. The sample can be added through a separate inlet, or by disconnecting the gas inlet. The solution must be stirred with a magnetic stirrer without damaging the ultrafilter. Some kinds of cells are more prone to leak than others. Perhaps the most important criterion for choosing a cell is its relative freedom from leaks, after proper assembly.

An important characteristic of the cell is to have a small void volume. Physically, the void volume is the volume from the surface of the membrane which is in contact with the solution to the tip of the solvent outlet tube. In practice, because of diffusion of small molecules during stirring, the void volume is somewhat different. One way to establish an experimental value for it is given in the algorithm (Fig. 2). The void volume can

[3] G. Weber, in "Molecular Biophysics" (B. Pullman and M. Weissbluth, eds.), p. 369. Academic Press, New York, 1965.

[4] F. W. Dalquist, this series, Vol. 48, p. 270.

[5] J. R. Cann, this series, Vol. 48, p. 299.

be reduced in a variety of ways, such as by using outlet tubes with capillary diameters.

The kind of membrane to be used must not bind either macromolecule or ligand to any appreciable extent. Binding of a macromolecule would occur only on the membrane surface of the filter. To test the membranes, a macromolecular solution 10-fold or more dilute than will be used in a titration is added to the ultrafiltration cell and without any applied pressure, it is allowed to stir for at least the length of time required to complete the experiment. The solution is removed and its volume measured to correct for concentration changes due to the passage of some solvent through the membrane. The corrected concentration is compared to the original one. The ligand may bind both to the membrane and to its matrix support, that is, the whole filter. Binding of ligand to the filter is determined conveniently by suspending the filter in a small volume of a highly diluted ligand solution in a covered dish or beaker. The concentration of ligand in the presence and absence of the filter is compared.

Decrease of the rate of ultrafiltration may be caused by concentration of the macromolecules onto the membrane surface. This is called membrane polarization. As the solvent passes through the membrane under pressure, the macromolecules cannot penetrate the membrane and concentrate near its surface. This can be minimized by adequate stirring. The cells cannot be immersed in a thermostat bath. Thus, to carry out studies at other than room temperatures, constant temperature rooms or constant temperature boxes must be used.

The whole experiment may be divided into four parts: (1) preparations for the titration; (2) the titration itself; (3) experimental operations after the titration; and (4) calculation of the results. In order to explain the details in these four parts, we summarize the overall procedure.

In our study,[2] we measured the binding of saccharides and metal ions to concanavalin A (Con A).[6] We used the following apparatus and conditions. An Amicon Model 8MC stirred cell was used with the microvolume accessory but without the volume-reducing insert. The solvent compartment of the cell was shut off. Also, the Amicon Model 3 was used. Outlet tubes with capillary diameters were employed. Thus, the void volume was reduced to 0.05 ml or less. PICG membranes from Millipore Corp. were employed, and there was no detectable binding of either protein or metal ions. In general, the flow was 0.05 to 0.1 ml/min at less than 1 psi pressure. The binding experiments were carried out at 23 to 25°.

The titration with ligand is discontinuous, or in steps. Usually, the

[6] Abbreviations used: Con A, concanavalin A; MAGlu, methyl α-D-glucopyranoside; tris, tris(hydroxymethyl)-amminomethane.

macromolecule and ligand of some initial concentration are incubated, and a known volume of the mixture is added to the cell. In our experiments 3–5 ml of the mixture was added. A summary of the procedure for titrating with ligand is as follows. The effluent fractions containing free ligand were collected in small 1- to 2-ml tubes that were stoppered after collection. The volumes of these fractions were determined by weight and the calculated density of the solutions. At each step of the titration with ligand, a fraction, called Gi, of approximately 0.3 ml was first collected to flush out the void volume. The amount of free ligand in these Gi fractions must be accounted for in the calculations. Two fractions of 0.5 to 1 ml were then collected and used for determining the free ligand concentration at that step. For the next step, the desired volume of stock ligand, and enough solvent to restore the protein solution to approximately its original volume, were added with microsyringes accurate to 0.002 ml. The solution was stirred for 1 to 5 min before collecting fractions again, at that step. Each titration consisted of four to five such steps; thus, during the titration, 12 or 15 fractions were collected.

At the end of the experiment, the protein solution was removed from the cell and its volume, protein, and total ligand concentration were determined. These measurements served as a check for a possible error that might have occurred during the titration. In addition, the far UV (200 to 250 nm) CD spectrum of the protein solution was measured, to determine whether any denaturation took place during the experiment.

In order to establish proper experimental conditions and investigate possible sources of error, several titrations of Con A with methyl-α-D-glucopyranoside (MAGlu) were carried out.[2] Part of the data is given in Fig. 1 and in the table. The fact that in some experiments the value of n, the maximum number of moles of ligand bound per mole of protein monomer, was greater than 1 was discussed and analyzed further mathematically.[2]

Preparations for the Titration

The concentrations of macromolecule and stock ligand to be used in the titration must be decided upon. This assumes some knowledge of the magnitude of the binding constant. If no such information is available, exploratory experiments with widely varying ratios of ligand to macromolecule ought to be carried out. It was pointed out[7] that in order to obtain reliable results, the concentration of the ligand binding sites should be within plus or minus an order of magnitude of the value of the dissociation constant of the complex. This criterion may be used for arriving at an

---

[7] R. E. Buller, W. T. Schrader, and B. W. O'Malley, *J. Steroid Biochem.* **7**, 321 (1976).

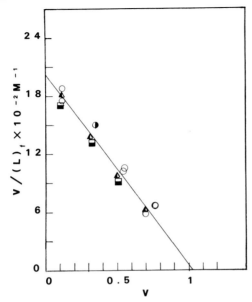

FIG. 1. Scatchard plot of MAGlu (L) binding to Con A, at 23 to 25°. ν is the moles of ligand (MAGlu) bound per mole of Con A monomer of M_r 25,500. Circles: data for titration No. 4, listed in the table; Squares: data for titration No. 5, listed in the table. The buffer for both of these titrations was 0.2 M NaCl, 0.05 M acetate buffer, pH 5.2. Triangles: Data for titration No. 7, listed in the table; the buffer was 0.2 M NaCl, 0.02 M Tris, pH 7.1. Open circles, squares, and triangles indicate a single point of a titration; half-filled symbols indicate two points of the same titration that were virtually identical. All solutions contained 0.5 mM Mn^{2+} and 1 mM Ca^{2+}; the Con A concentration varied between 0.2 and 0.5 mM.

optimum concentration of the macromolecule. The above reference[7] as well as other reviews[3–5] may be consulted for arriving at experimental conditions that would produce statistically reliable results. It should be pointed out that during the titration, usually fractions of a milliliter of the stock solution of ligand are added to raise the total concentration of ligand to some desired level, at each step. To arrive at the concentration of stock ligand to be prepared and the approximate volumes of it to be added at each step, one might proceed as follows. According to Eq. (3), the value of ν depends only on [L]. By using some value of an "overall" binding constant and for n, [L] is calculated for different desired values of ν. The values of concentration of the total ligand at each value of ν and the corresponding [L] are calculated by using the definition of ν in Eq. (3). To keep the calculations of such estimates simple, the amount of free ligand that would be removed at each titration step may be ignored.

ASSOCIATION OF METHYL α-D-GLUCOPYRANOSIDE WITH CONCANVALIN A

		Equation (4)		Equation (5)	
N^a	pH^b	$K_a \times 10^{-3} M^{-1}$	n	$K_a \times 10^{-3} M^{-1}$	n
1	5.2	1.56 ± 0.09	1.14 ± 0.04	1.74 ± 0.19	1.10 ± 0.05
2	5.2	1.70 ± 0.14	1.22 ± 0.05	1.70 ± 0.07	1.20 ± 0.04
3	5.2	2.19 ± 0.33	1.18 ± 0.09	1.83 ± 0.23	1.28 ± 0.12
4	5.2	1.93 ± 0.15	1.08 ± 0.04	2.09 ± 0.14	1.03 ± 0.05
5	5.2	2.00 ± 0.19	0.97 ± 0.04	2.00 ± 0.04	0.97 ± 0.02
6	5.2	1.59 ± 0.08	1.40 ± 0.02	1.58 ± 0.03	1.40 ± 0.03
7	7.2	1.98 ± 0.08	1.02 ± 0.02	1.88 ± 0.04	1.05 ± 0.02
8	7.1	2.14 ± 0.12	1.01 ± 0.03	2.32 ± 0.09	0.97 ± 0.02

[a] N, Number of the titration; each titration consisted of 8 to 10 measurements of free ligand.

[b] The buffer was 0.2 M NaCl, and either 0.02 M acetate at pH 5.2, or 0.01 M Tris at pH 7. The temperature was 23 to 25°. For each titration, the association constants and n, the maximum number of moles of ligand bound per Con A monomer of 25,500 molecular weight, were calculated by a linear least-squares solution of Eq. (4) or Eq. (5). Errors shown are standard deviations. In all studies, Mn^{2+} was 0.5 mM, Ca^{2+} was 1 mM, and Con A 0.2–0.5 mM.

The Titration

It would be helpful to prepare a table with appropriate column headings under which are entered the volumes of stock ligand and solvent to be added at each step and also of the ultrafiltrate volumes to be collected; blank columns are included in the table to record the actual volumes of ligand and solvent added and of the collected ultrafiltrates during the titration.

A high degree of accuracy is required. The volumes of the ultrafiltrates collected are determined by weight and the known density of the solvent. The small stoppered containers in which the ultrafiltrates were collected were weighed before and after the titration. Microsyringes were used to add the stock ligand and solvent at each step. Enough solvent was added at each step to restore the volume of the solution to nearly its original volume. The volume of the solution of the macromolecule added to the cell was determined either by using an accurate volumetric device such as a microsyringe, or by weight. The density of the macromolecule solution can be calculated by the formula

$$\rho = \rho_0 + (1 - \bar{v}\rho_0)*[P] \qquad (6)$$

where $\bar{v}$ and [P] are the partial specific volume in ml/g and macromolecule concentration in g/ml, respectively, and ρ and ρ_0 are the densities of the macromolecule solution and solvent, respectively.

A serious problem is a possible leak of the assembled cell. Before the titration we tested for leaks by adding some solvent without ligand or macromolecule and applying a little pressure. If the membrane is not positioned properly, the solvent will "filter" out immediately. If there were no leaks, we wiped the inside of the cell with some filter paper and proceeded with the titration. Whenever possible, the first few ultrafiltrate fractions ought to be checked for leakage of macromolecule. The volume of the solvent taken up by the filter must be accounted for in the calculations. A good value for this volume may be obtained by weighing an ultrafilter dry, wetting it, wiping off the excess solvent, and reweighing it.

The actual procedure for the titration was summarized above. An accurately measured volume of solution containing a known initial concentration of total macromolecule $[P_T]$ and total ligand $[L_T]$ is added to the cell. The solution must be stirred long enough for equilibrium to be reached. As already pointed out above, the rate of collecting the ultrafiltrate fractions does not disturb the equilibrium. The first three fractions are collected before adding any more ligand. Then stock ligand and solvent are added to restore the solution to nearly its original volume. After addition of titrant, the first fraction collected, called here a Gi fraction, is collected for the purpose of flushing out the void volume. The millimoles of ligand in this Gi fraction must be accounted for in the calculations. Without adding any more stock ligand, two more ultrafiltrate fractions are collected. If equilibrium had been reached before collecting the two fractions, the concentration of ligand should be the same in both of these fractions. The ligand concentration in these fractions is equal to the free ligand concentration in the cell. The cycle of adding ligand to the cell and collecting three fractions is repeated, usually three or four times.

It is not always necessary to collect two fractions at each step to determine the free ligand concentration; in some cases one such fraction might be sufficient. The two fractions serve as duplicates, and also to check if equilibrium was reached after adding more ligand. A possible time dependence of a reaction may affect the results of ultrafiltration, as follows. If binding were incomplete before collecting these two fractions, the concentration of free ligand should be greater in the first fraction. If little binding took place during the titration, no increased binding would be observed after successive additions of ligand. If the reaction were fast but irreversible, no free ligand would be detected until saturation of the binding sites had occurred.

To check for possible errors during the titration, one may determine the macromolecule and total ligand concentrations in the cell at the end of the titration. The volume of the final solution may be determined after withdrawing the solution from the cell using a syringe or pipet with a polyethylene or other plastic tip. Its measured volume should be approximately equal to (but usually less than) the calculated volume. The volume of the solution can be calculated by accounting for the volumes of ligand and solvent added and of the fractions collected. The final molarity of total ligand is estimated in a similar manner. If the algorithm is used, these values are printed in the last line of the output. The measured quantities should be fairly close to the calculated ones.

The volumes of the fractions are determined, preferably by weight, as described. Then the concentration of ligand in the fractions is determined by some suitable method. The DATA input in the algorithm requires a single entry for ligand added before collecting a Gi fraction. The molarity of ligand added at each Gi step is calculated by dividing the mmoles of stock ligand added by the sum of the volumes of stock ligand and solvent added.

Calculations of [L] and v and Use of the Algorithm

Although the calculations are simple, they are lengthy. A single error may invalidate the rest of the calculations. Thus, we have included a short algorithm in BASIC (Fig. 2) that could be used to calculate [L] and v. This is a "core" algorithm in that only the lines necessary for the calculations are included. It requires that quantities such as the volume and concentration of ligand added at a step have been precalculated. We shall comment on the main statements of the algorithm. This would also illustrate how to carry out the calculations without the aid of the algorithm.

There are three slightly different versions of the algorithm which depend on the state of the filter and the void volume, as follows.

Version 1. It is assumed that the filter is wet and the void volume is full of solvent without ligand. This condition would exist if the cell had been tested for leaks before adding the macromolecule solution; not all of the solvent is removed under pressure, so the filter is wet and the void volume is full of solvent. Excess solvent is wiped off from inside the cell. Version 1 is the one given in Fig. 2.

Version 2. The filter is wet when the cell is assembled but it is not tested for leaks. Thus, the void volume is empty. To produce Version 2, the following changes need be made to Version 1: (1) substitute line 35 with the line: 35 LET L = G − F; (2) substitute line 39 with the line: 39 LET R = A9*V9/(V9 − F).

Version 3. In this case the filter used is dry and the void volume is empty. To produce Version 3, substitute in Version 1 line 35 with the line: 35 LET L = G.

In all three versions, it is assumed that some known amount of ligand is initially present in the solution. It is also assumed that at least two ultrafiltrate fractions are collected before adding any more ligand. Three such fractions ought to be collected before adding any ligand, if duplicate fractions are collected to determine [L] at each step.

The data for the first two fractions are treated differently from the rest. The differences in the three versions occur in the calculations involving the first two fractions. In all three versions, the values of the initial conditions and for the first two filtrates are read in line 20. In line 30 are calculated P, the total millimoles of the macromolecule, which remain constant throughout the titration, and A, the total millimoles of ligand initially in the solution. Lines 35–40 calculate the concentrations in the cell "before collecting the 1st fraction." This means that the void volume, G, is full but that no fraction has been collected. In Version 1 this is true before separating any solvent and free ligand from the macromolecule solution. In Versions 2 and 3 this condition would be achieved if enough solvent with free ligand was drawn from the solution to fill the void volume.

We shall discuss first Version 1. In line 35, L is set to zero. L here does not stand for ligand; it is the volume of solvent that is drawn from the macromolecule solution to fill the void volume, to achieve the condition "before collecting the 1st fraction." In the other two versions it has a value other than zero. In line 38 the volume, H, of the solution is calculated. In Version 1, H is equal to the original volume added to the cell. The macromolecule concentration, D, is also calculated. In line 39, R stands for the concentration of free ligand in the cell, and is set equal to A8, the ligand concentration in the second fraction. Nothing is added before collecting the first two fractions, therefore A8 should be equal to [L] originally in the cell.

When the first fraction is collected, an equal volume of solvent with ligand is removed from the cell. However, the ligand concentration, A9, in the first fraction is less than that drawn from the cell. The reason is that the solvent without ligand in the void volume, G, is contained in the volume of the first fraction, V9. R may also be defined by LET R = A9*V9/(V9−G). The reasons for using A8 instead will be discussed with the calculations for the Gi fractions.

In line 40, Y is the total number of millimoles of ligand removed from the solution. In Version 1, none is removed up to that point. U is the total ligand concentration, which also remains unchanged up to that point; B is the concentration of ligand bound. The value of ν is given by X. Lines 350–380 print the calculated quantities.

This is Version 1. A format for entering the data is:

```
900 DATA N,T0,P0,V0,G,F
910 DATA V(1),A(1),V(2),A(2)
920 DATA V(I),A(I),0,0
930 DATA V(I),A(I),K(I),M(I)
```

In line 900 are entered: The number of ultrafiltrate fractions, N; the total initial molarity of ligand, T0; the total initial molarity of protein, P0; the initial total volume of the solution titrated, V0; a value for the void volume, G; a value for the volume of solvent trapped by the membrane, F. In line (910) are entered the volume, V9, and ligand concentration, A9, of the first fraction collected, and the same quantities, V8, A8, for the second fraction collected. Each subsequent DATA statement contains the values: V(I), A(I), K(I), M(I). K(I) is the total volume and M(I) the concentration of ligand of the titrant added. V(I) and A(I) are the volume and ligand molarity of the fraction collected after adding the titrant. If no titrant was added before collecting a fraction, then "0,0" are entered for K(I) and M(I). Enter 0, 0, 0, 0 as the final DATA line.

```
00010 DIM A(30), V(30),P(30),O(30),Z(30),Y(30),B(30)
00011 DIM C(30),X(30),S(30),T(30),K(30),M(30),Q(30),G(30)
00020 READ N,T0,P0,V0,G,F,V9,A9,V8,A8
00030 LET P=P0*V0:LET A=T0*V0
00035 LET L=0
00038 LET H=V0-L:LET D=P/H
00039 LET R=A8
00040 LET Y=R*L:LET U=(A-Y)/H:LET B=U-R: LET X=B/D
00050 LET V1=H-V9:LET P1=P/V1:LET Y1=Y+R*V9
00060 LET A1=(A-Y1)/V1:LET B1=A1-R: LET X1=B1/P1
00350 PRINT"CONCNS IN CELL, BEFORE COLLECTING THE 1ST FRACTION:",
00370 PRINT"TOTAL PROTEIN="; D,"TOTAL LIGAND=";U,"FREE LIGAND=";R
00380 PRINT"BOUND LIGAND";B,", NU = ";X
00390 PRINT "CELL CONCNS. AFTER COLLECTING 1ST FRACTION:"
```

FIG. 2. An algorithm in BASIC.

```
00400 PRINT "TOT. PROTEIN="; P1,"TOT. LIGAND=";A1,"FREE LIGAND=";R

00410 PRINT "BOUND LIGAND=";B1," NU=";X1

00450 LET V2=V1-V8:LET P2=P/V2:LET Y2=Y1+V8*A8:LET A2=(A-Y2)/V2

00500 LET B2=A2-A8: LET X2=B2/P2

00540 PRINT "CONCNS. IN CELL AFTER COLLECTING 2ND FRACTION:"

00550 PRINT "TOT. PROTEIN=";P2,"TOT. LIGAND=";A2,"FREE LIGAND=";A8

00560 PRINT "BOUND LIGAND=";B2," NU=";X2

00572 PRINT " N"," PROT. CONC.";" TOT. LIG C.";" FREE LIG. C.";

00575 PRINT " BOUND LIG. C.";" NU"

00578 FOR J=3 TO N

00579 READ V(J),A(J),K(J),M(J)

00580 NEXT J

00581 LET Q(2)=V2:LET P(2)=P2:LET Y(2)=Y2:LET T(2)=A2:LET B(2)=B2

00586 LET X(2)=X2:LET A(2)=A8

00590 FOR I=3 TO N-1

00592 IF K(I)=0 THEN GOTO 740

00600 LET Q(I)=Q(I-1)-V(I)+K(I):LET A=A+M(I)*K(I):LET P(I)=P/Q(I)

00620 LET Y(I)=Y(I-1)+V(I)*A(I+1):LET T(I)=(A-Y(I))/Q(I)

00630 LET C(I)=T(I)-A(I+1): LET O(I)=C(I)/P(I):PRINT I;"G NEXT A",

00660 PRINT P(I),T(I),A(I+1),C(I),O(I)

00680 LET G(I)=Y(I-1)+(V(I)*A(I)-G*A(I-1))*V(I)/(V(I)-G)

00690LETZ(I)=(A-G(I))/Q(I):LETS(I)=(V(I)*A(I)-G*A(I-1))/(V(I)-G)

00700 LET C(I)=Z(I)-S(I):LET O(I)=C(I)/P(I)

00730 PRINT I;"G DIFF",P(I),Z(I),S(I),C(I),O(I):GOTO 810

00740 LET Q(I)=Q(I-1)-V(I):LET P(I)=P/Q(I):LET Y(I)=Y(I-1)+V(I)*A(I)

00770LETT(I)=(A-Y(I))/Q(I):LETB(I)=(A-Y(I)-A(I)*Q(I))/Q(I)

00780 LET X(I)=B(I)/P(I)

00800 PRINT I;"AFTER",P(I),T(I),A(I),B(I);" ";X(I)

00810 NEXT I

1500 END
```

FIG. 2. (*continued*)

The quantities for "after collecting the 1st fraction" are calculated by lines 50–60. In line 50 the new volume, V1, is equal to the previous one, H, less the volume of the first fraction, V9. The value Y is increased by the millimoles removed, V9*R, to give Y1. The reason why R rather than A9 is used was explained above. In line 60 the new concentrations of total ligand, A1, bound ligand, B1, and v, X1, are calculated. Lines 390–410 print the calculated quantities.

Version 2 differs from Version 1 in lines 35 and 39. This version is for the case where the filter is initially wet but the rest of the void volume is empty. It should be noted that the volume of the filter that can be occupied by solvent is considered as part of the void volume, G. Thus, solvent of volume (G − F) is drawn from the cell to arrive at the condition "before collecting the 1st fraction," where F is the volume of solvent trapped in the membrane. Therefore, in Version 2, L is defined by the line: 35 LET L = G − F. The calculations in lines 38–60 are the same as in Version 1, except for line 39. In line 39, R stands for [L] in the cell and is calculated by multiplying the ligand concentration in the first fraction, A9, by the factor V9/(V9 − F); this accounts for the fact that a volume fraction F of V9 contained no ligand. Alternately, one could set R = A8, as in Version 1.

Version 3 differs from Version 1 in that the filter is initially dry and the void volume is empty. Only line 35 differs, where L is set equal to the void volume, G.

The rest of the calculations in all versions are identical. In line 450, the volume of the second fraction, V8, is subtracted from the solution volume, V1, to obtain the new volume, V2. The millimoles of ligand removed previously, Y1, is increased by adding to it the millimoles of ligand removed with the second fraction, V8*A8, to give Y2. Y2 is then subtracted from the original millimoles of ligand, A, and the new total ligand concentration, A2, is calculated. Lines 540–560 print the calculated quantities.

The output of the calculations that follow is printed in tabular form. The algorithm proceeds to perform one of two alternate series of calculations, depending on whether the fraction is a Gi one. This decision is made in line 592. If no titrant is added, that is, if it is not a Gi fraction, execution is directed to line 740. In this case, the concentration, A(l), of ligand in the fraction would be equal to [L] in the cell. In line 740, the new solution volume, Q(l), macromolecule concentration, P(l), and overall sum of millimoles of ligand removed, Y(l), are calculated.

It should be pointed out that throughout the algorithm, two quantities are used to account for the total millimoles of ligand. One of these is A. Initially, A is the total millimoles in the cell; the value A is increased every time ligand is added (see line 600). The other quantity is Y(l); Y(l) is increased by adding to it the millimoles of ligand removed with each fraction, as in line 740. The total millimoles of ligand present after collecting a fraction is the difference (A − Y(l)). See line 770. The calculations for a fraction which is not a Gi one are completed by calculating in line 770 the concentrations of total ligand, T(l), ligand bound, B(l), and in line 780 the value of ν, X(l).

The calculations for the Gi fraction begin on line 600. The volume Q(l − 1) is increased by the volume, K(l), of the ligand which is added before collecting the Gi fraction; from (Q(l) + K(l)) is subtracted the volume, V(l), of the collected Gi fraction to give the new solution volume, Q(l). The millimoles of ligand, A, is increased by the millimoles of ligand added, M(l)*K(l); then, the new macromolecule concentration, P(l), is calculated. Because of the uncertainty of the value of the void volume, the calculations that follow are carried out in two different ways.

The Gi fraction may be thought of as being the sum of two volume fractions, one of volume equal to the void volume, G, and of ligand concentration equal to A(l − 1); the remainder is of volume (V(l) − G) and of ligand concentration equal to A(l + 1), that is, the concentration in the fraction collected after the Gi fraction. However, all of the solution removed from the cell by collecting the Gi fraction contains free ligand of concentration A(l + 1). This is the proper value of [L]. With this in mind, we proceed to examine the two ways of calculating the data for the Gi fractions.

The first way is given in lines 620–660. In line 620, the ligand concentration, A(l + 1), in the fraction collected after the Gi fraction is used to calculate the millimoles of free ligand removed from the solution, V(l)*A(l + 1). The millimoles of ligand removed is added to the previous sum, Y(l − 1), to obtain the current value of Y, Y(l). This value for Y(l) is used in calculating the concentration of total ligand, T(l). In line 630 the concentration of bound ligand, C(l), is calculated by subtracting A(l + 1) from T(l). We consider this calculation more reliable than the one that follows, because it is relatively independent of the value of G. Thus, the value of Y(l) obtained here is used in the calculations of the fractions that follow.

The second way is given in lines 680–730. It should be noted that because the calculated values here are not used in any of the calculations that follow, variables different from the ones used above for the various quantities are defined. In this section of the algorithm for example, the equivalent of Y(l) is G(l). In line 680 the millimoles of ligand flushed out with the void volume is G*A(l − 1). If this is subtracted from the millimoles of ligand in the Gi fraction, V(l)*A(l), the result is the millimoles of ligand removed from the cell in a volume (V(l) − G). To find the millimoles in the volume removed from the cell, that is, V(l), the above number of millimoles is multiplied by the factor $V(l)/(V(l) - G)$. The resulting value is added to $Y(l - 1)$ to give $G(l)$. In line 690 the $G(l)$ value is used to calculate the total ligand concentration, $Z(l)$. In a similar manner the free ligand concentration, $S(l)$, is calculated. The values calculated are used mainly to check the accuracy of the value used for the void volume. If G is accurate, all quantities printed in the lines labeled "G NEXT A" and "G DIFF" should agree. If fairly large differences are observed between these two lines throughout the output, a new value for G should be entered and the data reprocessed. After the calculations for the Gi fraction are completed, the cycle is repeated via line 810.

A Comment on Analyzing the Results

Detailed analysis of complex systems has been dealt with in several studies.[3-5] Systems that fit a linear regression may be analyzed by the Scatchard equation, Eq. (4). A simple system where only two molecules of ligand bind per macromolecule, one with an association constant K_1 and the other with an association constant K_2, may be analyzed by one of the following equations[2]:

$$Y = 1/(K_1 + K_2) + XK_1K_2/(K_1 + K_2) \tag{7}$$

where $Y = [L](1/\nu - 1)$, and $X = (1 - 2/\nu)[L]$.

$$U = 1/(K_1K_2) + Z(K_1 + K_2)/(K_1K_2) \tag{8}$$

where $U = (2/\nu - 1)[L]^2$ and $Z = [L](1 - 1/\nu)$.

It has been pointed out[2] that in least-squares analysis, where $y = f(x)$, the dependent variable, y, should preferably be the quantity that is least reliable, if the error in y is being minimized. In the case where the complex is measured, the more appropriate equation would be Eq. (4). When free ligand is measured, as is the case here, the more appropriate equation would be Eq. (5).

Related Applications of Ultrafiltration

The effect of pH on ligand binding may be studied in an analogous manner. Instead of ligand, acid or base is added to the solution in the cell, and the resulting pH is measured with a thin combination electrode. In an exploratory study,[1] it was found that the association constant of manganous ion to Con A decreased sharply with decreasing pH. In pH studies, however, where dissociation of the complex may occur as the pH is changed, the rate of dissociation of the complex may be too slow and equilibrium may not be reached within the time of one step of the titration. In such cases, studies of increasing pH as well as decreasing pH are necessary.

Besides studies of titrations with ligand, ultrafiltration has been applied recently in clinical studies, such as in the determination of free thyroxine and triiodothyronine in serum[8] and in determining unbound cortisol in plasma.[9]

Besides equilibrium studies, it is possible to use ultrafiltration in kinetic studies. A limitation is that changes in the rate of binding must occur in a time span of the order of a minute. In such studies, macromolecule solution without ligand is loaded into the cell and stock ligand is added just before collection of ultrafiltrate. This kind of study is especially useful if titration data indicate that equilibrium is reestablished slowly.

Other Recent Methods

Diafilitration was the earliest method using permselective membrane filters to be applied to binding studies.[10] In diafiltration, ligand is added from a reservoir connected to the cell, simultaneous to the filtering process. Thus, the volume is kept constant. The addition of ligand continues until the concentration of ligand in the effluent is equal to that in the reservoir.

Another approach was described recently.[11] The experimental design is nearly identical to that in diafiltration, except that small fractions of effluent are collected while the concentration of ligand in the effluent is changing, that is, before a constant level of free ligand concentration is reached. Utilizing the establishment of ultrafiltration as a thermodynamically valid method, the authors derived equations applicable to the analysis of data by their method.[11] They also described the required modifications of the cell. This method might be called "continuous" ultrafiltration in contrast to the "discontinuous" one described here. The changes of ligand concentration must be small at each step; however, a large number of samples are collected during each experiment. There is no provision to establish that equilibrium has been reached at each step. This deficiency might be overcome by carrying out "discontinuous" ultrafiltration experiments at select ligand concentrations.

[8] J. Sophianopoulos, I. Jerkunica, C. N. Lee, and D. Sgoutas, *Clin. Chem. (Winston-Salem, N.C.)* **26**, 159 (1980).
[9] I. Jerkunica, J. Sophianopoulos, and D. Sgoutas, *Clin. Chem. (Winston-Salem, N.C.)* **26**, 1734 (1980).
[10] W. F. Blatt, S. M. Robinson, and H. J. Bixler, *Anal. Biochem.* **26**, 151 (1968).
[11] K. B. Roy and H. T. Miles, *Biochemistry* **21**, 57 (1982).

In our study,[1] we had assumed that the concentration of all the components remained uniform throughout the solution in the ultrafiltration cell, because the solution was stirred. Further theoretical exploration of the process shows that it is not necessary that the concentration remains uniform. Thus, the protein and total ligand may vary, as when using an unstirred device, and the concentration of free ligand will still be that at equilibrium. Various unstirred devices are available. Some are simply unstirred cells, where the effluent is pushed out by gas pressure. In others, the effluent is removed by centrifugation. At present, the possible sources of experimental error must be investigated with each device chosen. The choice of membrane filters available is restricted at present. Application of unstirred devices has been mainly in clinical assays.[12-14]

[12] I. Vlahos, W. MacMahon, D. Sgoutas, W. Bowers, J. Thompson, and W. Trawick, *Clin. Chem. (Winston-Salem, N.C.)* **28**, 2286 (1982).
[13] W. MacMahon, J. Stallings, and D. Sgoutas, *Clin. Biochem.* **16**, 240 (1983).
[14] W. MacMahon, J. Thompson, W. Bowers, and D. Sgoutas, *Clin. Chim. Acta* **131**, 171 (1983).

[20] Protein Chromatography on Hydroxyapatite Columns

By Marina J. Gorbunoff

The use of hydroxyapatite (HA) columns for protein chromatography was introduced by Tiselius *et al.* in 1956.[1] Its systematic study was undertaken by Bernardi,[2,3] who examined a variety of proteins using several solvent systems and worked out standard procedures for the operation of HA columns. He also proposed a mechanism for protein adsorption to and desorption from HA, since the principles of ion-exchange chromatography are not applicable to HA chromatography.[2,3]

In a study extending Bernardi's work,[4-6] a number of additional proteins with isoelectric points ranging from pH 3.5 to 11.0 have been examined, the relation between the ionic state of the HA column and protein retention has been scrutinized, and additional solvent systems have been

[1] A. Tiselius, S. Hjerten, and O. Levin, *Arch. Biochem. Biophys.* **65**, 132 (1956).
[2] G. Bernardi, this series, Vol. 22, p. 325.
[3] G. Bernardi, this series, Vol. 27, p. 471.
[4] M. J. Gorbunoff, *Anal. Biochem.* **136**, 425 (1984).
[5] M. J. Gorbunoff, *Anal. Biochem.* **136**, 433 (1984).
[6] M. J. Gorbunoff, *Anal. Biochem.* **136**, 440 (1984).

used to test the effect of the chemical nature of eluants upon the desorption process.[4] Furthermore, the roles of protein fine structure and of specific polar groups have been probed.[5] This last study included proteins in which individual polar groups were modified specifically, as well as those in which very extensive numbers of amino or carboxyl groups were modified with (1) retention of the same charge, (2) inversion of the charge, or (3) annihilation of the charge.[5] As a result of this work, a mechanism of protein–HA interaction was deduced,[6] which was based on three fundamental conclusions:

1. Adsorption and elution cannot be regarded as simple reversals of a single process.
2. Amino and carboxyl groups act differently in the adsorption of proteins to HA.
3. Elutions of basic and acidic proteins by different salts follow different mechanisms.

Adsorption of Proteins to Hydroxyapatite

Amino groups act in the adsorption of proteins to HA as the result primarily of nonspecific electrostatic interactions between their positive charges and the general negative charge on the HA column,[7] when the column is equilibrated with phosphate buffer:

$$HAPO_4^- \cdots H_3^+N-Prot$$

This is consistent with the observations that (1) retention of basic proteins (see Table I), polypeptides,[1,3] and amino acids[3] is controlled by the ionic state of the column, i.e., its net charge (negative in the phosphate cycle; neutral in the NaCl cycle; positive in the $CaCl_2$ or $MgCl_2$ cycle); (2) the lower the pH of the equilibrating phosphate buffer, the higher the molarity required for elution[3,4]; and (3) blocking of carboxyls strengthens the binding of both basic and acidic proteins to HA.[5] This electrostatic interaction is crucial to the binding of basic proteins. It is sufficient to block 5 out of 19 amino groups in α-chymotrypsin to make its retention marginal, while lysozyme with 7 out of 18 groups blocked is not retained on any column (see Table I).

Carboxyl groups act in two ways. First, they are repelled electrostatically from the negative charge of the column. Second, they bind specifi-

---

[7] The surface of HA crystals presents a mosaic of positive (calcium) and negative (phosphate) sites. Since HA columns are normally operated at pH 6.8 after extensive washing with phosphate buffer, the surface of the column can be regarded as negative due to partial neutralization of the positive calcium loci by phosphate ions.

TABLE I

INTERACTION OF PROTEINS AND PROTEIN DERIVATIVES WITH HYDROXYAPATITE COLUMNS IN VARIOUS CYCLES

Protein	Isoelectric point	Nominal net charge	Elution		Column retention
			NaCl	CaCl$_2$	
Lysozyme	10.5–11.0	+7	+	+	Not retained by MgCl$_2$ column; retained by NaCl and PO$_4$
Horse radish peroxidase	7.2		−	−	Not retained by NaCl and CaCl$_2$ columns; retained by PO$_4$ column only
Ovomucoid	3.8–4.5		−	−	Not retained by PO$_4$ column; retained on NaCl column
Trypsinogen	9.3	+7	+	−	Retained by all columns
Maleyllysozyme		−7	+	−	Retained by PO$_4$ and NaCl columns
Acetyltrypsinogen		−7	+	−	
Acetylchymotrypsin		∼ −2	+	−	Not, or very poorly retained by PO$_4$ columns; retained by NaCl columns
Succinylchymotrypsin		∼ −11		−	
Acetylchymotrypsinogen		−10		+	Not retained by PO$_4$ and NaCl columns; retained by MgCl$_2$ and CaCl$_2$ columns
Acetylribonuclease		−7			Not retained by any column
Acetyllysozyme		0–+1			
Lysozyme–AMS^a		+7	+	+	Not retained by NaCl columns; retained by PO$_4$ columns
Trypsinogen–AMS		+7	+	+	
Pepsinogen–AMS		−30			Not retained by NaCl columns; show some retention on PO$_4$ columns
STI–AMS		−11			

^{a} AMS, COOH → CONH$_2$CH$_2$SO$_3$H.

cally by complexation to calcium sites on the column, forming clusters of

$$[HACa—OOC–Prot]$$

This is consistent with the observations that (1) the relation between the ionic (net charge) state of the column and retention of acidic amino acids,[3] polypeptides,[3] and proteins (see Table I) is a mirror image of that of their basic counterparts; and (2) replacement of COOH by SO_3H, which does not affect the charge, prevents this complexation (see Table I). Thus acidic proteins must bind almost exclusively by this complexation, since their ability to bind to HA is lost on COOH → SO_3H replacement. This singularity of COOH–Ca complexation is demonstrated in a particularly striking manner by proteins which have clusters of carboxyls, such as trypsinogen and β-lactoglobulin.[4] A particularly striking example is trypsinogen (see Table II). Trypsinogen is a basic protein which displays retention and elution patterns which are a hydrid of those expected for basic and acidic proteins. The isoelectric point of trypsinogen is 9.3. This causes it to behave as a basic protein, except with respect to elution with $CaCl_2$. It contains, however, a cluster of four aspartate residues in the N-terminal of the molecule. Their blocking or transformation to SO_3H abolishes its inability to be eluted by 3 M $CaCl_2$ and renders it into a normal basic protein. It is interesting to note that this cluster of carboxyls constitutes one of the Ca^{2+} binding sites of this protein. Since retention on the column depends on the cooperative interaction of several carboxyls with HA, a sufficient density of carboxyls is required to generate statistically a cluster complementary to the calcium atoms immobilized in the HA crystals. Therefore, the inability of basic proteins to bind to HA once their

TABLE II
ELUTION OF TRYPSINOGEN FROM HYDROXYAPATITE[a]

| Derivative | Elution molarity | | | | | |
	NaPO$_4$(6.8)	NaF	NaCl	NaClO$_4$	CaCl$_2$	MgCl$_2$
Trypsinogen[b]	0.17	0.23	0.32	0.31	>3.0	0.1
Acetyltrypsinogen	0.004	0.004	0.5		>3.0	0.2
Trypsinogen–COX[c]	0.23	0.41	0.32	0.24	0.003	0.1
AMS–Trypsinogen[d]	0.13	0.18	0.18	0.24	0.001	0.001

[a] The loading columns were always in the pH 6.8 phosphate cycle, except for acetyltrypsinogen for which the column was in the chloride cycle.
[b] N-terminal peptide, Val-Asp-Asp-Asp-Asp.
[c] COX, COOH → $CONHNHCONH_2$.
[d] AMS, COOH → $CONH_2CH_2SO_3H$.

positive charge has been abolished must reflect their insufficient density of carboxyls.

Elution of Proteins from Hydroxyapatite

Basic proteins are eluted either as a result of normal Debye–Hückel charge screening, which operates in the elution by F^-, Cl^-, ClO_4^-, SCN^-, and phosphate, or by specific displacement by Ca^{2+} and Mg^{2+} ions which complex with column phosphates and neutralize their negative charges:

$$HAPO_4^- \cdots {}^+NH_3–Prot + CaCl_2 \longrightarrow HAPO_4–Ca + Cl^- + {}^+H_3N–Prot \quad (1)$$

The former mechanism is most strikingly illustrated by the separation of five isozymes of lactic dehydrogenase, where elution is related directly to charge.[8]

Acidic proteins are eluted by displacement of their carboxyls from HA calcium sites by ions which form stronger complexes with calcium than do carboxyls, e.g., fluoride or phosphate:

$$HACa–OOC–Prot + NaF \text{ (or } PO_4) \longrightarrow HACa–F + Na^+ + {}^-OOC–Prot \quad (2)$$

Since the formation constants for CaF or $CaPO_4 \gg CaOOC$, the displacement occurs at a rather low molarity of fluoride or phosphate. The ineffectiveness of Cl^- as eluant of acidic proteins is due to the fact that it does not form a complex with Ca^{2+} and, thus, cannot compete with the Ca–OOC complexes. The strength of the $CaPO_4$ bonds can be taken to be demonstrated by the unusually strong bonding to HA of phosphoproteins.[9] The behavior of nucleic acids need not contradict this, since they differ from phosphoproteins both in the state of their phosphate groups and their space distribution in the molecule.[3]

The ability of $CaCl_2$ and $MgCl_2$ to strengthen the bonding of acidic proteins to HA is due to the formation of additional bridges between protein carboxyls and column phosphate sites[3,6]

$$HAPO_4^- \cdots Ca^{2+} \cdots {}^-OOC–Prot$$

This can be used to advantage to retain acidic proteins on the column since these are eluted neither by 3 M $CaCl_2$ nor by 3 M $MgCl_2$. The difference in the formation constants of Ca–PO_4 and Mg–PO_4 provides a useful tool for the discrimination between acidic proteins and neutral proteins, i.e., those with isoelectric points between pH 5/5.5 and 8, since the last are eluted by $MgCl_2$ above 1 M, but not by $CaCl_2$.

[8] M. John and J. Schmidt, *Anal. Biochem.* **141**, 466 (1984).
[9] G. Bernardi and W. H. Cook, *Biochim. Biophys. Acta* **44**, 96 (1960).

Guidelines for the Use of HA Columns

The understanding of the principles of protein retention on and elution from HA columns has permitted to establish a set of rules for loading proteins on HA and eluting them from the columns. The choice of column cycle and ionic nature of eluting buffers must be dictated by the charge nature of the protein mixture to be fractionated, as well as the charge (positive, negative, or neutral) of the protein to be isolated. Keeping in mind that (1) basic proteins are not retained by $MgCl_2$ or $CaCl_2$ columns, while acidic proteins are strongly retained by these columns, as well as by NaCl columns, and (2) basic proteins are eluted by very low molarities of $MgCl_2$ and $CaCl_2$, while acidic proteins are not eluted by these solvents and neutral proteins are eluted only by $MgCl_2$ at 1 M, (3) that acidic proteins are not eluted by NaCl, while basic proteins are eluted by that solvent at molarities between 0.1 and 0.3, and (4) that acidic proteins are eluted by NaF and phosphate solvents at low molarity ($\sim$0.1 M), while basic proteins are eluted at molarities similar to those of NaCl, some general guidelines can be proposed.

Loading

Phosphate columns (0.001 M, pH 6.8) for mixtures of predominantly basic proteins, or if it is desired to retain basic proteins on the column, while possibly losing some acidic ones.

NaCl columns (0.001 M, unbuffered) for mixtures of predominantly acidic proteins (glycoproteins with loose structures in particular), or if in a mixture it is essential that all acidic proteins be retained on the column.

$MgCl_2$ or $CaCl_2$ columns (0.001 M, unbuffered) only for acidic proteins which do not bind to NaCl columns.

Washing

The same buffer as that on the column. For PO_4 columns this assures removal of most denatured proteins.

Elution

All elution procedures will give the following order of protein emergence from the column: basic > neutral > acidic. Chromatographic schemes can be developed on the basis of either step or gradient elution. Step elution can be carried out very quickly. The use of gradients, although more time consuming, can provide more refined separation. Since shallow gradients are to be preferred because of better resolution, the use of a gradient in place of step elution might prove at times impractical.

Three general Schemes can be proposed as points of departure. These will be set up either as a series of steps, gradients, or gradients combined with steps. The choice of a step or gradient will frequently be dictated by the intention of the experiment, whether a certain class of proteins is to be eluted in a batch without resolution, or if some resolution is desired.

Step elution	Gradient elution
	Scheme I
I. 0.005 M MgCl$_2$	0.001 to 0.005 M MgCl$_2$ gradient
to elute basic proteins	
II. 1.0 M MgCl$_2$	1.0 M MgCl$_2$ in a step
to elute proteins with isoelectric points between 5/5.5 and 8	
III. 0.3 M PO$_4$	0.01 to 0.3 M PO$_4$ gradient
to elute acidic proteins	
	Scheme II
I. 0.005 M MgCl$_2$	0.001 to 0.005 M MgCl$_2$ gradient
to elute basic proteins	
II. 1 M Na or KCl	0.01 to 1.0 M Na or KCl gradient
to elute proteins with isoelectric points of 7.0–7.6 and DNases[3]	
III. 0.3 M PO$_4$	0.01 to 0.3 M PO$_4$ gradient
to elute acidic proteins	
	Scheme III
I. 1 M NaCl or KCl	1 M NaCl or KCl in a step
to elute basic and neutral proteins, as well as DNases	
II. 0.3 M PO$_4$	0.01 to 0.3 M PO$_4$ gradient
to elute acidic proteins	

Replacement of the 1 M MgCl$_2$ step in Scheme I by a gradient is not likely to offer any advantages. The use of a gradient in place of the 1 M Na/KCl step is meaningful only in Scheme II, but not in Scheme III. Elution with a phosphate gradient should always be preceeded by washing with 0.001 M pH 6.8 phosphate buffer (two column volumes) to remove the salt, at high concentration, left from the preceding step. In the case of NaCl columns one can start with a 0.001/0.01 M PO$_4$, pH 6.8 step, which might cause elution of poorly bound acidic proteins. In general one should avoid the use of CaCl$_2$ eluants, unless very dilute, since these tend to cause plugging of the column.

For the purification of individual proteins a phosphate column and an appropriate gradient are sufficient. For basic proteins a NaCl gradient

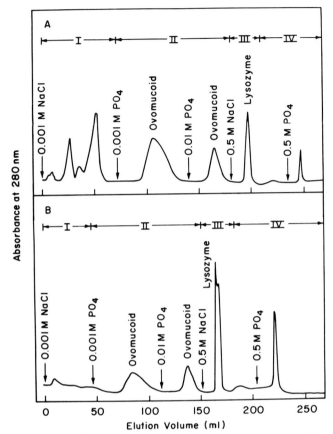

FIG. 1. Purification of two samples of commercial ovomucoid on hydroxyapatite columns. Zone I consisted of inactive material with maximal UV absorption at 260 nm; zone II contained the ovomucoid factions; zone III was lysozyme; zone IV contained the remainder of the impurities, trypsin- and chymotrypsin-active. (Reproduced from Ref. 10.)

(0.01–0.5 M) is one of choice, since it does not elute most neutral or acidic proteins. The following examples might serve as illustration.

1. The *purification of commercial ovomucoid*[10] which contains as impurities lysozyme, ovoinhibitor, conalbumin, and ovalbumin. The elution program is shown in Fig. 1 for two samples of the material. The procedure combines a NaCl column, washed with 0.001 M NaCl and stepwise elution: a 0.01 M PO$_4$, pH 6.8 step to elute ovomucoid (a glycoprotein having a loose structure), a 0.5 M NaCl step to remove basic proteins (lysozyme

---

[10] M. J. Gorbunoff, *J. Chromatogr.* **187,** 224 (1980).

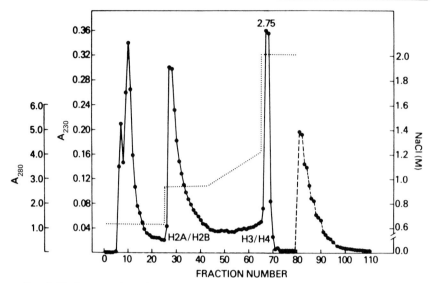

FIG. 2. Hydroxyapatite column chromatography of chromatin. Chromatin containing 34 mg of DNA in 0.63 M NaCl, 0.1 M potassium phosphate, pH 6.7, was loaded onto a 2.5 × 20 cm column, and eluted in 18-ml fractions at 60 ml/hr. The NaCl concentration of the running buffer is indicated by the dotted line. The concentration of potassium phosphate (pH 6.7) was maintained at 0.1 M until tube 79, then stepped to 0.5 M. The optical absorbance of the fractions was determined at 230 nm (solid line) and 260 nm (dashed line). Purified H2A + H2B and H3 + H4 were obtained by pooling fractions 26–42 and 66–69, respectively. (Reproduced from Ref. 11.)

and ovoinhibitor), and a 0.5 M PO$_4$ step to wash off other acidic impurities.

2. The *purification of histone pairs from chromatin* [11] is shown in Fig. 2. This procedure combined a NaCl step at 0.93 M to elute H2A and H2B, then a gradient to 1.20 M NaCl to wash off residual H2A and H2B, and finally a step a 2 M NaCl to elute the arginine-rich histones, followed by 0.5 M PO$_4$ to remove the DNA from the column.

3. The *purification of commercial 2× recrystallized papaya lysozyme* was carried out in both a phosphate and a NaCl gradient, as shown in Fig. 3. In both procedures, several basic impurities are eluted first, followed by the main peak at 0.24 M PO$_4$ and 0.25 M Cl$^-$. The impurities accounted for a considerable fraction of the total protein and the elution profile of crude, as well as 1× crystallized papaya lysozyme was identical to the 2× crystallized material. It is interesting to note that commercial highly puri-

[11] R. H. Simon and G. Felsenfeld, *Nucleic Acids Res.* **6,** 689 (1979).

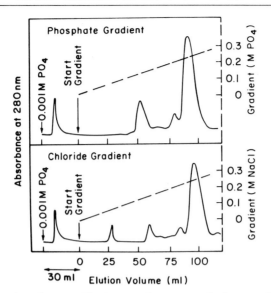

FIG. 3. Purification of recrystallized papaya lysozyme on hydroxyapatite columns. The protein was loaded on a 1 × 20 cm column in 0.001 M PO_4. The appropriate gradient was started after washing with 30 ml of the loading buffer.

fied crystalline soybean trypsin inhibitor was found by this technique to contain up to 20% impurity.

4. The *purification of tubulin* involved the use of a $MgCl_2$ column, washed with 0.005 M $MgCl_2$.[12] This was used primarily out of consideration of tubulin stability, since this highly labile protein is stabilized by Mg^{2+} ions. The procedure, shown in Fig. 4, involved washing with 0.005 M $MgCl_2$ which did not elute any protein, followed by a 0.001 to 0.3 M KPO_4 gradient. Three peaks were eluted. The first peak, which contained 15% of the protein, was identified as partially denatured tubulin which had lost its GTP. It became bound to HA probably because of the use of a $MgCl_2$ column. The second peak (85% of the protein) was pure tubulin which contained 2.0 GTP molecules per tubulin dimer. The third peak contained no protein. It was identified as pure GTP. It seems interesting to remark that when the loading column and washing buffer contained no magnesium ions, the protein came out in a single more diffuse peak. This may be due either to the washing through of the denatured tubulin or to the enhanced binding of the native GTP-containing tubulin to the column in the Mg^{2+} cycle, in view of the high affinity of the GTP for divalent cations.

[12] L. Grisham, M. J. Gorbunoff, B. Price, and S. N. Timasheff, unpublished results.

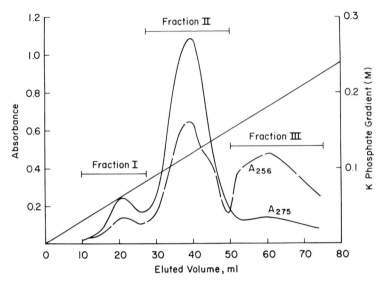

FIG. 4. Purification of Weisenberg tubulin on a hydroxyapatite column. The 1 × 20 cm column was prewashed with 95 ml of 0.005 M MgCl$_2$, 25 mg of protein was loaded on the column and washed with 30 ml of 0.005 M MgCl$_2$. A 0.001 to 0.3 M KPO$_4$ gradient was applied (50 g of each) and 2-ml fractions were collected. The eluted material was monitored at 275 and 256 nm. All operations were performed at 4° due to protein instability and to prevent aggregation at higher temperatures.

NOTE: It has been called to the attention of the author that the chromatographic behavior of commercial hydroxyapatite might differ widely between suppliers. All the studies reported here were performed with hydroxyapatite prepared in the laboratory by the Bernardi procedure[2] and stored in 0.001 M phosphate buffer. This material gave no trouble with flow rates unless it was several months old.

Acknowledgment

This work was supported in part by NIH Grant GM-14603.

[21] Measurement of Ligand–Protein Interaction by Electrophoretic and Spectroscopic Techniques

By Robert W. Oberfelder and James C. Lee

Introduction

Determination of the equilibrium binding constant and the stoichiometry for a protein–ligand interaction requires an appropriate choice of methodology. Each technique has advantages and disadvantages, as well as practical and theoretical limitations, and these must be considered when choosing a procedure. It is important to have a number of approaches available from which to choose, so that techniques can be used which will be compatible with the constraints placed on the system due to the nature of the protein and ligand of interest. Several procedures have been discussed in this series including equilibrium dialysis,[1] Hummel–Dryer chromatography,[2] Womack–Colowick rapid dialysis,[3] and countercurrent distribution.[4] In these methods the formation of a ligand–protein complex is detected by directly determining the difference in ligand concentrations. Indirect approaches utilizing absorbance spectroscopy, fluorescence spectroscopy, or electrophoresis may also be employed to perform binding studies. These studies use changes in the spectral properties of the protein or ligand to measure equilibrium binding constants. The procedures used to perform binding studies utilizing spectroscopy, electrophoresis,[5] and isoelectric focusing[6] will be discussed and the advantages and disadvantages intrinsic to these techniques will also be described.

Spectroscopic Methods

Any spectroscopic technique can be employed to monitor ligand–protein interaction. Utilization of spectroscopic change as a measure of the extent of interaction is based on the explicit assumption that the fractional

[1] U. Westphal, this series, Vol. 15, p. 762.
[2] G. K. Ackers, this series, Vol. 27, p. 441.
[3] F. C. Womack and S. P. Colowick, this series, Vol. 27, p. 464.
[4] G. Kegeles, this series, Vol. 27, p. 456.
[5] E. M. Ritzén, F. S. French, S. C. Weddington, S. N. Nayfeh, and U. Hansson, *J. Biol. Chem.* **249**, 6597 (1974).
[6] J. R. Cann and K. J. Gardiner, *Biophys. Chem.* **10**, 211 (1979).

METHODS IN ENZYMOLOGY, VOL. 117

saturation of binding sites by ligand is linearly proportional to the fraction change in the spectroscopic signal. Such an assumption is the weakest link in these indirect methods of monitoring ligand–protein interaction to obtain binding constants and stoichiometry.

Difference Spectroscopy

Theory. Binding of a ligand to a protein may result in a change in the environment surrounding either the ligand or amino acid residues in the protein. Such a change may perturb the electronic interactions of the ligand or the residues in the protein producing a shift in the absorbance spectrum.[7,8] The difference between the unperturbed and the perturbed spectrum is termed the difference spectrum which can be calculated according to the following equation.

$$A_x (P + L) - A_x (PL) = \Delta A_x \qquad (1)$$

where A_x is the absorbance at wavelength x. (P + L) indicates that the protein and ligand are separate, while (PL) indicates that mixing and presumably binding has taken place. ΔA_x is the difference in the absorbance of the two solutions at wavelength x. The difference spectrum is simply the expression of the ΔA_x values over the wavelength range studied.

Perturbation of the protein can most readily be detected in the 250- to 300-nm range for phenylalanine, tyrosine, and tryptophan. These residues when perturbed have difference spectra which are easily recognized. The spectral shifts may be a result of a direct interaction of the ligand in or around the binding site, or it may be a result of a conformational change distant from the binding site. Association and dissociation of the protein might also produce a spectral shift, but the binding of the ligand would have to be linked to the association or dissociation to affect a change that is ligand concentrations dependent.

The spectrum of the ligand may also be perturbed since it may be shifted from an aqueous environment to a hydrophobic one upon binding to the protein. Interaction of the ligand might also change the degree of ionization and thereby shift the spectrum. Thus, an observed spectral change can be caused by a wide variety of perturbations.

Examination of the observed difference spectrum should reveal a wavelength or a range at which ΔA_x is maximal. This wavelength will provide the maximum sensitivity in the observation of the change; hence, it should be employed in the studies to determine the effect of ligand

[7] P. Bennouyal and C. G. Trowbridge, *Arch. Biochem. Biophys.* **115**, 67 (1966).
[8] P. Cuatrecasas, S. Fachs, and C. B. Antinsen, *J. Biol. Chem.* **242**, 4759 (1967).

concentration on the spectral change. The maximum change in absorbance can then be determined by plotting $1/\Delta A_x$ versus $1/[L]$ and extrapolating to infinite ligand concentration. The y intercept will yield the maximum absorbance change, ΔA_{xm}. The fractional change in the absorbance $\Delta A_x/\Delta A_{xm}$ is assumed to be linearly proportional to the fractional saturation of binding sites, thus,

$$\frac{\Delta A_x}{\Delta A_{xm}} = \frac{\bar{\nu}}{n} \tag{2}$$

where $\bar{\nu}$ and n are the moles of ligand bound and the number of binding sites, respectively.

$\bar{\nu}$ may be determined by conducting another series of experiments. The increment in the ligand molar extinction coefficient due to binding, $\Delta\varepsilon = \varepsilon(\text{bound}) - \varepsilon(\text{free})$, can be determined from extrapolation of a titration of ligand solution of finite concentration with excess protein. Then aliquots of a protein solution are titrated with known total concentrations of ligands, the bound ligand concentration, $\bar{\nu}$, determined as $\Delta A_x/\Delta\varepsilon$ and the free ligand concentration estimated from the difference.

The binding isotherm can be expressed as

$$\bar{\nu} = \frac{n[L]_{\text{free}}}{K_d + [L]_{\text{free}}} \tag{3}$$

where $[L]_{\text{free}}$ and K_d are the free ligand concentration and the dissociation constant, respectively. The binding isotherm can also be expressed in the familiar Scatchard plot format as follows[9]

$$\frac{\bar{\nu}}{[L]_{\text{free}}} = \frac{n - \bar{\nu}}{K_d} \tag{4}$$

The slope of a plot of $\bar{\nu}/[L]_{\text{free}}$ versus $\bar{\nu}$ will yield a slope equal to $-1/K_d$.

An allosteric enzyme may be analyzed using difference spectroscopy as long as the relationship described by Eq. (2) is adhered to. The degree of cooperativity can be estimated since

$$\bar{\nu} = \frac{n [L]_{\text{free}}^{h}}{K_d + [L]_{\text{free}}^{h}} \tag{5}$$

where h is the Hill coefficient,[10] an indication of cooperativity in ligand binding.

Experimental Procedure. 1. The spectral data required for this procedure can be most readily obtained utilizing a double-beam spectrophotom-

[9] G. Scatchard, *Ann. N.Y. Acad. Sci.* **51**, 660 (1949).
[10] A. V. Hill, *J. Physiol. (London)* **40**, 190 (1910).

eter. Two tandem cuvettes will also be necessary to perform these experiments. The UV range is essential for the detection of absorbance changes in the protein and may be useful for detection of absorbance changes in the ligand as well. The visible range is useful in the detection of absorbance changes in the ligand or prosthetic groups associated with the protein. It is also useful to have the capacity to scan as a function of wavelength.

2. A stock protein solution should be prepared so that all of the solutions used in the experiments contain the same protein concentrations. Stock solutions of the ligand should be prepared using the ligand concentrations to be tested. All of these solutions should be prepared in the same buffer to avoid potential complications due to differing buffer compositions.

3. Of the protein stock 1.00 ml is added to one side of the two cuvettes and 1.0 ml of the ligand stock is added to the other side of both cuvettes. One cuvette will be employed as the sample and the other as the reference.

4. Scan the two unmixed cuvettes over a wavelength range encompassing both the range in which the protein and ligand absorb. This scan should be essentially flat, if the two cuvettes are well matched.

5. Mix the sample cuvette thoroughly, then scan again in order to detect spectral changes in the absorbance of the protein or the ligand. The spectral change may be either an increase or a decrease in the absorbance relative to the reference spectra.

6. Steps 3, 4, and 5 should be repeated for each ligand concentration to be tested.

7. Examination of the scans should reveal the wavelength at which the spectral change is maximal.

8. Utilizing the wavelength at which the perturbation was maximal for wavelength "x," the data should be plotted as $\bar{\nu}/[L]_{free}$ versus $\bar{\nu}$ so that the slope can be used to determine K_d in accordance to Eq. (4).

An illustration of the application of difference spectroscopy in monitoring ligand–protein interactions is shown in Fig. 1.[10a] The binding of 2-methoxy-5(2',3',4'-trimethoxyphenyl)-2,4,6-cycloheptatrien-1-one (MTC) to calf brain tubulin induces a difference spectrum. The magnitude of absorbance increment is proportional to the amount of MTC added. The binding isotherm is shown in Fig. 2. It is evident that the binding isotherm determined by difference spectroscopy resembles that determined by the equilibrium method, although the absolute values of $\bar{\nu}$ are lower.

[10a] J. M. Andreu, M. J. Gorbunoff, J. C. Lee, and S. N. Timasheff, *Biochemistry* **23**, 1742 (1984).

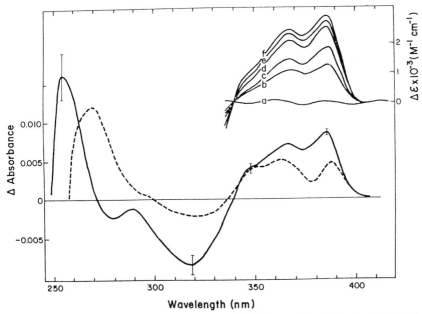

FIG. 1. Difference absorption spectra generated by the interaction of tubulin with MTC in $10^{-2}\ M$ phosphate, $10^{-4}\ M$ GTP (PG buffer) at pH 7.0 and 25°. (——) Difference spectrum of $3.1 \times 10^{-5}\ M$ MTC and $4.0 \times 10^{-6}\ M$ tubulin vs ligand and protein in separate solutions. The difference spectrum (– – –) generated by $1.6 \times 10^{-5}\ M$ tubulin is shown for comparison. The 350 to 400 nm spectra in the upper right hand corner were generated by $9.5 \times 10^{-6}\ M$ MTC with (a) no tubulin, (b) $6.9 \times 10^{-6}\ M$ tubulin, (c) $1.37 \times 10^{-5}\ M$ tubulin, (d) $2.72 \times 10^{-5}\ M$ tubulin, (e) $4.02 \times 10^{-5}\ M$ tubulin, and (f) $6.55 \times 10^{-5}\ M$ tubulin. (Reprinted with permission from Andreu et al.[10a] Copyright 1984 American Chemical Society.)

General Comments. The validity of the results of the difference spectroscopy experiments is entirely dependent upon the relationship described by Eq. (2). One should check the results by performing direct binding experiments to establish that the assumed relationship between the fractional spectral perturbation and the fractional saturation of the protein is correct. The apparent dissociation constant should be determined utilizing several different protein concentrations so that affects due to association–dissociation phenomena may be detected.

Fluorescence Spectroscopy

Theory. Formation of a ligand–protein complex may result in alteration in the fluorescence intensity of the aromatic amino acids of the protein, extrinsic probes or the ligand. Monitoring ligand–protein interaction by fluorescence has one complication in addition to those cited for differ-

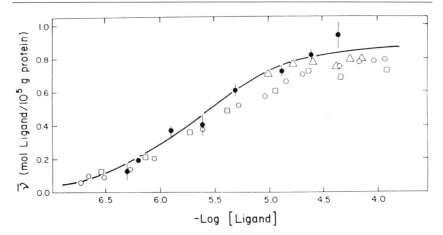

FIG. 2. Binding isotherm of MTC to tubulin; PG buffer pH 7.0, 25°. Solid circles (●) are column measurements at a protein concentration of $(0.5–1.2) \times 10^{-5} M$. Open symbols are binding measurements from ligand fluorescence at (□) $1.8 \times 10^{-6} M$, (○) $5.3 \times 10^{-6} M$, and (△) $3.6 \times 10^{-5} M$ protein. The solid line is a fit to the column measurements ($K_b = 4.8 \times 10^{-5}$ M^{-1}; $n = 0.88$) obtained from a Scatchard plot of these data. (Reprinted with permission from Andreu et al.[10a] Copyright 1984 American Chemical Society.)

ence spectroscopy, the inner filter effect. Fluorescence intensities are proportional to concentrations over a narrow range of optical densities. If the titrant in the measurement of ligand–protein interaction absorbs light at the excitation wavelength, then the intensity of the exciting light will decrease with each addition of the titrant, thus, an artifactual decrease in emission light intensity will be observed. Various methods have been proposed to correct for the inner filter effect and one of the simplest procedures is[11]

$$F_{corr} \cong F_{obs} \text{ antilog } \frac{OD_{ex} + OD_{em}}{2} \tag{6}$$

where F_{corr} and F_{obs} are the corrected and observed fluorescence intensity, respectively and OD_{ex} and OD_{em} are the optical density of the sample at the excitation and emission wavelengths, respectively. The best solution to inner filter effect is, however, to employ samples of low optical density, i.e., less than 0.05.

Having corrected for the inner filter effect, quenching of the intrinsic protein fluorescence by ligand can be employed to estimate the binding affinity. The maximal fluorescence quenching by excess ligand is mea-

[11] J. R. Lakowicz, "Principles of Fluorescence Spectroscopy." Plenum, New York, 1983.

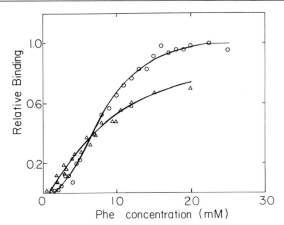

FIG. 3. The binding of Phe to rabbit muscle pyruvate kinase at pH 7.5 and 23°. The symbols and source of data are (△) equilibrium dialysis data (reproduced with permission from Kayne and Price[12a]) and (○) fluorescence data. (T. G. Consler and J. C. Lee, unpublished data.)

sured, then the fraction of sites occupied, θ, must be assumed to be equal to the fraction of the maximal quenching effect at a given total ligand concentration. The binding equilibrium constant can be determined employing the relationship[12] $\theta/(1 - \theta) = K_b [A]$, where [A] is the free ligand concentration calculated from the total ligand and protein concentrations and θ is the fraction of the protein in bound form, assuming a binding stoichiometry, if it has not been determined independently by other means.

Experimental Procedure. 1. To further limit the complication caused by inner filter effects small fluorescence cells of 0.5 × 0.5 cm or smaller can be used.

2. A stock ligand solution should be prepared. The concentration of the stock solution should ideally be about 100- to 500-fold that of the final ligand concentration to be tested. This will ensure minimal dilution of the protein solution, which should not be diluted by more than 5%.

An illustration of the application of fluorescence spectroscopy in monitoring ligand–protein interaction is shown in Fig. 3.[12a] The binding of L-phenylalanine to rabbit muscle pyruvate kinase induces a change in the intensity of the intrinsic protein fluorescence. Assuming the change in fluorescence intensity is directly proportional to ligand–protein complex formation, a binding isotherm can be defined. The sigmoidal nature of the

[12] S. S. Lehrer and G. D. Fasman, *Biochem. Biophys. Res. Commun.* **23**, 133 (1966).
[12a] F. J. Kayne and N. C. Price, *Arch. Biochem. Biophys.* **159**, 292 (1973).

binding isotherm indicates a significant degree of cooperativity in the binding of phenylalanine to the enzyme. Comparison of the fluorescence data with data obtained using equilibrium analysis (Fig. 3) shows that the two methods indicate different degrees of cooperativity. Clearly the isotherms are dependent upon the methods used to obtain them. Thus additional experiments must be performed in order to explain the discrepancy. It has been shown that pyruvate kinase undergoes a cooperative global structural change upon binding of phenylalanine, so the change in the fluorescence intensity may reflect both the structural change and ligand binding. These results illustrate the major limitation in using indirect methods to monitor ligand–protein interaction.

Electrophoretic Methods

A majority of methods available to monitor ligand–protein interaction rely on separation of the ligand–protein complex from free ligand. Thus, transport methods are involved, such as the chromatographic procedure of Hummel and Dreyer,[13,14] the gel exclusion procedure of Hirose and Kano,[15] and the sedimentation procedure of Steinberg and Schachman.[16] In this section, methods utilizing electrophoresis will be discussed.

Steady-State Polyacrylamide Gel Electrophoresis

Theory. Polyacrylamide gel electrophoresis at steady-state conditions can be employed to obtain reliable information on the equilibrium constant and stoichiometric relationship in ligand–protein interaction, as demonstrated by Ritzén et al.[5]

The principle utilized in this procedure is essentially the same as that of the chromatographic procedure of Hummel and Dreyer. The limitation is, however, that the ligand of interest should be noncharged so that the electric field does not affect the distribution of the ligand. The ligand is introduced into the polyacrylamide gel matrix by including it in the solution during the polymerization process, which should result in uniform distribution throughout the gel. As the protein migrates through the gel, it encounters a constant concentration of ligand and it will bind an amount of ligand which will be dependent upon the free ligand concentration stoichiometry and the binding constant. The amount of ligand bound to the protein will continue to increase until the free concentration of ligand

[13] J. P. Hummel and W. J. Dreyer, *Biochim. Biophys. Acta* **63**, 530 (1962).
[14] G. F. Fairclough and J. S. Fruton, *Biochemistry* **5**, 673 (1966).
[15] M. Hirose and Y. Kano, *Biochim. Biophys. Acta* **251**, 376 (1971).
[16] I. Z. Steinberg and H. K. Schachman, *Biochemistry* **5**, 3728 (1966).

in the region of the protein is equivalent to the initial free ligand concentration. Once this state has been achieved, there will no longer be any net exchange of ligand between the gel and the protein. This should lead to the observation of a peak in the ligand concentration for ligand associated with the protein, while a trough should be observed in the early portion of the gel. Addition of a large amount of protein may deplete the gel of ligand, so that net binding of the ligand occurs during the entire period of electrophoresis of the protein, thus making determination of bound and free ligand difficult. The attainment of equilibrium may be checked by determination of the concentration of ligand both in front of and behind the protein peak. The concentrations should be the same if equilibrium has been attained.

In order to determine the equilibrium binding constant, the concentration of the free ligand in the gel is required. The amount of ligand added to the solution prior to polymerization is known, and the total volume of solution is also known. The concentration of the ligand in the gel should be equivalent to the initial concentrations in the solution prior to polymerization. The free ligand concentration, C_L, should then be

$$C_L = L/V \tag{7}$$

where L is the amount of ligand added, and V is the total volume of liquid added. The calculated value can be verified experimentally. The density of the solution used in polymerizing the gel, excluding acrylamide should be measured. A slice of gel is then blotted free of excess moisture and weighed. The amount of ligand present in the slice (L) is then quantitated. The ligand concentration is

$$[L] = (L) \left[\frac{(S)}{G - AG} \right] \tag{8}$$

where $[L]$ is the concentration of the ligand, G is the total weight of the gel, A is the fraction of acrylamide in the gel by weight, and (S) is the density of the polymerization solution excluding acrylamide in grams/liter. It is assumed in this calculation that the ligand concentration is not appreciably affected by acrylamide, i.e., the ligand is neither bound by acrylamide nor preferentially excluded.

The amount of bound ligand must also be determined. The amount of ligand in each gel slice is measured so that one can assess the magnitude of the peak in the ligand concentration in the region of the protein. The free ligand concentration for each slice of gel can be determined from the region immediately in front of or behind the protein peak. The amount of ligand bound per mole of protein, $\bar{\nu}$, can be calculated using the following equation

$$\bar{\nu} = \frac{\Sigma L_B - \Sigma L_F}{p} \qquad (9)$$

where L_B is the total amount of ligand in the gel slices in the region of the protein, L_F is the amount of free ligand expected over the gel slices in which L_B was determined, and p is the amount of protein in the gel.

A range of free ligand concentrations must be tested and the values for $\bar{\nu}$ and $[L_F]$ determined. The data can then be plotted in the form of a Scatchard plot[9] to determine the stoichiometry and the equilibrium binding constant.

Experimental Procedure. 1. Preparation of the polyacrylamide tube gels is carried out according to the procedure described in detail by Davis,[17] omitting the stacking gel. Elimination of the stacking gel abolishes potential complication which might arise from the passage of the protein through zones containing two different hydrogen ion concentrations. A 1-cm space is left above the running gel so that the protein can be layered on top. The weight percent of the acrylamide can be varied to suit the protein, but 7% is generally satisfactory. The density of the running gel solution without the acrylamide should be determined so that the free ligand concentration can be calculated based upon Eq. (8). The ligand is included in the solution during polymerization of the running gel to ensure uniform distribution. The ligand must not be modified during the polymerization process, and this should be checked by extracting it from the gel followed by characterization by analytical procedures such as NMR, mass spectrometry, or high-pressure liquid chromatography.

The ligand used should be readily detectable in small quantities. Radioactive ligand provides such sensitivity.

2. The protein should be dissolved in buffer with 10% glycerol. The solution may also include the ligand at a concentration equivalent to that employed in the gel. Preequilibration of the protein with the ligand facilitates formation of the complex. The protein containing solution should be layered through the reservoir buffer and on top of the polymerized gel.

3. The electrophoresis apparatus should be water jacketed and the temperature of the system carefully controlled using a circulating constant temperature bath. Regulation of the temperature is important since the equilibrium constant may be significantly affected by temperature changes.

4. The time and current may be varied to match the time required for equilibrium to be achieved and to allow for the stability of the protein.

5. Once electrophoresis is concluded, the gel is removed from the tube as discussed by Davis[17] and sliced into small sections, approximately

[17] B. J. Davis, *Ann. N.Y. Acad. Sci.* **121**, 404 (1964).

2 mm in height, using a gel slicer. The gel slices can then be placed in one of the commercially available solubilizers such as Soluene 350 from Packard, then added to Insta-Fluor (Packard) and the amount of radioactive ligand determined using a liquid scintillation counter. The observed counts per minute are corrected for quenching and the amount of ligand in each slice determined from the specific radioactivity of the ligand. The gels should also be stained according to any of the procedures described by Wilson[18] to determine the location of the protein peak.

6. A plot of the moles of ligand versus the slice number should show a peak in the amount of ligand at the position of the protein. The peak should be flanked by regions in which the amount of ligand present in the slices is equivalent to the amount expected, based upon the original concentration of ligand in the gel. The electrophoresis time should be extended or the amount of protein loaded on the gel should be decreased until the flanking regions are clearly detectable.

7. A range of ligand concentrations should be tested utilizing a single protein concentration so that enough data can be amassed to determine the bound and free ligand concentrations. The free ligand concentration can be determined using Eq. (7) based on the initial ligand concentration in the solution used to prepare the running gel. Equation (8) should also be applied after determination of G, S, and L to check the free ligand concentration. The amount of bound ligand is then determined utilizing Eq. (9) after determination of ΣL_B in the region of the protein, ΣL_F and the total amount of protein added to the gel. Once the free and bound ligand concentrations have been determined, the data can be plotted in the form of a Scatchard plot.

General Comments. Results obtained using the steady-state polyacrylamide gel electrophoresis technique show good agreement with those determined by equilibrium methods.[5] The precision of the technique is estimated to be ±4%.

A potential advantage of this technique is that it is not necessary to employ a homogeneous protein sample. As long as the target protein can be resolved from the other proteins, a binding isotherm is estimated. Furthermore, simultaneous demonstration of several binding components is feasible in one sample under identical experimental conditions. Ritzén et al.[5] have demonstrated the simultaneous binding of dihydrotestosterone to albumin, corticosteroid-binding globulin and testosterone-binding globulin, as shown in Fig. 4. By this technique it is possible to measure binding to specific components over a wide range of binding affinities and to quantitate each binding component for which the binding constant is known.

[18] C. M. Wilson, this series, Vol. 91, p. 236.

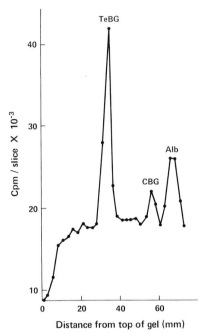

FIG. 4. Fractionation of prepubertal monkey serum by steady-state polyacrylamide gel electrophoresis. Serum diluted 1 : 10 was incubated for 16 hr at 0° with 10^{-8} M [^{3}H]dihydrotestosterone. One hundred microliters (corresponding to 0.7 mg of protein) was applied on a 6.5% polyacrylamide gel (10 × 70 mm) containing 2 nM [^{3}H]dihydrotestosterone. Steady state between association and dissociation after 3.5 hr of electrophoresis is shown by the equal levels of free radioactivity in front and behind the peaks. TaBG is testosterone-binding globulin. (Reprinted with permission from Ritzén *et al.*[5] Copyright 1974. The American Society of Biological Chemists, Inc.)

The technique can be successfully applied for quantitative measurements when the interaction between ligand and protein is in rapid equilibrium relative to the migration rate of the protein. To ascertain that this critical condition is fulfilled, it is necessary to conduct the experiment as a function of voltage across the gels. If the system is in rapid equilibrium, identical results should be obtained regardless of voltage.

Counterion Electrophoresis

Theory. The method of steady-state counterion electrophoresis was developed by Ueng and Bronner[19] to monitor the binding of ionic ligands to proteins. The basic theory is very similar to that for steady-state electrophoresis.

Experimental Procedure. 1. The ionic ligand is added to the lower

[19] T.-H. Ueng and F. Bronner, *Arch. Biochem. Biophys.* **197**, 205 (1979).

buffer chamber of a conventional disc electrophoresis apparatus,[17] while the protein sample is applied to the gel in the upper buffer chamber.

2. Under the influence of the applied electric field, the ionic ligand and protein will migrate in opposite directions, leading to complex formation.

3. A steady state is reached when the concentrations of ligand in front and behind the protein zone are identical.

4. The gel is sliced and the amount of ligand in each slice quantitated. Since in most cases the ligand employed is radioactive, it can be quantitated by counting.

5. Repeating the experiment as a function of ligand concentration, a binding isotherm can be established and the data analyzed to obtain the stoichiometry and apparent binding constant.

General Comment. This method has been applied to the study of calcium binding to rat intestinal calcium-binding protein,[19] as shown in Fig. 5. It is evident that there are two calcium binding components, band 1 and 2. A binding isotherm can be established for these two components by varying the amount of calcium in the buffer chamber. Figure 6A shows the linear relation between the amount of calcium added and the amount of calcium detected in the baseline. Figure 6B shows the binding isotherm for band 1 and band 2. The data were further analyzed by double reciprocal plots to yield apparent binding constants (Fig. 6C).

Cann and Fink[20] presented a theoretical analysis of this method and show that the apparent binding constant is highly dependent on protein concentrations. Thus, the apparent binding constant should be determined as a function of protein concentration. The value of the apparent binding constant at infinite dilution of protein is actually the product of a kinetic factor and the intrinsic binding data.

Isoelectric Focusing

Intrinsic ligand binding constants may be obtained for some protein–ligand interactions utilizing isoelectric focusing. This procedure is useful for interactions in which the binding of ligand results in a change in the isoelectric point of the protein, the theory of which has been discussed by Cann and co-worker.[6,21] Since the ligand free species can be distinguished electrophoretically from the liganded protein, the relative proportion of each can be determined. Utilizing a range of ligand concentrations, one should observe a ligand concentration-dependent shift in the position of the protein from the position of the unliganded species to that of the saturated protein. The position of the center of mass of the protein is assumed to be linearly proportional to the amount of ligand bound. Quan-

[20] J. R. Cann and N. H. Fink, *Biophys. Chem.* **21**, 81 (1985).
[21] J. R. Cann, *Biophys. Chem.* **11**, 249 (1980).

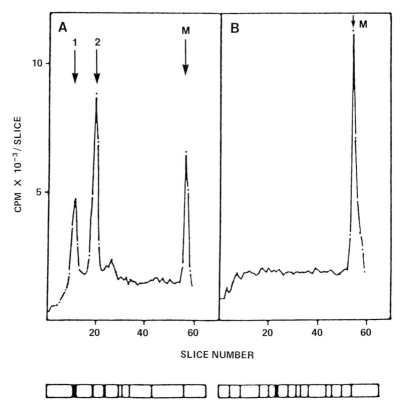

FIG. 5. Binding activity profiles and protein staining patterns of partially purified cal-cium-binding protein obtained from mucosal scrapings from animals on a low calcium (I) or a low calcium, vitamin D-deficient (I minus D) regimen. The specific activity was 33 nmol Ca_{bound}/mg protein from diet I animals (A) and 8 in material from animals on diet I minus D (B). Protein samples of 150 μg were electrophoresed with the anode buffer (700 ml) contain-ing 0.4 μM $CaCl_2$; 13,400 cpm ^{45}Ca/nmol; 0.25 ml of 0.014% bromphenol blue had been added to the cathode buffer (300 ml). Following electrophoresis gels were either stained or sliced (1 mm/slice) and the slices assayed for radioactivity. The protein patterns have been sketched from the destained gels. Protein migration was from left to right, toward the anode (increasing slice number); Ca^{3+} migrated from right to left toward the cathode. Arrows designate calcium-binding protein in band 1 and band 2 and the position of the marker dye (M, bromphenol blue). The baseline radioactivity of slices 36 to 50 (A) averaged 1480 cpm/slice; the binding activity in band 1 or band 3 was estimated to be 0.16 or 0.37 nmol Ca_{bound}, respectively, by dividing the radioactivity in the peak area above the baseline value by the specific radioactivity. Note the absence of bands 1 and 2 in the material from the D-deficient animals (B). (Reprinted with permission from Ueng and Bonner.[19] Copyright 1979, Aca-demic Press.)

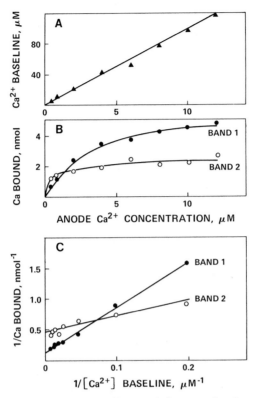

FIG. 6. Calcium binding of calcium-binding protein in counterion electrophoresis. Protein (200 μg), prepared from mucosal scrapings of animals on diet I, with a specific activity of 47 nmol Ca_{bound}/mg protein, was applied to the gel, with the anode buffer calcium concentration varied from 0.4 to 12 μM, while the radiospecific activity varied from 11,200 to 42,500 cpm ^{45}Ca/nmol. The cathode buffer contained 250 μl of 0.015% bromphenol blue. Following electrophoresis the gels were stained or sliced (0.8 mm/slice) and assayed. (A) Baseline calcium as a function of calcium concentration at the anode. Baseline calcium levels were estimated from the mean baseline counts divided by the buffer specific radioactivity. (B) Calcium content of bands 1 and 2 as a function of calcium at anode. The calcium content was estimated from the difference between baseline and peak area of bands 1 and 2. (C) Estimation of apparent dissociation constants. The graph was constructed from the values for baseline calcium (ordinate, A) and band calcium content (ordinate, B). (Reprinted with permission from Ueng and Bonner.[19] Copyright 1979, Academic Press.)

titative evaluation of the ligand concentration dependence of the position of the protein can be used to determine the equilibrium binding constant for the ligand.

Theory. The equilibrium between the protein, P, and the ligand, L, to form the complex, PL, is illustrated below:

$$P + L \rightleftharpoons PL$$
$$PL + L \rightleftharpoons PL_2$$
$$PL_{i-1} + L \rightleftharpoons PL_i$$
$$PL_{n-1} + L \rightleftharpoons PL_n$$

where n is the maximum number of ligand molecules bound. If binding of the ligand affects the electrophoretic mobility and isoelectric point of the protein, then the binding of each ligand should shift the isoelectric point of the complex by a constant increment in the velocity, ω. The intrinsic binding constant is designated K_0 which is assumed to be equal for each ligand, if multiple ligand molecules are bound per mole of protein.

The average velocity of the protein, $\bar{v}_p$, in the equilibrium mixture of P and PL_i is given as a function of position, x, in the following equation:

$$\bar{v}_p(x) = v_p(x) + \bar{\nu}\omega \tag{10}$$

where $v_p(x)$ is the velocity of the protein at position x while $\bar{\nu}$ is the amount of ligand bound per mole of protein. The ligand binding isotherm is described by the equation

$$\bar{\nu} = \frac{nK_0C_L(x)}{1 + K_0C_L(x)} \tag{11}$$

where $C_L(x)$ is the equilibrium concentration of the free ligand at position x.

Since the pH gradients used in isoelectric focusing experiments are generally linear, and the electrophoretic mobility of the protein varies linearly in the region of the isoelectric point, $\bar{V}_p(x)$ is assumed to be a linear function of x; hence

$$\bar{V}_p(x) = a - bx \tag{12}$$

Substitution of Eq. (12) into Eq. (10) results in

$$\bar{V}_p(x) = (a + \bar{\nu}\omega) - bx \tag{13}$$

Once the ligand–protein complex has reached equilibrium and moved to the constituent isoelectric point, $V_p(x) = 0$ so $x_0 = a/b$. Substitution into Eq. (13) yields

$$\bar{x} = x_0 + \bar{\nu}(\omega/b) \tag{14}$$

Substituting Eq. (11) into Eq. (14) leads to

$$\bar{x} = \frac{x_0 + nK_0C_L(\bar{x})}{1 + K_0C_L(\bar{x})} \, (\omega/b) \tag{15}$$

When the concentration of the free ligand at $\bar{x}$ approaches infinity, the observed isoelectric point approaches the position of the fully liganded protein, X_F. Equation (15) then reduces to

$$\omega/b = (X_F - X_0)/n \quad (16)$$

Substitution of Eq. (16) into Eq. (15) and rearrangement leads to the double reciprocal relationship

$$\frac{1}{\bar{X} - X_0} = \frac{1}{X_F - X_0} + \frac{1}{X_F - X_0} \frac{1}{K_0} \frac{1}{C_L(\bar{x})} \quad (17)$$

Plotting $1/(\bar{X} - X_0)$ versus $1/C_L(\bar{x})$ and extrapolation to infinite free ligand concentrations provides the value for $1/(X_F - X_0)$. Dividing the slope by this value results in the value for $1/K_0$ which is the reciprocal of the dissociation constant.

Systems may also be analyzed in which only one ligand is bound, and binding changes the frictional coefficient of the protein. In this case K_0 will be an apparent binding constant which depends upon the diffusion coefficients of the unliganded protein and the protein–ligand complex.

$$K_0' = K_0(D_{PL}/D_P) \quad (18)$$

K_0' is the apparent equilibrium constant, and D_{PL} and D_P are the diffusion coefficients of the complex and the unliganded protein, respectively. If the change in the diffusion coefficient is small, K_0' will be a good estimate of K_0.

Proteins which bind ligand cooperatively may be analyzed utilizing the isoelectric focusing technique, but in this case $C_L(x)$ must be replaced by $[C_L(x)]^h$ where h is the Hill coefficient. The validity of the analysis depends upon the assumption that the frictional coefficient of the protein does not change appreciably. The constant obtained for a cooperative interaction will require the acquisition of extensive binding and structural data to determine the significance of the experimentally derived equilibrium constant.

The stoichiometry cannot be obtained from the approach discussed thus far, but it may be estimated. If one can determine the amount of bound and free ligand present under saturating conditions for a known amount of protein, the molar ratio of ligand to protein will give an approximate value for the stoichiometry, as demonstrated by Park.[22]

Experimental Procedure. 1. The gels should be prepared as described by Wrigley[23] using 7.5% acrylamide and a pH range of Ampholines which

[22] C. M. Park, *Ann. N.Y. Acad. Sci.* **209**, 237 (1973).
[23] C. W. Wrigley, this series, Vol. 22, p. 559.

will encompass the pI of the protein as well as the pI of the protein–ligand complex.

2. The protein may either be included in the solution used to prepare the gels, or it may be loaded on top of the polyacrylamide gels after polymerization as discussed by Wrigley.[23]

3. The protein should then be focused in several different tubes. One of the tubes should be removed and the position of the protein, X_0, should be determined from a densitometric scan of the gel after staining.

4. Using the gels in which the protein has been prefocused, ligand should be added in the appropriate electrode compartment so that it will be driven through the region in which the protein focuses. The position of the protein after equilibration with the ligand, $\bar{x}$, should be determined from a densitometric scan of the stain gel. The value for $\bar{x}$ should be obtained using a range of ligand concentrations. A single set of gels may be used for the entire concentration range to be tested. The lowest concentration should be tested first, then one of the gels should be removed, additional ligand added to the electrode solution, and electrophoresis should be resumed. Several ligand concentrations may be tested in this manner, as long as the protein is stable under the conditions of the experiment for a sufficient length of time.

5. The gels may be stained as described by Wrigley[23] or Wilson[18] and then scanned. The values for X_0 and $\bar{X}$ can be ascertained by integration of the scans and determination of the position of the center of mass. The center of mass will be considered that point in the gel which has equal amounts of protein on either side of it. The position of the protein should be correlated with the pH gradient so that small changes in the length of the gel do not complicate the analysis.

6. The data should be plotted in a $1/(\bar{x} - x_0)$ versus $1/C_L(\bar{x})$ plot and the value for $1/(X_F - X_0)$ obtained by extrapolation to infinite $C_L(\bar{x})$. The value for the slope of the plot must then be divided by the value for $1/(X_F - X_0)$ to determine the dissociation constant.

7. The stoichiometry of the interaction may be determined utilizing radioactive ligand. The isoelectric focusing procedure should be performed using a concentration of ligand approaching saturation. After the protein and ligand have achieved equilibrium (the position of the protein is constant as a function of time), the gel should be removed from the apparatus and sliced. Each slice is then placed in a solubilizer and counted in a scintillation fluid. The counts should be corrected for quench, and the amount of ligand in each slice determined from the specific radioactivity of the ligand. The amount of ligand in the slices between the electrode compartment which originally contained the ligand, and x_0, should be constant and should represent the amount of free ligand associ-

ated with the protein. The difference between the total amount of ligand present in the slices containing protein and the amount of free ligand in these slices will be equal to the amount of bound ligand. Dividing the amount of ligand bound to the protein by the amount of protein in the slices will yield the stoichiometry.

8. The stability of the ligand in the electrode solution must be checked to assure that it is not degraded or modified.

9. The stability of the protein in the presence of Ampholines should also be checked for a period of time equal to or greater than the time required to perform the focusing experiments.

10. The temperature of the gel should be regulated over the course of the focusing experiment to avoid potential complications which might arise if the equilibrium binding constant is temperature dependent.

General Comments. The use of isoelectric focusing to determine the equilibrium binding constant for a protein–ligand interaction is based on several assumptions. (1) It must be assumed that the association and dissociation rates of complex formation are sufficiently fast to maintain equilibrium at every instant of electrofocusing. (2) The binding of the ligand must change the net charge of the protein, but must not significantly affect the frictional coefficient of the protein. (3) The binding of each ligand molecule is assumed to affect the electrophoretic velocity by the same increment.

A direct binding technique should be used to corroborate the results obtained with the isoelectric focusing technique. Several protein concentrations should be used in the determination of K_0 so that one can observe possible protein concentration effects on K_0.

This technique is limited to ligands which are ions and are capable of migrating through the pH gradient to a region beyond the region in which the protein is located. The ligand must not be bound by the Ampholines, since this will result in multiple equilibria which will be difficult to analyze. It is also essential that the Ampholines do not influence the affinity of the protein for the ligand.

This technique is potentially very powerful in studying ligand–protein interaction of a heterogeneous protein sample such as a solution of isozymes. The interaction between ligand and a specific isozyme can be studied quantitatively without having to purify the individual isozymes.

Acknowledgments

Supported by National Institutes of Health Grants NS-14269 and AM-21489.

[22] Measurement of Ligand Binding to Proteins by Fluorescence Spectroscopy

By LARRY D. WARD

In recent years, fluorescence spectroscopy has developed into a routine experimental procedure for the study of protein–ligand interactions. To study ligand binding by this method, one only requires that a change in quantum yield be consequent upon ligand binding, whether one is observing ligand fluorescence, intrinsic protein fluorescence, or fluorescence of a covalently or noncovalently bound fluorescent probe which is sensitive to ligand binding. Changes in quantum yield upon ligand binding are extremely common, being caused by a variety of different mechanisms such as sensitivity of a fluorescent ligand to environment,[1] energy transfer from protein to ligand causing quenching of protein fluorescence[2–4] or enhancement of ligand fluorescence,[5,6] a conformational change in ligand[7] on binding to the protein, or a ligand-induced conformational change of the protein.[8] Due to the indirect observation of ligand binding by such changes in fluorescent intensity, methods of data analysis differ from those which are used routinely to analyze binding data[9–12] obtained using methods such as equilibrium dialysis, where the free ligand concentration is measured directly. This chapter will describe methods that can be used to calculate the binding constants and stoichiometry of particular protein–ligand interactions from the dependence of the observed fluorescence of a protein–ligand mixture upon total ligand concentration. The relevant theory, approach, and possible pitfalls associated with the fluorescent measurement

[1] L. Stryer, *J. Mol. Biol.* **13,** 482 (1965).
[2] J. Andreu and S. N. Timasheff, *Biochemistry* **24,** 6465 (1982).
[3] R. A. Stinson and J. J. Holbrook, *Biochem. J.* **131,** 719 (1973).
[4] J. J. Holbrook, *Biochem. J.* **128,** 921 (1972).
[5] R. B. Martin, in "Calcium in Biology" (T. G. Spiro, ed.), p. 273. Wiley, New York, 1983.
[6] C.-L. A. Wang, R. R. Aquaron, P. C. Leavis, and J. Gergely, *Eur. J. Biochem.* **124,** 7 (1982).
[7] J. J. Holbrook, A. Liljas, S. J. Steindel, and M. G. Rossmann, in "The Enzymes" (P. D. Boyer, ed.), 3rd ed., Vol. 11, p. 191. Academic Press, New York, 1975.
[8] K. P. Kohse and L. M. G. Heilmeyer, Jr., *Eur. J. Biochem.* **117,** 507 (1981).
[9] G. Scatchard, *Ann. N.Y. Acad. Sci.* **51,** 660 (1949).
[10] F. W. Dahlquist, this series, Vol. 48, p. 270.
[11] L. W. Nichol and D. J. Winzor, in "Protein-Protein Interactions" (C. Frieden and L. W. Nichol, eds.), p. 338. Wiley, New York, 1981.
[12] A. V. Hill, *J. Physiol. (London)* **90,** iv (1910).

of interactions will be presented with respect to both perturbation of ligand fluorescence on binding, as well as the perturbation of the intrinsic fluorescence of acceptor on interacting with ligand.

Theory

Dependence of Fluorescence of Ligand Binding

Case (1): Perturbation of Ligand Fluorescence on Binding to Acceptor. It will be assumed for this case under the conditions of study that (1) the acceptor does not have appreciable fluorescence at the emission wavelength monitored or, if so, its fluorescence is not perturbed on ligand binding and (2) no nonspecific effects, such as inner filter effects,[13] are contributing to the observed fluorescence change.

Consider the situation where acceptor (A) has p sites for ligand (X), the appropriate equilibria for this interaction being described by the following:

$$[AX_{(i-1)}] + [X] \overset{K_i}{\rightleftharpoons} [AX_i], \qquad K_i = \frac{[AX_i]}{[AX_{i-1}][X]} \tag{1}$$

where $i = 1, 2 \ldots p$. On mixing acceptor A and ligand X, the change in fluorescence intensity due to interaction of ligand with acceptor is simply given by $\Delta F = F - F_0$ where F is the observed fluorescence of the protein–ligand mixture at a particular total concentration of ligand and F_0 is the fluorescence of ligand in the absence of acceptor at the same ligand concentration. The change in ligand fluorescence due to interaction with protein can be consequently described by

$$\Delta F = F - F_0 = \sum_{i=1}^{p} [AX_i](\phi_{AX_i} - i\phi_x) \tag{2}$$

where ϕ_{AX_i} and ϕ_x are the quantum yields of the individual acceptor–ligand complex and free ligand, respectively. On dividing ΔF by ΔF_{max}, where ΔF_{max} is the maximum fluorescent change observed when all acceptor-binding sites are saturated with ligand, it follows that

$$\frac{\Delta F}{\Delta F_{max}} = \frac{\sum_{i=1}^{p} [AX_i](\phi_{AX_i} - i\phi_x)}{[A_T](\phi_{AX_p} - p\phi_x)} \tag{3}$$

---

[13] C. A. Parker, "Photoluminescence of Solutions." Elsevier, Amsterdam, 1968.

where $[A_T]$ is the total concentration of acceptor in the system. If one assumes that each ligand bound by acceptor has a constant change in quantum yield, i.e.,

$$\phi_{AX_i} = i\partial\phi_x; \qquad \partial = \frac{\phi_{AX}}{\phi_x} \qquad (4)$$

where the symbol ∂ represents the enhancement of the quantum yield of ligand on binding to protein, it is easily shown that

$$\frac{\Delta F}{\Delta F_{max}} = \theta = \frac{\sum\limits_{i=1}^{p} i[AX_i]}{p[A_T]} \qquad (5)$$

Case (2): Perturbation of Intrinsic Acceptor Fluorescence on Ligand Binding. The relation described by Eq. (5) can also be shown to describe the dependence of the fluorescence change on total ligand concentration for this situation by a similar procedure, if it is assumed that each ligand attachment to protein causes a constant incremental change in the quantum yield of acceptor,

$$\phi_{AX_i} = \phi_A (1 + \Delta i) \qquad (6)$$

where Δ is the fractional change in acceptor fluorescence on binding of ligand expressed as a fraction of the quantum yield of acceptor. It is quite clear from Eq. (5) that provided the assumption is made of a constant change in quantum yield accompanying ligand binding for situations involving measurement of protein or ligand fluorescence, the ratio $\Delta F/\Delta F_{max}$ corresponds to the fractional occupancy of total acceptor binding sites (θ) by ligand, be they independent or nonindependent, heterogeneous or equivalent binding sites.

This assumption of a constant change in quantum yield accompanying ligand binding is obviously not always valid, particularly when studying a system with more than one class of sites. For situations in which binding to each class of sites results in a different change in quantum yield, it is extremely difficult to analyze binding data directly by fluorescence, unless the relative association constants describing the interaction between acceptor and ligand are sufficiently separated to allow the study of one in the absence of the other. If this is not the case, analysis of the dependence of ΔF on total ligand concentration requires direct comparison with binding data obtained by a method such as equilibrium dialysis in order to differentiate the contribution of each class of site to the overall change in fluorescence. In this chapter the simplest situation of equivalent and independent binding sites and a constant change in fluorescence on ligand binding will be discussed. Some causes and effects of deviations from this assumption of a constant change in quantum yield on ligand binding will be presented more fully in a later section.

The Klotz binding function (r) for equivalent and independent binding sites is expressed simply as[14]

$$r = \frac{\sum_{i=1}^{p} i[AX_i]}{[A_T]} = \frac{pK_A[X_f]}{1 + K_A[X_f]} \tag{7}$$

where K_A is the site, or intrinsic binding constant. By comparing Eq. (7) with Eq. (5) it is obvious that

$$\phi = \frac{r}{p} = \frac{K_A[X_f]}{1 + K_A[X_f]} \tag{8}$$

It should be noted that analysis of binding data according to Eq. (8) requires knowledge of $[X_f]$ and not $[X_T]$, which is known in fluorescent binding studies, and also that analysis of such data by conventional graphic procedures, such as double reciprocal and Scatchard plots, does not allow one to evaluate p, the total number of binding sites. In the following section methods will be presented which do allow the calculation of both the stoichiometry (p) and the intrinsic association constant describing the interaction from the dependence of ΔF on $[X_T]$.

Analysis of Fluorescent Binding Data

Estimation of ΔF_{max}. An accurate determination of θ is obviously dependent on an accurate estimate of ΔF_{max}, the fluorescence change observed when all ligand binding sites are occupied. Ideally, the measurement of ΔF_{max} can be obtained simply by titrating a known concentration of acceptor with saturating concentration of ligand, ΔF_{max} being the fluorescent intensity change obtained under these conditions. If it is not possible to saturate acceptor with ligand under the conditions of assay due to reasons such as a large inner filter effect at high ligand concentration, ΔF_{max} may be obtained by extrapolation procedures.

One such procedure is to titrate a solution of fixed total ligand concentration with an increasing concentration of acceptor. As the concentration of acceptor is increased, the fraction of ligand bound also increases allowing one to extrapolate a plot of $1/\Delta F$ vs $1/[A_T]$ to infinite protein concentration to obtain an estimate of ΔF_{max}. To obtain an accurate estimation of ΔF_{max} using this procedure the concentration of bound ligand should represent an appreciable fraction of the total ligand concentration and thus the concentration range of acceptor used should be in the region of the

[14] I. M. Klotz, *Arch. Biochem. Biophys.* **9**, 109 (1946).

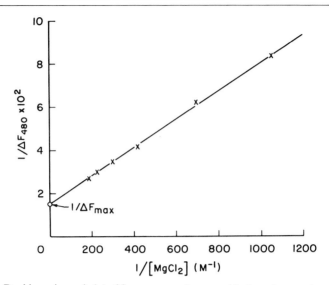

FIG. 1. Double reciprocal plot of fluorescence change on binding of magnesium to tubulin at 20° in 10 mM phosphate, 1 M sucrose, 1 mM EGTA, pH 7.0 by the ability of Mg^{2+} to enhance the fluorescence of noncovalently bound probe ANS. The concentrations of ANS and tubulin were 3×10^{-5} and 4×10^{-6} M, respectively. ANS was excited at 380 nm and emission measured at 480 nm; 5-nm excitation and emission slits were used.

dissociation constant describing the interaction. This approach should not be used if there is any dependence of the binding constant on the concentration of acceptor over the concentration range studied.[11,15,16] When studying systems where the total concentration of ligand ($[X_T]$) is much greater than the total acceptor concentration, ($[A_T]$), the free ligand concentration, ($[X_f]$), can be essentially equated with the total ligand concentration. Under such circumstances, F_{max} can be obtained from a conventional double reciprocal plot of $1/\Delta F$ vs $1/[X_T]$ by extrapolating to infinite concentration of ligand.[17] This is illustrated in Fig. 1 for the interaction of magnesium with calf brain tubulin, the binding of magnesium being monitored by its ability to perturb the fluorescence of the environmental probe, ANS. Use of this approach under conditions where bound ligand is contributing appreciably to the total concentration of ligand can result in erroneous estimates of ΔF_{max} as a linear relationship will not exist between $1/\Delta F$ and $1/[X_T]$ under such circumstances.

[15] L. W. Nichol, W. J. H. Jackson, and D. J. Winzor, *Biochemistry* **6**, 2449 (1967).
[16] G. C. Na and S. N. Timasheff, *Biochemistry* **19**, 1355 (1980).
[17] S. S. Lehrer and G. D. Fasman, *Biochem. Biophys. Res. Commun.* **23**, 133 (1966).

Calculation of K_A and p. When the total ligand concentration can be equated with the free ligand concentration as in Fig. 1 for the tubulin–Mg^{2+} interaction the dependence of ΔF on $[X_F]$ is described simply by Eq. (7) and it is thus a simple matter to calculate the binding constant describing the interaction directly from the double reciprocal plot. When this is not the case, two approaches can be used to calculate both K_A and p. The first and simplest approach is to use the graphic procedure of Stinson and Holbrook[3] where one can calculate both p and K_A directly. The second involves calculating by the method of continuous variation[18,19] the stoichiometry of the interaction where upon the free ligand concentration can be calculated directly from Eq. (9), assuming conservation of ligand.

$$[X_F] = [X_T] - p\phi[A_T] \tag{9}$$

On calculating $r = p\phi$ as a function of $[X_F]$, the data can be analyzed by conventional means[9–12] in order to evaluate the association constant describing the interaction. Stinson and Holbrook,[3] by relating the intrinsic association constant (K_A) to the concentrations of bound $(p\theta[A_T])$, and free binding sites $(p[A_T] - p\theta[A_T])$, and the concentration of free ligand $([X_T] - p\theta[A_T])$ showed with simplification and rearrangement that the following relationship exists for equivalent and independent binding sites:

$$\frac{1}{(1 - \theta)K_A} = \frac{[X_T]}{\theta} - p[A_T] \tag{10}$$

In this case, plots of $1/(1 - \theta)$ vs $[X_T]/\theta$ will give straight lines with slope of K_A and an intercept where $[X_T]/\theta = p[A_T]$ from which the stoichiometry of the interaction can be calculated. An illustrative plot of binding data according to this procedure is presented in Fig. 2 for the case of a system described by a K_A of 10^4 M^{-1} and $p = 1$ or $p = 2$. Accurate determination of p using this procedure requires one to use a concentration of protein at or about the dissociation constant describing the interaction. If too low a concentration of protein is used, the concentration of bound ligand will become very small relative to the total concentration of ligand and the intercept will not be easily distinguished from the origin as illustrated in Fig. 2 for $[A_T] = 1 \times 10^{-6}$ M. This procedure has obvious advantages over all others in that it relates the total ligand concentration with the fractional saturation directly and one can evaluate both the stoichiometry and binding constant directly from one plot.

The accuracy of the stoichiometry calculated by way of the plot of $1/(1 - \theta)$ vs $[X_T]/\theta$ can easily be tested using the method of continuous

[18] E. Asmus, *Z. Anal. Chem.* **183**, 321 (1961).
[19] P. Job, *Ann. Chim. (Paris)* **9**, 113 (1928).

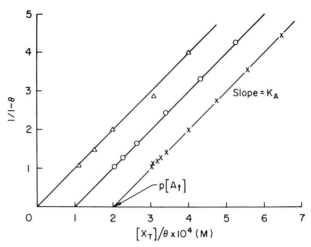

FIG. 2. Calculation of stoichiometry and association constant from fluorescent data by use of a plot of $1/(1 - \theta)$ vs $[X_T]/\theta$. The slope of the plot is the association constant (K_A) and the intercept the total number of binding sites $(p[A_T])$. Numerical examples showing the effect of varying the stoichiometry (p) and the acceptor concentration (A_T) for an interaction described by an association constant of $1 \times 10^4\ M^{-1}$. The respective curves are represented by ($\blacktriangle$) $p = 2$, $[A_T] = 1 \times 10^{-6}\ M$, ($\bigcirc$) $p = 1$, $[A_T] = 1 \times 10^{-4}\ M$, and ($\times$) $p = 2$, $[A_T] = 1 \times 10^{-4}\ M$.

variation.[18,19] Application of this method requires measurement of the fluorescence change (ΔF) of a series of acceptor–ligand mixtures under such conditions where the sum of the concentrations of acceptor and ligand is held constant, but the respective molar fraction of each is varied. The effects of the stoichiometry varying from 1 to 3 equivalent and independent binding sites on the form of curves generated by the continuous variation method for an acceptor–ligand interaction described by a K_A of $1 \times 10^4\ M^{-1}$ are given in Fig. 3. In this case, when the sum of the concentration of total ligand and total acceptor was kept constant at 3.0×10^{-4} M the maxima shifts progressively from a maximum at a mole fraction of 0.5 for a stoichiometry of 1 to lower mole fraction of tubulin as the stoichiometry increases. Stoichiometry can thus be calculated for an unknown system by comparison of experimental results of the continuous variation experiment with curves generated on the basis of particular stoichiometries and binding constants.[20,21] The exact position of the maxima for stoichiometries greater than one is dependent on the magnitude of the sum of the total ligand plus total protein concentrations relative to the

[20] C. Na and S. N. Timasheff, *Arch. Biochem. Biophys.* **182**, 147 (1977).
[21] V. Prakash and S. N. Timasheff, *J. Biol. Chem.* **258**, 1689 (1983).

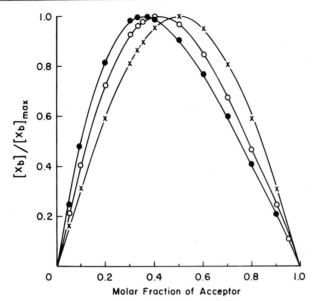

FIG. 3. Continuous variation titration of acceptor with ligand. Numerical examples showing the effect of varying the stoichiometry of interaction from $p = 1$ to 3 for an interaction described by an association constant of $1 \times 10^4 \ M^{-1}$ where the sum of the ligand and acceptor concentrations was kept constant at $3 \times 10^{-4} \ M$. The calculated curves for the binding stoichiometries are one (×), two (○), and three (●) ligands per macromolecule.

dissociation constant. If the sum of the concentrations is chosen such that it is much less than the dissociation constant, all curves will be centered at 0.5 within experimental error due to the low probability of any higher order complexes (AX_2, AX_3) being present, the predominant proportion of all complexes formed being AX. It is obvious that to obtain estimates of stoichiometry using this method for the case of equivalent and independent binding sites, one should work at least in the region of the dissociation constant of the interaction, higher if possible, in order to get a high proportion of higher order complex and thus maximum difference between the generated curves of varying stoichiometry.

The method of continuous variation as originally applied[18,19] was used to examine reactions of the type: $A + nX \leftrightarrows AX_n$, that is, situations where there are no intermediate species. In such cases there is no dependence of the position of the maximum on the relation of the dissociation constant to the sum of the concentration of reacting species.[18] In the scope of protein–ligand interactions, this would apply to the case of a highly cooperative binding or interaction of a protein with a polymerized form of ligand, such as a micelle. The quite distinctive behavior this system and that of equivalent and independent binding in continuous variation experiments makes this a useful tool for distinguishing these systems. In the limiting case where one is working at well above the dissociation constant at all concentrations one would expect the equivalent and independent binding case to give a similar curve to that for the cooperative situation. Once again, this method should not be used if there is any dependence of the binding constant on the concentration of acceptor over the region of interest.

Dependence of Fluorescence Intensity on Fractional Saturation

A constant change in fluorescence intensity on ligand binding, as is assumed in Eqs. (4) and (6), should give rise to a linear dependence of the observed fluorescence on the fractional occupancy of bound ligand (Fig. 4). Although fluorescent titrations are used routinely to estimate the binding of ligands to proteins, such a dependence is rarely investigated. Deviations from linearity in such a plot will obviously lead to quite erroneous

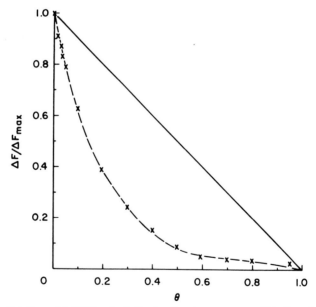

FIG. 4. Relation of $\Delta F/\Delta F_{max}$ to θ, the fractional occupancy of binding sites by ligand. The solid line describes the relationship that should exist if Eq. (4) or (6) is satisfied, i.e., a constant change in fluorescence intensity on binding of ligand. The curvilinear plot (b) was based on the results of Weber and Daniel,[24] for the interaction of ANS with bovine serum albumin.

estimates of K_A, p, and ΔF_{max} if data are analyzed according to the methods described. For equivalent and independent binding sites, this relationship should hold as long as any change in property of the fluorophore consequent on ligand binding is restricted to the region of bound ligand and thus does not affect the fluorescent properties of the other binding sites. Even for the situation in which binding to the various binding sites causes quite different changes in fluorescence, due to, for example, more tryptophans being located in the region of certain sites when measuring intrinsic protein fluorescence, a random distribution of ligand among the binding sites would require that the linear relationship shown in Fig. 4 should hold.

If the effect of ligand binding does affect the fluorescent properties of other binding sites, a nonlinear dependence of fluorescent change on fractional occupancy would result even for equivalent and independent binding sites. This type of effect could be caused by a conformational change on ligand binding affecting the fluorescent properties associated with consequent binding of ligand or via fluorescent energy transfer.[22,23] This last mechanism was shown to be consistent with the nonlinear quenching of the intrinsic protein fluorescence of lactate dehydrogenase on binding of NADH[4] and bovine serum albumin on binding of ANS,[24] as shown in Fig. 4. The high absorbance of NADH and ANS in the region of protein emission results in a large overlap integral and thus highly efficient energy transfer,[22,23] allowing NADH and ANS to quench in the region of other binding sites. In such cases changes in fluorescence with fractional occupancy are no longer linear as described by Eq. (6), but described by more complex formulations such as used by Holbrook[4] to describe the nonlinear dependence of lactate dehydrogenase fluorescence on ligand binding. This differing dependence of ΔF on θ for nonoverlapping and overlapping quenching volumes was exemplified in the case of the binding of Fe^{3+} and Cu^{2+} to transferrin, the binding of Cu^{2+} resulting in linear quenching and that of Fe^{3+} in nonlinear quenching.[25] This result was readily explained in terms of the smaller overlap integral of the transferrin–Cu^{2+} donor–acceptor pair with consequent nonoverlapping quenching volumes.

It is interesting to note that the dependence of ΔF on θ can be used to give information about the distribution of ligand within a protein population as ligands with overlapping quenching volumes for completely cooperative systems will give linear quenching due to the protein population

[22] T. Forster, *Ann. Phys. (Liepzig) [6]* **2**, 55 (1948).
[23] L. Stryer, *Annu. Rev. Biochem.* **47**, 819 (1978).
[24] G. Weber and E. Daniel, *Biochemistry* **5**, 1900 (1966).
[25] S. Lehrer, *J. Biol. Chem.* **244**, 3613 (1969).

being split into protein with all ligand binding sites occupied or with all binding sites unoccupied.[4,26] Such an approach was used by Weber and Daniel[24] to demonstrate a random distribution of ANS among its binding sites on bovine serum albumin.

Inner Filter Effects

Accurate analysis of binding data ideally requires the ligand concentration to be varied over a range of 0.1–10 times the dissociation constant describing the interaction. However, it is important to realize that when using the method of right angle illumination, where the fluorescent emission is measured at right angles to the incident beam, that the observed fluorescence is only proportional to the absorbance of the sample up to an optical density of 0.05 at the excitation wavelength.[27] This deviation from linearity results from the fact that the incident light must pass through a finite depth of solution before reaching the area viewed by the photomultiplier, the intensity of light reaching this region being effectively diminished by a factor 10^{-AL},[13] where A is the absorbance of the solution and L the distance through which the light must travel. It is obvious that failure to take into account this nonspecific absorption will lead to gross errors in the estimation of binding parameters if the ligand of interest is absorbing significantly at the excitation wavelength.

If the use of such procedures as working at an excitation wavelength away from the absorption maximum of ligand (see Fig. 5)[28,29] or the use of a short path length cell[28] does not allow one to obtain a concentration range of ligand where trivial absorption of the excitation beam is not significant, a variety of procedures have been presented in order to correct for these inner filter effects.[13,29–33] It should be pointed out that due to various assumptions in their formulation some of the correction procedures break down at quite low absorbance values and one should be certain that the correction procedure adopted is suitable under the conditions of the experiment. The simplest procedures to use are those in which a correction factor is calculated on the basis of the absorbance of

[26] A. J. Pesce, C.-G. Rosen, and T. L. Pasby "Fluorescence Spectroscopy." Dekker, New York, 1971.

[27] J. R. Lakowicz, "Principles of Fluorescence Spectroscopy." Plenum, New York, 1983.

[28] R. F. Chen and J. F. Hayes, *Anal. Biochem.* 13, 523 (1965).

[29] B. Birdsall, R. W. King, M. R. Wheeler, C. A. Lewis, Jr., S. R. Goode, R. B. Dunlap, and G. C. K. Roberts, *Anal. Biochem.* 132, 353 (1983).

[30] J. F. Holland, R. E. Teets, P. M. Kelly, and A. Timmick, *Anal. Chem.* 49, 706 (1977).

[31] C. Hélène, F. Brun, and M. Yaniv. *J. Mol. Biol.* 58, 349 (1971).

[32] M. L. Mertens and J. H. R. Kägi, *Anal. Biochem.* 96, 448 (1979).

[33] M. Ehrenberg, E. Cronvall, and R. Rigler, *FEBS Lett.* 18, 199 (1971).

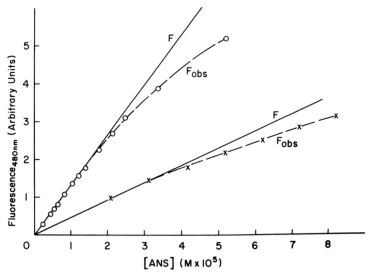

Fig. 5. Deviation from a linear dependence of fluorescence on the concentration of ANS when exciting at the absorption maximum (350 nm) (×) and removed from the absorption maximum (400 nm) (○).

the sample at the excitation wavelength and, in some cases, knowledge of the geometric arrangement of the optical system under which the fluorescence intensity is measured. The following is such a correction factor, as described by Parker[13] and Holland et al.[30]

$$\frac{F}{F_{IFT}} = C = \frac{2.303A\ (x_2 - x_1)}{10^{-Ax_1} - 10^{-Ax_2}} \tag{11}$$

where A is the absorbance of the sample at the excitation wavelength, x_1 and x_2 the distances from the front of the cell to the mask edges of the observation window, and F_{IFT} the observed fluorescence before application of the correction factor for inner filter effects. The range of absorbance over which this correction factor can be used seems to be limited by the assumption inherent in its derivation that the efficiency of detection of fluorescence is uniform across the observation window.[29,33] The range of absorbance over which this assumption is valid is dependent on the optical system of the fluorimeter in use. Under the conditions used by Holland et al.[30] this correction procedure was valid up to an absorbance of 2. However, before applying this correction procedure to ligand binding data, the range over which it is valid for the particular fluorimeter in use should be gauged by measuring the ability of it to correct for the nonlinearity of the concentration dependence of fluorescence of a suitable standard under the conditions to be used in the binding experiment.

The correction factor described by Eq. (11) can be simplified even further to a relation under which no knowledge of the geometric arrangement of the optical system is required. Such a correction procedure was described by Hélène et al.[31] who corrected for any change in the ratio of the fluorescence of acceptor in the absence of ligand to the fluorescence in the presence of ligand by application of the following correction factor:

$$C = \frac{A_T (1 - 10^{-A_0})}{A_0 (1 - 10^{-A})} \qquad (12)$$

where A_0 is the absorbance of acceptor alone and A_T is the total absorbance of the acceptor–ligand mixture. This correction procedure is valid only up to a total absorbance of 0.2–0.3 at the excitation wavelength[29,32] and will lead to gross errors if used at higher absorbance values.

As stated previously, these empirical formulations of correction factors are all ultimately limited by the assumption of a uniform efficiency of detection across the observation window. Two methods that have been presented to obviate this problem both involve curve fitting procedures which calculate the appropriate correction factors from the effect of increasing ligand concentration on the fluorescence of an appropriate standard.[29,33] Use of such a procedure is limited by the choice of a reliable standard that has identical absorption charcteristics to the fluorophore in question. When measuring ligand fluorescence an obvious choice of standard is the ligand itself. As illustrated in Fig. 5, correction factors can be calculated from the concentration-dependent fluorescence of the ligand under the conditions of assay by the curve fitting procedures[29,33] or by simply extrapolating the linear region of the curve such that the correction factor will equal F/F_{obs}.[27] N-Acetyltryptophanamide and tryptophan are two compounds which are commonly used as standards for proteins in calculating correction factors for inner filter effects. It should be emphasized that, as the excitation spectra of these compounds are not identical with those of most proteins, a region of the spectrum should be selected which is similar to the protein under study and the concentration of standard should be such as to have an identical absorbance at the excitation wavelength. This is particularly important if the ligand absorption profile is steep in this region, as this may lead to differential absorption of light across the slit width, and thus errors in the correction factors calculated using the standard compound if the excitation spectra of standard and protein are not identical. In such cases, a decreased excitation slit width can oviate some of this problem.[32]

The concentration of both protein and standard should be selected such that their observed fluorescence is directly dependent on their concentration in the absence of ligand. Correction factors can then be calculated from the dependence of the fluorescence on the concentration of ligand as described previously, or by simply using the ratio of the fluorescence of the protein to that of the standard at each concentration of ligand,[26] this ratio representing any real changes in protein fluorescence due to the specific interaction between protein and ligand.

A useful method of correcting for inner filter effects which obviates the use of standards by making use of a graphic extrapolation procedure of the protein–ligand binding data has been presented by Mertens and Kägi.[32] It was shown that under certain conditions the decrease in fluorescence of acceptor due to inner filter effects could be described by

$$\ln F_{IFT} = \ln F_0 - \varepsilon L[X_T] \tag{13}$$

where ε is the molar absorption coefficient of ligand and L the path length through which light must travel to the region observed by the photomultiplier. Such a relation indicates that a plot of $\ln F_{IFT}$ vs $[X_T]$ should be linear if all change in fluorescence is due to inner filter effect and thus when studying protein–ligand interactions this linear relation would be expected to apply at high concentration when all binding sites are saturated. If all ligand-binding sites can be saturated, this allows one to correct for inner filter effect at all ligand concentrations by extrapolation of the linear region of the curve. The application of this method is demonstrated in Fig. 6 for the binding of MgADP to creatine kinase as applied by

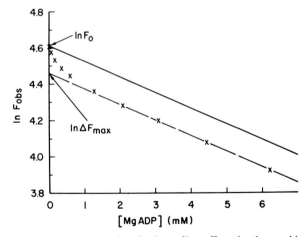

FIG. 6. Correction of fluorescent data for inner filter effects by the graphic extrapolation procedure of Mertens and Kägi[32] as applied by them to the binding of MgADP to creatine kinase. The $\ln F_{IFT}$ curve can be obtained by drawing a line parallel to the line describing the dependence of $\ln F_{obs}$ at high concentration, through F_0. ΔF, at each ligand concentration, is obtained from the difference between $\ln F_{IFT}$ and the F_{obs}.

Mertens and Kägi.[32] The validity of Eq. (13) is dependent on the assumption that the fluorescence originates from a point source.[33] The validity of this assumption, for the particular system under study, should be tested before application of this method by demonstrating that the dependence of the fluorescence of an appropriate standard on $[X_T]$ is indeed described by Eq. (13).

In conclusion, fluorescence analysis represents an easy way of studying protein–ligand interactions provided one can obtain an observable change in fluorescence due to ligand binding. The association constants describing the interaction can be obtained quite easily from conventional double reciprocal plots (Fig. 1) or plots of $1/(1 - \theta)$ vs $\theta/[X_T]$ (Fig. 2) for situations in which the concentration of bound ligand does not contribute appreciably to the total concentration of ligand. For this situation neither the plot of $1/(1 - \theta)$ vs $\theta/[X_T]$ nor the method of continuous variation[18,19] allows estimation of the stoichiometry of the interaction.

When bound ligand does contribute significantly to the total concentration of ligand both K_A and p can be obtained directly from a plot of $1/(1 - \theta)$ vs $\theta/[X_T]$. The method of continuous variation[18,19] can also be used to measure the stoichiometry. One should be aware of possible pitfalls in this analysis, such as nonlinearity of fluorescence change on fractional occupancy (Fig. 4) which can lead to gross errors in estimates of binding constants.[4] Possibly the greatest cause of error in these studies is failure to correct for inner filter effects. A variety of methods were presented which can correct for such trivial absorption of light, the method of choice being dictated by the conditions of study.

[23] Measurement of Metal–Ligand Distances by EXAFS

By Robert A. Scott

Introduction

Historical Preface

It has been only within the last decade that X-ray absorption spectroscopy (XAS) and the corollary technique of extended X-ray absorption fine structure (EXAFS) have become useful in examining biological systems. Although the phenomenon of X-ray absorption has been known since the

1920s,[1,2] a renaissance of interest occurred in the early 1970s when the Stanford Positron Electron Accelerating Ring (SPEAR) was first used as a source of synchrotron radiation. The unique characteristics (high intensity, brightness, broadband spectral distribution) of synchrotron radiation make it an ideal source for XAS and the Stanford Synchrotron Radiation Laboratory (SSRL) was established to make use of it. SPEAR was originally built to allow observation of subatomic particles created by collision of counterrotating bunches of positrons and electrons and thus, the first use of synchrotron radiation was "parasitic." SSRL currently receives 50% of SPEAR running for the "dedicated" production of synchrotron radiation. Other storage rings have also been used for synchrotron radiation research, but it is only recently (the last few years) that rings have been developed specifically for the production of synchrotron radiation.

The renaissance of XAS was also predicated on development of a theoretical basis for the EXAFS phenomenon, which came about during the early 1970s as well. The current theoretical understanding was sparked by the work of Sayers, Lytle, and Stern in 1970 on the single-scattering theory of EXAFS.[3-8]

Scope of This Chapter

It is the intended purpose of this chapter to discuss the methodology behind the use of EXAFS for determining metal–ligand environments (with specific application to metal sites in proteins and enzymes). This includes the collection, reduction, and analysis of X-ray absorption data. It does *not* include the theoretical basis of EXAFS. This will be covered only in outline, emphasizing those points required for an understanding of the methodology. In addition, no attempt has been made to survey the literature for applications of XAS to biological problems. Examples of such applications will be used only to illustrate the techniques discussed.

In addition, the edge and near-edge region of the X-ray absorption spectrum will be virtually ignored. This is not due to prejudice, since it is the author's opinion that studies involving the X-ray absorption edge

[1] W. Kossel, *Z. Phys.* **1,** 119 (1920).

[2] M. Siegbahn, "The Spectroscopy of X-Rays." Oxford Univ. Press, London and New York, 1925.

[3] D. E. Sayers, F. W. Lytle, and E. A. Stern, *Adv. X-Ray Anal.* **13,** 248 (1970).

[4] D. E. Sayers, F. W. Lytle, and E. A. Stern, *Phys. Rev. Lett.* **27,** 1204 (1971).

[5] D. E. Sayers, F. W. Lytle, and E. A. Stern, *J. Non-Cryst. Solids* **8–10,** 401 (1972).

[6] D. E. Sayers, F. W. Lytle, and E. A. Stern, *in* "Amorphous and Liquid Semiconductors," (J. Stuke and W. Brenig, eds.), p. 403. Taylor & Francis, London, 1974.

[7] E. A. Stern, *Phys. Rev. B: Solid State* [3] **10,** 3027 (1974).

[8] E. A. Stern, D. E. Sayers, and F. W. Lytle, *Phys. Rev. B: Solid State* [3] **11,** 4836 (1975).

(e.g., X-ray absorption near edge structure, XANES) will become increasingly important in providing electronic structural information. However, such techniques are still in their infancy compared to EXAFS and are much less amenable to discussion at this time.

Applicability of EXAFS

The XAS technique makes use of the fact that the energy dependence of the X-ray absorption coefficient (i.e., the X-ray absorption spectrum) of a material exhibits features characteristic of the elements contained in the material. For each element these features consist of a series of discontinuities (absorption edges) that occur at specific energies slightly affected by the electronic environment of the atom. In addition, if the atom is surrounded by a regular array of other atoms (e.g., in a crystalline matrix, or a homogeneous chemical compound), the absorption coefficient in the region of the spectrum just to higher energy of each absorption edge will exhibit oscillatory behavior, referred to as EXAFS. This spectral region contains structural information about the makeup of the "atomic neighborhood" of the absorbing atom. The EXAFS technique involves acquisition and analysis of data in the EXAFS spectral region in order to extract information concerning the local structural environment of the absorbing atom. Specifically, one can determine (1) the distance from the absorbing atom to other atoms; (2) the number of other atoms at a particular distance; and (3) the nature (i.e., atomic number, Z) of the other atoms.

The energy at which the X-ray absorption edge occurs is dependent upon the atomic number of the absorbing atom (see Fig. 1).[8a,8b] Thus, the range of elements available as absorbers in an EXAFS experiment is dictated by the availability of an adequate photon source in the appropriate energy region. For synchrotron radiation, the low-energy end of the range of available photon energies is affected by absorption of photons by window material, atmosphere, etc. With some special precautions (e.g., all He atmosphere), this places a lower limit of $\sim$2.5 keV on available photons which makes the EXAFS experiment difficult for absorbing atoms with $Z \leq 17$ (Cl). There is no effective upper limit on the elements that can be examined as absorbers, since elements with K edges above the available photon energy range have L edges within the available range. As Fig. 1 shows, the available photon energy range must extend to $\sim$25 keV to avoid gaps in the coverage of elements. This is usually not a problem

[8a] J. A. Bearden and A. F. Burr, *Rev. Mod. Phys.* **39**, 125 (1967).
[8b] W. H. McMaster, N. Kerr Del Grande, J. H. Mallett, and J. H. Hubbell, "Compilation of X-ray Cross Sections," UCRL-50174, Sect. II, Rev. 1. Natl. Tech. Inf. Serv., Springfield, Virginia, 1968.

with the current dedicated operation of synchrotron sources and the use of insertion devices such as wigglers.

For the most part, the transition metals are commonly used as absorbing atoms in EXAFS experiments on biological systems. Most typical is the use of EXAFS to examine the coordination environment of active sites in metalloproteins and metalloenzymes. Less typical but equally important are EXAFS studies of metal-binding sites (e.g., in calcium-binding proteins) and of appropriate atoms (e.g., As, Br, etc.) in inhibitors or substrates bound to (or near) metalloprotein active sites.

There are a few restrictions on the state of samples used in biological EXAFS studies. Physically, the sample may be a solution, frozen solution, lyophilized powder, microcrystalline solid, or single crystal. Chemically, one must take care to ensure as much homogeneity in the state of the metal site as possible. The EXAFS technique is notoriously poor at detecting heterogeneity in structural environments—only an average environment is found. A special case involves metalloenzymes with two or more atoms of the same metal in distinctive structural environments (e.g., cytochrome c oxidase, laccase, the iron sites in nitrogenase). In order to assign EXAFS features to individual sites, some kind of chemical perturbation must be used to alter the structure of one site while (presumably) not affecting the other(s). Generation of mixed valence states or "half-apo" protein is a possible approach to this problem.

Most recent EXAFS studies on biological systems have used the fluorescence excitation technique since its enhanced sensitivity makes it suitable for the lower concentrations accessible in biological samples. The sensitivity of the various fluorescence detection schemes dictates the lower concentration limit for EXAFS samples. This limit is dependent upon the atomic number of the absorbing atom and is currently ~0.5–1.0 mM at Fe and ~0.1–0.2 mM at Mo.

Theory

As already mentioned, it is not the purpose of this chapter to give a detailed discussion of the theory behind the use of the EXAFS technique for determination of metal environments. For those interested in the detailed mathematics, several treatments are available.[3–12] However, a certain amount of theory concerning the origin of EXAFS oscillations is

[9] C. A. Ashley and S. Doniach, *Phys. Rev. B: Solid State* [3] **11,** 1279 (1975).
[10] P. A. Lee and J. B. Pendry, *Phys. Rev. B: Solid State* [3] **11,** 2795 (1975).
[11] P. A. Lee and G. Beni, *Phys. Rev. B: Solid State* [3] **15,** 2862 (1977).
[12] P. A. Lee, *in* "EXAFS Spectroscopy. Techniques and Applications" (B.-K. Teo and D. C. Joy, eds.), p. 5. Plenum, New York, 1981.

necessary so that the reader can comprehend better the data analysis methodology.

X-Ray Absorption Spectrum

The shape of an X-ray absorption spectrum is controlled by the energies of dissociation of electrons from orbitals of the absorbing atom. There is a sharp discontinuity (an "edge") in the absorption spectrum in the region of these atomic ionizations. Dissociation of metal core ($n = 1$, 2) electrons usually requires photons in the X-ray range. Photodissociation of a 1s electron gives rise to the K-absorption edge while photodissociation of 2s or 2p electrons gives rise to the L edges. These events are

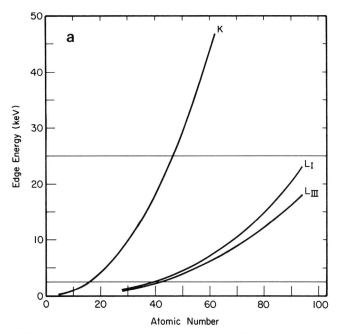

FIG. 1. X-Ray absorption edge energies. (a) The energy of the first inflection point of the elemental edge is plotted as a function of atomic number (Z) for K, L_I and L_{III} edges. A source giving off photons from ~2.5–25 keV allows coverage of all elements with $Z \geq 17$ (by either K or L_{III} edges). (b) Values of the energy of the first inflection point of the K and L edges for selected elements. The data are taken from the X-ray absorption energies tabulated by Bearden and Burr[8a] or from McMaster et al.[8b] In cases for which two values were available, the average was used. Values given in parentheses are those for which the two values disagreed considerably. In these cases, the values that fit better the trends in (a) were used.

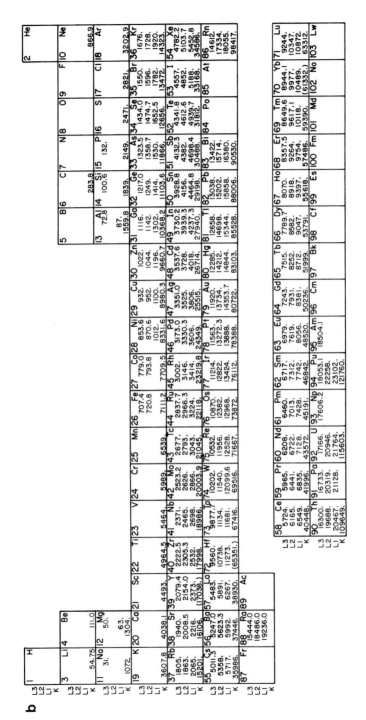

Fig. 1. (*continued*)

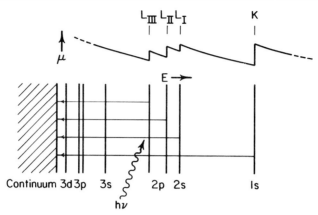

FIG. 2. General energy level diagram for X-ray absorption by an atom. The lower portion indicates the ionizations that give rise to the K, L_I, L_{II}, and L_{III} absorption edges indicated in the spectrum in the upper portion. The spectrum is plotted as absorption coefficient (μ) versus photon energy ($E \equiv h\nu$).

illustrated schematically in Fig. 2. Energies of K and L edges for selected elements are shown in the periodic table of Fig. 1.

We are interested specifically in the EXAFS region of the spectrum which starts ~50–100 eV beyond the absorption edge (see Fig. 3). In general, equivalent information is available from either K or L edge

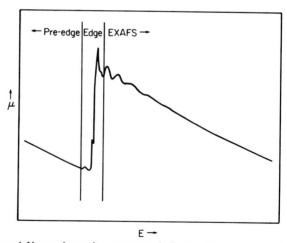

FIG. 3. General X-ray absorption spectrum indicating the energy regions of interest. Divisions between regions are poorly defined with the edge region entailing any features before, on, or just after the edge that are not analyzable as EXAFS.

EXAFS. L edge studies are generally done best using the L_{III} edge since the L_{II} EXAFS region is cut off early by the L_I edge and the L_I EXAFS region contains some remnant of L_{II} EXAFS. In any case, the EXAFS region contains oscillations on top of a smoothly varying background. Subtraction of this background (and normalization of the remainder by the size of the edge jump) yields the EXAFS data (the oscillations). It is this extraction of the EXAFS data and the subsequent analysis which are the subjects of this chapter.

Theoretical EXAFS Expression

A simple physical picture can be used to explain the oscillatory behavior of the EXAFS data. Upon absorption of an X-ray photon of energy E (where E is greater than E_0, the threshold energy for dissociation of the appropriate core electron), the atom is ionized, converting the bound electron into a photoelectron of wavelength λ. This wavelength is related to the photoelectron momentum, p, which is in turn related to the photoelectron kinetic energy [equal to $(E - E_0)$]. It proves convenient to define a photoelectron wave vector, $\mathbf{k}$, to use as the independent variable in displaying EXAFS data:

$$\mathbf{k} = 2\pi/\lambda = p/h = [2m_e(E - E_0)/\hbar^2]^{1/2} = [0.262449\ (E - E_0)]^{1/2} \quad (1)$$

The final expression holds for E and E_0 expressed in units of electron volts (eV) and $\mathbf{k}$ in units of $\overset{\circ}{A}^{-1}$.

Since no EXAFS is observed in the X-ray absorption spectrum of a monoatomic gas (e.g., Kr), it was postulated that the EXAFS oscillations arose from backscattering of this X-ray-induced photoelectron from the nuclear potentials of nearby atoms. Using this picture, the absorption coefficient would be proportional to the amplitude of the backscattered photoelectron wave at the absorbing atom. Since this amplitude is periodic (i.e., we can treat the photoelectron as a wave) and λ is inversely proportional to the square root of the kinetic energy $(E - E_0)$, the absorption coefficient would be expected to have an energy dependence proportional to the interference between the outgoing and backscattered photoelectron waves. Figure 4 gives a schematic illustration of this argument. At photon energy E_1 (Fig. 4a), the backscattered wave "arrives back" at the absorbing atom in phase giving constructive interference and a maximum in the absorption coefficient (Fig. 4c), whereas at photon energy E_2 (Fig. 4b), the interference is destructive giving a minimum in the absorption coefficient.

Thus, each scattering atom is expected to give rise to a periodic oscillation of the absorption coefficient and the EXAFS turns out to be a sum

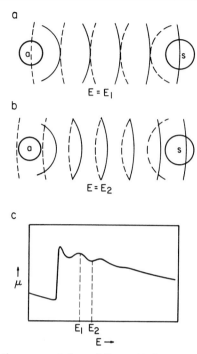

FIG. 4. Diagrammatic representation of the scattering processes that give rise to the EXAFS. (a) The outgoing and backscattered waves at energy E_1 giving rise to constructive interference and a maximum in μ. (b) The scattered waves at energy E_2 giving rise to destructive interference and a minimum in μ. [In both (a) and (b), the solid arcs represent the "peaks" of the outgoing wave from absorber a and the dashed arcs represent the "peaks" of the backscattered wave from scatterer s.] (c) The resulting spectrum indicating the values of μ at E_1 and E_2.

of damped, phase-shifted sine waves, one from each (set of) scatterer(s). This can be expressed mathematically in terms of the EXAFS (χ) as a function of the photoelectron wave vector (**k**):

$$\chi(\mathbf{k})\left(\equiv \frac{\mu - \mu_s}{\mu_0}\right) = \sum_s \frac{B_s N_s |f_s(\pi, \mathbf{k})|}{\mathbf{k}R_{as}^2}$$

$$\exp(-2\sigma_{as}^2 \mathbf{k}^2)\sin[2\mathbf{k}R_{as} + \alpha_{as}(\mathbf{k})] \quad (2)$$

The first equivalence describes the empirical extraction of $\chi(\mathbf{k})$ from the total absorption coefficient (μ) by subtraction of the smooth background (μ_s) and normalization to the free-atom absorption coefficient of the same edge height (μ_0). The theoretical expression was derived assuming single-scattering theory, but contains the extra factor, B_s. The summation over s is to be interpreted as a summation over all scatterers, s, or over all shells of scatterers, s. In this context, a "shell" of scatterers means a collection of scatterers (all of the same element) at (approximately) the same distance from the absorber. N_s is the number of scatterers in such a shell and R_{as} is the distance (or weighted average of the distances) of the scatterers in this shell from the absorber, a.

The exponential term is referred to as the Debye–Waller factor and it describes the damping (in **k**-space) of χ due to the relative mean square deviation (σ_{as}^2) in R_{as}. Actually, σ_{as}^2 results from the combination of two effects:

$$\sigma_{as}^2 = \sigma_{vib}^2 + \sigma_{stat}^2 \tag{3}$$

σ_{vib}^2 represents the dynamic contribution due to vibrational motion of atoms a and s. This may be calculated assuming a harmonic vibration of frequency ν.[13]

$$\sigma_{vib}^2 = (h/8\pi^2\mu\nu)\coth(h\nu/2kT) \tag{4}$$

μ is the reduced mass of the a–s system, T is the absolute temperature, and h, k are Planck's and Boltzmann's constants, respectively. σ_{stat}^2 represents the contribution from static disorder in the interatomic distances that make up the shell. This can be estimated for a shell containing m atoms at distance R_m and n atoms at distance R_n.[13,14]

$$\sigma_{stat} \approx \frac{(mn)^{1/2}}{(m + n)} |R_m - R_n| \tag{5}$$

This usually holds only for moderately disordered systems ($|R_m - R_n| \lesssim$ 0.1 Å).

The overall appearance of the EXAFS contribution from a shell is dependent upon the nature of the absorbing atom and (particularly) the scattering atom(s). In this dependence lies the main utility of the technique (i.e., the identification of the type of scattering atom from the shape of the EXAFS). This dependence is described by an inherent backscattering function which is broken up into amplitude and phase components in Eq. (2). The inherent backscattering amplitude is denoted as $B_s|f_s(\pi,\mathbf{k})|$. This amplitude is a function of **k** and is often referred to as the amplitude envelope. The phase component is denoted $\alpha_{as}(\mathbf{k})$, the inherent backscattering phase shift, which is also a function of **k**.

[13] B.-K. Teo, *Acc. Chem. Res.* **13**, 412 (1980).
[14] B.-K. Teo, R. G. Shulman, G. S. Brown, and A. E. Meixner, *J. Am. Chem. Soc.* **101**, 5624 (1979).

B_s is a factor which is best described as the product of two other factors:

$$B_s = S_s T_s \qquad (6)$$

S_s is a scale factor and T_s is a transmission factor. S_s is needed to correct errors in theoretically derived amplitude envelopes. T_s is usually assumed to be unity for "first shell" scatterers, but may be less than unity for longer distance scatterers. It may be used to correct for multiple scattering effects or for inelastic scattering losses of photoelectron amplitude. The latter effect is sometimes described by a factor of the form, $\exp(-R_{as}/\bar{\lambda})$, where $\bar{\lambda}$ is a mean free path for the photoelectron.

The types of information that are available from an EXAFS spectrum should be clear from the discussion so far. In each component "sine" wave there are basically three observables: frequency, amplitude, and phase. Measurement of the frequency ($\sim 2kR_{as}$) gives information about the absorber–scatterer distance, R_{as}. Measurement of the amplitude gives information about the number of scatterers, N_s, in the shell. Measurement of the phase [$\alpha_{as}(\mathbf{k})$] gives information about the type of atom doing the scattering. As already discussed, there is also information concerning the type of scatterer in the shape of the amplitude envelope [$B_s|f_s(\pi,\mathbf{k})|$]. In all of the above, the treatment has been oversimplified. A more detailed description must be delayed until the discussion of data analysis.

Experimental Techniques

One trigger event for the renaissance of XAS as a powerful structural technique was the advent of synchrotron radiation as an X-ray source for these experiments. Certainly, application of EXAFS to biological systems would not be possible without synchrotron radiation sources. In this section, the discussion will involve the components of a typical "X-ray absorption spectrometer" and the methodology of data acquisition and data analysis. Of particular interest are discussions of potential experimental difficulties and the inherent limitations of the EXAFS technique.

X-Ray Absorption Spectrometer

Figure 5 illustrates diagrammatically the general equipment components necessary to perform XAS measurements. These include a source of X rays, some means of energy resolving this radiation (one type of monochromator is shown, but dispersive elements may be used alternatively), and detection equipment. In addition, a means of control of spectrometer operation and data acquisition is required. In the case of XAS, a (dedicated) computer plays this role.

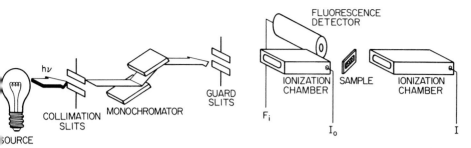

Fig. 5. The components of a general X-ray absorption spectrometer. The X-ray source is typically synchrotron radiation (not a light bulb). The monochromator shown is that described in Fig. 6b. The detection system can measure either transmission [$\mu x = \ln (I_0/I)$] or fluorescence excitation ($\mu x \propto \Sigma_i F_i/I_0$) spectra.

Source. For biological applications, the source requirement is for X-ray photons between ~2.5 and 25 keV. Other necessary features are a fairly flat spectrum (white radiation), high intensity (because of inherently dilute samples), and high collimation (so that small sample volumes can be used). The energy range requirement is fulfilled by three types of sources: conventional sealed-tube X-ray sources, rotating anodes, and synchrotron radiation.

Conventional X-ray tubes (e.g., a copper target source) produce a spectral distribution consisting of a Bremsstrahlung background with characteristic emission lines (e.g., copper K_α, K_β emissions). In order to scan the X-ray photon energy, the Bremsstrahlung background must be used and typical photon fluxes (through a 0.1-mrad^2 aperture, for example) are ~10^4 photons sec^{-1} eV^{-1} at 10 keV. In rotating anode X-ray tubes, the target is rotated (to increase the effective target surface area), allowing higher photon fluxes to be obtained (ca. one to two orders of magnitude higher than a similar conventional X-ray tube). The source spectrum still contains features characteristic of the target material. Thus, these sources give distinctly "nonflat" spectral distributions. This can be overcome partially by choosing the target material wisely. Their main disadvantage is their low photon flux, which has so far precluded their use with dilute metalloprotein samples.

SSRL was the first storage ring used as a synchrotron radiation source, but since then many more storage rings and synchrotrons have become available for XAS work. In the United States there are three main synchrotron radiation sources available for biological XAS: SSRL, the Cornell High Energy Synchrotron Source (CHESS) at Ithaca, New York, and the National Synchrotron Light Source (NSLS) at Brookhaven, Upton, Long Island. (As of this writing, the NSLS was still not fully commissioned.) In general, synchrotron radiation has the advantage of being fairly "white" (flat spectral distribution). It also displays extremely high brightness resulting from high intensity and high collimation. For a 0.1-mrad^2 aperture, fluxes can be easily $\gtrsim 10^{10}-10^{11}$ photons sec^{-1} eV^{-1} at 10 keV. With incorporation of insertion devices (e.g., wigglers), these fluxes can be increased by as much as two to three orders of magnitude. It is clear from this discussion that synchrotron radiation is the "source of choice" for biological XAS work.

Distribution of synchrotron radiation beamtime to the user community is done for the most part by proposal. A prospective user submits a proposal to perform specific experiments, requesting a certain amount of beamtime. A system of external peer review is used to evaluate the scientific impact of the work and the competence of the investigator(s). Based on the reviews, a rating is assigned (at SSRL by a Proposal Review Panel) to each proposal and this rating is used to decide on scheduling of beamtime. At SSRL in 1983, about 60% of the proposals that were rated and requested beamtime actually received beamtime.

At the NSLS (and also at SSRL), a different mode of operation is also being used. This involves the establishment of a Participating Research Team (PRT), usually made up of both academic and industrial concerns. The PRT is usually requested to make both a scientific and financial commitment to setting up an experimental station at the facility, in return for which they are guaranteed a certain percentage of the beamtime available on that station. In all cases to date, some fraction of the time is also set aside for peer-reviewed proposals. The most recent example of the PRT approach is the construction of beam line VI at SSRL jointly by Exxon, Lawrence Berkeley Laboratory (LBL), and SSRL. This beam line was commissioned in spring, 1984.

Monochromator. In recording XAS data, the normal procedure involves collecting data points at discrete photon energies, one at a time. This procedure requires monochromatization of the white source radiation. In general, this is accomplished by taking advantage of Bragg reflection from a certain diffraction plane of a single crystal:

$$n\lambda = n \left(\frac{hc}{E}\right) = 2d \sin \theta \tag{7}$$

where n is the order of the reflection, λ is the wavelength of the diffracted ray, E is the energy of the diffracted ray, d is the d-spacing of the crystal lattice planes, θ is the angle between the Bragg planes and the incident ray, h is Planck's constant, and c is the speed of light. For energies in the proper region (2.5 keV $\leq E \leq$ 25 keV) and reasonable incident angles (θ), the optimum d-spacing is ~1.5–3.5 Å (for use of a first order reflection, $n = 1$). Typical crystal planes used for X-ray monochromators include

Si[111], Si[220], Si[400], Ge[111], Ge[220], Ge[311], etc. Silicon proves to be a good choice since large single crystal specimens of silicon are available for cutting and polishing.

Flat crystal monochromators make use of crystals polished parallel (or nearly parallel) to a set of Bragg planes. Several different arrangements have been tried (at one time or another). These are illustrated in Fig. 6a–d. The simplest arrangement involves a single flat crystal rotated during the scan to different incident angles (θ). This arrangement is clumsy since the experiment (sample and detectors) has to track the reflected beam during the scan. Figure 6b shows a conventional double crystal monochromator (as used at SSRL, for example) in which the incident beam is Bragg reflected from two flat crystals in succession resulting in a reflected ray parallel to the incident ray. The crystals are mounted on a goniometer at fixed separation and the goniometer is rotated by computer-controlled stepping motors to effect a scan. As a result of the fixed intercrystal separation, the vertical position of the beam changes during the scan. This is usually compensated by moving a stepping motor-controlled table (with the experiment) vertically to track the reflected beam. The advantage of using this monochromator lies in its ease of construction and operation.

There are two possible arrangements that can avoid movement of the reflected beam during a scan (at the expense of extra complexity in the mechanisms). The double–double crystal monochromator is shown in Fig. 6c. In this case, two pairs of fixed-separation crystals are rotated in tandem in opposite directions. Figure 6d shows the variable-separation double crystal monochromator (the Golovchenko design). During a scan, both crystals are rotated and kept parallel to each other while one is also moved horizontally (on an air bearing) so that the reflected ray is always at the same vertical position. The last two monochromators are both "fixed-in, fixed-out" monochromators. For any of the monochromators that need two (or more) crystals with parallel Bragg planes, some sort of tuning mechanism is required to assure alignment. In the current double crystal design at SSRL, a piezoelectric stepper on one crystal is used for tuning.

Figure 6e illustrates a bent-crystal focusing monochromator of a design often used with laboratory sources (e.g., rotating anodes) to enhance the photon intensity collected from the source.[15–18] Increases of as much as three orders of magnitude in photon flux have been reported.[15] In a

[15] G. S. Knapp, H. Chen, and T. E. Klippert, Rev. Sci. Instrum. 49, 1658 (1978).
[16] G. G. Cohen, D. A. Fischer, J. Colbert, and N. J. Shevchik, Rev. Sci. Instrum. 51, 273 (1980).
[17] R. Haensel, AIP Conf. Proc. 64, 73 (1980).
[18] S. Khalid, R. Emrich, R. Dujari, J. Shultz, and J. R. Katzer, Rev. Sci. Instrum. 53, 22 (1982).

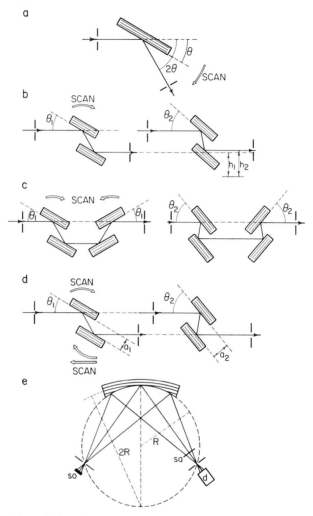

FIG. 6. Various designs for crystal X-ray monochromators that make use of Bragg reflection. Each diagram illustrates the diffracting crystal(s) in cross section (the lines indicate the Bragg planes) and typical ray-tracings. (a) A simple flat crystal monochromator. (b) The common double crystal (fixed separation) monochromator. (c) A double–double crystal (fixed separation, fixed in/fixed out) monochromator. (d) The Golovchenko double crystal (variable separation, fixed in/fixed out) monochromator. (e) A bent crystal monochromator using the Johansson geometry (d, detector; sa, sample; so, source). This latter type of monochromator is typically used with a laboratory X-ray source.

scan, as the bent crystal is rotated by θ, the sample and detector are rotated about the same pivot by 2θ.

It is of interest to mention here the importance of resolution (spectral bandwidth) of the various monochromators already discussed. For EXAFS data, a spectral bandwidth of 10–20 eV is sufficient since EXAFS oscillations occur over $\gtrsim$50 eV. However, if one is interested in X-ray absorption edge structure as well, ~2 eV resolution is required. There is *always* a trade-off between resolution and intensity. For flat crystal monochromators, the resolution is given by

$$\Delta E = E(\Delta\theta) \cot \theta \qquad (8)$$

convolved with the rocking curve of the crystal reflection being used. In Eq. (8), θ is the Bragg angle at photon energy E and $\Delta\theta$ is the spread in θ for rays passing through the monochromator. For synchrotron radiation, $\Delta\theta$ is defined by the vertical collimation of the radiation entering the monochromator (see Fig. 5). For a typical high-resolution XAS scan at SSRL, the vertical aperture might be ~1 mm at 20 m from a 1-mm-high source point. This gives $\Delta\theta \approx 0.1$ mrad and for a Bragg reflection with a sufficiently narrow rocking curve (e.g., Si[220]), $\theta = 18.84°$ at 10 keV and Eq. (8) gives $\Delta E \lesssim 3$ eV ($\Delta E \lesssim 2$ eV at 8 keV). If the vertical aperture is opened to 2 mm, a 2-fold increase in photon flux is available in exchange for a 2-fold decrease in resolution. A typical bent-crystal focusing monochromator might give resolution of ~10–15 eV[15] due to the broad rocking curve of the bent crystal and the nonideal source image (i.e., it is not a line source). This can be improved somewhat (again accompanied by lower photon flux) by inserting a "flag" pointed toward the center of the bent crystal (the flag would be moved up in Fig. 6e). This allows only a smaller cone of rays to be incident on and reflected from the crystal.

In some experiments utilizing synchrotron radiation, the beam is first reflected off a focusing mirror before passing through the monochromator. These mirrors are generally bent toroidal mirrors (Pt-coated fused quartz, for example) which are designed to collect a larger portion of the beam and focus it into a spot at the sample, increasing the brightness by a significant factor. It should be realized that such focusing mirrors degrade the resolution of the photon energy substantially. Figure 7 shows that this is a result of the mirror converting the horizontal divergence of the beam into vertical convergence which then contributes to $\Delta\theta$ in Eq. (8). Often, horizontal collimation of the beam (*before* the mirror) by the use of a vertically positioned "V"-slit is used to improve resolution (again, at the expense of photon flux).

X-Ray Detectors. The absorption coefficient, μ, of a sample of thickness x can be measured using the relationship

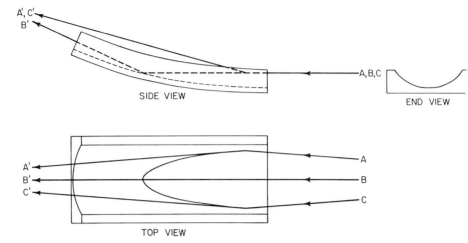

SIDE VIEW

END VIEW

TOP VIEW

FIG. 7. Diagram of the effect of a toroidally bent focusing X-ray mirror on the energy resolution of synchrotron radiation. For this diagram, the X-ray beam incident on the mirror (rays A, B, C) is assumed to have only horizontal divergence. The effect of the mirror is to convert this horizontal divergence to vertical convergence in the reflected beam (rays A', B', C').

$$I = I_o \exp(-\mu x) \qquad (9)$$

where I_0 is the intensity of X rays incident on the sample and I is the intensity transmitted. In general, measurement of the product μx is sufficient since the data usually are normalized to the edge jump. The detection equipment must then be able to give a quantity proportional to I_0 and one proportional to I. The transmission technique measures I_0 and I directly and $\ln(I_0/I)$ yields a quantity proportional to μx. Indirect techniques involve measurement of processes that are a result of the absorption of a photon rather than measurement of I directly. The fluorescence technique involves measurement of X-ray fluorescence (e.g., K_α emission for K-edge absorption) and this technique is used almost exclusively in biological applications.

In the transmission and fluorescence techniques, detectors that generate output proportional to X-ray intensities are required. Two types are used: pulse (photon counting) detectors and dc (integrating) detectors.[19] Measurement of I_0 requires a resonably transparent detector so that the beam still strikes the sample. This is usually accomplished with a gas ionization chamber, a dc detector filled with an appropriate gas that can

[19] E. A. Stern, AIP Conf. Proc. **64**, 39 (1980).

be ionized by the X rays. The current thus generated across two charged plates is amplified and measured to be proportional to the X-ray intensity. The gas (e.g., He, Ne, N_2, Ar, etc.) is chosen to absorb an optimum fraction (~20–30%) of the radiation. For transmission work, the I detector should be chosen to absorb all the remaining photons. A larger ionization chamber often serves the purpose since it is generally useful to allow some residual radiation to pass through the I detector (for alignment purposes or for a simultaneous energy calibration measurement).

For detection of fluorescent photons, pulse detectors are commonly used (due to their better sensitivity at low count rates). In addition, pulse detectors allow some amount of energy discrimination (a factor important in fluorescence excitation measurements). In some cases, energy discrimination can also be achieved by using appropriate filters and this method has been used with both pulse detectors and gas ionization chambers in fluorescence detection experiments. Pulse detectors include gas proportional counters, scintillation counters, gas scintillation counters, and solid state detectors.

Data Acquisition

Transmission. Most biological samples (i.e., metalloproteins, metalloenzymes) are spectroscopically dilute and, as a result, fluorescence excitation techniques are used normally to collect XAS data. However, every EXAFS analysis requires data on "model compounds" for extraction of phase and amplitude information. (In this context, a model compound is defined as any structurally characterized metal complex with an appropriate absorber–scatterer pair.) XAS data on such compounds are almost always collected on solid samples using the transmission technique.

Figure 5 illustrates the simple setup required for transmission XAS data collection: a monochromator, an I_0 ionization chamber, some means to hold the sample, and an I ionization chamber. The sample usually rests on a remotely controlled y-z translation stage (defining x as the beam propagation direction) for easy alignment of the sample with the beam. The ionization chambers are filled with an inert gas chosen so that a certain fraction of the X-radiation is absorbed by the I_0 detector. The optimum fraction depends on the sample thickness and one can calculate the optimum sample thickness (x_{opt}) for best signal-to-noise ratio (S/N) to be[20]

[20] B.-K. Teo, *in* "EXAFS Spectroscopy, Techniques and Applications" (B.-K. Teo and D. C. Joy, eds.), p. 13. Plenum, New York, 1981.

$$\mu_t x_{opt} \approx 2.5 \qquad (10)$$

where μ_t is the total absorption coefficient of the sample (just above the absorption edge of interest). (Note that the measurement of $\mu_t x$ is given by

$$\mu_t x = \ln \left(\frac{I_0}{I} \frac{I'}{I_0'} \right) \qquad (11)$$

where I_0' and I' are measurements in the absence of the sample.) Given such a sample, the optimum fraction of the radiation absorbed by I_0 is 20–25%.[19,20] One can (crudely) adjust this fraction by putting different gases in the I_0 ionization chamber for different photon energies. It proves convenient to use N_2 for photon energies between ~6 and 12 keV and Ar for higher energies.

The optimum sample thickness in Eq. (10) should be considered a maximum—other experimental effects can lower the expected S/N. Such effects include beam instability, incorrect sample positioning, heterogeneous sample thickness (e.g., pinholes, etc.), harmonic contamination of the X-ray beam, and monochromator "glitches." The last two effects are treated in detail later. It is necessary to be extremely careful about making up solid samples, making sure the material is homogeneously ground and (usually) pressed. One can adjust $\mu_t x$ by using different thickness sample holders or by diluting the sample with an inert solid containing only low-Z atoms (BN, Li_2CO_3, and $LiBF_4$ are commonly used).

In most XAS measurements, the data are collected at discrete energies, selected by stepping motor positioning of the monochromator crystals to some Bragg angle, θ. It is necessary to use variable intervals (in stepping motor pulses, proportional to θ) since often much smaller steps are used across the edge. In the EXAFS region, both variable intervals and variable integration times are used often to compensate for the eventual k^n weighting of the processed EXAFS data. Thus, the scan is set up to collect data points equally spaced in k and the integration times are longer in the high-k region (to give constant S/N throughout the EXAFS region after multiplication by k^n). In practice, the total desired scan time precludes perfect compensation so that most reported EXAFS data still exhibit more noise in the high-k region.

The time used for collection of one spectrum (scan time) can vary anywhere from ~10 min to a few hours. The overhead time for (computer-controlled) movement of the monochromator stepping motor makes shorter scan times less efficient and (at SSRL) an effective lower limit is probably ~15–20 min. Due to spurious beam instabilities and the hazards associated with attempting to include shortened scans in the overall average, the upper limit of scan time is a function of the frequency of the beam

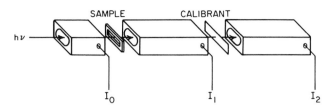

FIG. 8. Detector arrangement for measurement of internal calibration. The sample absorption is given by $\mu x = \ln(I_0/I_1)$ and the radiation that gets through the I_1 detector is used to measure the calibrant absorption, $\mu x = \ln(I_1/I_2)$.

problems. Since it is standard practice to discard the entire scan once a beam problem corrupts the data, if they are more frequent, the scan time should be shortened so that more complete scans can be acquired. At SSRL, an effective upper limit of ~1 hr seems reasonable, and an average scan time of ~30 min is common.

Energy calibration of the monochromator is generally accomplished by recording a high-resolution spectrum of the edge region of the appropriate metal foil. The inflection point of the absorption edge is used then for a single-point calibration. This is accomplished best by the internal calibration procedure illustrated in Fig. 8. A third ionization chamber is located behind the second with the calibrant between. Three measurements are recorded (I_0, I_1, I_2). A small amount of radiation passes through I_1, through the calibrant, and into I_2. The absorbance of the sample [$\mu_t x = \ln(I_0/I_1)$] is recorded simultaneously with the absorbance of the calibrant [$\mu_c x_c = \ln(I_1/I_2)$]. Thus each scan of the sample is calibrated internally simultaneously with data acquisition. This can be done for fluorescence excitation data as well, providing at least some radiation passes through the sample.

Fluorescence. For concentrated samples, the transmission technique is adequate. However, for spectroscopically dilute samples (as in most biological applications), the fluorescence excitation technique is the preferred method of data acquisition. In practice, for first-row transition metals, the lower limit of concentration for which transmission XAS is feasible is probably ~10 mM.[21]

The fluorescence XAS technique takes advantage of the fact that a certain fraction of the atoms from which a K-shell (1s) electron has been dissociated (by absorption of an X-ray photon) relax by emitting a fluorescent photon (see Fig. 9a). The fluorescent yield is a monotonically increasing function of atomic number. For example, K_α fluorescent yields

[21] R. G. Shulman, P. Eisenberger, and B. M. Kincaid, *Annu. Rev. Biophys. Bioeng.* **7**, 559 (1978).

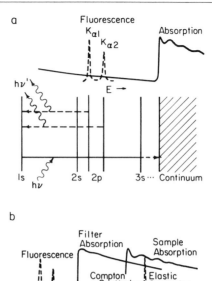

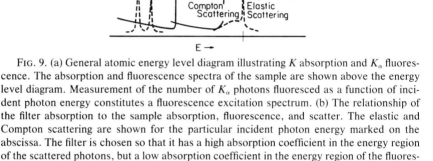

FIG. 9. (a) General atomic energy level diagram illustrating K absorption and K_α fluores-cence. The absorption and fluorescence spectra of the sample are shown above the energy level diagram. Measurement of the number of K_α photons fluoresced as a function of inci-dent photon energy constitutes a fluorescence excitation spectrum. (b) The relationship of the filter absorption to the sample absorption, fluorescence, and scatter. The elastic and Compton scattering are shown for the particular incident photon energy marked on the abscissa. The filter is chosen so that it has a high absorption coefficient in the energy region of the scattered photons, but a low absorption coefficient in the energy region of the fluores-cent photons. This is typically an elemental filter with atomic number one less than the absorbing atom (a "$Z - 1$" filter).

for Fe and Mo are 0.347 and 0.764, respectively.[20] The data collection proceeds in the same manner as for transmission XAS with the mono-chromator scanned through the energy range containing the sample ab-sorption edge and EXAFS. However, instead of monitoring the transmit-ted intensity, the number of K_α photons emitted is measured (usually by photon counting techniques). This quantity is directly proportional to the number of photons absorbed and thus contains the same information. I_0 is still measured with a partially transparent ionization chamber and the signal of interest is F/I_0 (where F is the measured fluorescence signal). The advantage of the fluorescence technique for dilute (or thin) samples comes about due to the difference in energy of the K_α photons making up the signal and the photons making up the background (the elastic and

Compton scattered photons at higher energies). This is illustrated in Fig. 9b. The fluorescence technique gains sensitivity by making use of various scatter rejection schemes (i.e., being able to energy discriminate against scattered photons). This can be accomplished either by using detector systems with high-energy resolution or by choosing an appropriate low-pass filter which absorbs selectively the higher energy scattered photons (Fig. 9b). For study of an element with atomic number Z, an appropriate filter can be made from a thin foil or deposited layer of the element with atomic number $Z-1$.

There are four basic types of detectors that have been used for fluorescence-detected XAS work: scintillation counters,[22–25] solid-state detectors,[26] gas-filled detectors (either ionization chambers or proportional counters),[19] and the so-called "barrel monochromator."[27,28] Scintillation counters use a material that can convert photon-produced photoelectrons into electronic excitation energy which is discharged by emission of a visible or near-UV photon. A photomultiplier tube behind the scintillator detects this fluorescence. The most commonly used scintillators are NaI(Tl) or certain plastics. The advantage of scintillation counters is that they can reach fairly high count rates and they are inexpensive enough that arrays of detectors can be built[24,25] to cover a large solid angle (as high as $\sim20\%$ of 4π) around the sample. The maximum count rate available to any photon counting detector is determined by its characteristic deadtime, τ. This is defined as the minimum time (Δt) which must separate adjacent pulses for each of them to be detectable. If $\Delta t < \tau$, the two pulses are counted as only one. Therefore, the observed count rate (N') is related to the actual count rate (N) by

$$N' = N(1-N\tau) \tag{12}$$

Equation (12) holds up to $N\tau \approx 0.4$.[19] Thus, *with* appropriate deadtime corrections, a detector will yield reliable count rates up to N_{max} where

$$N_{max} \approx 0.4/\tau \tag{13}$$

[22] R. G. Shulman, P. Eisenberger, B.-K. Teo, B. M. Kincaid, and G. S. Brown, *J. Mol. Biol.* **124**, 305 (1978).
[23] E. A. Stern and S. M. Heald, *Rev. Sci. Instrum.* **50**, 1579 (1979).
[24] S. P. Cramer and R. A. Scott, *Rev. Sci. Instrum.* **52**, 395 (1981).
[25] J. C. Phillips, *J. Phys. E* **14**, 1425 (1981).
[26] J. Jaklevic, J. A. Kirby, M. P. Klein, A. S. Robertson, G. S. Brown, and P. Eisenberger, *Solid State Commun.* **23**, 679 (1977).
[27] J. B. Hastings, *in* "EXAFS Spectroscopy. Techniques and Applications" (B.-K. Teo and D. C. Joy, eds.), p. 205. Plenum, New York, 1981.
[28] M. Marcus, L. S. Powers, A. R. Storm, B. M. Kincaid, and B. Chance, *Rev. Sci. Instrum.* **51**, 1023 (1980).

For NaI (Tl) scintillation counters, the deadtime is ~ 1 μsec and $N_{max} \approx 4 \times 10^5$ cps. With plastic scintillators, the deadtime can be about an order of magnitude shorter[19] yielding another factor of 10 in maximum count rate. The main disadvantage of scintillation counters is the low energy resolution (no better than $\sim 50\%$ at 7 keV) which is not sufficient to separate K_α emission from scattering. Therefore, scintillation detector systems are generally used with appropriately designed filter assemblies.[23,24]

Solid-state detectors use semiconductor materials (Si or Ge) as ionization detectors. In some cases, to reduce the dark current of the semiconducting material, impurities are added (Li is most common) and the detector is operated at liquid nitrogen temperature. These detectors (most commonly SiLi or intrinsic Ge) have very good energy resolution (perhaps ~ 100 eV at 7 keV), but can operate only at a maximum of $\sim 20,000$ cps.[23] They also tend to be expensive and this (plus the liquid nitrogen dewars) makes it prohibitive to build an array to collect a significant solid angle of fluorescence.

The barrel monochromator[27,28] achieves its high-energy resolution by using Bragg reflection from graphite or LiF crystals to monochromatize the emitted radiation and using a Rowland circle geometry to focus a cone of rays from a point source (the sample) to a point on the detector (a solid-state detector was used in both reported systems). The crystals are formed (as a mosaic) into a concave "barrel" geometry in order to accomplish the focusing with monochromatization of the proper photon energy range (e.g., the FeK$_\alpha$ emission line). The main advantage of this system is its high-energy resolution, although this seems to be outweighed by the disadvantage of the small solid angle subtended ($\sim 1.5\%$ of 4π[23]) and the fact that a new crystal arrangement must be developed for use at each energy (e.g., one for Fe, one for Cu, etc.).

As with scintillation counters, gas-filled detectors (such as ionization chambers) have very poor energy resolution. Thus, for use as fluorescence detectors, a filter is virtually required (for dilute samples). In general, thicker filters give better scatter rejection but a point of diminishing returns is reached soon. This is due to the presence of filter fluorescence, produced by the absorption of scattered photons. As more scattered photons are absorbed by the filter, more filter fluorescence photons are produced and reach the detector, effectively generating more filter-transmitted scatter. This effect can be overcome partially by using baffles (known as Soller slits) which prevent a substantial amount of filter fluorescence (emitted in *all* directions by the filter) from reaching the detector. The Soller sites are designed to allow photons that originate at the sample to pass all the way to the detector unimpeded. Thus, depending upon the profile of the sample "viewed" by the detector, the design of appropriate Soller slits can be quite complex.

The optimal arrangement for fluorescence detection studies on low-Z metals (e.g., Fe, Cu) has the sample at 45° to both the incident radiation and the fluorescence detector with the detector in the horizontal plane[29] (see Fig. 5). This not only allows the fluorescence detector to see an optimal projection of sample surface area, but also is the geometry which gives the minimum amount of background scatter. With an array of detectors, one packs the detectors around this particular orientation to collect as much solid angle as possible.[24] For XAS studies of higher-Z metals (e.g., Mo), one can use a sample with greater pathlength (up to ~1 cm of H_2O at 20 keV), still allowing transmission of some radiation for internal calibration, yet allowing more surface area of the sample to be "viewed" by the fluorescence detectors.

In general, the same (dedicated) minicomputer which controls the XAS experiment also takes care of data acquisition. Data from ionization chambers start out as current and go through a current preamplifier and a voltage-to-frequency converter in order to be recorded as cps. For scintillation counters, the individual pulses need to be amplified (by fast delay line amplifiers, for example) and then go through single channel analyzers (that allow discrimination against harmonic scatter) before being recorded as cps. For applications that use multiple detector arrays, the outputs of all the detectors can be summed in an analog fashion to give one signal (assuming the computer interface can handle the overall count rate), but for statistical reasons it is best to collect data from each channel (detector) separately and use an appropriate weighting scheme.[30]

Each channel will have a different S/N (depending on the specific electronic components and on the geometrical location of the detector) and in order to avoid overemphasizing the data from a channel with large scatter-to-fluorescence ratio, one should use a weighting scheme based on estimated S/N. The S/N for each channel can be estimated before data collection is begun by examining the actual data to get an estimate of the background and fluorescence counts and then assuming that photon statistics are obeyed. Figure 10 shows a hypothetical spectrum [of $(F/I_0)_i$ versus energy] which might be observed for the ith fluorescence channel. The signal of interest (e.g., the EXAFS) is assumed proportional to the size of the edge jump (measured as cps and normalized to I_0) and the noise

[29] D. R. Sandstrom and J. M. Fine, *AIP Conf. Proc.* **64**, 127 (1980).
[30] R. A. Scott, S. P. Cramer, R. W. Shaw, H. Beinert, and H. B. Gray, *Proc. Natl. Acad. Sci. U.S.A.* **78**, 664 (1981).

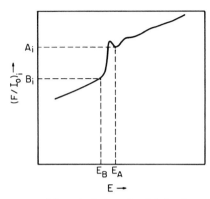

FIG. 10. Measurements used for calculation of weighting factors for fluorescence excitation spectra. The proper weighting factor can be calculated to be proportional to the relative edge jump $[\Delta_i' = (A_i - B_i)/A_i]$.

is estimated (by photon statistics) as proportional to the square root of the total count rate (again normalized to I_0). For the ith channel, one need make only two measurements (see Fig. 10): a count rate (B_i) at an energy (E_B) just before the edge of interest and a count rate (A_i) at an energy (E_A) just after the edge of interest. Then the signal (S_i) and noise (N_i) for the ith channel are estimated as

$$N_i \propto A_i^{1/2} \tag{14}$$
$$S_i \propto (A_i - B_i) \equiv \Delta_i \tag{15}$$

The proper weighting factor (W_i) is proportional to the square of the S/N but also must include normalization of the data in the ith channel to unit edge jump (this is so that channels with equal S/N contribute equally to the weighted average edge jump):

$$W_i \propto \frac{(S_i/N_i)^2}{\Delta_i} = \frac{\Delta_i}{A_i} \equiv \Delta_i' \tag{16}$$

Thus, the proper weighting factors are just proportional to the relative edge jumps (Δ_i') in each channel. The properly weighted fluorescence $(\langle F \rangle)$ is then calculated from the fluorescence in each channel (F_i) by

$$\langle F \rangle = \frac{\Sigma_i W_i F_i}{\Sigma_i W_i} \tag{17}$$

With the aid of the computer, the proper weighting may be done at the time of data acquisition. (This saves storage space since only one number is stored per energy.) However, in some cases, fluorescence detectors at different locations respond differently to beam or monochromator instabilities (see below). Since it is difficult to know the extent of such behavior at data acquisition, it may be advantageous to postpone the proper averaging until the individual channels can be examined in detail.

Experimental Difficulties. The EXAFS oscillations that contain the desired structural information are a small modulation of a large signal. Measurement of EXAFS data is thus very sensitive to minor defects in sample or beam integrity. The mechanics of collecting EXAFS data is straightforward—the art is to be aware of possible artifacts and to know how to circumvent them. This section discusses a number of potential pitfalls, their causes, and possible solutions.

Especially for biological systems, sample integrity is of utmost importance. One must be able to demonstrate that the state (usually, oxidation state of the active site) of the sample is the same after EXAFS data collection as it was before. It is also desirable to have some evidence that this state existed *during* data collection (and was not just regenerated when the X rays were turned off). One solution is to do "on-line" monitoring of the sample in the X-ray beam.[31] This solution is complicated and expensive. It should suffice to perform the experiment at cryogenic temperatures, check the appropriate spectroscopic "handles" (usually EPR, but also perhaps optical reflectance) before and after data collection (also at cryogenic temperatures), and show that the XAS spectrum (both edge and EXAFS) does not change over the course of data collection.

The problem, of course, is that X-radiation is ionizing—hydrated electrons are easily produced by irradiation of aqueous solutions. It has been demonstrated that free radical EPR signals are observed in aqueous cytochrome *c* oxidase solutions after X-irradiation.[31–33] The danger exists that these hydrated electrons may diffuse to the metal active site(s), causing "photoreduction" or "radiation damage." Low temperatures (≤ 200 K) seem to alleviate the problem, since diffusion of the hydrated electrons to the metal sites at these temperatures becomes slower than radical recombination. At 77 K, in cytochrome *c* oxidase samples, the radiation-induced radicals disappear completely without significant reduction of the EPR-detectable metal sites in ~2 weeks.

Homogeneity of the physical state of the sample is also important if one is to avoid artifacts. This is of less importance in fluorescence data

[31] B. Chance, P. Angiolillo, E. K. Yang, and L. Powers, *FEBS Lett.* **112**, 178 (1980).
[32] G. W. Brudvig, D. F. Bocian, R. C. Gamble, and S. I. Chan, *Biochim. Biophys. Acta* **624**, 78 (1980).
[33] R. A. Scott, *in* "The Biological Chemistry of Iron" (H. B. Dunford, D. H. Dolphin, K. N. Raymond, and L. C. Sieker, eds.), p. 475. Reidel Publ., Boston, Massachusetts, 1982.

collection. For example, frozen solutions (not glasses) can *usually* be examined by fluorescence with no problems, but transmission experiments on such physically heterogeneous samples are not possible. For transmission work, any portions of the sample that are more transparent to X rays (e.g., thin spots, bubbles, pinholes, cracks) will increase the I ionization chamber reading above what it should be yielding a lower $\ln(I_0/I)$ quantity. Assume that some fraction, f, of the sample is thinner than the rest (of thickness x) by an amount, δ (i.e., the thin spots have thickness $x - \delta$). Then the measured signal would be

$$(\mu x)_{\text{meas}} = \ln(I_0/I) = \mu x - \ln\{1 - f[1 - \exp(\mu\delta)]\} \tag{18}$$

[instead of $\ln(I_0/I) = \mu x$, for a homogeneous sample of thickness x]. This signal is no longer proportional to μx, the result being that the normalized EXAFS amplitude will be underestimated. For example, if $\mu x = 2.5$ [the optimum thickness given in Eq. (10)], 1% of the illuminated surface area of the sample consists of pinholes (or cracks) with $\delta = x$, and the edge-to-background ratio is 4, the EXAFS amplitude would be underestimated by ~4%. Worse yet, if the experiment does not track the vertical motion of the beam perfectly during the scan, the percentage of radiation that "leaks" through the pinholes will change, giving rise to small discontinuities in the data.

For fluid solution samples, these problems can be overcome by assuring that the solution is well mixed and that no bubbles are in the path of the X-ray beam. For solid samples, the material should be extensively ground with a mortar and pestle and pressed into a spacer (usually with Mylar or Kapton windows) to form a flat, homogeneous pellet.

Even with perfectly homogeneous samples, transmission experiments can be affected by beam artifacts. With synchrotron radiation, these can be of two types: (1) spurious, time-dependent changes in the synchrotron radiation beam; and (2) artifacts introduced or emphasized by the optics (i.e., the monochromator). The first category includes sudden changes in beam intensity, beam density, or beam position. Any of these are manifested as a sudden drop in I_0 reading. These discontinuities are usually reflected in the measured $\ln(I_0/I)$ quantity due to the fact that I_0 and I usually do not ratio perfectly (vide infra). Such artifacts are infrequent, are easy to spot, and are usually out of the experimenter's control.

The second category of beam artifacts are a direct result of the double flat crystal monochromators used for XAS. The presence of n in the Bragg relationship, Eq. (7), results in contamination of the reflected beam (supposedly just the fundamental, $n = 1$) with higher harmonics ($n = 2, 3,...$). For example, a monochromator set to reflect 10 keV X rays also reflects 20 keV X rays (along precisely the same path). Of course, the absorption coefficient of the sample for the higher energy radiation is much lower than for the fundamental. Thus, most of this harmonic radiation "leaks" through the sample (it is often called leakage radiation) causing the same type of distortion of the EXAFS data as leakage of the fundamental through pinholes. (This is usually referred to as the "thickness effect" since the distortion is worse for thicker samples.[34-38]) If the detectors could discriminate against this leakage radiation, it would not be a problem, but ionization chambers have very poor energy resolution. Thus, it is typically the presence of this leakage radiation which causes I_0 and I to ratio imperfectly. This in turn causes spurious beam fluctuations to show up in $\ln(I_0/I)$.

Another artifact introduced by the double crystal monochromators is the so-called monochromator "glitch." In addition to the set of Bragg planes being used for fundamental reflection, there are many others that may reflect some of the fundamental intensity in another direction at precise energies in the scan. This is registered as a sharp dip in the I_0 reading which occurs at the same energy all the time (i.e., every scan) and is referred to as a "glitch." Again, due to imperfect ratioing (from leakage radiation), this sharp "glitch" shows up as a positive spike in the $\ln(I_0/I)$ data.

It is clear from the above discussion that leakage radiation can introduce subtle distortions to the resulting EXAFS data that must be minimized. To avoid distortions due to the thickness effect, one should be sure to make samples that are thinner or more dilute so that the distortions are not as large. However, this is merely treating the symptoms. The preferred method of solving all such problems is to reduce or eliminate the leakage radiation present in the beam. The most direct method of harmonic discrimination (assuming that the harmonic is the source of leakage radiation) is adjustment of the source spectrum to decrease the *relative* amount of higher energy photons present. For laboratory sources, one can just turn down the power on the X-ray tube, while for synchrotron radiation, one must choose to run at lower electron energy (shifting the source spectrum to lower critical energy), or turn down a wiggler magnet field (if applicable). In any case, some sacrifice in fundamental intensity is necessary.

[34] S. M. Heald and E. A. Stern, *Phys. Rev. B: Solid State* [3] **16**, 5549 (1977).
[35] D. M. Pease, L. V. Azaroff, C. K. Vaccaro, and W. A. Hines, *Phys. Rev. B: Condens. Matter* [3] **19**, 1576 (1979).
[36] E. A. Stern, S. M. Heald, and B. A. Bunker, *Phys. Rev. Lett.* **42**, 1372 (1979).
[37] E. A. Stern, B. A. Bunker, and S. M. Heald, *Phys. Rev. B: Condens. Matter* [3] **21**, 5521 (1980).
[38] E. A. Stern and K. Kim, *Phys. Rev. B: Condens. Matter* [3] **23**, 3781 (1981).

Another method is to choose a set of Bragg planes for which the harmonic reflection is forbidden. For example, with Si[111] crystals, the even order harmonics (e.g., Si[222], Si[444]) are nearly forbidden. This method is not always available for a particular experiment. One of the most commonly used methods of harmonic discrimination is monochromator "detuning." This method takes advantage of the relatively narrow rocking curve of the harmonic (compared to the fundamental). Detuning refers to rotating one of the monochromator crystals (with respect to the other) slightly out of parallel alignment. Detuning slightly off the rocking curve maximum will reduce drastically the amount of harmonic reflected while affecting only slightly the reflected fundamental intensity.

A slightly more drastic approach (to achieve harmonic discrimination) is to condition the source spectrum before the radiation reaches the monochromator by the use of X-ray reflecting optics. In the X-ray region, total external reflection from a material occurs only at glancing angles (below some critical angle).[39] This critical angle depends on photon energy; the higher the energy, the smaller the critical angle. Thus, with proper choice of material and incident angle, one can design an X-ray mirror which will reflect the fundamental but absorb the harmonic. One can use a flat mirror provided a material with adequate optical flatness is available (float glass has been used for this purpose), Platinum- (or gold-)coated fused quartz has been used in mirrors designed for X-ray focusing applications. Although the mirror shown in Fig. 7 was designed to gather a wide horizontal acceptance and focus it into a bright spot at the sample, a secondary effect is very good harmonic discrimination (when the harmonic is above the cutoff energy defined by the angle of the mirror).

In the absence of leakage radiation, distortion of the EXAFS data (i.e., underestimation of EXAFS amplitudes) can also be caused by detector nonlinearity. For transmission data, the I ionization chamber observes a signal which varies over a much wider dynamic range (due to the sample absorption edge) than the I_0 signal. If the I detector has a nonlinear response, the effect will resemble closely the leakage radiation effect (before the edge, the I reading will be high, in a nonlinear region, and will be measured as smaller than it really is; after the edge, the I reading decreases into a linear region and is measured correctly). Fortunately, ionization chambers exhibit extremely linear response, but if other kinds of detectors are used, their linearity should be checked (against an ionization chamber, for example).

Fluorescence detectors can also exhibit nonlinear behavior [as described by their inherent deadtime as in Eq. (12)]. Thus, if fluorescence

[39] R. C. Gamble, *AIP Conf. Proc.* **64**, 113 (1980).

detectors are operated near their high count rate limit, distortions will occur again since the fluorescence signal will get considerably larger after the sample absorption edge. Deadtime corrections can correct this problem partially but only up to a certain limit [Eq. (13)]. The best solution is to use fluorescence detectors well below their saturation limit.

Just as with leakage radiation, detector nonlinearity will result in imperfect ratioing of I_0 and I. Along with this comes susceptibility to monochromator "glitches," spurious beam fluctuations, etc. For an ideal XAS experiment, one should assure that all detectors are operating in a linear range and that effective harmonic discrimination is used.

Data Reduction and Analysis

Collecting the data is only half of what is required to use EXAFS successfully to determine the structure of a metal site. Reduction and analysis of the data are at least as important and can be as susceptible to artifacts as data acquisition. It is convenient to divide the data workup into two steps: (1) data reduction involves extraction of the EXAFS data from the raw XAS data; (2) data analysis involves extraction of the structural information from the EXAFS data (by curve-fitting).

Data Reduction. The raw data as acquired consist of I_0, I, and possibly F (fluorescence) data as a function of monochromator motor steps. (If the internal calibration method was used, another "I_2" column corresponding to the third ionization chamber will also be present.) If more than one fluorescence detector was used, the fluorescence data may be in separate columns (channels), F_i. If this is the case, the F_i should be properly averaged within each scan (each data file) before any further processing occurs. The proper averaging consists of weighting the fluorescence channels appropriately based on their individual S/N (as discussed above). However, it is often useful to carry each F_i column individually through the data reduction and compare the EXAFS of each column to determine whether there are F_i columns that are too noisy to keep in the average. Once this is determined, the proper weighted average of the selected F_i columns can be done and the data reduction can proceed with the averaged fluorescence data, $\langle F \rangle$.

The next sept in data reduction (or the first step if transmission data is being used) is to convert monochromator motor steps into electron volts (eV). A single-point calibration is generally used to perform this conversion. With the external calibration procedure, separate scans of the appropriate metal foil should be recorded before and after the sample data. The $\ln(I_0/I)$ data from these scans give the calibrant edges which must be analyzed to locate the inflection point. With the internal calibration method, each sample data file has the calibrant spectrum as $\ln(I/I_2)$. In either case, the inflection points are located by finding all of the downgoing zero-crossings in the second derivative of the edge spectrum. The first inflection point in each metal edge corresponds to the tabulated energies in the periodic table of Fig. 1. Assuming that the motor step positions of all the calculated inflection points are approximately equal, they can be averaged to give the calibration point, S_c (in motor steps). The energy, E_c, corresponding to this is taken from the table (Fig. 1) and the energy, $E(i)$, for the ith point in each spectrum is calculated by

$$E(i) = hc/2d \sin \theta(i) \tag{19}$$

$$\theta(i) = \theta_c + [S(i) - S_c]/\Delta \tag{20}$$

$$\theta_c = \sin^{-1}(hc/2dE_c) \tag{21}$$

where d is the monochromator crystal d-spacing, $S(i)$ is the motor step value for the ith data point, and Δ is the number of motor steps per degree of monochromator crystal rotation.

Usually, at the same time the energy scale is generated a $\mathbf{k}$-scale ($\mathbf{k}$ is the photoelectron wave vector, in $\mathring{A}^{-1}$) and the appropriate ratioed data can also be generated. $\mathbf{k}(i)$ is calculated for each $E(i)$ using Eq. (1). In order to perform this calculation, one must choose a threshold energy, E_0, which is not a trivial matter. For a particular absorber, E_0 is expected to depend upon the valence state, but it cannot be chosen to match the inflection point of the sample absorption edge because bound state transitions may influence this quantity. The usual approach is to arbitrarily pick an E_0 for a particular absorber or use the measured inflection point and then allow E_0 to vary (redefining the $\mathbf{k}$-scale) during curve-fitting. The ratioed data will be $\ln(I_0/I)$, (F_i/I_0), or $(\langle F \rangle/I_0)$. Once these ratioed data are generated, plots of each data file (each scan) can be examined to determine whether any need to be discarded (e.g., if they were not finished, too noisy, or contained beam artifacts). The remaining good scans can then be averaged together and the EXAFS data will be extracted from this averaged XAS spectrum.

The next step in data reduction is the subtraction of the background from the edge and EXAFS. For transmission data, this background is due to absorption by all the other (lower atomic number) elements in the sample plus residual absorption due to the lower energy edges of the element being studied (e.g., if the Cu K edge is being examined, the Cu L edges will contribute to the background). For fluorescence data, the background consists of any scattered intensity not removed by the scatter rejection scheme (e.g., the filters). In general, the procedure is to con-struct a smooth curve (a polynomial of low order, $-2 \leq n \leq 3$) which mimics the background as well as possible.

Four methods may be useful. (1) Measure the background using an appropriate sample and fit it with a polynomial. This works well for fluorescence data for which a sample of buffer can mimic the scatter background reasonably well. For transmission data, however, it is extremely difficult to generate a sample containing all the appropriate elements except the metal. (2) Fit the sample data in the preedge region and extrapolate it through the edge and EXAFS regions. This may be successful with either transmission or fluorescence data, but caution must be exercised since the extrapolation is over a wider energy range than is actually fit and it is not guaranteed to mimic the background. (3) Fit the sample data in the EXAFS region (starting ~50 eV above the edge) with a polynomial and subtract a constant from it to force it through the data just before the edge. This approach can work for either transmission or fluorescence data. The polynomial is not guaranteed to mimic the background well, but it will level out the data in the EXAFS region, making the spline fit easier (vide infra). (4) For transmission data, one can use the Victoreen formula[40] to estimate the background (given the elemental composition of the sample). The Victoreen formula is given by

$$\mu = C\lambda^3 - D\lambda^4 \qquad (22)$$

where λ is the X-ray wavelength and C, D are tabulated coefficients.[40] Each element has two sets of coefficients (one set for before the element's absorption edge, one for after) and the calculated Victoreen must be summed for each element in the sample, then multiplied by a factor, γ (related to the unknown thickness and concentration of the sample), to force it through the sample data just before the edge.

The selection of background to subtract may seem rather arbitrary. It is important to realize that the exact nature of this curve is not critical to the data reduction. In principle, the raw data [denoted F_0, but could be $\ln(I_0/I)$ or (F/I_0)] can be written as

$$F_0 = \gamma\mu + F_{back} \qquad (23)$$

where γ is a proportionality constant (related to concentration, sample thickness, fluorescence yield, etc.) and F_{back} represents the background. Ideally, the procedure described above would find a perfect model of F_{back} and subtraction would give the signal, F_1:

[40] C. H. MacGillavry and G. D. Rieck, eds., "International Tables for X-Ray Crystallography," Vol. III, p. 171. Kynoch Press, Birmingham, England, 1968.

$$F_1 \equiv (F_0 - F_{\text{back}}) = \gamma\mu \tag{24}$$

Ultimately, we want to extract the EXAFS data (χ), which is theoretically represented by

$$\chi \equiv \frac{\mu - \mu_0}{\mu_0} \tag{25}$$

where μ_0 is the absorption expected for a free atom. In our ideal case, γ is known and the next step in data reduction would involve subtracting and normalizing F_1 with the function $\gamma\mu_0$:

$$F_2 \equiv (F_1 - \gamma\mu_0)/\gamma\mu_0 = \chi \tag{26}$$

Unfortunately, in the real world, F_{back} can never be perfectly modeled and γ is an unknown. The procedure described above yields some imperfect approximation to F_{back} (defined as F'_{back}) and subtraction yields

$$F'_1 \equiv (F_0 - F'_{\text{back}}) = \gamma\mu + \Delta F \tag{27}$$

where $\Delta F \equiv (F_{\text{back}} - F'_{\text{back}})$. In this nonideal case, the background is not completely removed and the next step which should involve subtraction of the scaled atomic absorption coefficient $(\gamma\mu_0)$ will not work.

However, the appropriate function to subtract can be found by simply fitting the EXAFS region (after background subtraction) to a smooth curve, μ_s. This is usually accomplished by the use of a polynomial spline. Typically, the EXAFS region is broken up into two or three spline regions and each region is fit with a polynomial (of third order, for example). The spline criteria are that the polynomials meet at the region endpoints with equal value and equal slope. This calculation of μ_s is found to be generally useful for both transmission and fluorescence data. Care must be exercised so that (1) μ_s does not track any EXAFS oscillations, thus eliminating them; and (2) there is no low frequency curvature left in the EXAFS data after subtraction of μ_s. Point (1) is usually not a problem as long as the spline polynomial order is low enough (third order is almost always sufficient) and the number of spline regions is not too large (no more than four regions are needed). Point (2) causes a problem only when μ_s does not fit the "background" curvature in the EXAFS region. This is the reason for the F'_{back} subtraction; it gives an F'_1 which is reasonably level (except for the EXAFS oscillations) and usually allows the spline to fit the overall curvature more precisely. It is also possible to check for an improper spline fit by looking for artifactual low-frequency peaks in the Fourier transform (vide infra).

The assumption in the above discussion is that the spline fit takes care of both the atomic falloff and the residual background:

$$\mu_s = (\gamma\mu_0 + \Delta F) \tag{28}$$

and the spline subtraction then yields

$$F_1' - \mu_s = \gamma(\mu - \mu_0) \tag{29}$$

All that remains is to normalize the data to the atomic falloff and the result will be the EXAFS. The proper normalization function (N) is the atomic falloff [as modeled by the Victoreen formula, Eq. (22)] scaled so that the atomic absorption has an edge jump identical to the F_1' data:

$$N = \gamma\mu_0 \tag{30}$$

$$\chi = (F_1' - \mu_s)/N \tag{31}$$

In general, χ is treated as a function of $\mathbf{k}$, the photoelectron wave vector, rather than a function of energy [the theoretical EXAFS expression, Eq. (2), is written as a function of $\mathbf{k}$]. $\chi(\mathbf{k})$ is severely damped in $\mathbf{k}$-space, yet much information is contained in oscillations at high $\mathbf{k}$. $\chi(\mathbf{k})$ is usually weighted by some power of $\mathbf{k}$ ($\mathbf{k}^3$ is now almost universal) in displayed EXAFS and in curve-fitting to compensate for this damping. [Note that some EXAFS spectra were and may still be displayed as $\mathbf{k}^n\chi(\mathbf{k})$ with $n = 0, 1, 2$. Care must be exercised to know what is being plotted before comparisons of literature spectra can be done.]

Data Analysis. The data reduction procedures summarized above have succeeded in extracting the EXAFS data, $\chi(\mathbf{k})$, from the raw XAS data. The data analysis procedures to be discussed in this section are designed to interpret these EXAFS data in terms of structural information concerning the metal site. Data analysis may be treated on different levels: qualitative information may be obtained by examining the Fourier transform (FT) of the EXAFS data; quantitative information may be extracted by the use of curve-fitting techniques. Both of these approaches will be described here.

EXAFS data are normally displayed in $\mathbf{k}$-space and since $\mathbf{k}$ has units of Å^{-1}, a Fourier transform will yield transform data in the inverse space, having units of Å. This is real space (with units of distance) but the transform is *not* a true radial distribution function. Instead, the inverse space will be defined here as R'-space, where R' is related to R (the true distance from the absorbing atom) by

$$R' = R + \alpha_1 \tag{32}$$

where α_1 is a (linear) phase shift. In general, the FT is carried out using

$$F(R') = \int_{\mathbf{k}_{min}}^{\mathbf{k}_{max}} \mathbf{k}^n\chi(\mathbf{k})\exp(2i\mathbf{k}R')d\mathbf{k} \tag{33}$$

The resultant $F(R')$ is a complex function of R' with both real and imaginary components, k_{min} and k_{max} are the limits used to select the $\chi(k)$ data to be included in the transform (k_{min} is usually chosen to be ~ 4 Å^{-1}), and k^n reflects the weighting of the $\chi(k)$ data as already discussed (usually, n = 3). Of course, with discrete $\chi(k)$ data, the integral of Eq. (33) must be done numerically using, for example, a Simpson's rule approach.[41]

Since the $\chi(k)$ function is assumed to be only real, the $F(R')$ function is required to be a symmetric function. However the only physically useful data are those for which $R' > 0$. Typically only the modulus of the FT is plotted (as shown in Fig. 11). This is given by Eq. (34):

$$|F(R')| = \{[F_{re}(R')]^2 + [F_{im}(R')]^2\}^{1/2} \qquad (34)$$

Each term in the EXAFS summation of Eq. (2) gives rise to a characteristic sine wave, which in turn gives rise to a single peak in the FT. Thus, observation of a peak in the $F(R')$ data suggests the presence of a shell of atoms in the EXAFS. Care needs to be taken since small peaks appear in the FT due to truncation of the $\chi(k)$ data and due to noise. Thus, only FT peaks larger than some baseline may be interpreted properly as true shells of atoms. This baseline may be approximated by using the height of peaks in the "high-frequency" (large R') portion of the $|F(R')|$ data (see Fig. 11).

For a shell of atoms at distance R_{as} from the absorbing atom, $|F(R')|$ will exhibit a peak centered at $R' = R'_{as}$, where R'_{as} is related to R_{as} by Eq. (32). The phase shift (α_1) occurs because the EXAFS expression [Eq. (2)] is *not* a sum of simple sine waves. In particular, $\alpha_{as}(k)$ is k dependent and may be approximated as a polynomial in k:

$$\alpha_{as}(k) \approx a_0 + a_1 k + a_2 k^2 + \cdots \qquad (35)$$

Thus, the sine term of Eq. (2) may be written

$$\sin[(2R_{as} + a_1)k + a_0 + a_2 k^2 + \cdots]$$

and the frequency is really given by ($2R_{as} + a_1$). In the FT described by Eq. (33), the 2 has been factored out and thus $R'_{as} \approx R_{as} + a_1/2$ is the peak position in the $|F(R')|$ data ($\alpha_1 \approx a_1/2$). For many different types of scatterers, α_1 takes on values ranging from ~ -0.5 to ~ -0.2 Å.

If each shell gave rise to a simple sine wave in $\chi(k)$, one would also expect delta functions in the $|F(R')|$ data. Thus, the k dependence of α_{as} also contributes to broadening of the FT peaks. The peak widths contain an additional contribution from the overall shape of the amplitude envelope for that shell. This comes both from the inherent backscattering

[41] P. A. Stark, "Introduction to Numerical Methods," p. 202. Macmillan, London, 1970.

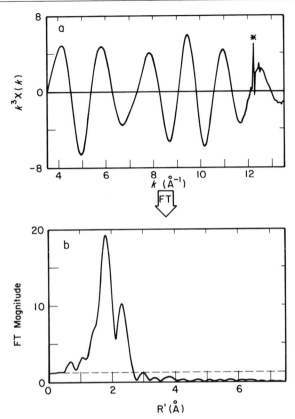

FIG. 11. Example of Fourier transformation. (a) Fe EXAFS data of $(NEt_4)_2$ [$Fe_4S_4(t$-BuS$)_4$ (the sample was a gift from K. Hagen and R. H. Holm), plotted as $k^3\chi(\mathbf{k})$ vs $\mathbf{k}$. The beat pattern indicates more than one shell of atoms around the Fe. The asterisk marks a monochromator glitch. (b) The result of Fourier transformation [using Eq. (33)] of the Fe EXAFS data in (a) over the range $\mathbf{k} = 3.5–13.5$ Å^{-1}. Mathematically, the two major FT peaks observed are the result of the beat pattern in (a). Physically, the peak at 1.85 Å is assignable to Fe–S scattering and the peak at 2.34 Å is due to Fe–Fe scattering. The dashed line indicates the approximate noise level in this FT.

amplitude, $B_s|f_s(\pi,\mathbf{k})|$, and from the Debye–Waller factor, $\exp(-2\sigma_{as}^2\mathbf{k}^2)$. Larger σ_{as} values give rise to broader FT peaks, so that data collection at lower temperatures gives rise to smaller σ_{as}, yielding sharper FT peaks. In some instances, FT peaks at high R' can be brought out of the noise by lowering the temperature of data collection.

Given the ability to account for variations in σ_{as} from one sample to another, the area of an FT peak can be used as a measure of N_s, the number of scatterers in the shell. It is often difficult, however, to glean any independent information concerning the value of σ_{as} and determinations of N_s are therefore susceptible to large errors. Thus, FT peak heights (or areas) can be used only for crude estimations of relative numbers of atoms. Some methods of curve-fitting analysis will be described below to address this problem.

One other utility of Fourier transformation should be mentioned. This involves a technique known as Fourier filtering. A certain region of $F(R')$ may be selected and backtransformed to **k**-space. If the selected region includes all of the important FT peaks, but not the high-R' region of $F(R')$ (which presumably consists solely of noise), then the result of backtransformation is the $\chi(\mathbf{k})$ data with all the high-frequency noise filtered out. Alternatively, the selected portion of $F(R')$ may contain a single peak, in which case the backtransformed data consist of the contribution of a single shell to $\chi(\mathbf{k})$. In this manner, it may be possible to extract out the individual "sine" waves that make up the $\chi(\mathbf{k})$ data. Initial curve-fitting may then be performed on this single shell. Figure 12 shows an example of each of these filtering techniques.

For curve-fitting analysis, it is necessary to know the amplitude and phase of the photoelectron backscattering function for the particular absorber–scatterer pair under consideration. In Eq. (2), these are represented by $B_s|f_s(\pi,\mathbf{k})|$ and $\alpha_{as}(\mathbf{k})$, respectively. Different methods exist for obtaining these functions and these methods will be discussed in detail below. One of these methods involves complex backtransformation[21,42] of a single peak in the $F(R')$ data. Selection of the appropriate portion of $F(R')$ is usually done by multiplying $F(R')$ by a window function consisting of zeroes outside the R' range of interest. The window function is typically unity in the desired R' region, but is preferably shaped so that a smooth transition from unity to zero occurs on each side of the selected FT peak. A typical shape function is a half-Gaussian and the window shown in Fig. 12c is a function made up of two half-Gaussians, each with 0.15 Å HWHM (half width at half maximum).

If we define $F_w(R')$ to be $F(R')$ multiplied by the appropriate window function, the complex backtransform may be described by the integral

$$\mathbf{k}^n\chi'(\mathbf{k}) = \frac{2}{\pi} \int_{R'_{\min}}^{R'_{\max}} F_w(R')\exp(-2i\mathbf{k}R')\,dR' \tag{36}$$

In this expression, $\chi'(\mathbf{k})$ represents the filtered EXAFS (a complex quantity) and $R'_{\min}$ and $R'_{\max}$ are selected to include all of the nonzero points in the windowed $F_w(R')$ data. The resultant $\chi'(\mathbf{k})$ data are typically useful only over the original **k**-space region used for the forward transform ($\mathbf{k}_{\min}$

---

[42] P. A. Lee, P. H. Citrin, P. Eisenberger, and B. M. Kincaid, *Rev. Mod. Phys.* **53**, 769 (1981).

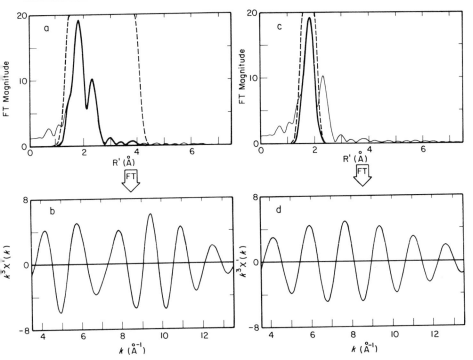

FIG. 12. Examples of Fourier filtering. In both (a) and (c), the dashed line is the filter window and the bold line is the product of this window and the Fourier transform, i.e., the data to be backtransformed. The results of transformation [using Eq. (36)] of the data in (a) and (c) are given in (b) and (d), respectively. The $\chi'(\mathbf{k})$ data in (b) are the result of "noise filtering" of the $\chi(\mathbf{k})$ data of Fig. 11a (note the absence of the monochromator glitch in Fig. 11a). The $\chi'(\mathbf{k})$ data in (d) are the result of extraction of the Fe–S shell (the 1.85 Å FT peak) from (c).

to $\mathbf{k}_{max}$). The real and imaginary components of $\chi'(\mathbf{k})$ over this region can be used to form the following functions:

$$A(\mathbf{k}) = |\chi'(\mathbf{k})| = \{[\chi'_{re}(\mathbf{k})]^2 + [\chi'_{im}(\mathbf{k})]^2\}^{1/2} \tag{37}$$

$$\Phi(\mathbf{k}) = \tan^{-1}[\chi'_{im}(\mathbf{k})/\chi'_{re}(\mathbf{k})] \tag{38}$$

These functions are components of an EXAFS expression of the form

$$\chi'(\mathbf{k}) = A(\mathbf{k})\sin[\Phi(\mathbf{k})] \tag{39}$$

An example of extraction of $A(\mathbf{k})$ and $\Phi(\mathbf{k})$ is given in Fig. 13. Comparison of Eq. (39) with Eq. (2) indicates that $A(\mathbf{k})$ is related to the backscattering amplitude and $\Phi(\mathbf{k})$ to the backscattering phase of the absorber–scatterer pair giving rise to the selected FT peak:

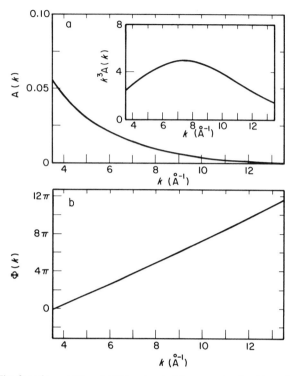

FIG. 13. The functions $A(\mathbf{k})$ and $\Phi(\mathbf{k})$ extracted [by complex backtransform, Eqs. (37), (38)] from the Fe–S shell (the data in Fig. 12c and d). The inset to (a) is the $\mathbf{k}^3$-weighted $A(\mathbf{k})$ function which constitutes the amplitude envelope of the $\mathbf{k}^3\chi'(\mathbf{k})$ data of Fig. 12d.

$$A(\mathbf{k}) = \frac{N_s B_s |f_s(\pi,\mathbf{k})|}{kR_{as}^2} \exp(-2\sigma_{as}^2 \mathbf{k}^2) \tag{40}$$

$$\Phi(\mathbf{k}) = 2\mathbf{k}R_{as} + \alpha_{as}(\mathbf{k}) \tag{41}$$

[One should note that, in practice, the complex backtransform technique for determining $A(\mathbf{k})$ is unique, but the determination of $\Phi(\mathbf{k})$ by Eq. (38) is subject to some ambiguity. $\Phi(\mathbf{k})$ should be a monotonically increasing function, dominated by $2\mathbf{k}R_{as}$, but the $\tan^{-1}$ function has discontinuities. Thus, the calculated $\Phi(\mathbf{k})$ data must have $n\pi$ added to it periodically. Also, the form of the Fourier transform equations used herein introduces a sine/cosine ambiguity so that $\pm\pi/2$ must be added to the calculated $\Phi(\mathbf{k})$, depending on the quadrant $\Phi(\mathbf{k})$ starts in, for $\chi'(\mathbf{k})$ to mimic the $\chi(\mathbf{k})$ data properly.]

The most important part of EXAFS data analysis is the curve-fitting of the EXAFS data, which allows interpretation of the EXAFS oscillations in terms of the actual structure around the absorbing metal atom. The general curve-fitting approach involves hypothesis of a local structural environment (involving types and numbers of atoms at given distances from the metal), followed by calculation of the EXAFS spectrum expected for this proposed structure, and comparison of this simulated EXAFS with the observed data. Typically, some sort of optimization procedure is used to alter the details of the proposed structure until the simulated EXAFS best fits the observed data according to a least squares criterion. Relating this to the EXAFS expression of Eq. (2), selection of the scatterer type (i.e., atomic number, Z) corresponds to generating the functions $B_s|f_s(\pi,\mathbf{k})|$ and $\alpha_{as}(\mathbf{k})$ (by methods yet to be described). Then, for each shell, initial values for the parameters R_{as}, N_s, and σ_{as} are chosen and the optimization consists of varying any or all of these parameters until the best least-squares fit is obtained. In some optimization procedures, the $\mathbf{k}$-scale may also be optimized by varying a parameter called ΔE_0 (for each shell) which corresponds to a difference in the ionization threshold energy for different shells.

The difference among the various methods of curve-fitting originate in different methods for generating the necessary amplitude and phase functions $[B_s|f_s(\pi,\mathbf{k})|$ and $\alpha_{as}(\mathbf{k})]$ for a given absorber–scatterer pair. Two basic methods are available: *ab initio* calculation of theoretical amplitude and phase functions and empirical measurement of amplitude and phase functions using measured EXAFS data on structurally characterized ("model") compounds.

Theoretical amplitude and phase functions are tabulated (for most elements) from $\mathbf{k} \approx 4.0–15.0\ \text{Å}^{-1}$ by Teo and Lee.[43] The phase functions depend upon the E_0 value used to extract the EXAFS data, so that curve-fitting with theoretical functions usually also involves optimization of ΔE_0. Although the theoretical phase functions seem to match experiment fairly well (given the proper choice of E_0), the theoretical amplitude functions often predict the EXAFS to be larger than observed (by a factor of ~2) and thus, require the use of the scale factor, S_s, of Eq. (6). Teo and co-workers have developed a modified curve-fitting approach known as the FABM (fine adjustment based on models) method[44] for using structurally characterized compounds to calculate appropriate values for S_s (and ΔE_0).

The empirical approach to generating amplitude and phase functions

[43] B.-K. Teo and P. A. Lee, *J. Am. Chem. Soc.* **101**, 2815 (1979).
[44] B.-K. Teo, M. R. Antonio, and B. A. Averill, *J. Am. Chem. Soc.* **105**, 3751 (1983).

for a particular absorber–scatterer pair involves using only data from structurally characterized (model) compounds. Two methods are available for doing this. The first involves parameterization of the functions and was pioneered by Hodgson and coworkers.[45] The particular parameterized amplitude and phase functions used are given by

$$B_s|f_s(\pi,\mathbf{k})| = c_0 \exp(-c_1\mathbf{k}^2)/\mathbf{k}^{c_2} \qquad (42)$$
$$\alpha_{as}(\mathbf{k}) = a_0 + a_1\mathbf{k} + a_2\mathbf{k}^2 + a_{-1}\mathbf{k}^{-1} \qquad (43)$$

[In the $\alpha_{as}(\mathbf{k})$ expression, usually *either* a_2 or a_{-1} are set to zero.] With this method, EXAFS data from a model compound are usually curve-fit with R_{as} and N_s for the shell fixed at the known values, and the c_i and a_i parameters are optimized. These parameter values are then fixed for use of the amplitude and phase functions in optimizations on data from a compound of unknown structure.

An alternate method for extracting amplitude and phase functions from model compounds involves the Fourier filtering and complex backtransform techniques already discussed. Equations (40) and (41) indicate that knowledge of R_{as} (for the model compound) allows calculation of $\alpha_{as}(\mathbf{k})$ from $\Phi(\mathbf{k})$ (generated in the complex backtransform) and knowledge of R_{as}, N_s, and σ_{as} allows calculation of $B_s|f_s(\pi,\mathbf{k})|$ from $A(\mathbf{k})$. Thus, except for the fact that σ_{as} is not available generally, one need make no assumptions about the functional form of the amplitude and phase functions [as required by Eqs. (42) and (43)] to calculate them as discrete functions empirically.

One of the main problems with any of the curve-fitting approaches described is the general lack of knowledge concerning the Debye–Waller σ_{as} for any particular compound. As already discussed, this term describes both static and dynamic variations in R_{as} within a particular shell. A crystallographic structural determination allows calculation of the static contribution [by Eq. (3)], but the dynamic contribution (due to thermal vibrations of the a–s bond) is only accessible for very symmetric compounds through use of vibrational frequencies and a normal coordinate analysis.[46] Curve-fitting analysis of EXAFS data on simple compounds can yield in principle σ_{as} values (by optimization of R_{as}, N_s, and σ_{as}) but the values of N_s and σ_{as} are highly correlated causing such a determination to be very error prone. These problems with σ_{as} have con-

[45] S. P. Cramer, K. O. Hodgson, E. I. Stiefel, and W. E. Newton, *J. Am. Chem. Soc.* **100**, 2748 (1978).
[46] S. J. Cyvin, "Molecular Vibrations and Mean Square Amplitudes." Elsevier, Amsterdam, 1968.

tributed significantly to the inability of practitioners of EXAFS to predict coordination numbers with accuracy any better than $\sim \pm 25\%$.

Described below is a hybrid method of curve-fitting analysis used with some success in the author's laboratory which attempts to use the strengths of both the theoretical and empirical approaches to circumvent some of the problems mentioned above. This method has some characteristics in common with the FABM method[44] but differs principally in that empirically derived amplitude and phase functions are used for the ultimate curve-fitting.

In outline, the method consists of (1) establishing a **k**-scale for both model and unknown data (by E_0 adjustment) which is compatible with the theoretical functions for the particular a–s pair; (2) assuming that the overall shape of the theoretical amplitude function is correct and using this to estimate a Debye–Waller factor for the model data; (3) using the known structural parameters and the estimated Debye–Waller factor for the model to extract (by complex backtransform) empirical amplitude and phase functions derived from the model; and (4) using these functions to perform curve-fitting on unknown data.

An example will serve to illustrate the method. The EXAFS data displayed in Fig. 11a are from a [4Fe–4S] cluster analog of Fe–S protein active sites. As an example, the Fe–S shell of this model compound will be used to generate amplitude and phase functions for curve-fitting. These functions should then be useful in fitting Fe–S shells in data from compounds of unknown structure. The first step is to Fourier filter just the Fe–S contribution as illustrated in Fig. 12c and d. This complex backtransform also generates the $\Phi(\mathbf{k})$ function (Fig. 13b) for this shell [by Eq. (38)]. Using the known average Fe–S distance from crystallographic data ($R_{as} = 2.280$ Å[47]), an empirical phase function [$\alpha_{as}(\mathbf{k})$] may be extracted from $\Phi(\mathbf{k})$ using Eq. (41). The $\alpha_{as}(\mathbf{k})$ calculated will depend upon the value of E_0 chosen to define the **k**-scale of the original EXAFS data (and to a lesser extent on the reverse FT window limits) and the empirical $\alpha_{as}(\mathbf{k})$ functions for three different choices of E_0 are plotted in Fig. 14. Also plotted in Fig. 14 is the theoretical phase function for Fe–S as calculated by Teo and Lee.[43] In order to place the empirically derived amplitude and phase functions to be used in this method on the same **k**-scale as the theoretical functions, E_0 is chosen as the value that yields empirical and theoretical phase functions that match as closely as possible. Figure 14

---

[47] The structural dimensions of the [4Fe-4S] core of [Fe$_4$S$_4$-(t-BuS)$_4$]$^{2-}$ were assumed to be the same as those for [Fe$_4$S$_4$(SC$_6$H$_5$)$_4$]$^{2-}$, taken from J. M. Berg and R. H. Holm, *in* "Iron-Sulfur Proteins" (T. G. Spiro, ed.), Metal Ions in Biology, Vol. 4, Chapter 1. Wiley, New York, 1982.

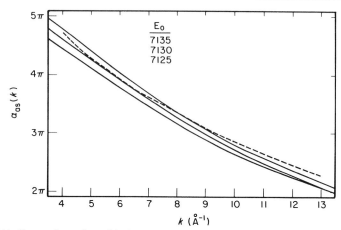

FIG. 14. Comparison of empirical and theoretical phase functions $[\alpha_{as}(\mathbf{k})]$. The empirical phase functions (solid lines) were extracted from the $\Phi(\mathbf{k})$ data of Fig. 13b as described in the text, with the original $\chi(\mathbf{k})$ data (Fig. 11a) extracted using different E_0 values (as indicated). The theoretical phase function is interpolated (using a third order polynomial) from values calculated by Teo and Lee in Ref. 43. This type of comparison is used to estimate the E_0 (7135 eV for this example) which places the observed data on the same $\mathbf{k}$-scale as that used in generation of the theoretical functions.

suggests that $E_0 = 7135$ eV is a reasonable value for the Fe–S interaction in this example.

Once the $\mathbf{k}$-scale is defined, the next step involves estimation of the Debye–Waller σ_{as} value for the model data. This consists of curve-fitting the model data using the empirical phase function determined above and the theoretical amplitude function given by Teo and Lee.[43] R_{as} and N_s are fixed at the known values and B_s and σ_{as} are optimized (while ΔE_0 is fixed at 0). For our example, this curve-fitting gives optimized values of $B_s = 0.354$ and $\sigma_{as} = 0.053$ Å. With the assumption that the shape (but not necessarily the absolute magnitude) of the theoretical amplitude function is correct, we can use this optimized value of σ_{as} as a physically meaning-ful one. Even if the assumption is invalid, this Debye–Waller σ_{as} can be used for comparison in fits of data from compounds of unknown structure.

The next step consists of using this optimized σ_{as} along with the known R_{as} and N_s to extract an empirical amplitude function $[B_s|f_s(\pi,\mathbf{k})|]$ from $A(\mathbf{k})$ using Eq. (40). $[A(\mathbf{k})$ was calculated in the complex backtrans-form of Fig. 12c and d using Eq. (37).] With this, both empirical phase and empirical amplitude functions have been extracted from the model data

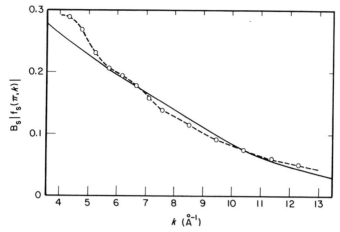

FIG. 15. Comparison of empirical and theoretical amplitude functions $[B_s|f_s(\pi, \mathbf{k})|]$. The empirical amplitude function (solid line) was extracted from the $A(\mathbf{k})$ data of Fig. 13a as described in the text. The theoretical amplitude function (○) consists of $|f_s(\pi, \mathbf{k})|$ values tabulated by Teo and Lee in Ref. 43 and $B_s = 0.354$ (see text). The points (○) indicate the tabulated values and the dashed line is calculated by interpolating between the points using a third-order polynomial.

and are now available for use in other curve-fitting. Figure 15 shows the extracted empirical amplitude function from the example and compares this function with the theoretical amplitude function as calculated by Teo and Lee.[43] [The theoretical amplitude function constitutes $|f_s(\pi,\mathbf{k})|$ and this is multiplied by the best fit scale factor, $B_s = 0.354$, for display in Fig. 15.]

In using these empirical amplitude and phase functions for curve-fitting single shells of data from a compound of unknown structure one has the option of varying R_{as}, N_s, and σ_{as}. (For multishell fits, ΔE_0 may require adjustment for some of the shells.) One problem often encountered is the high degree of correlation of N_s and σ_{as}. One method of circumventing this problem involves allowing N_s to take on only integer values. (For example, separate optimizations may be performed fixing N_s at a different integer value for each one.) One may then use the optimized value of σ_{as} (along with the "goodness-of-fit" indicator from the least-squares optimization) to determine which fit is best. In order to judge reasonably the values of σ_{as} one must be able to assess how physically reasonable a σ_{as} is. The utility of the value calculated for σ_{as} in the model compound is thus to have a basis for judging σ_{as} values obtained in various fits on compounds of unknown structure.

Accuracy of Results

Whatever specific method is used for EXAFS data analysis, there are inherent limitations to the accuracy of the results obtained. These can be summarized best as error limits for the determination of the three quantities of interest: R_{as}, N_s, and Z (atomic number). The error in R_{as} is dependent upon the ability to measure the frequency of the EXAFS oscillations and also slightly dependent upon the proper choice of E_0. With any of the curve-fitting techniques described, a lower bound on this error is probably $\sim \pm 0.02$ Å.

Errors in predicted N_s values come from various sources. First, distortions in the EXAFS amplitudes resulting from improper data collection (e.g., the thickness effect) will result in a systematic underestimation of N_s. For spectroscopically dilute samples, noise in the EXAFS data can combine with spurious beam artifacts (e.g., glitches) to cause random errors in N_s determination. (This is less of a problem in R_{as} determinations.) Probably most important is the contribution of the Debye–Waller factor to the overall EXAFS amplitude. Since independent knowledge of the proper value of σ_{as} is very difficult to come by, misestimation of this factor contributes to large errors in N_s. In the past, these factors have all contributed to errors in calculated N_s values as large as $\pm 30\%$. Although adequate testing has not yet been performed, it is hoped that the hybrid method of data analysis described above will be able to improve this error.

The atomic number, Z, of the scattering atom(s) is determined indirectly by finding which phase and amplitude functions fit the data best. Therefore, the error in Z is dependent on our ability to distinguish among phase and amplitude functions for adjacent elements in the periodic table. A typical error limit on Z is assumed to be $\Delta Z \approx \pm 1$. This means, for example, that C, N, and O are difficult to distinguish since their phase and amplitude functions all look fairly similar. This error limit applies to relatively low-Z scatterers. For metal scatterers, it should be increased probably to $\Delta Z \approx \pm 2$.

Conclusion

This chapter was written with one purpose in mind: broader exposure of the scientific community to the methodological details of the EXAFS technique will allow more critical examination of published data and eventually result in higher quality work and more acceptance of EXAFS as a useful structural technique. As with any infant technique, expectations have run the gamut: EXAFS is not the savior of bioinorganic and biophysical chemists, but it does have a niche to fill. Any potential (or current) EXAFS practitioners must be aware of both the strengths *and* the limitations of the technique. It is hoped that this chapter has made some small contribution toward a more rational use of the EXAFS technique in the study of biological systems.

[24] Velocity Sedimentation Study of Ligand-Induced Protein Self-Association

By George C. Na and Serge N. Timasheff

A large number of protein reactions are regulated through the reversible stoichiometric interactions with small molecules or ligands. Within this wide spectrum of protein reactions, the reversible association and dissociation of protein subunits have drawn considerable research interest. The biological significance of such interactions is evident. Subunit associations are ubiquitous in key enzymes regulating metabolic pathways, in the formation of cellular organelles, and in regulatory and transporting proteins.

This chapter is devoted to the methodology of studying ligand-induced protein self-associations using the velocity sedimentation technique. The type of study described here is aimed at characterizing macromolecular associations in terms of their stoichiometries and equilibrium constants, as well as at elucidating ligand bindings to macromolecules and the equilibrium linkages between the ligand binding and the self-association reactions. The discussion is restricted to fast and reversible self-association systems. Special emphasis is given to the experimental considerations and the methods of data analysis. Theoretical treatments are minimized and limited to providing a basic conceptual framework for the data analysis. There have been several excellent articles in previous volumes of this series and elsewhere dealing with similar subjects.[1-3] The readers should consult them for references.

This chapter is divided into four sections. The first section consists of a general consideration of the experimental approach in the velocity sedimentation of ligand-induced self-association system. The second section

[1] L. M. Gilbert and G. A. Gilbert, this series, Vol. 48, p. 195.

[2] J. R. Cann, this series, Vol. 48, p. 299.

[3] D. J. Winzor, *in* "Protein-Protein Interactions" (C. Frieden and L. W. Nichol, eds.), p. 129. Wiley, New York, 1981.

describes preliminary diagnoses of a ligand-induced self-association using velocity sedimentation. It can be used by those whose interest is limited to a qualitative understanding of a ligand-induced self-association system. The third section of the chapter deals with more quantitative and in-depth analyses of the velocity sedimentation data including model fitting. The last section details linked function analyses of the association systems. The methods described throughout the chapter are derived from our experience on a number of self-association systems, particularly the self-associations of the microtubule protein tubulin, which are used as examples throughout the chapter. The methods, however, should be generally applicable to other systems within the confinements of a fast, reversible self-association.

The ligand-induced dissociations of proteins into their subunits, although as interesting as the ligand-induced association, are not discussed in the present chapter. As far as the experimental approach and data analysis are concerned, they should be quite similar to the ligand-induced self-associations. One should be able to analyze such systems by making minor modifications of what is described in this chapter.

Velocity Sedimentation

Numerous review articles have been published dealing with both the theoretical and experimental aspects of velocity sedimentation. To avoid redundancy, only a few points which are less frequently mentioned and are immediately pertinent to ligand-induced self-association systems will be emphasized here.

Advantages and Limitations of Velocity Sedimentation

Many physical techniques are available for probing protein self-association. The choice of velocity sedimentation is based on its advantages, namely, that it can generate a wealth of information on the association stoichiometry and equilibrium constant through a few experimental runs and within a short time span. The shapes of the velocity sedimentation boundaries of a ligand-induced self-association are frequently diagnostic of the ligand–protein affinity. The short duration of a velocity sedimentation experiment often makes it the best choice for unstable proteins which will deteriorate if long hours of study are needed. On the other hand, it is true that the data analysis, and thus the conclusions derived from velocity sedimentation, are usually not as rigorous as for some of the other available techniques, for instance equilibrium ultracentrifugation. This, as will become evident in the latter part of this chapter, is due mostly to the fact that in addition to the molecular weight, the size and shape of a macromolecule affect its hydrodynamic properties and, thus, both influence the sedimentation velocity of the macromolecule. This drawback sometimes can turn into an advantage when information on the size and shape of the macromolecule is desired. Notwithstanding these considerations, the ease of the experiment makes it a good choice as a preliminary probe of a system before more rigorous and time-consuming studies are undertaken.

The magnitude of the association constant measurable by velocity sedimentation is limited and these limitations should be a factor to be considered before adopting the technique. Frequently, the limiting factor is the concentration range of macromolecules that can be handled by the instrument. With schlieren optics, the protein concentration needed to obtain a usable sedimentation boundary usually lies in the range of 2 to 20 mg/ml. When using interference optics, the applicable concentration range can be extended approximately one order of magnitude at the lower end of the limit. With the optical scanner, this range depends on the extinction coefficient of the protein and it frequently can be extended down to 20–100 μg/ml.[4,5] Considering a typical protein with a molecular weight of 1×10^5, the measurable range corresponds to molar concentrations of 2×10^{-7} to 2×10^{-4} M. Since self-association reactions are best examined at or near 50% completion, this means that velocity sedimentation is suitable for measuring macromolecular interactions with association constants in the range of 5×10^3 to 5×10^6 M^{-1}, corresponding to standard free energy changes of -5 to -9 kcal/mole. The actual range is flexible to a certain extent and is dependent on the molecular weight, the association stoichiometry, and the existence of cooperativity. However, if a self-association is much stronger than the upper limit, the system will essentially behave as a nondissociating system in velocity sedimentation. Likewise, if an association is much weaker than the lower limit, it will essentially look like a nonassociating system in velocity sedimentation. These detection limits have two implications. First, experimentally, a strongly associated species frequently can be considered as nondissociating within the concentration range used in a velocity sedimentation study of its further associations. This is exemplified by the microtubule protein, tubulin. The α-β heterodimer of tubulin has an association constant of 1.25×10^6 M^{-1}.[6] Within the protein concentration ranges employed in the

[4] H. K. Schachman and S. J. Edelstein, *Biochemistry* **5**, 2681 (1966).
[5] L. K. Hesterberg and J. C. Lee, *Biochemistry* **20**, 2974 (1981).
[6] H. W. Detrich, III and R. C. Williams, Jr., *Biochemistry* **17**, 3900 (1975).

velocity sedimentation studies of the magnesium and vinblastine-induced tubulin self-associations,[7–10] the dissociation of the α-β dimer is negligible. Second, one should not lose sight of the fact that the *in vivo* concentration of a given protein may not coincide with the concentration range that is suitable for velocity sedimentation study. Consequently, there will be significant self-association reactions that are not measurable by velocity sedimentation and conversely there will also be self-association reactions measurable by velocity sedimentation that are not really biologically significant.

Fast and Reversible Self-Association

As stated earlier, this chapter deals only with fast and reversible self-association systems. The velocity sedimentation of extremely slow or irreversible association systems can be interpreted in a straightforward manner and will not be discussed here. In fact, such systems frequently lend themselves better to other methods of attack, such as column gel filtration. Those systems with intermediate reaction rates, i.e., a reaction half time ranging from approximately 1 min to an hour, cannot be studied quantitatively by velocity sedimentation. It is, therefore, essential to ascertain that the reaction of interest is a fast and reversible one before proceeding with velocity sedimentation studies, as described below. The kinetics of a macromolecular association reaction are usually best determined by using a spectroscopic technique, such as the monitoring of the light scattering or fluorescence of the macromolecule. For the purpose of establishing that the reaction half time is short enough for velocity sedimentation studies, a fast-kinetic device, such as temperature jump or stopped-flow attachment, is usually not necessary. The reversibility of the reaction can be tested by determining whether the same equilibrium is reached from both the association of protomers and the dissociation of polymers. In the case of ligand-induced association systems, this can be achieved by either increasing or decreasing the solution ligand concentration, as described in the sample preparation section below.

Strong and Weak Ligand-Induced Self-Associations

In velocity sedimentation studies, ligand-induced self-association has been classified into strong ligand-induced and weak ligand-induced. The terms "strong" and "weak" used here refer to the ligand–protein affinity

[7] R. P. Frigon and S. N. Timasheff, *Biochemistry* **14,** 4559 (1975).
[8] R. P. Frigon and S. N. Timasheff, *Biochemistry* **14,** 4567 (1975).
[9] G. C. Na and S. N. Timasheff, *Biochemistry* **19,** 1347 (1980).
[10] G. C. Na and S. N. Timasheff, *Biochemistry* **19,** 1355 (1980).

and not to the protein self-association. Such classification has emerged out of the observation that ligand-induced self-associations with different ligand affinities display characteristic sedimentation boundaries and demand different experimental approaches. The affinity of the ligand in the present context is actually gauged against the total concentration of the macromolecule used. A strong ligand-induced self-association comprises systems with apparent ligand binding constants, K_x^{app}, greater than the reciprocal of the molar concentration of the macromolecule by one order of magnitude or more. Likewise, weak ligand-induced self-associations can be taken as those systems where K_x^{app} is less than the reciprocal of the molar concentration of the macromolecule by two orders of magnitude or more. Taking a macromolecule with a molecular weight of 1×10^5 at a concentration of 10 mg/ml, the reciprocal of the macromolecular concentration is $10^4\ M^{-1}$. Thus, K_x^{app} must be $10^5\ M^{-1}$ or higher to be considered a strong ligand-induced association system and $10^3\ M^{-1}$ or lower to be considered a weak ligand-induced association system. If the K_x^{app} value falls in between these two extremes, the system belongs to neither category and will be referred to as intermediate ligand-induced self-association hereon. The need for such classifications and the different experimental approaches called for, for these systems, will be detailed later.

Sample Preparation and Maintenance of a Constant Ligand Concentration

In quantitative velocity sedimentation studies of a ligand-induced self-association, the data analysis dictates that the sedimentations be carried out under a constant free ligand concentration. This is because the equilibrium constants of a ligand-induced self-association system measured by velocity sedimentation are only apparent values, due to the fact that the velocity sedimentation technique cannot differentiate between macromolecules with identical sedimentation coefficients, but with different numbers of bound ligand molecules. This can be illustrated by the simple ligand-induced dimerization reaction shown below:

$$2A + 2X \underset{}{\overset{k_1}{\rightleftharpoons}} AX + A + X \underset{}{\overset{k_1}{\rightleftharpoons}} 2AX \underset{}{\overset{k_2}{\rightleftharpoons}} A_2X_2 \qquad (1)$$

where the macromolecule A binds the ligand X and two liganded macromolecules then associate to form a dimer. We adopted the convention of using lower case k's for the intrinsic or microscopic equilibrium constants and upper case K's for the apparent or macroscopic equilibrium constant. Equilibrium constants are usually expressed in units of M^{-1}. Certain calculations lend themselves better to the use of the equilibrium constants in units of ml/mg. They are denoted by the same k's or K's with the superscript prime. k_1 and k_2 are the respective intrinsic association constants. The apparent dimerization constant of the reaction can be expressed as

$$K_2^{app} = \frac{[(AX)_2]}{([A] + [AX])^2} = \frac{k_2}{[1 + 1/(k_1[X])]^2} \tag{2}$$

where [X] is the concentration of the free or unbound ligand. The numerator contains all the dimer species present in the solution whereas the denominator contains all the monomer species in the solution. Assuming that the intrinsic or microscopic equilibrium constants, k_1 and k_2, remain unchanged, the apparent equilibrium constant will be dependent only on the free ligand concentration. Thus, one can maintain a constant apparent association constant only by working at a constant free ligand concentration.

For a weak ligand-induced self-association, no extra care needs to be taken to achieve a constant free ligand concentration. Since the concentration of the ligand added to the solution will be much higher than the concentration of the macromolecule, the free ligand concentration will not be significantly different from the total ligand concentration. This is exemplified by the magnesium ion-induced tubulin self-association.[7,8] Since the magnesium ion-tubulin binding constant is $10^2 \, M^{-1}$, the study was carried out within the range of magnesium ion concentrations of 10^{-3} to $1.6 \times 10^{-2} \, M$. This is much higher than the protein concentration used. Consequently, the zweition was introduced into the protein solution simply by adding small aliquots of a high concentration stock solution.

For a strong or intermediate ligand-induced self-association, a constant free ligand concentration can be maintained by equilibrating the protein at a known concentration of the ligand, using a method such as gel filtration or equilibrium dialysis.[9,10] The gel filtration method is more suitable for handling proteins or ligands that are available only in small quantities. For unstable proteins, the short duration of gel filtration is also advantageous over the equilibrium dialysis method. Usually a column with dimensions of 1×10 cm, packed with a gel such as Sephadex G-25 is sufficient for equilibrating a 20-mg protein sample. To eliminate the effect of temperature variation, it is best to jacket the column and regulate its temperature by a circulating water bath. Before applying the protein, the column should be washed thoroughly with the experimental buffer containing the ligand. Small aliquots of 0.5 ml can be collected from the column and used directly in the velocity sedimentation experiments. Aliquots collected in different regions of the eluting peak usually can provide a range of protein concentrations. Since the ligand binding of such a system is dependent on the total protein concentration, once the ligand equilibration is achieved there should be no further attempt to adjust the protein concentration by dilution with the experimental buffer, since this would release bound ligand and disturb the solution free ligand concentration. An indication of satisfactory ligand equilibration is given by the identity of sedimentation boundaries obtained from fractions in the leading edge of the eluting peak and in the trailing edge of the peak, provided that they are of the same protein concentration. Since the transport of macromolecules in a gel column is similar to that in velocity sedimentation, it is likely that depletion of free ligand at the trailing edge of the eluting peak will occur for strong ligand-induced self-associations. If, in the particular gel used, the partition coefficients of the protein species vary with the degree of polymerization, the above ligand depletion effect could be significant at low ligand concentrations. This is usually indicated by consistently different sedimentation boundaries obtained from aliquots taken at the leading edge and the trailing edge of the eluting peak, despite the fact that they are of the same protein concentration, and that efforts, such as increasing the size of the column and decreasing the rate of elution, have been made to achieve equilibration. Under such conditions, one can either use only fractions in the leading edge of the peak where ligand depletion is expected to be less severe, or switch to equilibrium dialysis for achieving the ligand–protein equilibration.

An advantage of using gel filtration or equilibrium dialysis to equilibrate the protein with the ligand is that the ligand concentration is increased gradually rather than abruptly as in the introduction of aliquots of concentrated stock solution of the ligand. Frequently, this can avoid the formation of irreversible aggregates.

Another measure which can sometimes effectively prevent the deterioration of the protein sample is to introduce the ligand to the protein solution at the earliest stage of the sample preparation. Many proteins, including allosteric enzymes, are known to be stabilized by their ligand effectors.

Selection of Optics

As mentioned earlier, the concentration range of the protein and the optics of the centrifuge to be used depend on the magnitude of the self-association constant to be determined. Schlieren optics are suitable for high protein concentrations up to 20 mg/ml, while interference optics are applicable in the concentration range below 5 mg/ml. The applicable concentration range for the optical scanner depends on the extinction coefficient of the protein at the particular wavelength used. For a wide range of coverage, more than one type of optics should be employed. The readability of the schlieren boundaries usually deteriorates below 4 mg/ml of protein; at that point, the interference optics should take over. If the optical scanner is used in a system where the ligand of interest also absorbs light, one must choose the scanning wavelength judiciously to assure that the boundaries obtained correspond to that of the protein and not the ligand.

Calculation of the Weight-Average Sedimentation Coefficient

For rigorous quantitative analysis of velocity sedimentation data, the second moment of the sedimentation boundary should be used in the calculation of the weight-average sedimentation coefficient. This is particularly true for those strongly skewed or bimodal boundaries where the second moment could differ substantially from the apex of the boundary. Traditionally, schlieren boundaries recorded on glass plates are measured manually with a microcomparator, such as the Nikon Model-6C. To simplify the plate measurement, the manual type micrometers on the microcomparator can be replaced with an electronic type, such as the Elk Model 9200 precision digital positioner. As one turns the positioner, it sends out digital pulses which can be counted by a digital counting device. An LED then displays the digital count directly. The x-interval between two readings should depend on the sharpness of the boundary. For a boundary that spans approximately 0.5 cm on the plate, an interval of 10–20 μm per reading can be used outside the apex of the boundary and 5 μm per reading is usually sufficient around the apex. The second moment of the boundary can be calculated from these readings through numerical integration, using the trapezoidal rule:

$$\bar{r}^2 = \frac{\int_{r_m}^{r_p} r^2 \frac{dn}{dr}\, dr}{\int_{r_m}^{r_p} \frac{dn}{dr}\, dr}$$

$$\cong \frac{\sum_{i=1}^{n} \left[\left(r^2 \frac{dn}{dr} \right)_{i+1} + \left(r^2 \frac{dn}{dr} \right)_i \right] \frac{r_{i+1} - r_i}{2}}{\sum_{i=1}^{n} \left[\left(\frac{dn}{dr} \right)_{i+1} + \left(\frac{dn}{dr} \right)_i \right] \frac{r_{i+1} - r_i}{2}} \tag{3}$$

where $\bar{r}$ is the second moment of the boundary, r is the radius, and (dn/dr) is the refractive index gradient obtained from the schlieren boundary. The integration should be carried out from a point in the supernatant to a point in the plateau. For sedimentation boundaries obtained through either interference optics or with an optical scanner, the second moment can be calculated according to

$$\bar{r}^2 = \frac{\int_{r_m}^{r_p} r^2 dc}{\int_{r_m}^{r_p} dc} \cong \frac{\sum_{i=1}^{n} \frac{\Delta c_i}{2} (r_{i+1}^2 + r_i^2)}{\sum_{i=1}^{n} \Delta c_i} \tag{4}$$

where c is the weight concentration of the macromolecule. With interference optics, since the vertical distance between two fringes represents a fixed concentration increment, the measurement of the second moment can be simplified by taking the x axis readings where the hairline of the microcomparator intersects the fringes. A programmable hand calculator is sufficient for the numerical integration. However, the entire process of data acquisition and analysis can be greatly simplified if the digitized results, either from a microcomparator or directly from an electronic optical scanner, are processed directly by a computer. From the second moments of the boundaries, the weight-average sedimentation coefficients can be calculated, converted to the condition of water at 20°, and plotted against the total protein concentration. It is important to emphasize here that the weight-average sedimentation coefficient obtained is a rigorous measure of the sedimentation of the macromolecules in the plateau region.[11] This fact has two implications. First, this sedimentation coefficient is independent of the protein or ligand concentration gradient present at the boundary. Second, one should use the protein concentration at the plateau region in the above plot. If the sedimentation coefficients were obtained from boundaries that have traveled a substantial distance from the miniscus, it is prudent to take into consideration the radial dilution effect and calculate the actual protein concentration by using the average of the second moments used in calculating the sedimentation coefficient.

Qualitative Diagnosis of a Ligand-Induced Self-Association

Velocity Sedimentation in the Absence of the Ligand Effector

When confronted with a protein without any prior knowledge of its self-association properties, the initial work involves a characterization of the velocity sedimentation of the protein in the absence of the ligand effector, i.e., usually in a buffer that contains the minimum ingredients necessary to maintain the pH and the integrity of the protein. The protein under such conditions can be taken as the "ground state" of the self-association reaction. First of all, the protein ought to be examined for its

[11] R. J. Goldberg, *J. Phys. Chem.* **57,** 194 (1953).

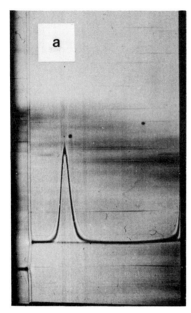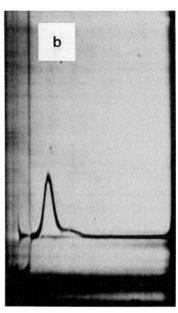

FIG. 1. Typical velocity sedimentation patterns of a homogeneous globular protein in its native state (a) and its partially, but irreversibly self-aggregated state (b). The pictures shown are those of calf brain tubulin (8 mg/ml) in 0.01 M NaP$_i$, 10^{-4} M GTP, pH 7.0 buffer sedimented at 60,000 rpm.

homogeneity through analysis of the shape of the sedimentation boundary. A homogeneous globular protein ought to display a single peak with its trailing edge slightly sharper than the leading edge, as exemplified by the sedimentation boundary of calf brain tubulin in a phosphate buffer shown in Fig. 1a. The sedimentation boundaries should also be closely examined both during rotor acceleration and throughout the run to assure that the protein has neither dissociated into smaller species nor aggregated irreversibly during the course of the run. The presence of higher molecular weight aggregates is usually indicated by the appearance of fast moving shoulders or spikes. This is exemplified by Fig. 1b, showing a velocity sedimentation boundary of the same protein sample as Fig. 1a, except that it had undergone a small degree of self-aggregation. Such examination is, of course, only complementary to other techniques, such as gel electrophoresis and equilibrium ultracentrifugation, which are inherently more effective in probing the homogeneity of a protein. Subsequently, the weight-average sedimentation coefficient, $\bar{s}_{20,w}$, of the protein ought to be determined as a function of the total protein

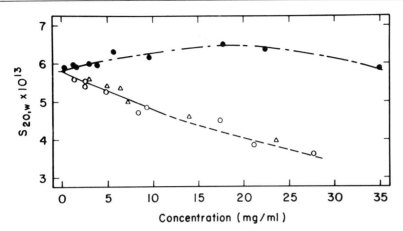

FIG. 2. Concentration dependence of the sedimentation coefficient of a globular protein under nonassociating conditions. The results shown are those of tubulin in 0.01 M NaP$_i$, 10^{-4} M GTP, pH 7.0 buffer. The straight line through the data is the linear least-squares fit, whereas the dashed portion indicates deviation from linearity at high protein concentrations. The data represented by the filled circles were obtained in the same buffer with the addition of 0.1 M NaCl. The slight increase of the latter sedimentation coefficients with protein concentration suggested the presence of incipient self-association. (Taken from Frigon and Timasheff.[7])

concentration. For a homogeneous protein under nonassociating conditions, the sedimentation boundaries are usually symmetrical enough so that the apex of the boundary can be used in place of the second moment in the calculation of the sedimentation coefficient. The $\bar{s}_{20,w}$ can usually be fitted by the linear equation[12]

$$\bar{s}_{20,w} = \bar{s}^0_{20,w}(1 - gc_t) \tag{5}$$

where $\bar{s}^0_{20,w}$ is the weight-average sedimentation coefficient extrapolated to zero protein concentration, g is the hydrodynamic nonideality constant, and c_t is the total protein concentration in mg/ml. For globular proteins, the value of g usually ranges from 0.01 to 0.04 ml/mg and it increases with increasing asymmetry of the protein.[12,13] The positive value of g means that, for nonassociating proteins, $\bar{s}_{20,w}$ decreases with increasing protein concentration, and an increase of $\bar{s}_{20,w}$ with increasing protein concentration can be taken as an indication of the presence of a self-association reaction. These are exemplified by the sedimentation coefficients of calf brain tubulin in a phosphate buffer shown in Fig. 2. The

[12] H. K. Schachman, "Ultracentrifugation in Biochemistry." Academic Press, New York, 1959.
[13] J. L. Trujillo and W. C. Deal, Jr., *Biochemistry* **16**, 3098 (1977).

data below 10 mg/ml can be fitted well by Eq. (5) using a g value of 0.018 ml/mg. Introduction of 0.1 M NaCl into the buffer caused a slight increase of $\bar{s}_{20,w}$ with the protein concentration suggesting the presence of a weak self-association.

Velocity Sedimentation in the Presence of the Ligand Effector

In the past two decades, extensive studies using computer simulation technique have generated a wealth of information on the characteristic shapes of sedimentation boundaries for self-association systems with various association stoichiometries and ligand affinities.[7,9,14–23] Today, based on these computer simulation results, one can reach a qualitative diagnosis of an association system, including the macromolecular stoichiometry, the presence or absence of cooperativity in the self-association, and obtain an approximation of the ligand–protein affinity by simply analyzing the characteristic shapes of a few sedimentation boundaries. Such preliminary analysis of a self-association system is based primarily on two important features of a reaction boundary:

Asymmetry of the Boundary. As mentioned earlier in this section, a homogeneous nonassociating macromolecule should display a single peak with its trailing edge slightly sharper than the leading edge, due to the hydrodynamic nonideality.[16] Upon introduction of a ligand effector, the occurrence of protein self-association is usually reflected by the sedimentation boundary becoming asymmetrical, with its trailing edge extended. The degree of such a change usually increases with the self-association stoichiometry. Self-associations of lower stoichiometries usually display less skewed boundaries. They are exemplified by the magnesium ion-induced tubulin self-association shown in Fig. 3. At low magnesium ion concentrations, small oligomers dominate and the sedimentation boundaries are only slightly skewed with the trailing edge becoming more diffused than the leading edge. As the weight fractions of the higher polymers increase, the sedimentation boundaries become increasingly

[14] D. J. Cox, *Arch. Biochem. Biophys.* **112**, 249 (1965).
[15] D. J. Cox, *Arch. Biochem. Biophys.* **112**, 259 (1965).
[16] D. J. Cox, *Arch. Biochem. Biophys.* **119**, 230 (1967).
[17] D. J. Cox, *Arch. Biochem. Biophys.* **129**, 106 (1969).
[18] D. J. Cox, *Arch. Biochem. Biophys.* **142**, 514 (1971).
[19] D. J. Cox, *Arch. Biochem. Biophys.* **146**, 181 (1971).
[20] R. R. Holloway and D. J. Cox, *Arch. Biochem. Biophys.* **160**, 595 (1974).
[21] J. R. Cann and W. B. Goad, "Interacting Macromolecules." Academic Press, New York, 1970.
[22] J. R. Cann and W. B. Goad, *Science* **170**, 441 (1970).
[23] J. R. Cann and W. B. Goad, *Arch. Biochem. Biophys.* **153**, 603 (1972).

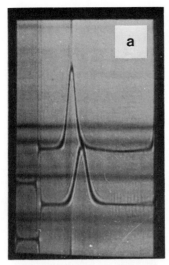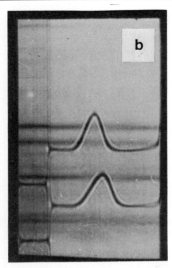

FIG. 3. Typical velocity sedimentation patterns of a weak isodesmic stepwise self-association. The results shown are those of magnesium ion-induced tubulin self-association. The speed was 60,000 rpm. The protein concentration was approximately 8 mg/ml; (a) 41 min after reaching speed; upper, no magnesium; lower, 0.0027 M MgCl$_2$; (b) 38 min after reaching speed; upper, 0.0055 M MgCl$_2$; lower, 0.0082 M MgCl$_2$. (Taken from Frigon and Timasheff.[7])

skewed.[20] This is also exemplified by the high concentration vinblastine-induced isodesmic indefinite self-association of tubulin shown in Fig. 4.

Modality of the Boundary. The most frequently encountered velocity sedimentation boundaries are usually either unimodal or bimodal. For a fast reversible self-association system, the bimodality of the sedimentation boundary usually indicates either the presence of cooperativity in the macromolecular self-association or the induction of the self-association by the strong binding of a ligand. The former phenomenon was first described by Gilbert[24,25] and is referred to as a Gilbert system, whereas the latter one was first reported by Cann and Goad[21–23] and is termed a Cann–Goad type sedimentation boundary.

Since a Gilbert type self-association involves some kind of cooperativity, it contains an association step or steps with a more negative free energy change than the rest of the association reaction. A moderately cooperative self-association frequently involves the formation of a looped structure as the end product. The incorporation of the last monomer to

[24] G. A. Gilbert, *Discuss. Faraday Soc.* **20**, 68 (1955).
[25] G. A. Gilbert, *Proc. R. Soc. London, Ser. A* **250**, 377 (1959).

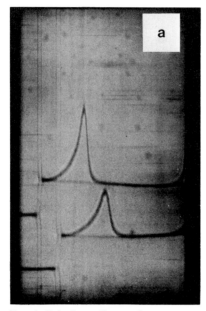
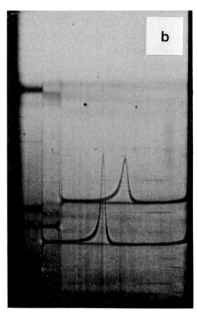

FIG. 4. Velocity sedimentation patterns of an isodesmic indefinite self-association. The results shown are those of the vinblastine-induced tubulin self-association at 60,000 rpm. The vinblastine concentrations were (a) 2×10^{-4} M and (b) 2.5×10^{-4} M. The protein concentrations were 6.7 mg/ml for the top pattern of (a), 8.3 mg/ml for the bottom pattern of (a), 8.5 mg/ml for the top pattern of (b), and 6.8 mg/ml for the bottom pattern of (b). (Taken from Na and Timasheff.[9])

enclose the looped structure is energetically more favorable because it is accompanied by the formation of two bonds instead of one.[7,8] A strongly cooperative self-association can be envisioned as resulting from multibody collisions which involves no intermediate species. In reality, the probability of such an event is small, particularly in the formation of polymers of high stoichiometries. In all likelihood, certain intermediate species do exist, but their quantities are so small they cannot be detected by existing physical techniques. In any event, the differentiation of the association pathways must rely on kinetic studies and is beyond the scope of the equilibrium studies described here. A dimerization system has only one association step which apparently would not allow the existence of any cooperativity in the self-association and it, therefore, cannot belong to a Gilbert system.

Excellent examples of the Gilbert system are provided by the tetra-

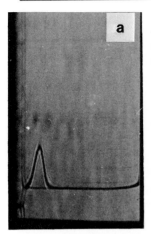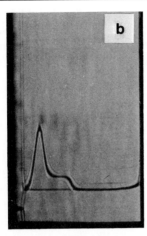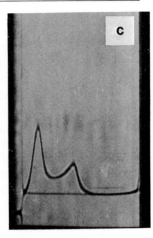

FIG. 5. Velocity sedimentation patterns of a Gilbert system. The results shown are those of magnesium ion-induced tubulin self-association to form a double ring structure of 26 ± 2-mer. The pictures were taken 21 min after reaching the speed of 48,000 rpm. The magnesium ion concentration was 0.01 M, whereas the protein concentrations were (a) 4.7, (b) 10.4, and (c) 15.5 mg/ml, respectively. (Taken from Frigon and Timasheff.[7])

merization of β-lactoglobulin,[26] the self-association of myosin,[27] the magnesium ion-induced self-association of tubulin into a double ring structure containing 26 ± 2 monomers[7,8] and the subunit association of phosphofructokinase.[5] In a Gilbert self-association system, as exemplified by the magnesium ion-induced tubulin self-association shown in Fig. 5, the sedimentation boundaries are unimodal at low protein concentrations, where usually stepwise isodesmically associated species predominate. Once the protein concentration is increased above a certain threshold value, the cooperative step for the formation of the favorable end product sets in and a fast moving peak begins to emerge. From there on, the size of the slow moving peak should remain constant. Further increases in protein concentration only serve to enlarge the fast moving peak. Therefore, in theory, the Gilbert system is very similar to the cooperative helical polymerization formulated by Osawa and Kasai,[28] except that in the former case the polymerization stops after the formation of the energetically favorable looped structure, whereas in the latter case the self-association continues on in a helical manner.

[26] R. Townend, R. J. Winterbottom, and S. N. Timasheff, *J. Am. Chem. Soc.* **82,** 3161 (1960).
[27] R. Josephs and W. F. Harrington, *Biochemistry* **7,** 2834 (1968).
[28] F. Oosawa and M. Kasai, *Biol. Macromol.* **5,** 261 (1971).

In the Cann–Goad systems, the bimodality of the sedimentation boundary is caused by the strong, stoichiometric ligand–protein interaction which leads to the generation of a concentration gradient of free ligand across the sedimentation boundary.[21–23] The formation of the concentration gradient can be easily understood by considering a simple asymptotic sedimentation of a ligand-induced monomer–dimer self-association system. At the beginning of the sedimentation, the monomers and dimers are in equilibrium with each other and with the free ligand. After application of a centrifugal field for a short interval, some of the macromolecules are pelleted at the bottom of the cell. For a self-associating system, the mass ratio of dimer to monomer pelleted is higher than the mass ratio of dimer to monomer found in the solution. This is simply because the dimers sediment faster than the monomers. According to the Wyman linkage theory, for a unit weight of macromolecule, more ligand is bound to the dimers than to the monomers.[29] Consequently, the molar ratio of the ligand bound to the macromolecules that pellet at the cell bottom is higher than that found in the plateau solution. This extra amount of ligand pelleted must be taken from the region centripetal to the macromolecule boundary, since in the plateau area neither the macromolecule nor the ligand changes its concentration other than as a result of radial dilution. Within the boundary, reequilibration of monomers to dimers does occur, with the binding of some free ligand. A repetition of the process, i.e., the transport of the macromolecule and the maintenance of the reaction equilibrium throughout the boundary, serves essentially as a ligand pump which constantly removes unbound ligand from the region centripetal to the boundary to the cell bottom and, thus, generates a free ligand concentration gradient across the boundary. By the same token, a similar free ligand gradient across the boundary will also be generated for a strong ligand-induced dissociation at certain intermediate ligand concentrations. However, instead of depletion, one expects to observe enhancement of the free ligand concentration centripetal to the boundary. A quantitative account of why a free ligand concentration gradient can lead to a bimodal sedimentation boundary will be given in the next section where computer simulation of the sedimentation boundary is discussed.

An excellent example of the Cann–Goad system is provided by the low concentration vinblastine-induced self-association of tubulin shown in Fig. 6. It is characteristic of the Cann–Goad system that, if the self-association is not a cooperative one, biomodal sedimentation boundaries are observed at intermediate ligand concentrations whereas unimodal sedimentation boundaries are observed at either low or high ligand con-

[29] J. Wyman, *Adv. Protein Chem.* **19**, 224 (1964).

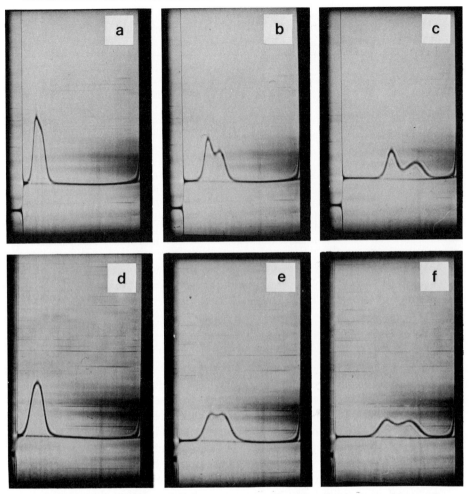

FIG. 6. Typical velocity sedimentation patterns of a Cann–Goad system. The results shown are those of tubulin in the presence of 1×10^{-5} M vinblastine. Pictures (a–c) were taken at 16, 32, and 64 min after reaching the speed of 60,000 rpm. Pictures (d–f) were taken from a separate run of the same sample except at 32, 64, and 104 min and at the speed of 48,000 rpm. (Taken from Na and Timasheff.[9])

centrations. Notice also that the bimodal sedimentation boundary of a Cann–Goad system first emerges from the meniscus as a single peak which then resolves into a bimodal one after sedimenting a distance from the meniscus. This contrasts with the Gilbert system where the boundary emerges from the meniscus as a bimodal one. This difference

TABLE I
MODALITY OF VELOCITY SEDIMENTATION BOUNDARIES OF VARIOUS LIGAND-INDUCED
PROTEIN SELF-ASSOCIATIONS

Ligand affinity	Association	
	Isodesmic	Cooperative with favorable end-product
Weak	Unimodal	Low protein conc.: unimodal High protein conc.: biomodal (Gilbert system)
Intermediate	Low ligand conc.: unimodal	Low ligand and low or high protein concs.: unimodal
Strong	Intermediate ligand conc.: biomodal (Cann–Goad system)	High protein conc. intermediate or high ligand conc.: bimodal (trimodal?) (due to ligand gradient and/or cooperativity)
	High ligand conc.: unimodal	High protein and high ligand concs.: bimodal (due to cooperativity)

can be used to differentiate the two systems. It is also evident from Fig. 6d–f that the establishment of the Cann–Goad type bimodal sedimentation boundary is dependent on the maintenance of a stable free ligand concentration gradient across the boundary and is strongly dependent on the rotor speed. At low rotor speed, the ligand concentration gradient tends to diffuse out, leading to poorly resolved bimodal boundaries.

Table I presents a flow chart of the procedures to analyze the velocity sedimentation boundaries and to establish a qualitative characterization of a ligand-induced self-association. In carrying out the velocity sedimentation study, the most useful solution variables are the protein and ligand concentrations. In a Gilbert type cooperative self-association, at a given ligand concentration, the self-association should be examined over a range of protein concentrations. One expects to observe unimodal sedimentation boundaries at low protein concentrations and bimodal boundaries at high protein concentrations. The critical concentration at which the fast moving peak starts to emerge should decrease with increasing ligand concentration. Thus, if bimodality cannot be observed, it may be because the critical concentration is too high. One should then increase the ligand concentration in order to lower the critical concentration to

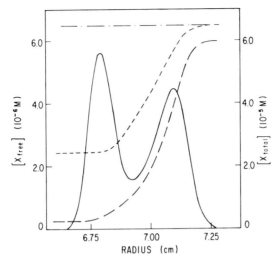

FIG. 7. Computer simulated Cann–Goad bimodal sedimentation boundary for a one-ligand-mediated monomer–dimer–trimer–tetramer isodesmic self-association. The total protein concentration was 1×10^{-4} M, the total ligand concentration was 6×10^{-5} M, and the initial free ligand concentration was 6.5×10^{-6} M. The microscopic equilibrium constants were 1.8×10^{4} M^{-1} for the binding of the ligand to the monomer and 1.8×10^{5} M^{-1} for the association between liganded macromolecules. The symbols are (——) protein concentration gradient, (–––) total ligand concentration, (----) free ligand concentration, and (–·–·–) initial free ligand concentration before centrifugation (G. C. Na, J. R. Cann, and S. N. Timasheff, unpublished).

within a measurable range. A Cann–Goad type strong ligand-induced self-association can be identified by examining the sedimentation boundaries over a range of ligand concentrations. With the Cann–Goad bimodal sedimentation boundaries, if the ligand being examined has UV or visible absorption that does not overlap with the protein absorption, one should be able to use the electronic optical scanner to measure the total concentration of the ligand across the entire cell. Figure 7 depicts a computer simulated Cann–Goad type bimodal sedimentation boundary together with the distribution of the total and free ligand concentrations for a monomer–dimer–trimer–tetramer isodesmic self-association. There are two salient features to be noted. First, although the protein boundary is bimodal, the ligand boundary is only unimodal. Apparently there is little ligand bound to the protein in the slow moving peak. Second, there is a strong depletion of free ligand centripetal to the boundary. The free ligand concentration found in the supernatant is only 38% of the original free ligand concentration before the centrifugation. On the other hand, this depletion comprises only a small percentage of the total ligand concentra-

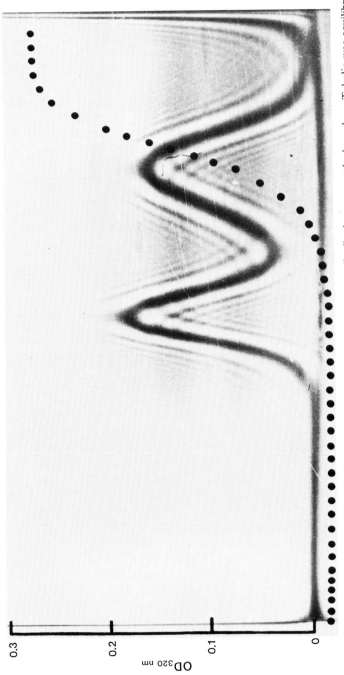

Fig. 8. Bimodal sedimentation boundary of the tubulin and vinblastine concentration distributions across the boundary. Tubulin was equilibrated with 1×10^{-5} M vinblastine, 0.01 M NaP$_i$, 10^{-4} M GTP, pH 7.0 buffer. An aliquot from the equilibrating column, containing 16.6 mg/ml tubulin, was centrifuged at 60,000 rpm in a double-sector cell with the reference sector filled with the equilibrating buffer. The schlieren picture was taken at a 65° bar angle 80 min after reaching speed. The dotted line depicts the photoelectric scan at 320 nm obtained simultaneously with the schlieren picture. (Taken from Ref. 10.)

tion in the plateau region. A simple demonstration of the presence of a total ligand concentration gradient across the boundary is apparently not sufficient evidence for a Cann–Goad system, since even a simple binding system will show such a gradient of total ligand. For a Cann–Goad system, the ligand concentration in the supernatant must be lower than the original free ligand concentration used in equilibrating the protein. These characteristics are completely born out in the bimodal sedimentation boundary of the vinblastine-induced tubulin self-association shown in Fig. 8.

By changing these solution variables and comparing the resulting boundaries with the table, one should be able to reach a preliminary diagnosis of the system of interest. Initially, in probing the effect of a particular ligand, the presence of other ligand effectors should be avoided in order to minimize complications. Other variables such as the concentration of the hydronium ion and the solution temperature should also be kept constant. After the initial characterization, one can then manipulate the latter variables to obtain information, such as the effects of different ligands and the enthalpy change of the association.

One can notice that there is a type of self-association, shown in Table I, which we are not aware of having either been encountered in a real system or having been studied theoretically. It is a cooperative self-association induced by a strong ligand binding reaction, i.e., a Cann–Goad-type Gilbert system. One would expect it to possess characteristics both of the Cann–Goad system and of the Gilbert system. It is yet uncertain whether a trimodal sedimentation boundary can emerge at certain high protein concentrations and intermediate ligand concentration. This question still awaits to be answered by computer simulation studies.

Quantitative Analysis of Ligand-Induced Self-Association

A reversible self-association system can be expressed by the following generalized equations:

$$A + A \rightleftharpoons A_2 \qquad K_2 = \frac{[A_2]}{[A]^2}$$

$$A_2 + A \rightleftharpoons A_3 \qquad K_3 = \frac{[A_3]}{[A_2][A]}$$

$$\vdots \qquad \qquad \vdots \qquad \qquad (6)$$

$$A_{n-1} + A \rightleftharpoons A_n \qquad K_n = \frac{[A_n]}{[A_{n-1}][A]}$$

where A_n and $[A_n]$ denote the n-mer and its molar concentration. K_n's are the association constants for the formation of successive bonds between monomer and $(n - 1)$-mer. For such a self-association, the mass distribution of the protein among various polymers is reflected in a functional dependence of $\overline{MW}$, the weight-average molecular weight of the macromolecule, on the total protein concentration and the association parameters:

$$\overline{MW} = \frac{\sum\limits_{i=1}^{n} c_i MW_i}{c_t} \tag{7}$$

where c_i is the concentration of i-mer in mg/ml, c_t is the total protein concentration, and MW_i is the molecular weight of the i-mer. In velocity sedimentation, the mass distribution is reflected by a similar dependence of the weight-average sedimentation coefficient on the total protein concentration and the association parameters:

$$\bar{s} = \frac{\sum\limits_{i=1}^{n} c_i s_i}{c_t} = \frac{\sum\limits_{i-1}^{n} c_i s_i^0 (1 - g_i c_t)}{c_t} \tag{8}$$

where s_i and g_i are the sedimentation coefficient and hydrodynamic nonideality constant of the i-mer, respectively. Thus, by fitting the experimentally determined weight-average sedimentation coefficients with theoretical curves, calculated from various association models, one can rule out certain association schemes and determine the association stoichiometry and equilibrium constants that can describe the system best.

Another method for studying quantitatively the velocity sedimentation of a ligand-induced self-association system is through the computer simulation of the sedimentation boundaries. As mentioned earlier, the sedimentation boundaries of systems undergoing different mechanisms of association often possess strikingly different characteristics. As a result, a comparison of the experimental sedimentation boundaries with those calculated with a computer can also lead to information about the mechanism of the association reaction. These two methods of velocity sedimentation analysis are described in detail in the following section.

Probing the Association Mechanism through Analyzing $\bar{s}_{20,w}$

The selection of theoretical models to fit the experimental data is guided by the preliminary boundary analyses described previously. Further hints as to the association mechanism can be obtained from the dependence of the $\bar{s}_{20,w}$ on total protein concentration. For a stepwise self-association without any cooperative event, the weight-average sedimentation coefficient should show a simple parabolic type dependence on total protein concentration.[9,10] On the other hand, if a cooperative step or steps are present in the self-association, a sigmoidal curve should be expected.[5,7,8] The degree of inflection of the sigmoidal curve should be directly related to the cooperativity whereas the number of inflections is related to the number of cooperative steps present. After examining the plot of $\bar{s}_{20,w}$ versus the protein concentration, preferably at several free ligand concentrations, one can set up a theoretical association model. Following the association model, the concentration of each i-mer can be calculated by numerically solving the following polynomial using the Newton–Ralphson rule:

$$c_t = \sum_{i=1}^{n} c_i = \sum_{i=1}^{n} K_2' K_3' \cdots K_i' c_1^i \tag{9}$$

where c_t is the total protein concentration and c_1 is the concentration of the monomer, both being in mg/ml; MW is the molecular weight of the monomer; K_i' is the apparent association constant between a monomer and an $(i-1)$-mer in units of ml/mg. It is related to K_i in the unit of M^{-1} by the equation

$$K_i' = \frac{i}{(i-1)\text{MW}} K_i \tag{10}$$

It is important to recognize here that the number of self-association schemes which can be devised is limited only by one's imagination. Thus, in fitting experimental data with theoretical models, the rule to observe is to stay with the simplest model that can accommodate the data. On the practical side, the solution of Eq. (9) can be greatly simplified if one can minimize the number of variables present. In the generalized expression of a self-association system, shown in Eq. (6), each of the self-association constants can assume any value. If there is no sign of cooperativity in the self-association, one should start with a simple stepwise isodesmic self-association, i.e., one should assign the same equilibrium constant to each step of the self-association. For a moderately cooperative association system, one may assign a higher valued association constant to the cooperative step.[7,8] For a strongly cooperative system, one may even set all the intermediate association constants to zero, which is essentially equivalent to assuming the absence of all intermediate species.[5] If the self-association proceeds indefinitely and the equilibrium constants for each step of the self-association are identical, then Eq. (9) converges to a simple binomial

$$c_t = \frac{c_1}{[1 - (K_2 c_1 / \text{MW})]^2} \tag{11}$$

and the concentrations of monomer and each i-mer can be obtained easily:

$$c_i = i(K_2/MW)^{i-1}c_1^i \tag{12}$$

In the case of a cooperative self-association, both the association scheme and data analysis can be simplified substantially if it is possible to assign a single equilibrium constant for all the association steps prior to the formation of the favorable end product. The mass distributions of the protein among the various polymers for the above two association schemes can be calculated. As depicted in Fig. 9A, in the isodesmic indefinite type of self-association, the small oligomers dominate at low protein concentrations; as the concentration increases, the weight fractions of higher polymers gradually increase. Also, as shown in Fig. 9B, at a given protein concentration, if the self-associations is weak, the monomer is the dominating species and the weight fractions of the polymers decrease with

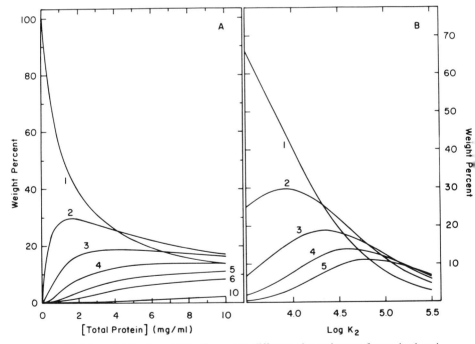

FIG. 9. Mass distribution of tubulin among different size polymers for an isodesmic, indefinite self-association. (A) Dependence on the total protein concentration; $K_2^{app} = 5.2 \times 10^4 \, M^{-1}$. (B) Dependence on the polymerization constants; total protein concentration = 10 mg/ml. The number next to each curve is the corresponding degree of polymerization. (Taken from Ref. 9.)

increasing degrees of polymerization. As the association becomes stronger, the trend reverses itself, i.e., higher molecular weight species become dominant. On the other hand, for a Gilbert system, such as the magnesium ion-induced tubulin self-association shown in Fig. 10, below the critical concentration for the formation of the favorable end polymer, the weight distribution of the protein is similar to the isodesmic association described above. Above the critical concentration, the concentration of the favorable end product of the self-association, in this case a 26-mer, departs from zero and increases linearly whereas those of the intermediate species remain essentially constant.

Once the concentrations of all i-mers are obtained, they are put into Eq. (8) to calculate the theoretical weight-average sedimentation coefficient as a function of total protein concentration. The data analyses are usually carried out under the assumption that the binding of the low-molecular weight ligand does not affect the sedimentation coefficient of the macromolecule, either through a slight change of the molecular weight or a change of the structure of the macromolecule. In Eq. (8), one needs to know s_i, the sedimentation coefficient of the i-mer. In a few instances, the sedimentation coefficient of the end polymer can be derived by extrapolating the sedimentation coefficients obtained at high protein concentration and at a saturating ligand concentration. This is particularly true for the Gilbert-type strongly cooperative self-association system where a certain end polymer will dominate at high protein concentrations.[5,7,8] However, in most cases the sedimentation coefficients of individual polymers are unavailable and are not readily measureable. To pro-

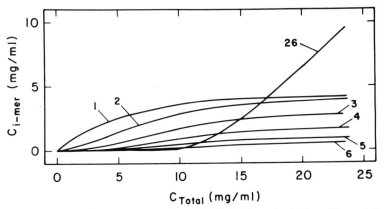

FIG. 10. Variations in the concentrations of individual species in the self-association of tubulin as a function of the total protein concentration in the presence of 0.008 M MgCl$_2$. The numbers indicate the degree of polymerization of each associated species. (Taken from Ref. 7.)

ceed with the calculation, it becomes necessary to use theoretical s_i values obtained by assuming a certain physical shape of the polymer. The simplest and most frequently used model is a spherical one for which the sedimentation coefficient of the i-mer can be calculated from

$$s_i^0 = i^{2/3} s_1^0 \tag{13}$$

where s_i^0 is the reduced sedimentation coefficient of the i-mer and s_1^0 is the reduced sedimentation coefficient of the monomer. More complicated and elaborate models can be used, if one has some knowledge of the physical size and shape of the polymers obtained through other methods, such as electron microscopy. In this regard, study of the morphology of the polymers using an electron microscope can provide valuable information not only on the association stoichiometry but also on the size and shape of the polymer species. Examples of such an approach can be found in the magnesium-induced tubulin self-association into a double ring structure,[7,8] as well as in several other multisubunit proteins.[30,31] In some cases, the association product was observed vividly under the electron microscope permitting its sedimentation coefficient to be deduced by the Kirkwood theory.[32] The theoretical values were found to be consistent with the experimental values obtained from extrapolation.

Another unknown factor in Eq. (8) is g_i, the hydrodynamic nonideality constant of the i-mer. As mentioned earlier, the value of g appears to increase with the asymmetry of the protein. Currently, there is no solid formulation of g as a function of the size and shape of a macromolecule. The usual practice is to assign the value of g_1 to all g_i. The uncertainties in the quantities of s_i and g_i must undoubtedly lead to some errors. Such errors may stay within the experimental uncertainty for those systems where the polymers are symmetrical in shape and have low association stoichiometries, but may become significant for those systems that form asymmetrical polymers of high association stoichiometries. In the latter case, these uncertainties could prevent an unequivocal deduction of the association stoichiometry and equilibrium constant.

An association model fitting is carried out through nonlinear least-squares fitting of the experimental $\bar{s}_{20,w}$. This technique is well illustrated by the magnesium-induced formation of the tubulin double-ring structure and the vinblastine-induced isodesmic indefinite self-association of tubulin. In the magnesium-induced self-association, the sedimentation boundaries shown in Fig. 5 displayed all the characteristics of the Gilbert system described above. The weight-average sedimentation coefficients depicted

[30] P. R. Andrews and P. D. Jeffrey, *Biophys. Chem.* **11**, 49 (1980).
[31] P. D. Jeffrey and P. R. Andrews, *Biophys. Chem.* **11**, 61 (1980).
[32] J. G. Kirkwood, *J. Polym. Sci.* **12**, 1 (1954).

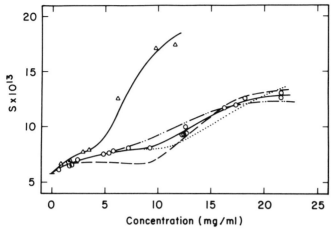

FIG. 11. Theoretical fitting of the concentration dependence of the weight-average sedimentation coefficient, $\bar{s}_{20,w}$, of tubulin under self-associating conditions in the presence of Mg^{2+}. ($\bigcirc$) 0.008 M $MgCl_2$; ($\triangle$) 0.016 M $MgCl_2$. The lines are least-squares fits to the data according to the models described in Ref. 7. (Taken from Ref. 7.)

in Fig. 11 showed sigmoidal dependence on the total protein concentration and could be fitted well with an initial isodesmic self-association which then terminated at the 26-mer which formed with a more negative free energy change. In contrast, the vinblastine-induced tubulin self-association, characterized by the strongly skewed sedimentation boundaries of Fig. 6, suggested an isodesmic indefinite self-association mechanism. Further proof of the latter association mechanism was obtained from the parabolic dependence of the weight-average sedimentation coefficients on the protein concentration, as depicted in Fig. 12, and the good fit of the data by the theoretical curves calculated according to the isodesmic indefinite mechanism at various vinblastine concentrations.

Computer Simulation of Velocity Sedimentation Boundaries

The method of numerical calculation of velocity sedimentation boundaries using an asymptotic approach was first introduced by Gilbert three decades ago.[24,25] Since then, the numerical simulation technique has been greatly refined and used fruitfully in the deduction of the mechanisms of biomacromolecular self-association. Basically, there are two types of simulation methods. The first method is for those systems where no ligand gradient is formed across the boundary. They include both weak ligand-induced self-associations and strong ligand-induced self-associations at saturating ligand concentrations. With a constant ligand concentration

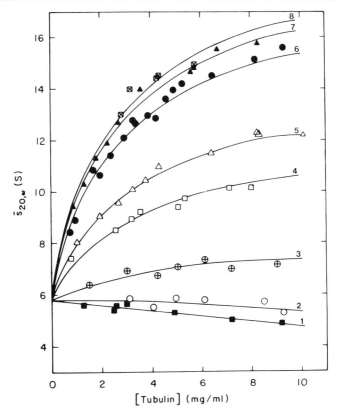

FIG. 12. Weight-average sedimentation coefficients ($\bar{s}_{20,w}$) of tubulin determined as a function of total protein concentration. Tubulin was equilibrated with different concentrations of vinblastine [(1) PG (0.01 M NaP$_i$; 10^{-4} M GTP, pH 7.0) only; (2) 1 × 10^{-6} M; (3) 5 × 10^{-6} M; (4) 2.5 × 10^{-5} M; (5) 5 × 10^{-5} M; (6) 1 × 10^{-4} M; (7) 2 × 10^{-4} M; (8) 5 × 10^{-4} M] in PG buffer, using both a batch and a flow column of Sephadex G-25 gel. Aliquots (0.5 ml) were collected from the column and used directly for ultracentrifugation without dilution to avoid changes in free vinblastine concentration. Solid lines are least-squares fittings of the experimental data by the isodesmic, indefinite, self-association model. (Taken from Ref. 9.)

throughout the cell, the simulation can be carried out using a constant apparent association constant. For a Cann–Goad bimodal sedimentation boundary, where a gradient of free ligand concentration does form around the boundary, the apparent association constant will vary throughout the boundary. For the latter systems, one should use the second method, which is designed to keep track of the free ligand concentration across the boundary. For those interested in carrying out the numerical simulation

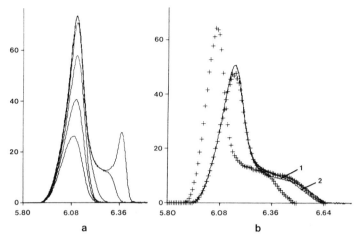

Fig. 13. Simulated sedimentation velocity profiles of magnesium-induced tubulin self-association in a sector-shaped cell at 48,000 rpm. (a) The sedimentation time was 1880 sec. A sharp initial boundary was set at 5.884 cm. The ordinate is the concentration gradient calibrated in units of mg/ml/cm and the abscissa indicates radial position in cm from the center of rotation. The protein concentrations corresponding to the shown patterns are 4.0, 6.0, 8.0, 10.0, 12.0, and 14.0 mg/ml. The association parameters are described in Ref. 7. (b) The calculated boundary was depicted by the solid curve. The experimental concentration gradient profile (+), shown at an early sedimentation time of 1260 sec was used as the initial boundary data. The theoretical profiles were then calculated for an additional sedimentation time of 720 sec and the curves are shown compared with an experimental profile (+) at 1980 sec. (Taken from Ref. 7.)

study, there have been two excellent review articles dealing with details of the methods.[33,34]

Given a theoretical self-association model, the computer simulation technique is capable of predicting boundary shapes and showing how they change in response to variations of certain molecular or association parameters. This is exemplified by Figs. 13a and 14A where, starting from synthetic hypersharp boundaries, simulated sedimentation boundaries were generated for a Gilbert system and an isodesmic indefinite self-association system. The qualitative agreement of the characteristic shapes of the simulated boundaries with the experimental boundaries found in the magnesium ion and vinblastine-induced tubulin self-associations has led to the initial confirmation of the association schemes. Fur-

[33] D. J. Cox, this series, Vol. 48, p. 212.
[34] D. J. Cox and R. S. Dale, in "Protein-Protein Interactions" (C. Frieden and L. W. Nichol, eds.), p. 173. Wiley, New York, 1981.

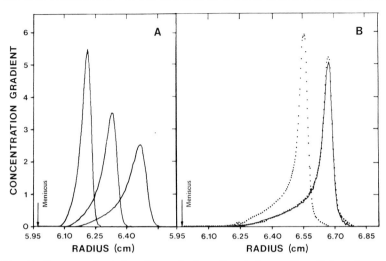

FIG. 14. Computer simulation of the velocity sedimentation profiles of tubulin in a sector-shaped cell. The ordinates for both parts A and B are in random units, whereas the abscissa represents the radial distance from the center of rotation. (A) The simulation process was started with an initial sharp boundary of 10 mg/ml set at 6.09 cm from the center of rotation. It was then subjected to simulated sedimentation at 52,000 rpm according to the isodesmic, indefinite self-association model. The resulting schlieren sedimentation patterns were calculated for 510, 964, and 1475 sec of centrifugation, respectively. (B) The simulation process was started after 32 min of sedimentation. The dotted line patterns are experimental schlieren patterns, after 32 and 40 min of sedimentation after reaching speed, of tubulin (12 mg/ml) equilibrated with 5×10^{-5} M vinblastine, 0.01 M NaP$_i$, 10^{-4} M GTP, pH 7.0 buffer. The early sedimentation pattern was used as the initial boundary for computer simulation according to the isodesmic, indefinite self-association model. The simulated sedimentation was allowed to proceed for 8 min. The resulting simulated sedimentation pattern (solid line) is to be compared with the second experimentally obtained sedimentation pattern. (Taken from Ref. 9.)

ther substantiations of the mechanisms were obtained through point-by-point comparison of the experimental boundaries with the simulated ones as shown in Figs. 13b and 14B. The latter simulations were carried out by using as the initial boundary an experimental boundary obtained from the same run at an earlier time.

The computer simulation studies, useful as they are, usually cannot generate new quantitative insights into the system other than those that have already been obtained from the analysis of the weight-average sedimentation coefficients. To begin with, in order to carry out the computer simulation, one needs to have on hand the complete functional dependence of $\bar{s}_{20,w}$ and $\bar{D}_{20,w}$ on the total protein concentration. In other words, the analysis of the weight-average sedimentation coefficient must precede the boundary simulation. The diffusion coefficient, another unknown factor, can be measured experimentally using techniques such as light scattering. However, in practice, it is usually calculated from the weight-average sedimentation coefficient using the Svedberg equation. In the latter case, additional uncertainties will be introduced which further complicate the fitting process. If one accepts the Svedberg equation in calculating the diffusion coefficient, there should be a one-to-one correspondence between a given shape of $\bar{s}_{20,w}$ versus total protein concentration plot and a specific shape of sedimentation boundary. Several salient features of the sedimentation boundaries can be understood better if one recognizes these relationships. For instance, both the Gilbert cooperative self-association and the Cann–Goad-type strong ligand-induced self-association can give bimodal sedimentation boundaries. This is because both systems show a sigmoidal dependence of their weight-average sedimentation coefficient on total protein concentration. In the Gilbert system, the sigmoidal dependence arises from the presence of cooperativity in the self-association. In the Cann–Goad system, it arises from the generation of a concentration gradient of free ligand across the boundary. This is exemplified by the vinblastine-induced tubulin self-association, as shown in Fig. 12. At a given concentration of the free ligand, $\bar{s}_{20,w}$ increases parabolically with increasing protein concentration. However, if a concentration gradient of free ligand is present across the boundary, as shown in Fig. 7, the actual dependence of $\bar{s}_{20,w}$ on the protein concentration across the boundary will no longer follow one of the solid curves but will shift across several curves following the lower ones at low protein concentration and the higher curves at high protein concentration. If the free ligand concentration gradient is sharp, the actual dependence of $\bar{s}_{20,w}$ on protein concentration will become sigmoidal. Also, in this particular system, the self-association plateaus at saturating ligand concentrations. Thus, at a given protein concentration the increment of $\bar{s}_{20,w}$ per unit increase of ligand concentration is maximal at intermediate ligand concentrations. As a result, the actual dependence of $\bar{s}_{20,w}$ on the total protein concentration gradually loses its sigmoidal character at both high and low ligand concentrations and consequently unimodal sedimentation boundaries are observed. Furthermore, for a Cann–Goad system, since the sigmoidal dependence of $\bar{s}_{20,w}$ on the protein concentration results from the concentration gradient of the free ligand across the boundary, the sedimentation boundary first emerges as unimodal and then starts to resolve into a bimodal one only after a period of sedimentation sufficient to allow the establishment of the free ligand concentration gradient.

Probing the Multiple Equilibria of a Ligand-Induced Self-Association

In the preceding section, we demonstrated that the self-association constant of a ligand-induced self-associating system obtained from the velocity sedimentation studies is an apparent one and it is a function of the ligand activity. In this section, we will discuss the method of analyzing this function to probe the multiple equilibria of a ligand-induced protein self-association. From the multiple equilibria, one can then derive information, such as the stoichiometry and association constant of the ligand-binding site(s) linked to the self-association reaction and the mechanism through which the ligand-binding reaction brings about the self-association of the macromolecule. This method of data analysis is illustrated below by using a ligand-induced dimerization reaction as an example. Assuming that both the ligand binding and self-association reactions are fast and reversible and thus both are constantly in equilibrium, the molecular species present in the solution depends on the association pathway. Considering the following association pathways:

pathway I

$$
\begin{array}{ccc}
2A + 2X \xrightleftharpoons{k_1} AX + A + X \xrightleftharpoons{k_1} 2AX \\
K_4 \Updownarrow \qquad\qquad\qquad K_2 \Updownarrow \\
A_2 + 2X \xrightleftharpoons{2K_3} A_2X \xrightleftharpoons{k_3/2} A_2X_2
\end{array}
\tag{14}
$$

pathway II

$$
\begin{aligned}
A + X &\xrightleftharpoons{k_1} AX \\
AX + A &\xrightleftharpoons{k_1} AXA \\
AX + X &\xrightleftharpoons{k_1} XAX \\
XAX + A &\xrightleftharpoons{k_1} XAXA \\
XAXA + X &\xrightleftharpoons{k_1} XAXAX
\end{aligned}
\tag{15}
$$

where A denotes the macromolecule and X denotes the ligand, the self-associations designated as pathway I and pathway II in Eq. (14) have previously been given the names of ligand-mediated association and ligand-facilitated association, respectively.[35] In the ligand-mediated association, the ligand binding to the macromolecule precedes the self-association reaction, whereas in the ligand-facilitated association, the ligand

[35] G. Kegeles and J. R. Cann, this series, Vol. 48, p. 248.

TABLE II

ssociation mechanism	K_2^{app}		$\left(\dfrac{\partial \ln K_2^{app}}{\partial \ln[X]}\right)_{T,P,a_{j\neq x}}$
I	$\dfrac{[(AX)_2]}{([A]+[AX])^2}$	$\dfrac{k_2}{[1+1/(k_1[X])]^2}$	$\dfrac{2}{1+k_1[X]}$
II	$\dfrac{[A_2]+[A_2X]+[A_2X_2]}{[A]^2}$	$k_4(1+2k_3[X]+k_3^2[X]^2)$	$\dfrac{2k_3[X]+2k_3^2[X]^2}{1+2k_3[X]+k_3^2[X]^2}$
I + II	$\dfrac{[A_2]+[A_2X]+[A_2X_2]}{([A]+[AX])^2}$	$k_4\dfrac{1+2k_3[X]+k_3^2[X]^2}{(1+k_1[X])^2}$	$\dfrac{2k_3[X]+2k_3^2[X]^2}{1+2k_3[X]+k_3^2[X]^2} - \dfrac{2k_1[X]}{1+k_1[X]}$
Crosslink	$\dfrac{[A_2X]+[A_2X_2]+[A_2X_3]}{([A]+[AX]+[AX_2])^2}$	$\dfrac{k_1^2[X]}{(1+k_1[X])^2}$	$\dfrac{1-k_1[X]}{1+k_1[X]}$

binding follows the association reaction. If the self-association can proceed through both pathways I and II then it is called a combined ligand-mediated and facilitated mechanism. In the self-association reaction described by Eq. (15), the ligand is divalent and it serves essentially as a cross-linker between two macromonomers. This self-association mechanism will be referred to hereon as a cross-linking mechanism. For these self-association mechanisms, Table II lists the apparent dimerization constants measured by the velocity sedimentation technique, both in terms of the constituent macromolecular concentrations, and in terms of the microscopic equilibrium constants and ligand activities. Figure 15A presents plots of the logarithm of these apparent dimerization constants versus the logarithm of the free ligand concentration. This plot will be referred to as the Wyman plot, since it is essentially a reflection of the Wyman linked function[29,36]:

$$\left(\frac{\partial \ln K}{\partial \ln[X]}\right)_{T,P,a_{j\neq x}} = \left(\frac{\partial m_x}{\partial m_p}\right)_{T,P,m_{j\neq p}} - \left(\frac{\partial m_x}{\partial m_r}\right)_{T,P,m_{j\neq r}}$$

$$= (\bar{X}_{product} - \bar{X}_{reactant}) - \frac{[X]}{[W]}(\bar{W}_{product} - \bar{W}_{reactant}) \qquad (16)$$

$$\cong \bar{X}_{product} - \bar{X}_{reactant} = \Delta\bar{X}$$

where m_x, m_r, m_p are the molal concentrations of the ligand, the reactant, and the product, respectively. [W] is the molal concentration of water in the solution, [X] is that of free ligand; $\bar{W}_{reactant}$ and $\bar{W}_{product}$ are the hydrations of the reactant and product, and $\bar{X}_{reactant}$ and $\bar{X}_{product}$ are the respec-

[36] C. Tanford, J. Mol. Biol. 39, 539 (1969).

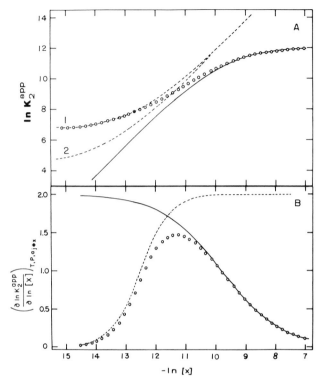

FIG. 15. Wyman plots of the apparent dimerization constant for the ligand-mediated, ligand-facilitated, and ligand-mediated plus facilitated mechanisms of self-association. (A) ln K_2^{app} versus ln[X]. The solid line is for ligand-mediated self-association, where $k_1 = 1.8 \times 10^4 \, M^{-1}$ and $k_2 = 1.8 \times 10^5 \, M^{-1}$. The two dashed lines are for ligand-facilitated self-association, where the intrinsic equilibrium constants used are (1) $k_3 = 2.7 \times 10^5 \, M^{-1}$ and $k_4 = 8 \times 10^2 \, M^{-1}$, and (2) $k_3 = 8.5 \times 10^5 \, M^{-1}$ and $k_4 = 80 \, M^{-1}$. The open circles are for the ligand-mediated plus facilitated mechanism of self-association where $k_1 = 1.8 \times 10^4 \, M^{-1}$, $k_2 = 1.8 \times 10^5 \, M^{-1}$, $k_3 = 2.7 \times 10^5 \, M^{-1}$, and $k_4 = 8 \times 10^2 \, M^{-1}$. (B) $(\partial \ln K_2^{app}/\partial \ln[X])_{T,P,a}$ versus ln[X]; the symbols have the same meaning as in (A). (Taken from Ref. 10.)

tive average ligand bindings by the reactant and the product. The Wyman linked function states that $\Delta \bar{X}$, the slope of the above double logarithmic plot, should be equal to the difference between the average ligand bound to the product and the average ligand bound to the reactant. For the above described association mechanisms, these derivatives are listed in the last column of Table II and their values are depicted in Fig. 15B. It is evident that $\Delta \bar{X}$ is dependent on the ligand concentration and the association mechanism. For the ligand-mediated self-association defined above, the slope of the double-logarithm plot at low ligand concentration approaches two, the stoichiometry of the ligand in the self-association reaction. It approaches zero at saturating ligand concentration. On the other hand, for a ligand-facilitated self-association, the slope approaches zero at low ligand concentration but reaches two at high ligand concentration. For the combined ligand-mediated and facilitated self-association, $\Delta \bar{X}$ approaches zero at both the high and low ligand concentration ends, while, between these two extremes, it reaches a maximal value which is a function of the two ligand binding constants, k_1 and k_3, and is less than two, the ligand stoichiometry in the dimerization reaction. For the cross-linking mechanism of self-association, the Wyman plot shows an initial increase followed by a decrease of ln K_{app} with increasing free ligand concentration. This is because, at high ligand concentration, the ligand binding sites tend to become saturated, which drives the macromolecule to its monomeric state. Thus, by plotting ln K_{app} versus ln[X] and fitting the data with theoretical curves calculated for different association schemes, one can deduce the mechanism of linkage as well as the intrinsic equilibrium constants of the ligand-induced self-association reaction.

The procedure of the data analysis is as follows. One should focus on the apparent association constant of a single association step at a time. For an isodesmic type self-association, the conclusions derived from analyzing a single apparent association constant should be applicable to all the association steps with the same association constant. For a given association mechanism, list out all of the macromolecular species stipulated by the mechanism and express the apparent association constant accordingly. This is followed by nonlinear least-squares fitting of the experimentally obtained apparent association constants with the equation. Examples of such a linkage analysis are provided by the calcium and hydronium ion effects on hemocyanin subunit association,[37] and the magnesium-induced and vinblastine-induced tubulin self-associations.[7-10] In order to derive the microscopic equilibrium constants, it is necessary that the apparent association constant be determined over a wide range of ligand concentrations, preferably at least one order of magnitude from both ends of half-saturation, i.e., from one-tenth to 10 times of $1/K_x^{app}$, where K_x^{app} is the apparent ligand binding constant. This is exemplified by the vinblastine-induced tubulin self-association shown in Fig. 16A where curve fitting of the Wyman plot provided the microscopic equilibrium constants and indicated that the isodesmic indefinite self-association of tubulin proceeds through a one-ligand-mediated mechanism.[10]

A word of caution is needed regarding the interpretation of the linked function analysis. It is often said that an equilibrium study cannot be used

---

[37] K. Morimoto and G. Kegeles, *Arch. Biochem. Biophys.* **142**, 247 (1971).

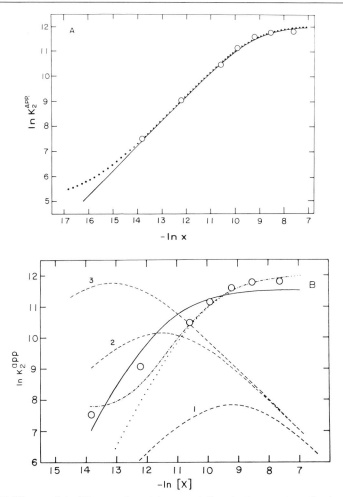

FIG. 16. Wyman plot of the experimental apparent dimerization constants for the vinblastine-induced self-association of tubulin and their least-squares fittings by different reaction mechanisms. The experimental results are depicted by the open circles. (A) Least-squares fittings by one-ligand-mediated mechanism (——) and one-ligand-mediated and facilitated mechanism ($\cdots$). (B) Least-squares fittings by two-ligand-mediated mechanism at [vinblastine] $\geq 1 \times 10^{-6}\ M$ (——) and at [vinblastine] $\geq 2.5 \times 10^{-5}\ M$, ($\cdots$) and two-ligand-mediated and facilitated mechanism (---). The dashed lines in (B) are the Wyman plots for self-association via a cross-linking mechanism. The intrinsic equilibrium constants, k_1, used in this case are (1) 1×10^4, (2) 1×10^5, and (3) $5 \times 10^5\ M^{-1}$. (Taken from Ref. 10.)

to prove a reaction pathway. Indeed, what the Wyman linked function analysis provides is a free energy map of the macromolecular states and the manner in which it is affected by the free ligand concentration. At equilibrium, the distribution of the macromolecules among the various states should follow the Boltzmann equation. This molecular distribution is a static picture and not a dynamic one. It tells nothing about how the macromolecules are converted from one state to another. However, the equilibrium distribution of macromolecule as a function of ligand activity can often be used as evidence to rule out certain reaction pathways. This usage of the linked function is best expressed in the words of J. Wyman[29]: "Although by itself the principle, being thermodynamic, has nothing to say about mechanism, nevertheless it provides a powerful touchstone for exploring the implications of any proposed mechanism as well as for establishing the consistency of underlying observations at a phenomenological level."

Concluding Remarks

In the past two decades, both the theory and methodology of velocity sedimentation have made such significant advances that it has evolved into a powerful tool for the examination of macromolecular self-associations. Computer simulation of velocity sedimentation boundaries has contributed significantly to this and it should continue to guide future developments. On the theoretical front, our knowledge is slowly gaining ground in the area of predicting the sedimentation coefficient and hydrodynamic nonideality based on the size and shape of a macromolecule. It is evident that further development in this area is needed in order to strengthen the technique to allow fully quantitative characterization of macromolecular associations. In this context, it is certain that the potential of the velocity sedimentation technique can be greatly enhanced by using it in conjunction with other physical methods, such as electron microscopy, light scattering, and X-ray diffraction.

Acknowledgment

This work was supported in part by NIH Grants (to SNT) CA-16707 and GM-14603.

Reference to brand or firm names does not constitute endorsement by the U.S. Department of Agriculture over others of a similar nature not mentioned.

[25] Measurement and Analysis of Ligand-Binding Isotherms Linked to Protein Self-Associations

By GEORGE C. NA and SERGE N. TIMASHEFF

In the preceding chapter [24], we examined the method of studying ligand-induced protein self-associations using the velocity sedimentation technique.[1] Toward the end of the chapter, we discussed how to utilize the apparent self-association constants derived from the velocity sedimentation study and analyze them as a function of the ligand activity to probe the multiple equilibria of the system. In a ligand-induced protein subunit association system, the equilibria of the ligand-binding reaction and the self-association reaction are linked. Consequently, the apparent self-association constant and the apparent ligand-binding constant should be mutually dependent. A number of theoretical studies have shown that in linked self-association systems the shape of the ligand-binding isotherm depends on the macromolecular concentration, the intrinsic (microscopic) equilibrium constants, and the mechanism of the equilibrium linkage. These isotherms obtained at different protein concentrations and plotted in the Scatchard format could be concave either upward or downward and their Y intercepts could either converge or diverge.[2-5] Therefore, by analyzing the ligand-binding data and fitting it with theoretically calculated isotherms, one can also probe the multiple equilibria of the system. These two different approaches, although probing the same multiple equilibria, utilize different experimental methods. One measures the macromolecule self-association while the other determines the ligand binding. They are, therefore, independent of each other with regard to the nonidealities present in the experimental methods and the assumptions invoked in the data analyses. Consequently, by taking both approaches to examine a linked system and obtaining consistent results, one can gain much stronger confidence in the conclusions derived. Furthermore, as will become evident later in this chapter, these two approaches frequently shed light on different aspects of a linkage and complement each other well. On the phenomenological level, it is important to recognize that the cooperativity between a ligand-binding reaction and a self-association

[1] G. C. Na and S. N. Timasheff, this volume [24].
[2] J. R. Cann and N. D. Hinman, *Biochemistry* **15**, 4614 (1976).
[3] L. W. Nichol and D. J. Winzor, *Biochemistry* **15**, 3015 (1976).
[4] J. R. Cann, this series, Vol. 48, p. 299.
[5] R. F. Stein, *Mol. Cell. Biochem.* **31**, 5 (1980).

METHODS IN ENZYMOLOGY, VOL. 117

reaction can lead to curvilinear ligand-binding isotherms that are similar to those found in either allosteric or heterogeneous ligand-binding systems. Such similarities have led to frequent misinterpretations of ligand-binding data in the literature. In this chapter, we discuss the methods of diagnosing ligand-induced protein self-associations and probing the multiple equilibria of such systems through the measurement and analysis of their ligand-binding isotherms.

Measurement of Ligand–Protein-Binding Isotherms

There are countless numbers of methods available for the measurement of ligand–protein bindings. Many of them do not meet the criteria to be used in linked self-association systems. We shall examine some of the reasons why they do not, and then describe the experimental protocol of a batch gel filtration method which we have used successfully that can be adapted easily to other linked systems.

One can classify the methods of measuring ligand–protein-binding isotherms into two major categories, indirect measurement and direct measurement. In an indirect measurement, one monitors a specific change in a physical or chemical property of either the protein or the ligand that ensues from the ligand–protein interaction. This can be, for instance, a shift in the absorption spectrum, quenching of the fluorescence spectrum, or a change in NMR relaxation time. Such methods have the advantage that the measurements can be carried out without prior separation of the ligand from the macromolecule. However, they are subject to the serious drawback that, in order to calculate the amount of the bound ligand, one usually has to make the assumption that the characteristic property change is linearly proportional to the quantity of the ligand bound to the macromolecule. The validity of this assumption is frequently difficult to verify independently. It is very likely an oversimplification when dealing with proteins with multiple ligand-binding sites or those with the ligand-binding reaction linked to the subunit association reaction. This difficulty essentially rules out the use of indirect measurements in the rigorous and unambiguous determination of ligand binding in linked systems.

It is frequently easier to measure the total concentration of a ligand in a solution through monitoring a physical or chemical property of the ligand such as radioactivity or light absorption. With this type of approach, the property of the ligand to be monitored should be independent of the interaction of the ligand with the macromolecule, a condition that is usually not difficult to establish. However, in order to obtain directly the concentration of both the bound and free ligand, one usually has to separate the macromolecule from the free ligand. Thus, the method of separating the unbound ligand from the macromolecule, instead of the actual measurement of the ligand activity itself, often becomes the issue. The selection of the method of separation is dictated by two requirements. First, the separation must be carried out without perturbing the chemical equilibrium of the system, i.e., the results must reflect the true equilibrium of the reaction. Second, it must allow precise control of the protein concentration so that a ligand-binding isotherm can be determined under a constant protein concentration. Molecular transport methods, such as ultracentrifugation, size exclusion chromatography, and electrophoresis, have been used frequently to separate the unbound ligand from macromolecules. For instance, one can precipitate macromolecules by centrifugation, leaving only the free ligand in the supernatant. Such an approach has been demonstrated by Steinberg and Schachman, using an analytical ultracentrifuge equipped with an optical scanner to monitor the distribution of the ligand across the centrifuge cell.[6] However, as described in the preceding chapter,[1] in fast reversible ligand-induced protein associations, due to the differences in the sedimentation coefficient and ligand affinity between the monomer and the polymers, the ligand concentration found in the supernatant could differ substantially from the equilibrium concentration. This renders the method unacceptable for the present purpose. This difficulty with velocity sedimentation can be circumvented by using instead equilibrium ultracentrifugation. That method, however, is not suitable for protein concentration-dependent studies. Furthermore, the low sensitivity of the UV absorption optics often places a limit on the magnitude of the binding constant that can be measured by the instrument. In zonal methods such as the Hummel–Dreyer column gel filtration, the results are usually expressed as ligand binding across the complete protein peak instead of at a given protein concentration.[7] It is, therefore, not suitable for protein concentration-dependent studies. Paper discs impregnated with ion-exchange resins also have been used to separate the unbound ligand from the protein. This method requires extensive washing of the disc after the adsorption of the protein to the ionic resin to remove the unbound ligand. Since the bound ligand can be released and the protein concentration can be changed during the washing, the method is unacceptable unless one can demonstrate that the rates of the ligand and the protein association/dissociation are both very slow.

The methods that seem most suitable for the determination of the ligand binding of linked systems are those that use the principle of size exclusion to separate, yet maintain an equilibrium between the macromolecule and the free ligand. For instance, with certain ions, one can use

---

[6] Z. Steinberg and H. K. Schachman, *Biochemistry* **5**, 3728 (1966).
[7] J. P. Hummel and W. J. Dreyer, *Biochim. Biophys. Acta* **63**, 530 (1962).

specific electrodes which employ semipermeable glass or membranes that allow the determination of the free ion activity in the presence of the macromolecule. Two other methods that use the same principle are equilibrium dialysis and batch gel filtration. The technique of equilibrium dialysis has been discussed in detail in earlier volumes of this series and will not be discussed here.[8,9] The batch gel filtration method is, in principle, quite similar to equilibrium dialysis. It uses a porous gel instead of a semipermeable membrane to maintain chemical and diffusional equilibria between two separate phases. The external phase corresponds to the mobile phase in column gel filtration and contains both the macromolecule and the ligand. The internal phase corresponds to the stationary phase in column gel filtration and contains only the free ligand.[9-11] Compared with equilibrium dialysis, the batch gel filtration method usually requires less material and the diffusional equilibrium can be reached within a shorter period of time. These could be significant advantages when dealing with precious and/or unstable proteins. Furthermore, the batch gel filtration method allows more precise control of the protein concentration than does equilibrium dialysis. However, as in the case of equilibrium dialysis, the method is sometimes complicated by the adsorption of ligand, usually aromatic ones, to the gel matrix which requires corrections in data analysis.[10-13] Nevertheless, parallel studies have shown that the ligand-binding results obtained from the batch gel filtration method are consistent with those obtained from other techniques.[10,14] The following is an experimental protocol of the batch gel filtration method that we have used successfully in the determination of the binding of the anticancer drug vinblastine to tubulin.[16] One should be able to adopt it easily for other linked systems, and it is of course applicable to systems which do not exhibit linkage.

Determination of Ligand–Protein Binding by Batch Gel Filtration

Preliminary Considerations. Before embarking on the measurement of ligand–protein bindings, one should first experiment with different types of gels. They frequently show different degrees of ligand adsorption due

[8] L. C. Craig, this series, Vol. 11, p. 870.
[9] P. McPhie, this series, Vol. 22, p. 23.
[10] P. Fasella, G. G. Hammes, and P. R. Schimmel, *Biochim. Biophys. Acta* **103**, 708 (1965).
[11] W. H. Pearlman and O. Crépy, *J. Biol. Chem.* **242**, 182 (1967).
[12] J.-C. Janson, *J. Chromatogr.* **28**, 12 (1967).
[13] B. Gellote, *J. Chromatogr.* **3**, 330 (1960).
[14] M. Hirose and Y. Kano, *Biochim. Biophys. Acta* **251**, 376 (1971).
[15] G. C. Wood and P. F. Cooper, *Chromatogr. Rev.* **12**, 88 (1970).
[16] G. C. Na and S. N. Timasheff, in preparation.

to their different chemical characteristics. In addition to the commonly used gels, such as the polyacrylamide-based BioGel (Bio-Rad Chemical Co.)[17] and Dextran-based Sephadex (Pharmacia Chemical Co.), the silicate-based porous glass beads (Pierce Chemical Co.) often display less ligand adsorption and deserve consideration. It should be cautioned, however, that glass beads in the simple form often absorb proteins, particularly those with higher charge densities. This difficulty can be partially alleviated by choosing the carbohydrate-coated glass beads and using a higher ionic strength buffer. With a given type of gel, the ligand adsorption usually decreases with increasing pore size of the gel.[12–14] Consequently, it is best to select the largest pore size gel that still excludes the protein. The adsorption of the ligand to the gel is often time dependent. To minimize the adsorption, the gel can be swollen in the absence of the ligand so that the ligand of interest is allowed to be in contact with the gel for a short period of time, just sufficient to achieve diffusional and chemical equilibria. Adsorption is not the only factor that makes the partition coefficient of the ligand, defined as the ratio of the ligand concentration in the internal phase of the gel to that in the external phase, deviate from the ideal value of one. If the size of the ligand molecule is substantially larger than that of water, the partition coefficient of the ligand could be less than one. This is apparently due to the exclusion of the ligand from a portion of the internal volume of the gel that is accessible to water molecules. At low ligand concentrations, the adsorption of the ligand to the gel, if it occurs, is usually the dominant factor and the partition coefficient becomes higher than one. At high ligand concentrations, the amount of the adsorbed ligand plateaus and becomes insignificant relative to the total ligand. The size exclusion effect, if any, may emerge to reduce the partition coefficient of the ligand to less than one. Again, to account for the nonideality, due either to the adsorption of the ligand to the gel matrix or to the exclusion of the ligand from a portion of the internal volume of the gel, the data have to be corrected against the partition coefficient of the ligand determined in the absence of the protein.

Experimental Protocol. Borosilicate culture tubes (12 × 75 mm) are soaked in Nochromix solution (Godax Lab., Inc.) and then rinsed thoroughly before use. The washed glass tubes should show little solution adherence to the test tube wall which may affect the accuracy of the measurements. Our experience has shown that plasticwares, both test tubes and disposable pipettes, frequently absorb aromatic ligands and should be avoided. In the method described below, the volumes of the

---

[17] Reference to brand or firm name does not constitute endorsement by the U.S. Department of Agriculture over others of a similar nature not mentioned.

aliquots are based on BioGel P-100. Both the weight of the gel and the volumes of the aliquots can be modified, scaled either up or down depending on the particular gel used and the availability of the material under investigation.

Forty milligrams of the dried gel is weighed in each tube using a semimicroanalytical balance. Of the experimental buffer 0.5–0.6 ml is then added to each tube. The buffer contains all the ingredients except the protein and the ligand of interest. The volume of the buffer is approximately 20% higher than the internal volume of the gel to ensure complete swelling of the gel. The tubes are tightly sealed with parafilm and the gel is allowed to swell for 20 hr at room temperature. At the end of the swelling, 0.2 ml of the experimental buffer containing a known concentration of the ligand is added and mixed gently. This is followed immediately by 0.2 ml of the protein stock solution, which had been preequilibrated with the same experimental buffer without the ligand through either column or batch gel filtration or equilibrium dialysis. The complete mixture is incubated at a constant temperature with gentle shaking to avoid splashing the solution to the test tube wall. The time interval of incubation is predetermined to be sufficient for achieving both chemical and diffusional equilibria, usually within 10 min. After the incubation, the tubes are left to stand still to allow the gel to settle by gravity to the bottom. Aliquots are then carefully removed from the supernatant for measurements of the ligand and protein concentrations. Controls are treated identically as the samples. They contain the same mixture except either the protein or the ligand solution is replaced by additional experimental buffer. The former controls give the partition coefficient of the ligand, while the latter ones give the excluded volume of the protein, both with respect to the particular gel used. Higher degrees of precision and accuracy can be achieved if all of the aliquots are measured gravimetrically with a semimicroanalytical balance. Weights of the aliquots can be converted to volumes using the density of the buffer solution which can be determined accurately with a precision densimeter.[18] The presence of the protein and the ligand at low concentrations should have little effect on the solution density and can be ignored in the calculation. The internal or excluded volume of the gel is also determined by using Blue Dextran 2000 (Pharmacia Chemical Co.). Three milliliters of Blue Dextran 2000 at concentrations ranging from 0.02 to 0.2% is added to each tube containing 100 mg of the dried gel. The gel is incubated for 20 hr at room temperature. Aliquots are then taken from the supernatant for measurement of the Blue Dextran concentration using its UV absorbance at 260 nm. If the gel is properly chosen, the excluded

[18] J. C. Lee, K. Gekko, and S. N. Timasheff, this series, Vol. 61, p. 26.

volume obtained using Blue Dextran 2000 should not differ by more than 5% from the value obtained with the protein.

Data Calculation. The excluded volume per unit weight of the gel, $\bar{v}_i$, can be calculated from

$$A_e/A = v_t/v_e \tag{1}$$
$$v_t = v_e + v_i \tag{2}$$
$$\bar{v}_i = v_i/g \tag{3}$$

where A and A_e are the absorbances at 260 nm of Blue Dextran 2000 in the stock solution and the external phase, respectively, v_t is the total volume of the solution, v_e is the external volume, v_i is the internal volume, and g is the weight of the gel. Once $\bar{v}_i$ is determined, for a given weight of the gel, the internal and external volumes can be calculated from Eq. (3). Following the law of conservation of mass, [X_i], the concentration of the ligand in the internal volume of the gel, can be calculated from

$$[X_i] = \frac{(X) - [X_e]v_e}{v_i} \tag{4}$$

where (X) is the total amount of the ligand added and $[X_e]$ is the concentration of the ligand in the external volume. In an ideal case, the partition coefficient of the free ligand should be equal to one. However, because of

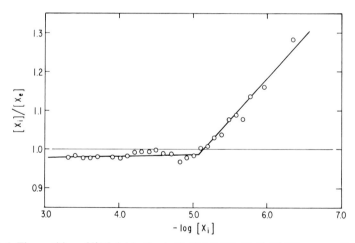

Fig. 1. The partition of [³H]vinblastine in BioGel P-100. BioGel P-100 was swollen for 24 hr. [³H]Vinblastine was then added and the mixture was incubated at 20° for 1 hr before aliquots were removed for radioactive scientillation counting. Data were least-squares fitted to two straight lines, from 10^{-7} to 10^{-5} M vinblastine and from 10^{-5} to 10^{-3} M vinblastine.

the factors described above, for many ligands it may deviate from one. The nonideality can be corrected using the partition coefficient of the free ligand determined in the absence of the protein as a function of the ligand concentration. Figure 1 shows the partition coefficient of vinblastine, a plant alkaloid and anticancer drug, in the presence of BioGel P-100. Above $1 \times 10^{-5} M$ of ligand, the partition coefficient has a rather constant value of 0.98 ± 0.01. Below $1 \times 10^{-5} M$, the adsorption of the ligand to the gel caused a linear increase of the partition coefficient as a function of the logarithm of the ligand concentration within the gel. By assuming that these partition coefficients are not affected by the presence of the protein, one can obtain $[X_f]$, the concentration of the free ligand in the external volume, from $[X_i]$ by interpolating from the predetermined partition coefficients of the free ligand. Once $[X_f]$ is known, one can calculate the concentration of the ligand bound to the protein in the external volumes, $[X_b]$:

$$[X_b] = [X_e] - [X_f] \tag{5}$$

Analyses of Ligand-Binding Isotherms

Data Acquisition

Protein self-association, being an intermolecular event, is dependent on protein concentration. Consequently, a ligand-binding reaction linked to a protein self-association reaction must also be dependent on the protein concentration. This contrasts with the cases of heterogeneous, but independent ligand-binding sites on a protein and allosteric ligand binding in which free energy coupling takes place only intramolecularly between different ligand-binding sites. Although both of these types of ligand binding can display curvilinear binding isotherms, neither should be dependent on the macromolecular concentration. Thus, the first step toward establishing a ligand-induced self-association system is to demonstrate that the ligand-binding isotherms obtained are dependent on the total protein concentration. Subsequent to this, a more rigorous and quantitative linkage analysis can be carried out, if one has on hand several ligand-binding isotherms that span one order of magnitude or more of the protein concentration. The range of protein concentration used should depend on the self-association constant and the linkage mechanism. For quantitative analysis, the binding isotherm obtained at the lowest protein concentration should be close to that of the intrinsic binding, whereas that obtained at the highest protein concentration should manifest substantial deviation from the intrinsic binding.

Data Plotting

The ligand binding data obtained are best analyzed by the Scatchard plot[19] ($\bar{X}/[X_f]$ vs $\bar{X}$, where $\bar{X} = [X_b]/[A_t]$; $[X_f]$ and $[X_b]$ are the concentration of the free and bound ligand, respectively, and $[A_t]$ is the molar concentration of the macromolecule in terms of the monomer) rather than the double-reciprocal plot ($\bar{X}^{-1}$ vs $[X_f]^{-1}$) or the saturation plot ($\bar{X}$ vs $[X_f]$). When dealing with ligand bindings that are linked thermodynamically to self-associations, the Scatchard plot is superior to the other two plots in its ability to diagnose different equilibrium linkage mechanisms and to allow easy derivation of the intrinsic association constants.[2] The reason lies in the intrinsic differences between these plotting methods. In the Scatchard plot, the ligand binding constant can usually be extracted from either the slope or the Y intercept of the data.[20] Since ligand binding isotherms of linked self-association systems are usually curvilinear, the ligand binding constants can be more easily and accurately derived from the Y intercept rather than the slope of the curve. By using the Y intercept of the Scatchard plot, the data analysis is focused at the region where the ligand concentration approaches zero. At diminishing ligand concentrations, the Y intercept of the Scatchard plot can be expressed in simpler terms of the intrinsic equilibrium constants and the macromolecular concentration and is, therefore, more amenable to analysis. On the other hand, with a double-reciprocal plot, the binding constant is derived from either the slope of the data or the extrapolated value on the X axis. The extrapolation can be carried out only when the isotherm is linear. Consequently, for curvilinear isotherms which prevail in the ligand-induced association systems, one has to resort to the slope of the plot to derive the intrinsic equilibrium constants. The latter process requires additional manipulation of the data and is less accurate. Furthermore, as shown in the latter part of this chapter, different linkage mechanisms of ligand-induced self-association manifest most pronounced differences in their apparent ligand bindings in the region where the association-linked binding site is less than half saturated. Such differences appear more distinctly and dramatically in the Scatchard plot than in the other two plots.[2]

Although one can use the same Scatchard plot and extrapolate the data to the X axis to obtain the ligand-binding stoichiometry, the validity of the latter value depends on the linearity of the data. Since the Scatchard plots of self-association-linked ligand bindings are almost never linear, one should be cautioned against this practice. In most cases, it is better to plot the same data in the form of a saturation plot ($\bar{X}$ vs log $[X]$)

[19] G. Scatchard, *Ann. N.Y. Acad. Sci.* **51**, 660 (1949).
[20] I. M. Klotz and D. L. Hunston, *Biochemistry* **10**, 3065 (1971).

which allows direct visualization of the degree of saturation of a given binding site or class of binding sites and is therefore more reliable than the Scatchard plot in deriving the ligand binding stoichiometry.[21]

No matter how the data are plotted, there is no way of assuring complete saturation of a protein by a given ligand. Consequently, the range of the ligand concentration used in a binding study must be decided through other means. The best guideline obviously can be found in a complementary study of the macromolecular self-association reaction, i.e., determination of the range of ligand concentrations that exerts a clearly observable effect on the equilibrium of the protein self-association reaction and measurement of the ligand–protein binding in that range, plus maybe one order of magnitude at each end. Ligand bindings that take place outside this range are independent of and irrelevant to the self-association reaction.

Model Fitting

Ligand-binding isotherms in the form of Scatchard plots can be analyzed quantitatively by fitting them with theoretically calculated isotherms. The characteristic shape of the ligand-binding isotherm is usually more uniquely dependent on the linkage mechanism and is generally insensitive to the macromolecular association stoichiometry, i.e., if one varies the macromolecular stoichiometry but maintains the same linkage mechanism, the resulting calculated isotherms usually change only quantitatively without losing their characteristic shape. Consequently, one can frequently deduce the linkage mechanism simply from the shape of the binding isotherm and without any knowledge of the stoichiometry of the macromolecular association. However, for a rigorous quantitative analysis, it is imperative to determine the stoichiometry of the macromolecular association through other means, such as analytical ultracentrifugation or light scattering, so that the exact theoretical ligand-binding isotherms can be calculated and used to fit the data. To aid the process of model selection and isotherm calculation, we present below the calculation of ligand-binding isotherms for the four different ligand-induced self-associations examined in the preceding chapter while discussing the linked function analysis of the apparent macromolecular self-association constant.[1] Each of these systems manifests a ligand-binding isotherm with a characteristic shape. They can be used by the readers, when confronted with an unknown system, as an initial guideline in selecting linkage mechanisms to fit the data.

---

[21] I. M. Klotz, *Science* **217,** 1247 (1982).

In the following sections, we will use lower-case k's for the intrinsic (microscopic) equilibrium constants and upper-case K's for the apparent (macroscopic) equilibrium constants. Although only dimerization reactions are shown, each linkage mechanism can be extended to any macromolecular association stoichiometry. For a ligand-induced self-association, the binding of the ligand to the macromolecule, irrespective of the reaction mechanism, can be expressed in the Scatchard format as

$$
\frac{\bar{X}}{[X_f]} = \frac{[X_b]}{[A_t][X_f]} = \frac{\displaystyle\sum_{i=1}^{n} \sum_{j=1}^{m} j[A_iX_j]}{[A_t][X_f]}
\tag{6}
$$

where $[X_b]$ and $[X_f]$ refer to the concentration of the bound and free ligand, respectively. $[A_t]$ is the total molar concentration of the macromolecule in terms of the monomer and $[A_iX_j]$ is the molar concentration of the i-mer containing j bound ligand. Two somewhat different approaches are available for the calculation of the theoretical ligand-binding isotherms. In the first approach, the concentration of each i-mer containing j bound ligand is expressed in terms of $[A_f]$, the free monomer concentration, $[X_f]$, the free ligand concentration, and the intrinsic equilibrium constants. Given the total protein concentration, the free ligand concentration, and the intrinsic association constants, one can calculate $[A_f]$, the free monomer concentration. For a dimerization reaction and an isodesmic indefinite self-association, the calculation is simple because it only requires the solving of a binomial equation. For associations with other stoichiometries, the values of $[A_f]$ can be obtained by solving the equation numerically using the Newton–Raphson rule.[22] Given $[A_t]$, $[A_f]$, $[X_f]$, and all the intrinsic equilibrium constants, the summation of Eq. (6) gives the Scatchard binding isotherm. In the case of an indefinite self-association, the summation of Eq. (6) at each free ligand concentration proceeds with increments of i until a major portion of the total protein, e.g., 99.9% is accounted for.

An alternative approach to obtain the theoretical isotherms is to calculate the total concentration of each i-mer containing all the liganded species with the use of the apparent self-association constant. From the total concentration of each i-mer, the amount of the ligand bound to the i-mer usually can be calculated. Both of these methods of calculating the ligand binding isotherms are demonstrated below.

Ligand-Mediated Self-Association. This association mechanism was defined as one in which the binding of the ligand to monomers precedes the macromolecular self-association. The original definition permits the

[22] M. E. Magar, "Data Analysis in Molecular Biology." Academic Press, New York, 1973.

polymerization between a liganded and an unliganded monomer.[23] We further stipulate here that each and every monomer must first bind the ligand before it can participate in the self-association reaction. A one-ligand-mediated dimerization is depicted by the following equation:

$$\begin{array}{cc} A + X \rightleftharpoons AX & k_1 \\ 2\,AX \rightleftharpoons A_2X_2 & 2k_2 \end{array} \qquad (7)$$

If the linkage mechanism is extended to an isodesmic indefinite one, then the apparent self-association constant for each step of the polymerization is identical. For a one-ligand-mediated isodesmic self-association, i.e., self-association induced by the binding of one ligand per monomer,[1] the apparent association constant for each step of the reaction can be expressed as

$$K_2 = \frac{[A_2X_2]}{([A] + [AX])^2} = \frac{2k_1^2 k_2 [X_f]^2}{(1 + k_1[X_f])^2} \qquad (8)$$

where $[X_f]$ is the concentration of the free ligand and k_1 and k_2 have the same definitions as in Eq. (7). Since all polymers must have their association-linked ligand binding sites occupied, one can set $j = i$ and $m = n$ in Eq. (6) and obtain

$$\frac{\bar{X}}{[X_f]} = \frac{\displaystyle\sum_{i=1}^{n} ik_1^i (2k_2)^{i-1} [A_f]^i [X_f]^{i-1}}{[A_t]} \qquad (9)$$

where $[A_f]$ is the concentration of unliganded monomers and $[A_t]$ is as defined above in Eq. (6). Consequently,

$$\lim_{[X_f] \to 0} \frac{\bar{X}}{[X_f]} = k_1 \qquad (10)$$

which means that the Y intercept of a Scatchard plot of this type of linked system will have the value of k_1 regardless of the total protein concentration used.

Figure 2 depicts the Scatchard binding isotherms of a one-ligand-mediated isodesmic indefinite self-association at different protein concentrations. Similar characteristic ligand-binding isotherms were found in the ligand-mediated self-associations with lower macromolecular stoichiometries or with higher ligand stoichiometries.[2,3] There are several salient features in the ligand-binding isotherms of such systems which allow their easy identification. They approach a straight line near zero protein concentration but become increasingly concave downward as the protein

[23] J. R. Cann and G. Kegeles, *Biochemistry* **13**, 1868 (1974).

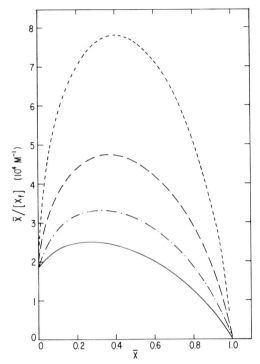

FIG. 2. Scatchard plots of the ligand binding of a one-ligand-mediated isodesmic indefinite self-association. The protein concentrations are $1.6 \times 10^{-5}\ M$ (———), $3.4 \times 10^{-5}\ M$ (-··-), $7.8 \times 10^{-5}\ M$ (— —), and $2.3 \times 10^{-4}\ M$ (--). The intrinsic association constants used are $k_1 = 1.8 \times 10^4\ M^{-1}$ and $k_2 = 9 \times 10^4\ M^{-1}$.

concentration increases. This is because the ligand binding is enhanced by the self-association reaction and, thus, increased at higher protein concentration. As the protein concentration approaches zero, the self-association diminishes and the apparent ligand binding approaches the intrinsic ligand binding. This is mathematically equivalent to setting the self-association constant k_2 to zero. At higher protein concentrations, a maximum can be found in the curvilinear isotherms between ligand stoichiometries of 0.3 and 0.5 for the one-ligand-mediated system. As predicted by Eq. (10), all binding isotherms show the same Y intercept k_1, the intrinsic ligand-binding constant for the ligand-binding site.

The ligand-binding isotherms of ligand-mediated self-association systems can be analyzed by first determining k_1, the ligand-binding constant of the monomer, from the Y intercept of the isotherms, and subsequent fitting of the isotherms to obtain k_2, the association constant between two liganded monomers.

Ligand-Facilitated Self-Association. A ligand-facilitated self-association is defined as one in which the ligand binds only to the polymers. A one-ligand-facilitated dimerization is depicted by the following equations:

$$
\begin{array}{lll}
2A \rightleftharpoons A_2 & & 2k_4 \\
A_2 + X \rightleftharpoons A_2X & & k_3 \\
A_2X + X \rightleftharpoons A_2X_2 & & k_3
\end{array} \tag{11}
$$

The model assumes an identical ligand-binding constant, k_3, for the two association-linked sites on the dimer. For such a self-association scheme, the apparent self-association constant can be expressed by

$$
K_2 = \frac{[A_2] + [A_2X] + [A_2X_2]}{[A]^2} = 2k_4 + 2k_3k_4[X_f] + 2k_3^2k_4[X_f]^2 \tag{12}
$$

Let us consider a similar ligand-facilitated isodesmic indefinite self-association having the same intrinsic association constants k_3 for each ligand binding site and k_4 for each step of the self-association. The weight concentration of each i-mer, c_i, can be calculated by using the equation

$$
c_t = \frac{c_1}{[1 - (K_2c_1/MW)]^2}, \qquad c_i = i \left(\frac{K_2}{MW}\right)^{i-1} c_1^i \tag{13}
$$

where MW is the molecular weight of the monomer. The concentration of the ligand bound to the macromolecule can be obtained by

$$
[X_b] = \sum_{i=2}^{n} \frac{c_i}{MW} \frac{k_3[X_f]}{(1 + k_3[X_f])} \tag{14}
$$

Figure 3 shows the Scatchard binding isotherms for a ligand-facilitated isodesmic indefinite self-association obtained by using Eqs. (12), (13), and (14). The model assumes $k_4 = 9 \times 10^2 \, M^{-1}$, a weak self-association of the monomer in the absence of the ligand. Similarly to the ligand-mediated self-association, the binding isotherms for the ligand-facilitated self-association are also concave downward. However, their Y intercepts increase with increasing protein concentration, instead of converging to a single point as do those of the ligand-mediated association systems. Consequently, at high protein concentrations, the isotherm loses its downward curvature. The amount of the ligand bound to the macromolecule can also be expressed as

[24] G. Kegeles and J. R. Cann, this series, Vol. 48, p. 248.

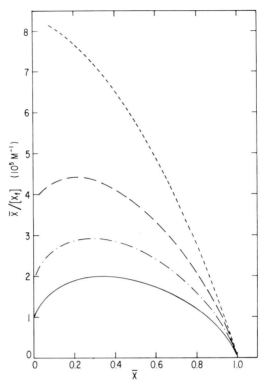

FIG. 3. Scatchard plots of ligand binding of a one-ligand-facilitated isodesmic indefinite self-association. The macromolecular concentrations increase from the bottom to the top and are the same as in Fig. 2. The intrinsic association constants used are $k_3 = 1.8 \times 10^6 \, M^{-1}$ and $k_4 = 9 \times 10^2 \, M^{-1}$.

$$[X_b] = \sum_{i=2}^{n} \sum_{j=1}^{i} j [A_i X_j]$$

$$= \sum_{i=2}^{n} \sum_{j=1}^{i} \frac{i!}{(i-j)!j!} j \, k_3^{\,j} (2k_4)^{i-1} [A_f]^i [X_f]^j \qquad (15)$$

and the Y intercept of the Scatchard plot for this self-association scheme should be

$$\lim_{[X_f] \to 0} \frac{\bar{X}}{[X_f]} = \frac{\displaystyle\sum_{i=2}^{n} i k_3 (2k_4)^{i-1} [A_f]^i}{[A_t]} \qquad (16)$$

At low protein concentrations, if k_4 is small, $[A_f] \cong [A_t]$ and the Y intercept can be expressed by

$$\lim_{[X_f]\to 0} \frac{\bar{X}}{[X_f]} = \sum_{i=2}^{n} ik_3(2k_4)^{i-1}[A_t]^{i-1} \qquad (17)$$

Figure 4a shows the Y intercepts of the Scatchard plots of the above linkage mechanism calculated from Eq. (16) and plotted against the total protein concentration. They show a curvilinear dependence on the total protein concentration and the curvature increases with increasing values of the intrinsic association constants. It is evident that the Y intercept approaches zero at low protein concentration. This result is a consequence of the assumption that the monomers which prevail at low protein concentration do not bind ligand.

A quantitative analysis of the ligand binding isotherms of ligand-facilitated self-association systems can be carried out by first analyzing the Y intercepts of the Scatchard isotherms as a function of the total protein concentration to obtain estimates of k_3 and k_4, followed by point-by-point fitting of the binding isotherms to determine quantitatively the intrinsic association constants.

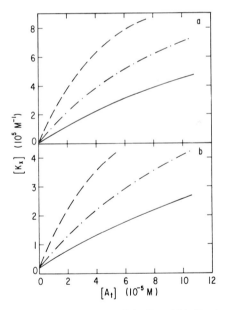

FIG. 4. Y intercept of the Scatchard plots of the ligand-binding isotherms of linked self-association systems as a function of total protein concentration. (a) Ligand-facilitated isodesmic indefinite self-association. The intrinsic association constants are $k_3 = 1.8 \times 10^6$, and $k_4 = 9 \times 10^2$ (——), 1.8×10^3 (—·—·—), and 3.6×10^3 (———). (b) Ligand-mediated and facilitated self-association. The intrinsic association constants are $k_1 = 1.8 \times 10^4$, $k_3 = 1.8 \times 10^6$, and k_2 and k_4 are 9×10^4, 9×10^2 (——); 1.8×10^5, 1.8×10^3 (—·—·—); and 3.6×10^5, 3.6×10^3 (———).

A system that behaves like a ligand-facilitated dimerization can be found in the interaction of the peptide hormone oxytocin with bovine neurophysin II.[25] Equilibrium ultracentrifugational studies showed that the protein undergoes a weak dimerization even in the absence of the ligand. The introduction of oxytocin shifted the existing equilibrium toward the dimeric state. The ligand-facilitated mechanism of association is also indicated by the downward curvature of the Scatchard ligand binding isotherms and the increase of the Y intercept of the binding isotherm with increasing protein concentration.

Ligand-Mediated and Facilitated Self-Association. This mechanism is defined as a combination of the above two mechanisms, i.e., ligand can bind to both monomers and polymers and both liganded and unliganded monomers can participate in the self-association reaction. A one-ligand-mediated and facilitated dimerization reaction is shown by

$$
\begin{array}{ccccc}
A_2 + 2X & \overset{k_1}{\rightleftharpoons} & XA_2 + X & \overset{k_3}{\rightleftharpoons} & A_2X_2 \\
\downarrow\uparrow & & \downarrow\uparrow & & \downarrow\uparrow \\
2A + 2X & \overset{k_1}{\rightleftharpoons} & AX + A + X & \overset{k_1}{\rightleftharpoons} & 2AX \\
\downarrow\uparrow & & \downarrow\uparrow & & \downarrow\uparrow \\
A_2 + 2X & \overset{k_3}{\rightleftharpoons} & A_2X + X & \overset{k_1}{\rightleftharpoons} & A_2X_2
\end{array}
\tag{18}
$$

In addition to the self-associations defined by Eqs. (7) and (11), the association between a liganded and an unliganded monomer is permitted. The model assumes an asymmetry in the self-association, i.e., when two monomers self-associate, the association constant is dependent on the liganding state of only one of the two monomers. The association constant is assumed to be k_2 if that monomer is liganded and k_4 if it is not liganded. This model can simulate a system where the monomer, upon binding a ligand, undergoes a conformational change and acquires a stronger propensity to associate. For such a system, the apparent dimerization constant can be expressed as

$$
\begin{aligned}
k_2 &= \frac{[A_2] + [A_2X] + [A_2X_2]}{([A] + [AX])^2} \\
&= \frac{2k_4 + k_1k_2[X_f] + k_1k_4[X_f] + 2k_1^2k_2[X_f]^2}{(1 + k_1[X_f])^2}
\end{aligned}
\tag{19}
$$

If the reaction scheme is extended to a stepwise isodesmic indefinite self-association, the ligand bound to the macromolecule can be expressed as

[25] P. Nicolas, M. Camier, P. Dessen, and P. Cohen, *J. Biol. Chem.* **251**, 3965 (1976).

$$[X_b] = \sum_{i=1}^{n} k_1{}^i(2k_2)^{i-1}[A_f]^i[X_f]^i + \sum_{i=2}^{n}\sum_{j=1}^{i} \frac{(i-1)!}{(i-j)!(j-1)!} k_1{}^j(2k_2)^{j-1}(2k_4)^{i-j}$$

$$[A_f]^i[X_f]^j + \sum_{i=2}^{n}\sum_{j=1}^{i} \frac{(i-1)!}{(i-j-1)!j!} k_1{}^j(2k_2)^j(2k_4)^{i-j-1}[A_f]^i[X_f]^j \quad (20)$$

The first term of Eq. (20) represents those macromolecules with their self-association linked sites fully occupied. This is contributed by the ligand-mediated self-association. The second and third terms consist of those macromolecules with their self-association-linked sites unsaturated. They are contributed by the rest of the association mechanisms. The second term covers those polymers with a liganded monomer at the starting terminal of the polymer whereas the third term contains those polymers with an unliganded monomer at that position.

One can also take the alternate approach to calculate the concentration of the bound ligand, i.e., first calculate the total concentration of each i-mer using the apparent self-association constant. There are two classes of ligand binding sites on the i-mer, those with a ligand affinity k_1 and those with a ligand affinity k_3. The ligand affinity of a given site depends on whether it has been coupled to the self-association reaction. For the model described above, with the exception of one terminal site, all ligand-binding sites on the polymer are linked to the self-association reaction and have a ligand-binding constant of k_3. The terminal site not linked to the self-association should have a ligand binding constant of k_1. Consequently, the total ligand bound to the macromolecule can be expressed as

$$[X_b] = \sum_{i=1}^{n} \frac{C_i}{iMW}\left[\frac{(i-1)k_3[X_f]}{(1+k_3[X_f])} + \frac{k_1[X_f]}{(1+k_1[X_f])}\right] \quad (21)$$

Figure 5 shows the theoretical binding isotherms of a one-ligand-mediated and facilitated isodesmic indefinite self-association system. They are concave upward and their Y intercepts increase with increasing protein concentration. In a manner similar to the ligand-facilitated self-association, the analysis of the ligand-binding isotherm of the ligand-mediated and facilitated self-association can be started by first obtaining estimates of the intrinsic equilibrium constant through analyzing the protein concentration dependence of the Y intercept of the Scatchard plot. According to Eq. (20)

$$\lim_{[X_f]\to 0} \frac{\bar{X}}{[X_f]} = \frac{k_1[A_f] + \sum_{i=2}^{n} k_1(2k_4)^{i-1}[A_f]^i + \sum_{i=2}^{n}(i-1)k_1(2k_2)(2k_4)^{i-2}[A_f]^i}{[A_t]}$$

$$(22)$$

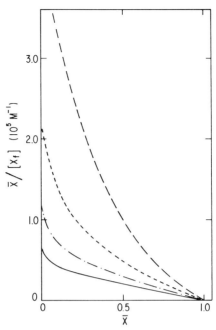

FIG. 5. Scatchard plots of the ligand binding of a one-ligand-mediated and facilitated isodesmic indefinite self-association. The macromolecular concentrations increase from the bottom to the top and are the same as Fig. 2. The intrinsic association constants used are $k_1 = 1.8 \times 10^4 \ M^{-1}$, $k_2 = 9 \times 10^4 \ M^{-1}$, $k_3 = 1.8 \times 10^6 \ M^{-1}$, and $k_4 = 9 \times 10^2 \ M^{-1}$.

Figure 4b shows the Y intercept of the Scatchard plot for such a linked system calculated according to Eq. (22) as a function of the total protein concentration. If $[A_t]$ is small, all terms in Eq. (22) higher than first order in $[A_f]$ can be ignored, and Eq. (22) becomes

$$\lim_{[X_f] \to 0} \frac{\bar{X}}{[X_f]} = k_1\{1 + (2k_2 + 2k_4)[A_t]\} \qquad (23)$$

Thus one can obtain estimates of k_1, k_2, and k_4 from the Y intercept and the slope of Fig. 4b. As the values of the intrinsic association constants k_2 and k_4 increase, the dependence of the Y intercept of the Scatchard plot on the total protein concentration becomes increasingly curvilinear and the data must be analyzed through nonlinear least-squares fitting technique. The validity of these estimated values of the intrinsic association constants can be tested subsequently through least-squares fittings of the experimental binding isotherms with the theoretical ones.

An example of the ligand-mediated and facilitated self-association can be found in the interaction of the anticancer drug vinblastine with the microtubule protein, tubulin. Velocity sedimentation studies of the self-association showed that the apparent self-association constant plateaued at high ligand concentrations, suggesting the presence of the ligand-mediated association.[26,27] At lower ligand concentrations, the Wyman plot showed a slope of approximately one, suggesting that the binding of one ligand is required for the association between two monomers. Studies of the vinblastine binding isotherms have shown that they are concave upward which rules out the ligand-mediated association as the only association mechanism.[28] In fact, the experimental isotherms can be fitted well by a theoretical model where association between a liganded monomer and an unliganded monomer is permitted.[16,29] Further proof of the linkage mechanism was derived from the complete consistency of both the stoichiometry and the intrinsic association constants obtained from ligand-binding and self-association studies.

Ligand-Induced Cross-Linking Self-Association. A protein self-association induced by ligand cross-linking is defined by Eq. (24) where the ligand serves as a bridge connecting two macromonomers[3,27]:

$$
\begin{array}{lll}
\text{A} & + \text{X} \rightleftharpoons \text{AX} & k_1 \\
\text{AX} & + \text{A} \rightleftharpoons \text{AXA} & k_1 \\
\text{AXA} & + \text{X} \rightleftharpoons \text{AXAX} & k_1
\end{array} \tag{24}
$$

$$
\vdots \qquad \vdots \qquad \vdots
$$

Each monomer has two ligand-binding sites. During its self-association, each macromonomer binds only the ligand and not another monomer. For such a ligand-induced association, the apparent self-association constant can be expressed as[27]

$$
K_2 = \frac{[\text{A}_2\text{X}] + [\text{A}_2\text{X}_2] + [\text{A}_2\text{X}_3]}{([\text{A}] + [\text{AX}] + [\text{AX}_2])^2} = \frac{k_1^2[\text{X}_f]}{(1 + k_1[\text{X}_f])^2} \tag{25}
$$

If the association proceeds indefinitely, the concentration of ligand bound to the macromolecule can be calculated from

$$
[\text{X}_b] = \sum_{i=1}^{n} \left[\frac{i-1}{i} \frac{c_i}{\text{MW}} + \frac{2}{i} \frac{c_i}{\text{MW}} \frac{k_1[\text{X}_f]}{(1 + k_1[\text{X}_f])} \right] \tag{26}
$$

where the first term represents the ligands that bridge two macromono-

[26] G. C. Na and S. N. Timasheff, *Biochemistry* **19**, 1347 (1980).
[27] G. C. Na and S. N. Timasheff, *Biochemistry* **19**, 1355 (1980).
[28] J. C. Lee, D. Harrison, and S. N. Timasheff, *J. Biol. Chem.* **250**, 9276 (1975).
[29] G. C. Na and S. N. Timasheff, *Abstr. Pap., 180th Meet., Am. Chem. Soc. Abstr. Biol.* p. 170 (1980).

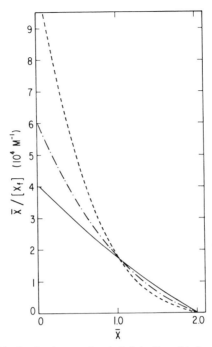

FIG. 6. Scatchard binding isotherms of an indefinite ligand-induced cross-linking of macromolecules. The details of the association mechanism are described by Eq. (24). The intrinsic ligand-binding constant to the monomer is 1.8×10^4. The protein concentrations are $1.6 \times 10^{-5}\ M$ (——), $7.8 \times 10^{-5}\ M$ (-··-), and $2.3 \times 10^{-4}\ M$ (———).

mers and the second term represents the ligand bound to the two terminal monomeric units not involved in bridging two monomers. Figure 6 shows the binding isotherms of such a linkage system at different protein concentrations. They are concave upward and intersect with each other at the stoichiometry of one. In fact, this is where the apparent self-association constant reaches its maximum when plotted against the free ligand concentration. Since the polymers must dissociate in order to have two ligands bound to each monomer, the apparent self-association constant peaks at this ligand concentration and then drops off at higher ligand concentrations as shown in the preceding chapter.[1] Such a behavior has been observed in the mercuric chloride-induced association of human mercaptalbumin, suggesting that this protein is cross-linked by the divalent ligand.[30]

[30] H. Edelhoch, E. Katchalski, R. H. Maybury, W. L. Hughes, Jr., and J. T. Edsall, *J. Am. Chem. Soc.* **75**, 5058 (1953).

The dotted curve of Fig. 6 can be used as a good example of the danger in determining ligand stoichiometry from a Scatchard plot alone. In this curve the Y coordinate descended more than 80% from the Y intercept over the region in which the ligand binding increased from zero to one ligand per monomer. Consequently, without looking at the saturation plot, it is very tempting to stop the experiment at a stoichiometry near one, linearly extrapolate the data, and reach the erroneous conclusion that there is only one ligand binding site per monomer.

Concluding Remarks

The multiple equilibria in a ligand-induced self-association can be studied by measuring and analyzing either the apparent macromolecular association or the apparent ligand binding. These two approaches, one discussed in the preceding chapter of this volume[1] and the other in this chapter, each have its own advantages and drawbacks, and they complement each other well. For instance, probing the ligand binding through analysis of the apparent self-association constant as described in the preceding chapter gives information only on the self-association-linked ligand binding site(s) and not the self-association-independent site(s). Information regarding the latter ligand-binding site(s) can be obtained only from direct ligand-binding measurements as described in this chapter. Conversely, in a ligand-binding study of the linkage mechanism, the shape of the ligand-binding isotherm is usually not sensitive toward the stoichiometry of the macromolecular association and thus cannot be used to obtain that information. For a thorough and detailed characterization of a linked system, it is essential to follow both approaches, i.e., first examine the mechanism of the macromolecular association reaction using a method such as analytical centrifugation and then analyze the ligand-binding isotherms as described in the present chapter. Since these two approaches measure two entirely different molecular events, they should employ different physical techniques. Consequently, both the encountered nonidealities and the invoked assumptions are different. By taking both approaches in examining a given system and obtaining consistent results, one can gain much stronger confidence in the conclusions derived. Another advantage which is gained sometimes from the complementary use of different physical techniques is that they focus frequently at different regions of the binding isotherm. For a given system, the sensitivity of one approach may pick up where that of the other starts to drop off. For instance, in the linked function analysis of the apparent macromolecular self-association shown in the preceding chapter, self-association is stronger at higher ligand concentrations. Consequently, the results obtained in the region where the association-linked ligand binding site is half to fully saturated are usually more reliable. If the apparent self-association constant plateaus at saturating ligand concentrations, it suggests the presence of a ligand-mediated association mechanism. The presence of a ligand-facilitated self-association should cause the apparent association constant to level off at low ligand concentration. However, if the apparent self-association constant does not level off at the lowest ligand concentrations examined, one still cannot rule out the presence of a weak ligand-facilitated pathway accompanying the ligand-mediated pathway that is beyond the detection by the available technique. Contrary to the self-association study, probing the linkage mechanism through analysis of the ligand binding as shown in this chapter focuses on the region where the association-linked ligand binding site is less than half saturated. It is in this region that different linkage mechanisms manifest most pronounced differences in the shape of their binding isotherms, and good quality data of ligand binding often can be obtained to allow differentiation of these linkage mechanisms. For instance, the question whether the ligand-mediated self-association is the sole association mechanism can be most easily tested by examining if the Y intercepts of the Scatchard binding isotherms converge to a single point.

There are instances where the self-association approach, rather than the ligand binding one, allows a more unambiguous differentiation of the linkage mechanisms. For example, the ligand-binding isotherms of the ligand-induced cross-linking of macromolecules exhibit certain similarities to those of the ligand-mediated and facilitated self-association. Given a certain amount of experimental uncertainties, it becomes difficult to differentiate these two linkage mechanisms on the basis of binding isotherms. However, the apparent self-association constants of the two systems show completely different dependences on the ligand concentration, one reaching a maximum and then dropping off, while the other plateaus at high ligand concentrations.

It should be noted that in the self-association study of a ligand-mediated and facilitated system, the value of the intrinsic self-association constant k_2 was obtained at saturating ligand concentration.[1] It should, therefore, correspond to the self-association constant between two fully liganded monomers. On the other hand, in the ligand-binding study of the same system, described above, the value of the intrinsic self-association constant is obtained at ligand concentrations approaching zero. Thus, in Eq. (23), k_2 corresponds to the self-association constant between a liganded monomer and an unliganded monomer. Although in the linkage mechanism defined by Eq. (18), the values of these two constants differ only by a factor of two, due to a hypothetical statistical factor, in a real case these two constants can be totally independent of each other.

The linkage mechanisms discussed in this chapter assume that there is no self-association-independent ligand-binding site on the macromolecule. If self-association-independent ligand-binding sites do exist on the macromolecule, their contribution to the total ligand bound can be easily accounted for and added to the above equations. Assuming that there are n_0 such association-independent sites with an equivalent association constant of k_0, their contribution to the total ligand bound can be expressed as

$$[X_b] = \frac{n_0 k_0 [X_f]}{(1 + k_0 [X_f])}$$ (27)

In the actual data analysis, if the ligand affinity of these sites is weaker than that of the self-association-linked site, the binding of the ligand to the association-independent site will follow the association-dependent site and appear at the right hand side of the Scatchard plot. This will interfere very little with the analysis of the linked ligand-binding site. This is the case in the vinblastine-induced tubulin self-association system.[16,29] However, if the self-association-independent site has an affinity that is much stronger than that of the self-association-linked site, the ligand binding to the former site will precede the latter one and appear at the left-hand side of the Scatchard plot. This could complicate the data analysis.

Last, it is important to emphasize that the methods described in this chapter are intended only for equilibrium systems. It is prudent to test the attainment of equilibrium through either a kinetic measurement or by ascertaining that the same results can be obtained from either the binding of the free ligand or the dissociation of the bound ligand.

Author Index

Numbers in parentheses are footnote reference numbers and indicate that an author's work is referred to although the name is not cited in the text.

Subject Index

A